页岩气开发理论与实践

（第一辑）

贾爱林　位云生　编著

科学出版社
北　京

内 容 简 介

页岩气开发是非常规气藏开发的重要组成部分。本书论文分为综合类、开发地质类、气藏工程类、生产应用类，汇总了一批国内页岩气开发领域专家的最新研究成果与心得，以及部分国外页岩气开发的最新进展，可为国内页岩气开发提供理论参考和方法借鉴。

本书可供从事页岩气开发的科研人员使用，也可以作为高等院校相关专业师生的参考用书。

图书在版编目(CIP)数据

页岩气开发理论与实践．第一辑／贾爱林，位云生编著．—北京：科学出版社，2018.1

ISBN 978-7-03-055690-5

Ⅰ.①页… Ⅱ.①贾… ②位… Ⅲ.①油页岩资源-油气田开发-文集 Ⅳ.①P618.130.8-53

中国版本图书馆 CIP 数据核字（2017）第 292538 号

责任编辑：王 运 姜德君／责任校对：张小霞

责任印制：肖 兴／封面设计：铭轩堂

科学出版社 出版

北京东黄城根北街16号

邮政编码:100717

http://www.sciencep.com

中国科学院印刷厂 印刷

科学出版社发行 各地新华书店经销

*

2018年1月第 一 版 开本：787×1092 1/16

2018年1月第一次印刷 印张：20 1/2

字数：486 000

定价：238.00 元

（如有印装质量问题，我社负责调换）

前　言

我国页岩气资源丰富，技术可采资源量达31.6万亿m^3［据美国能源信息署（EIA）预测］，仅次于美国。同时我国也是世界上除美国、加拿大之外，少数获得页岩气开发突破的国家之一，特别是近几年，我国页岩气产业发展迅猛，年产量从2014年的13亿m^3快速增长至2016年的78亿m^3，成为继美国、加拿大之后，第三大页岩气生产国。在国内非常规气藏开发中，页岩气产量首次超过了地面煤层气的产量，成为继致密气之后，第二大非常规气藏类型。

我国页岩气的跨越式发展，得益于四川盆地及其周缘长宁—威远、昭通、涪陵三个国家级页岩气示范区海相页岩气开发技术的长足进步。“十二五”期间，在政府大力投资和补贴政策的激励下，通过一系列技术攻关和开发试验，形成了埋深3500m以浅的海相页岩气藏储层评价、优快钻井、体积压裂、产能评价和开发技术优化五大主体技术系列：高演化强改造复杂山地地球物理资料解释及储层综合评价技术、水平井地质工程一体化的优快钻井技术、以大排量滑溜水+低密度支撑剂+可溶桥塞+拉链式压裂模式为主的水平井体积压裂工艺技术、页岩气体积压裂水平井概率性产能评价技术、水平井及裂缝参数综合优化技术。这些技术的试验与应用，提高了单井产量，降低了开发成本，奠定了我国页岩气规模开发的技术基础，并对占我国南方海相页岩气可采资源量2/3、埋深在3500m以深的页岩气的成功开发，以及北方大面积的陆相及海陆过渡相页岩气资源开发技术的突破提供了借鉴和示范作用。

《页岩气开发理论与实践》系列论文集，立足于我国页岩气勘探开发所取得的成果，收集整理了近年来具有代表性的学术论文，按照综合类、开发地质类、气藏工程类、生产应用类进行分类编辑，结集出版，希望在总结页岩气勘探开发理论和技术成果的同时，对这些理论和技术成果的推广应用发挥积极作用，并对新理论、新技术的探索有所启发，从而对我国页岩气勘探开发技术的进一步发展起到推动作用。

目　录

一、综　合　类

二、开发地质类

三、气藏工程类

四、生产应用类

一、综 合 类

海相页岩气开发评价关键技术进展

贾爱林　位云生　金亦秋
（中国石油勘探开发研究院）

摘要：以四川盆地及其周缘下古生界龙马溪组和筇竹寺组页岩为对象，总结中国海相页岩的地质和储集层特征，结合"十二五"期间页岩气开发取得的理论和技术成果，建立了中国页岩气开发评价关键技术指标体系和静动态参数及经济指标相结合的页岩气井综合分类标准。针对页岩气井压裂后形成的复杂缝网特征，联合分形介质和连续介质理论刻画复杂裂缝系统，建立页岩气全过程渗流生产模型，并根据输入参数特征构建概率分布模型，利用蒙特卡洛随机模拟方法，预测不同概率下的气井生产动态指标；同时提出水平井段及主裂缝参数、开发井距和生产制度等关键开发参数优化的具体思路和方法。中国页岩气大规模效益开发在地质理论、渗流机理及产能评价方法、开发技术政策及经济效益方面仍然面临许多亟待解决的难题。

关键词：页岩气；储集层特征；产能评价；水平井长度；裂缝参数；开发井距；生产制度

页岩气主要以吸附或游离状态赋存于细粒泥岩或页岩中。自20世纪90年代起，得益于良好的政策环境和持续的技术突破，美国天然气工业迅猛发展，已在古生界—新生界多套海相页岩层系实现大规模商业开发。根据美国能源信息署（EIA）历年页岩气产量数据[1]，2015年页岩气产量达到$4294\times10^8m^3$，占美国天然气总产量的46.1%。中国页岩气分布面积广，有利层系众多，可采资源量达$12.85\times10^{12}m^3$，居世界前列[2]，其中海相页岩气占主体。目前已在四川盆地及其周缘的下古生界龙马溪组和筇竹寺组获得重大突破，并于涪陵、长宁—威远、富顺—永川、昭通等地启动页岩气产能建设，初步进入规模开发阶段。本文通过对中美页岩气开发条件的对比，回顾页岩气开发实践和技术进步，总结近年来页岩气开发评价的关键技术进展，以期为中国页岩气工业的发展提供参考。

1　中国海相页岩气地质与储集层特征

1.1　中国海相页岩气地质特征

与美国主要页岩气田如Barnett、Marcellus、Haynesville、Eagle Ford等相比，中国海相页岩气田具有以下典型地质特征。

（1）中国南方海相页岩埋藏较深。下寒武统筇竹寺组、上奥陶统五峰组—下志留统龙马溪组海相页岩比北美地区大多数页岩沉积年代更为古老，因而页岩地层埋深普遍较大，只有盆地内局部隆起区和盆地边缘地带埋藏较浅。以筇竹寺组页岩为例，盆地边缘埋深2000m左右，盆地内部埋深大于8000m。埋深大一方面增加了开发的难度和风险，对开发工艺提出了更高的要求，另一方面，随着深度的增加地层压力变大，页岩气资源丰度显著提高。

（2）中国海相页岩构造复杂，断裂发育。古生界两套富有机质页岩沉积以后，经历了中生代—新生代一系列的构造演化，对页岩气的保存产生重要影响。构造运动在改造盆地形态的同时会形成天然裂缝系统。天然裂缝能提高富有机质页岩的渗流能力,是页岩气从基质孔隙流向井底的重要通道之一,同时,也是页岩气体积压裂形成复杂缝网系统的必要条件。

(3)目前中国开发的海相页岩气多位于盆地边缘埋深相对较浅的区域,这些区域以山地和丘陵为主,沟壑纵横。崎岖的地表环境、脆弱的生态条件,给井场、管道建设和压裂施工带来困难,增加了页岩气整体开发的难度。

1.2 中国海相页岩气储集层特征

与常规油气储集层相比,海相页岩储集层在甜点规模、气体赋存状态、储集空间类型和评价参数等方面都具有非常明显的特征。

中国南方海相优质页岩储集层主要位于深水陆棚亚相,地势平缓、分布范围广,古环境、古气候和古地理因素在较大区域范围内保持一致,储集层厚度平面分布稳定,变化率极小。与致密气有效砂体呈透镜状零散分布特点[3]相比,优质页岩具有“大甜点”的分布特征,单个甜点的范围在几十至上百平方千米。

页岩中的天然气一部分是以游离状态储存在孔隙空间中,另一部分呈吸附状态赋存于有机质和黏土矿物的表面。单位体积内的游离气含量与孔隙度和含气饱和度密切相关,而吸附气含量取决于页岩的吸附能力,影响因素主要包括有机质含量、温度和压力等,目前普遍采用 Langmuir 等温吸附模型对吸附气含量进行计算。

页岩储集空间主要含两种类型:基质孔隙和裂缝,前者包含黏土矿物间微孔、脆性矿物晶间孔、次生溶蚀孔以及有机质孔,孔隙大小多为纳米级,一般为 2 ~ 200nm;后者根据裂缝的大小分为宏观裂缝和微观裂缝。研究表明,页岩基质孔隙度与总有机碳含量和成熟度有关[4]:随着总有机碳含量的增加,孔隙度逐渐增大,当总有机碳含量大于 5% 以后,孔隙度随总有机碳含量增加而增大的趋势变缓,甚至有降低的趋势,原因在于富含有机质的页岩抗压能力弱,有机质孔隙不易保存;页岩成熟度与微观孔隙结构间的关系较为复杂,随着成熟度的增加,有机质热解生烃后形成的微孔数量也逐渐增多,但在此过程中,黏土矿物中比表面积大的蒙脱石逐渐向伊/蒙混层转化,并最终形成伊利石和绿泥石,使矿物间微孔的比表面积和孔隙体积大大降低。页岩中的宏观裂缝在构造运动频繁的区域(如盆地边缘造山带)比较发育,而脆性矿物含量较高的地层中更容易发育微裂缝。

页岩储集层特征决定了页岩气必须采用以水平井和体积压裂为主体的开发技术。因此,在对储集层进行评价时,除需要考虑含气量、压力系数、优质页岩厚度等传统评价参数外,还需对影响储集层可压性和压裂效果的工程地质条件进行评价,包括岩石弹性模量、泊松比、水平两向应力差、天然裂缝发育程度等。

2 中国海相页岩气开发现状

“十二五”期间中国页岩气勘探开发取得重大突破,特别是 2014 ~ 2015 年一系列重要理论突破、体制创新和技术进步助推中国页岩气实现初步规模商业开发。截至 2015 年年底,全国共完钻海相页岩气评价井 198 口,开发水平井 393 口,压裂投产 267 口,建成产能 $77\times10^8m^3$,并在重庆涪陵、四川长宁—威远、滇黔北昭通建成 3 个海相页岩气示范区。

通过“十二五”示范区的攻关试验，甜点区优选、优快钻井和体积压裂技术取得长足进步，实现了由无效资源变为有效产量的技术跨越，使单井综合投资由早期的1亿元降至7000万元左右，推动了中国页岩气产量从2014年的$12.8\times10^8m^3$快速增加至2015年的$44.6\times10^8m^3$，初步实现了产量规模增长。“十三五”期间，如何将有效产量变为规模效益产量是页岩气开发的主攻方向。

针对中国南方海相页岩气的实际特点，中国石油天然气集团公司、中国石油化工集团公司探索出“井位部署平台化、钻井压裂工厂化、采输设备撬装化、组织管理一体化”的高效勘探开发“四化”模式，同时形成了“国际合作、国内合作、风险作业、自营开发”4种页岩气规模效益开发体制机制，促进了技术进步，实现了降本增效。

3 页岩气开发评价关键技术指标体系

3.1 页岩气井靶体位置优选指标

页岩气的资源特征和开发方式决定了甜点区评价必须考虑地质甜点与工程甜点双重因素。优选含气量、厚度和压力系数作为主要地质参数，其中含气量和厚度是页岩储集层产能潜力的直观表征；作为自生自储的页岩气资源，压力系数既可影响含气量，又可反映地层保存条件的优劣。选取脆性指数、水平两向主应力差和裂缝发育程度作为主要工程参数，其中脆性指数是地层可压性的重要参考指标，储集层岩石脆性指数越高，可压性越强，压后形成的缝网更为复杂。储集层水平两向主应力差越小，越容易形成复杂缝网。

根据中国蜀南地区地质甜点区与工程甜点区的主要参数与试气产量的对应关系，形成页岩气井靶体位置优选的关键指标界限（表1），为水平井靶体位置的选择与优化提供依据。

表1 水平井靶体位置优选界限值

地质参数	界限值	工程参数	界限值
含气量/(m^3/t)	≥3.5	脆性指数/%	≥50
优质页岩厚度/m	≥3.0	水平两向主应力差/MPa	≤10
压力系数/f	≥1.5	裂缝发育程度	发育

3.2 页岩气水平井分类评价指标

页岩气水平井开发评价参数众多，为了简化页岩气开发过程中的评价指标体系，并准确反映气井产能的大小，本文通过关联性分析优选了优质页岩厚度、含气量、压力系数、压裂加砂量和返排率5个主要指标。这些地质指标和工程指标相互较为独立，且整体能够反映页岩气井的生产能力，利用其综合系数作为评价页岩气井的分类指标。综合系数表达式为

$$\psi=\ln\frac{(h/h_a)(C/C_a)(\beta/\beta_a)(S/S_a)}{\eta/\eta_a}$$

式中，C为含气量，m^3/t；h为优质页岩厚度，m；S为压裂加砂量，t；ψ为综合分类系数；β为压力系数，f；η为返排率，%；下标a表示区域平均。

建立评价标准的具体做法如下：首先以内部收益率12%和8%作为分类界限值，将页岩

气井分为Ⅰ类、Ⅱ类、Ⅲ类(表2)；再根据气价1.335元/m^3(不含税，含补贴0.3元/m^3)得到单井综合投资5500万元时，内部收益率12%和8%对应的单井最终累积产量的界限值；根据图1(a)可知Ⅰ类、Ⅱ类、Ⅲ类井对应的综合分类系数的界限值；再根据图1(b)可得到Ⅰ类、Ⅱ类、Ⅲ类井的试气产量的界限值。最终形成采用静态综合分类系数、试气产量、单井累积产量和内部收益率相统一的页岩气井综合分类标准(表2)。

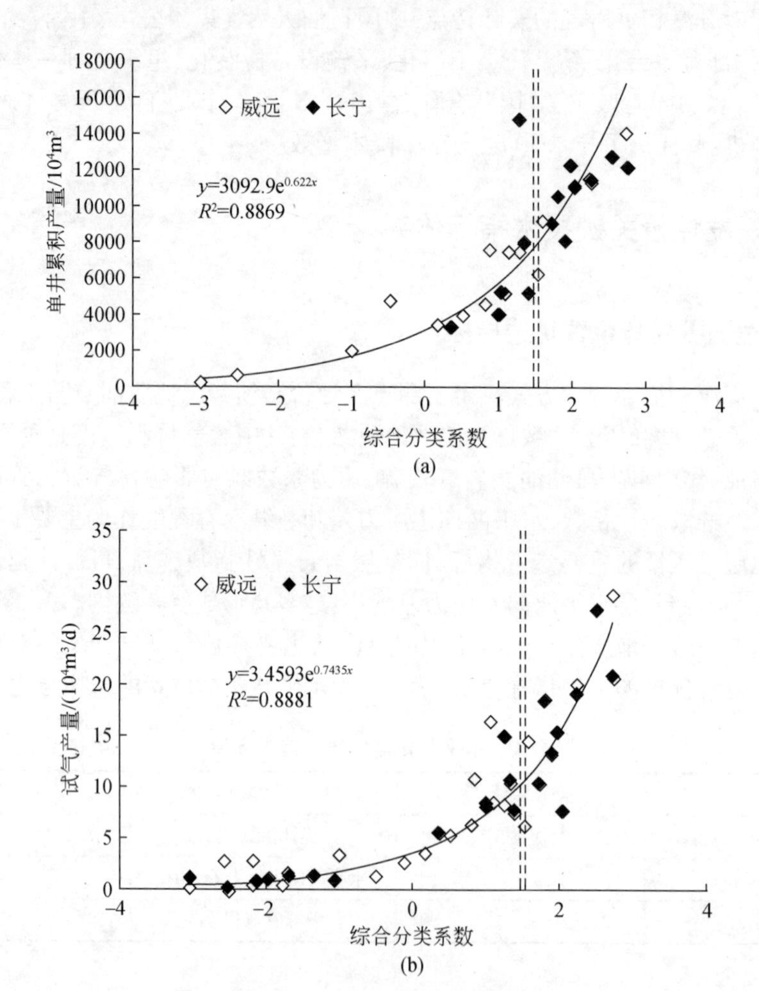

图1　威远、长宁地区气井试气产量和单井累积产量与综合分类系数的关系

表2　页岩气井综合分类标准

气井类型	综合分类系数	试气产量/($10^4m^3/d$)	单井累积产量/10^4m^3	内部收益率/%
Ⅰ	>1.603	>9.62	>8380	>12
Ⅱ	1.456~1.603	8.64~9.62	7651~8380	8~12
Ⅲ	<1.456	<8.64	<7651	<8

4 页岩气产能评价技术

页岩储集层具有纳米级孔隙、微米级裂缝的孔-缝结构,与体积压裂后形成的毫米级裂缝系统,共同组成“人造页岩气藏”复杂的缝网系统。由于页岩储渗空间尺度差异较大,气体流动形式包括渗流和解吸扩散两种方式,流动状态与常规气藏有着本质的区别,因此基于达西流的渗流模型已不能描述其复杂的流动形式。

4.1 模型建立

针对页岩气的流动特点,从微尺度解吸渗流机理分析出发,研究吸附气解吸原理、多尺度流动机制和从基质到裂缝系统的扩散过程等对气井生产动态的影响,从机理上厘清页岩气流动的物理本质。以考虑解吸、非达西流动等效应的拟压力控制方程为基础,利用复合线性流构建了多裂缝模型。同时用连续模型描述基质、天然微裂缝,用离散模型描述大尺度主裂缝,用分形模型[5-8]描述缝网区域内流动空间的复杂性(图2),建立完整的多段体积压裂水平井不稳定渗流数学模型。

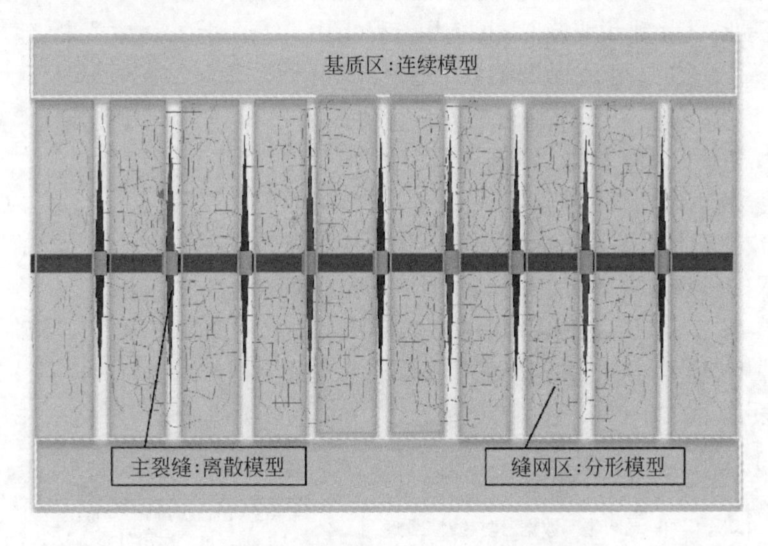

图2 多段体积压裂水平井物理模型

页岩储集层体积压裂形成缝网的条件通常包括压裂层水平两向主应力差小于5MPa、天然微裂缝发育、压裂液大排量泵入等。北美地区2006年以来的现场实践证明,相对常规储集层,页岩储集层形成缝网的条件更为苛刻,目前页岩气水平井压裂形成的裂缝仍以主裂缝为主[9]。而中国页岩储集层水平两向主应力差更大,体积压裂形成真正的有效缝网更难,因此本文建立的物理模型是以主裂缝为主的体积改造模型。

针对体积压裂形成的人工裂缝与天然裂缝耦合渗流系统,引入分形介质和连续介质理论刻画裂缝系统的跨尺度效应,形象表征裂缝系统的空间分布模式,建立起更准确的全过程渗流生产模型,从而更为合理地模拟气井生产动态,取得更好的生产数据拟合效果,进而获得更为准确的单井开发指标和生产动态预测。

4.2 概率性产能评价方法

以多段体积压裂水平井不稳定渗流数学模型为核心，结合页岩气生产数据，形成页岩气井生产动态分析和产能评价方法，以获得合理的开发指标参数。但页岩气井达到拟稳态生产的时间很长[10]，如 Barnett 页岩气田经改造后地层渗透率为 $0.1\times10^{-3}\mu m^2$ 时气井达到拟稳态时间为 2.3a，而渗透率为 $0.001\times10^{-3}\mu m^2$ 时则需要 230a[11]，因此，难以利用拟稳态数据准确获取模型中的基础参数。同时早期生产数据规律性不明显，很多流动状态无法清晰地反映出来，导致解析模型中的未知参量个数(地层、裂缝等参数)大于约束方程个数(特征流动段)，解释结果存在多解性。而且不同的解释结果会对产量预测产生很大影响，生产历史越短、预测期越长，参数影响越大。因此，根据各生产特征参数的特点，借助压裂设计、微地震监测、实验室测量等辅助信息，确定裂缝间距、有效渗透率、主裂缝半长等关键参数的取值范围，构建相关输入参数的概率分布模型，采用蒙特卡洛随机模拟方法，获得不同分布概率条件下的指标可信域。本文采用 80% 的可信度区间，即 $P_{10}\sim P_{90}$(可信域范围，计算结果中累积概率大于 10%、小于 90% 对应值的可能性)，一般同时给出 50% 的可信度 P_{50} 对应的值。另外，根据概率性结果的数学期望及期望的上下限假设，可获得最终的参数结果取值范围。

以线性流动阶段为例，说明该方法的使用流程(图 3)：①通过动态数据分析，建立线性回归方法，获得可靠参数组合的约束方程，即有效渗透率与裂缝半长的参数组合($x_f K^{0.5}$ 等于常数)。②根据概率统计分析给出如地层厚度、渗透率等基础性参数的概率分布模型[图 4(a)]。需要强调的是这里的渗透率下限值为岩心测试结果，上限值根据探测半径公式确定[12]。③基于随机模拟原理(即蒙特卡洛随机模拟方法)，对渗透率参数进行大量随机抽样，每次抽样结果均与约束方程结合，用以计算对应的未知参数。④按从小到大顺序重新整理计算结果(单井控制储量)，形成未知参数的概率分布结果及对应的可信域[图 4(b)]。

用获得的 P_{10}、P_{50}、P_{90} 三种概率下的渗透率及单井控制储量值为参数基础，利用线性流模型预测气井生产动态，对气井未来的生产动态及单井累积产量进行风险量化评估(图 5)。根据国内已投产的 270 口多段体积压裂水平井的生产数据分析，平均单井累积产量 P_{50} 为 $(0.6\sim1.0)\times10^8 m^3$。

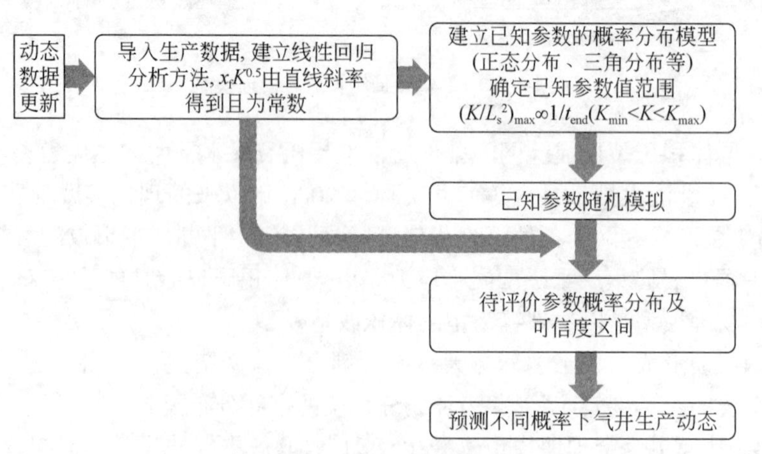

图 3 概率性产能评价方法使用流程

K. 渗透率，$10^{-3}\mu m^2$；L_s. 裂缝间距，m；t_{end}. 线性流结束时间，d；x_f. 裂缝半长，m

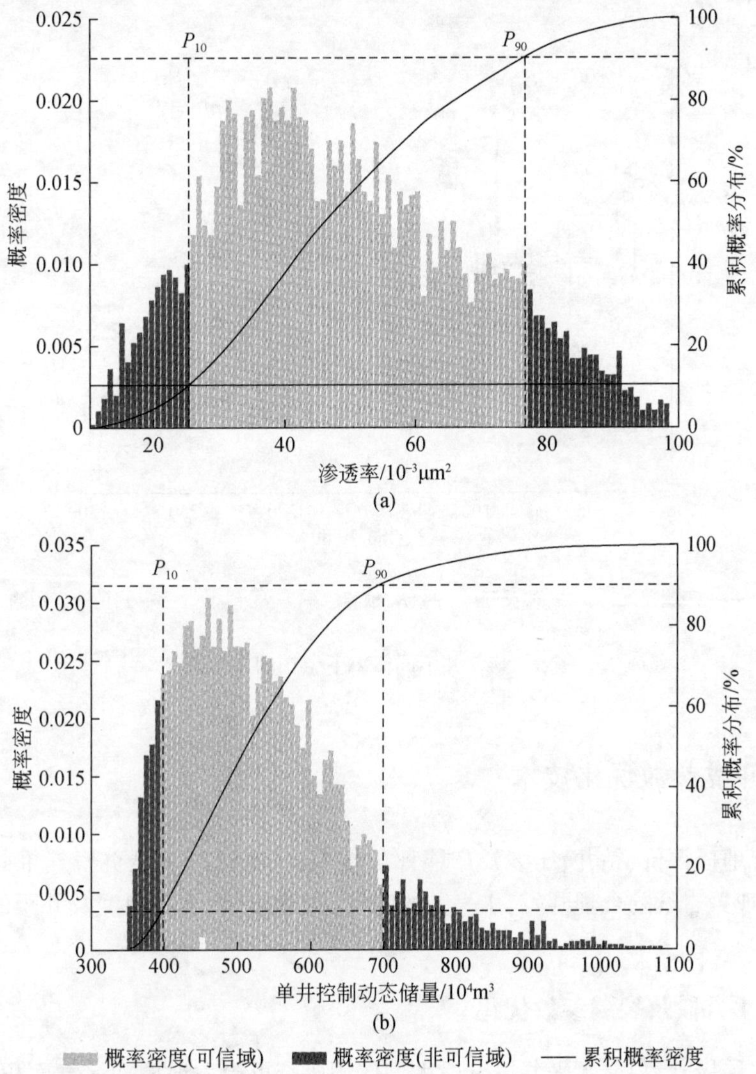

图4 渗透率与单井控制动态储量累积概率分布及可信度分布范围

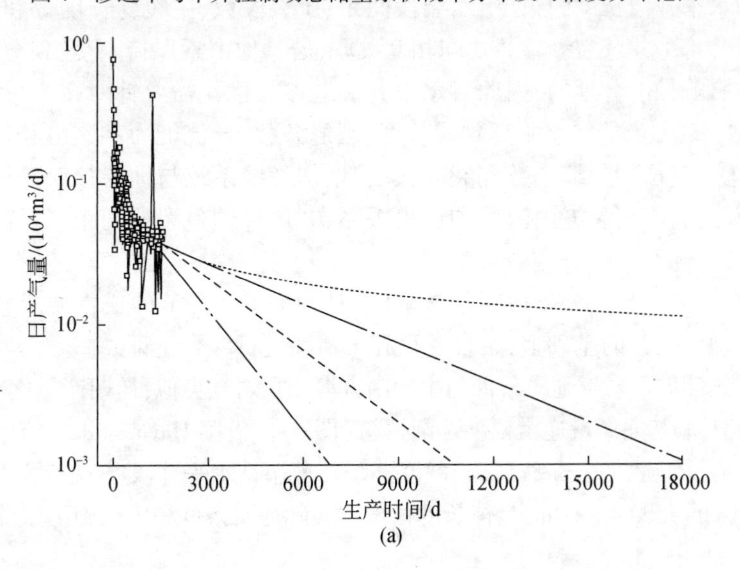

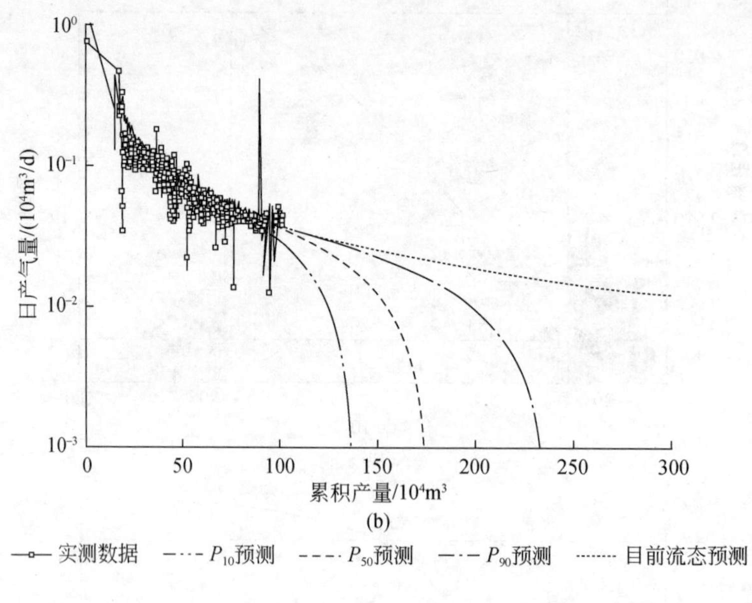

图 5　某气井不同概率下生产动态预测

5　页岩气开发参数优化技术

通过综合地质评价、钻井、压裂及产能评价技术，基本可以满足页岩气单井开发的技术需求，但要实现页岩气的合理开发，水平井段长度、裂缝布局、生产制度及井距的优化问题至关重要。

5.1　水平井段及裂缝参数优化

页岩储集层极其致密，水平井段分段大型体积压裂可极大地增加泄气面积，提高单井产量。在一定的水平井段长度约束下，压裂段间距、压裂段数、主裂缝长度等裂缝参数直接影响水平井产量。水平井及裂缝参数之间相互关联，若其他因素保持不变，仅分析某一个因素变化对产量的影响是不合理的，因此对水平井裂缝参数优化需要针对多个影响因素同时做整体优化[13,14]。气井产能受水平压裂段长度、裂缝条数、导流能力和裂缝长度等影响显著，优化的思路：①通过增加裂缝条数和裂缝长度，增加裂缝系统与地层接触面积；②通过调整裂缝有限导流能力，平衡裂缝内流入和流出关系；③通过调整裂缝间距、裂缝与封闭边界相对位置，降低裂缝相互干扰。

另外，页岩气开发目前处于边际效益阶段，对各个工程环节的成本要求较高，水平井及裂缝参数应在理论研究的基础上，结合工程施工情况进行综合论证。

美国 Haynesville 页岩气地质条件与中国南方海相页岩气类似，初期产量逐年递增，但随水平井段长度的增加，初期产量递增幅度逐渐变缓(图 6)。借鉴 Haynesville 页岩气的生产经验，综合考虑目前实际作业能力、工程风险和经济效益，遵循钻水平井段“一趟钻”[15]原则，推荐建产期水平井段长度 1500 ~ 1600m，具体长度可根据实际地质条件和工程施工条件适当调整。

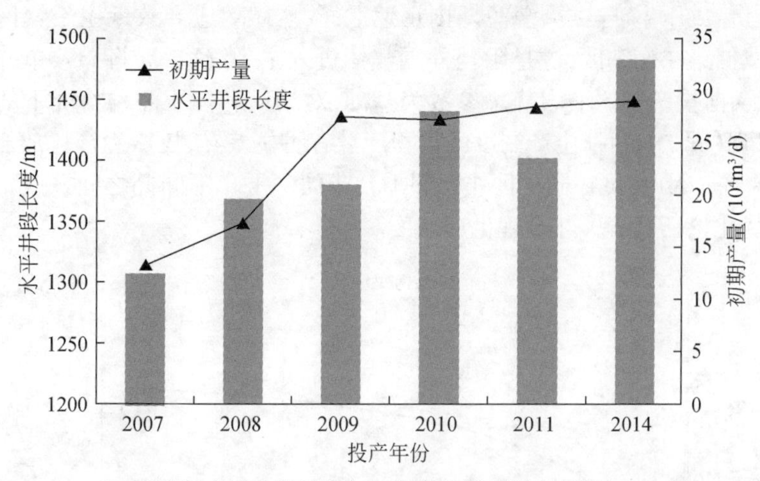

图6 美国 Haynesville 页岩气水平井段长度与初期产量随时间变化图(2012 年、2013 年无数据)

5.2 开发井距优化

目前页岩气水平井采用平台化布井、"工厂化"钻井和压裂以及大规模连续作业。若开发井距偏大,井间的剩余储量可能永远留在地下;若开发井距偏小,将降低单井累积产量,影响开发效益。中国页岩气井生产时间相对较短,早期开发井距的确定主要依据微地震监测结果。但目前实际试采动态监测结果表明,微地震监测确定的开发井距明显偏大[16]。由表3可见,与美国四大页岩气开发区块相比,中国蜀南地区页岩气开发区块平均单段支撑剂用量明显偏少,由于页岩储集层本身水平层理的限制,形成的裂缝长度也较小,合理的井距也应该小一些,但蜀南地区目前平均井距更大,对比来看,蜀南地区目前开发井距具备进一步缩小的空间。可采用理论优化、干扰试井和生产动态数据分析方法,结合现场试验,综合优化开发井距。

表3 中国蜀南与美国典型页岩气开发区块井距对比

区块	水平井段长度/m	井控面积/km^2	平均井控面积/km^2	平均井距/m	单段支撑剂用量/t
Barnett	1219	0.24~0.65	0.45	280	129.7
Haynesville	1402	0.16~2.27	0.50	260	162.3
Marcellus	1128	0.16~0.65	0.42	260	181.2
Eagle Ford	1494	0.32~2.59	0.60	300	112.6
中国蜀南	1448	0.36~1.10	0.65	400~500	97.5

5.3 生产制度优化

美国 Barnett、Marcellus 页岩气采用放压、大压差的生产方式,而 Haynesville 页岩气由于地层压力高,考虑页岩薄层状储集层强压敏效应,气井采用控压限产的方式进行生产。中国石油页岩气开发区块普遍压力较高,目前长宁和威远采用放大压差生产,昭通区块采用控压限产方式生产,两个示范区的压裂工艺不同,因此无法进行定量对比。

从理论上讲，放压生产，主裂缝短期内迅速泄压，压敏效应导致近主裂缝区渗透率急剧下降，快速形成储集层伤害区，过早阻挡外围气体进入主裂缝系统，将导致单井累积产量减少(图 7)。以当前投产井的平均静态参数为基础，对放压和控压两种情况下的气井产量进行数值模拟发现(图 7)，两种生产制度的累积产量有明显差异，控压生产比放压生产累积产量提高 28% 左右。因此，可在同一区块相邻平台或同一平台不同井之间开展不同生产制度对比试验，以优化论证页岩气井合理生产制度。

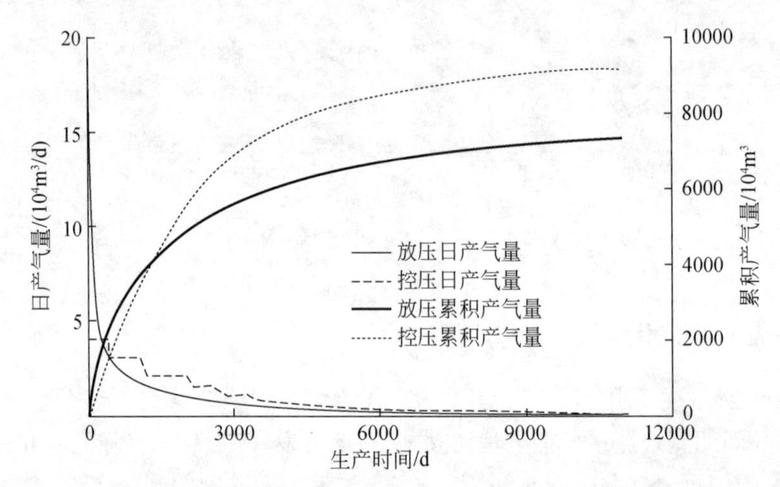

图 7　考虑应力敏感时放压与控压条件下产量模拟结果对比

6　页岩气开发前景与技术方向

页岩气作为一种新的非常规气资源，具有巨大的勘探开发潜力。根据国家油气专项中国石油页岩气项目组新一轮资源评价结果，预测中国南方海相页岩气可采资源量为 $8.82\times10^{12}m^3$，其中川东涪陵地区资源量达 $2.10\times10^{12}m^3$，川南—黔北长宁、威远、富顺—永川、昭通 4 个有利区资源量为 $4.37\times10^{12}m^3$。随着地质理论认识的提升和工艺水平的进步，预计到 2020 年，中国页岩气年产量将达 $300\times10^8m^3$[2]，到 2030 年页岩气将可能成为中国产量最大的天然气资源类型。但针对中国特殊的地质背景和当前低油价的国际形势，中国页岩气大规模效益开发仍然面临许多亟待解决的难题。

中国页岩气地质理论还不完善，小层间及区域间的储集层对比研究有待进一步深入。虽然当前的研究揭示了页岩气成藏的基本规律和特征，但往往将页岩视作均质储层，尚未有效研究不同层位、不同区域间的差异，不能有效指导开发储量计算和产能接替区的优选。

基于体积改造的人工缝网和页岩气纳米级孔隙的渗流机理及产能评价方法有待进一步完善。页岩中的孔隙以纳米级为主，纳米效应导致气体流动状态不同于常规储集层，尤其是在体积压裂后形成的复杂缝网系统中，多尺度流动空间和纳米效应共同影响，使气体流动空间的描述和渗流规律的表征高度复杂化，给产能评价带来了巨大挑战。

页岩气关键开发技术政策尚未定型。开发技术政策是油气田科学、高效开发的最关键问题，如开发井距影响气藏的最终采收率，生产制度影响单井累积产量等。这些关键技术政策的确定目前仅基于理论分析和模拟论证，或缺少实际的现场试验数据验证和支撑，或现场的测试手段不适应导致测试数据误差大。理论与现场实际相结合,开展页岩气关键开发技术政策综合论证意义重大。

从经济效益来看,中国页岩气钻井和压裂成本虽然有了较大幅度的下降,但仍然偏高,需要进一步降低单井综合开发成本。结合当前气价,要达到12%的内部收益率,单井综合投资应降低到5000万元以下。需要从技术和管理两方面进一步提高和创新,尽快将资源优势转化为效益产量优势。

7 结论

与美国相比,中国页岩气埋藏深、构造和地表条件复杂,增加了页岩气开发的难度。目前中国3500m以浅的页岩气已经投入初步规模开发,正在向更深的区域进军,具备更广阔的开发前景。

动静态参数及经济指标相统一的页岩气井综合分类评价标准已基本形成,需要进一步更新完善。考虑页岩气井压裂后多尺度空间及多种流动状态的水平井不稳定渗流数学模型和概率性的产能评价方法趋于完善。

水平井段长度、裂缝布局、生产制度及开发井距的优化问题是实现页岩气整体合理开发的关键,理论研究已经基本成熟,亟待现场试验验证。

页岩气开发理论的建立、开发评价方法的深化及开发成本的控制,是页岩气规模开发阶段的核心任务。

参考文献

[1] Energy Information Administration. Shale in the United States[EB/OL]. https://www. eia. gov/energy_in_brief/article/shale_in_the_ united_states. cfm[2016-07-26].

[2] 中国地质调查局. 中国页岩气资源地质调查报告[R]. 北京:中国地质调查局,2014.

[3] 贾爱林,郭建林,何东博. 精细油藏描述技术与发展方向[J]. 石油勘探与开发,2007,34(6):691-695.

[4] 梁兴,张廷山,杨洋,等. 滇黔北地区筇竹寺组高演化页岩气储层微观孔隙特征及其控制因素[J]. 天然气工业,2014,34(2):18-26.

[5] Chang J C, Yortsos Y C. Pressure-transient analysis of fractal reservoirs[J]. SPE Reservoir Evaluation & Engineering,1990,5(1):31-38.

[6] 同登科,陈钦雷,廖新维,等. 非线性渗流力学[M]. 北京:石油工业出版社,2002.

[7] Raghavan R, Chen C. Fractional diffusion in rocks produced by horizontal wells with multiple, transverse fractures of finite conductivity[J]. Journal of Petroleum Science and Engineering,2013,109(9):133-143.

[8] Raghavan R, Chen C. Fractured-well performance under anomalous diffusion[J]. SPE Reservoir Evaluation & Engineering,2013,16(3):237-245.

[9] Wong S W. Geomechanics of multiples fractures in horizontal wells[R]. Brisbane: Society of Petroleum Engineers,2015.

[10] Anderson D M, Nobakht M, Moghadam S, *et al.* Analysis of production data from fractured shale gas wells[R]. SPE 131787,2010.

[11] Yu W, Sepehrnoori K. Optimization of multiple hydraulic fractured horizontal wells in unconventional gas reservoirs[R]. SPE 164509,2013.

[12] 王军磊,贾爱林,甯波,等. 基于拟时间函数的气井不稳定生产数据分析[J]. 天然气工业,2014,

34(10):1-7.

[13] Valko P P, Economides M J. Heavy crude production from shallow formations: Long horizontal wells versus horizontal fractures[R]. SPE 50421, 1998.

[14] 王军磊,贾爱林,位云生,等. 有限导流压裂水平气井拟稳态产能计算及优化[J]. 中国石油大学学报(自然科学版),2016,40(1):100-107.

[15] 张东晓,杨婷云. 美国页岩气水力压裂开发对环境的影响[J]. 石油勘探与开发,2015,42(6):801-807.

[16] 朱维耀,亓倩,马千,等. 页岩气不稳定渗流压力传播规律和数学模型[J]. 石油勘探与开发,2016,43(2):261-267.

Progress in Key Technologies for Evaluating Marine Shale Gas Development in China

Jia Ailin Wei Yunsheng Jin Yiqiu

(Research Institute of Petroleum Exploration and Development, PetroChina)

Abstract: The Lower Paleozoic Longmaxi and Qiongzhusi Formation shale plays were taken as research objects to summarize the geologic and reservoir characteristics of marine shale gas in China. Based on the theory and technology achievements made during "the twelfth five-year plan", a key index system was established to evaluate shale gas well production effect in China, and a comprehensive classification was consequently presented by integrating geological statistical indexes and economic evaluation indexes; to characterize the complex fracture network formed by fracturing, a generalized analytical model was established by incorporating fractal and continuum geometry theory to capture the transient behavior of such fractured horizontal well throughout production life. An associated probabilistic analysis, based on the Monte Carlo simulation was presented by linking a probabilistic worksheet with the analytical model, and then used to determine a range of possible outcomes (i. e., EUR, decline rate and production time). Meanwhile, an optimization approach for development parameters, including fractured horizontal segment, fracture geometry (i. e., fracture length, conductivity, number), well spacing and drawdown management was advanced. There are still many unresolved problems on geologic theory, flow mechanism and productivity evaluation method, development technical policy and economic benefits, which limit large-scale and high-efficiency exploitation of shale gas resource in China.

Key words: shale gas; reservoir characteristics; productivity evaluation; horizontal section length; fracture parameter; development well spacing; production system

1 Introduction

Shale gas, which is a new kind of energy resource, mainly exists in fine-grained mudstone or shale in absorbed and free states. Benefiting from the favorable policy environment and continuous technological breakthroughs since the 1990s, American shale gas industry has made great progresses; shale gas in multiple sets of commercialized development in marine shale formations from Paleozoic to Cenozoic has been achieved developed commercially. According to the latest statistics from Energy Information Administration (EIA)[1], shale gas production of United States (US) increased amounted to 429.4 billion cubic metres meters in 2015, which occupies 46.1% of America's total gas output. Shale gas was widely distributed in China, numerous favorable strata, and recoverable available resources of 12.85 trillion cubic meters, ranking among the top in the world[2], in which, those strata mainly are marine shale. At present, great advancement has been achieved in the lower Paleozoic Longmaxi and Qiongzhusi shale formations in Sichuan Basin and its peripheral regions. A series of productivity constructions in Fuling, Changning-Weiyuan, Fushun-Yongchuan and Zhaotong have been initiated, marking the preliminary stage of large scale development of shale gas. This paper compares the shale gas developing conditions between China and United States, reviews the developing practice and technological innovations and gives a summary of progress in key technologies of evaluating shale gas development in the past few years, in the hope of providing instructive references for the development of China's shale gas industry.

2 Geologic and reservoir characteristics of marine shale gas in China

2.1 Geologic characteristics of marine shale gas in China

Compared with Barnett, Marcellus, Haynesville and Eagle Ford shale gas fields in the United States, China's shale gas plays have the typical geologic characteristics below.

(1) Deep burial depth of China's marine shale. The Lower Cambrian Qiongzhusi Formation and Upper Ordovician Wufeng Formation-Lower Silurian Longmaxi Formation are older than most shale formations in Northern America, so they have larger burial depth in general except for local uplifts in the Sichuan Basin and its edge. For example, Qiongzhusi Formation is about 2000m deep at the basin edge, but over 8000m deep inside the basin. Large burial depth on one hand adds difficulty and risk to shale gas development, and set higher requirements on development processes, on the other hand, with the increase of depth, formation pressure increases, and abundance of shale gas increases significantly too.

(2) Complex geologic structure and abundant faults of marine shale formations in China. After deposition, the two shale formations of Paleozoic have undergone a series of tectonic evolution from Mesozoic to Cenozoic which had important effect on shale gas preservation. While changing the geometry of the basin, tectonic movements can cause the formation of natural fracture system. Natural fracture can improve permeability of organic-rich shale, and is one kind of pathways for shale gas to move from matrix pores to well, and is the necessary condition for forming complex fractured systems by artificial fracturing.

(3) At present, the developed marine shale gas fields with shallow burial depth in China are located in the low mountains and hills around the verge of Sichuan Basin, where the ground surface is rugged. Rugged landscape and vulnerable ecological environment cause difficulties in constructing drilling sites and gas pipelines, adding difficulty to the overall shale gas development.

2.2 Reservoir characteristics of marine shale gas in China

Compared with conventional oil and gas reservoirs, marine shale gas formations have distinct differences in "sweet spot" scale, gas occurrence state, reservoir space type and evaluation parameters.

The high quality marine shale reservoirs in southern China mainly deposited in deep shelf environment, where the terrain was gentle and broad, and the paleoclimate, paleoenvironment and paleogeography were consistent in a large scope, so the thickness of reservoir is stable on the plane. Different from the lenticular scattered distribution of tight gas effective reservoir sand bodies, the scale of "sweet spots" of high quality shale ranges from dozens to hundreds of square kilometers[3].

Part of the shale gas accumulates in the pore space in free state and the other part is absorbed on the surface of organic matter and clay minerals. The free gas content per unit volume is controlled by porosity and gas saturation, while the absorbed gas content depends on the absorbability of shale, which is affected mainly by organic content, temperature and pressure. Langmuir isotherm model is commonly used to calculate the absorbed gas content currently.

Shale reservoir space is classified into two types, matrix pore and fracture. Matrix pores include clay or brittle mineral inter-crystalline pores, secondary dissolution pores and organic matter pores, and are mostly nano-scale with sizes of 2 ~ 200nm. Fractures can be divided into macro and micro fractures according to their size. Researches have shown that matrix porosity is related to its total organic carbon (TOC) content and maturity. The matrix porosity increases with the rise of TOC[4], this trend, however, slows down or even reverse when TOC content is over 5%, this is because the organic-rich shale is low in pressure resistance, therefore, the organic matter pores are difficult to preserve. The relationship between maturity and micro-pore structure of shale has been proved complex, with the rise of maturity, micro-pores formed by hydrocarbon generation of organic matter increase gradually, but in this process, montmorillonite with large specific surface area, changes into illite-montmorillonite mixed-layer and finally illite and chlorite, making specific surface area and porosity of micro-pores between mineral grains decrease remarkably. Macro-fractures are more developed in regions with complicated geological structure (for example, orogenic belt at basin edge), while micro-fractures are more easily to form in formations with high content of brittle minerals.

Horizontal well drilling and artificial fracturing are the main technologies used for shale gas development due to the properties of shale gas reservoirs. Besides gas content, pressure coefficient, thickness of high quality shale and other traditional evaluation parameters, some engineering geological conditions should be taken into consideration in shale reservoir evaluation. Such engineering geological conditions include rock elastic modulus, poisson's ratio, difference of horizontal stresses and development of fractures, which determine the reservoir fracability and fracturing effect.

3 Development status of China's marine shale gas

Shale gas exploration and development in China have made great breakthrough during "the twelfth five-year plan" period. A series of major theories, systematic innovations and technology progresses fueled commercial development of China's shale gas from 2014 to 2015. By the end of 2015, 198 evaluation wells, 393 horizontal wells of marine shale gas had been finished, 267 wells had been fractured and put into production, with the productivity of $77\times10^8 m^3$, and three marine shale gas demonstration areas, Fuling, Changning-Weiyuan and Zhaotong had been built.

The selection of "sweet spot", optimized and accelerated drilling and volume artificial fracturing technologies have made major progress through research and test during "the twelfth five-year plan", which has supported the transition of invalid resource to effective production, lowered the total investment of single well from 100 million Yuan in the early days to 70 million Yuan now, and brought about the rapid increase of shale gas production from $12.8\times10^8 m^3$ in 2014 to $44.6\times10^8 m^3$ in 2015. How to translate effectively production into scale production is the most important research direction in the thirteenth five year plan period.

According to the characteristics of marine shale gas in southern China, CNPC and SINOPEC have come up with the four modes of efficient exploration and development, "well deployment on wellpad, factory drilling and fracturing, skid-mounting of production and storage equipment, and integrated organization and management". Meanwhile, they have also worked out "domestic cooperation, international cooperation, risk operation, proprietary development" four kinds of scale development regimes for shale gas, which have promoted the technical progress, lowered cost and increased efficiency.

4 Key technological index system for shale gas development evaluation

4.1 Index of shale gas target selection

The characteristics of shale gas resource and its development mode make it necessary to consider the "geological sweet spots" and "engineering sweet spots" in "sweet spot" evaluation. Shale gas content, thickness and pressure coefficient of shale formations are selected as the principle geological parameters, in which shale gas content and thickness directly reflect the potential of shale reservoir, while pressure coefficient can affect gas content and reflect the preservation condition of shale reservoir. Brittleness index, difference of horizontal stresses and development of fractures are taken as the main engineering parameters, in which brittleness index is an important indicator of the reservoir fracability, in general, the higher the brittleness index, the higher the fracability of the reservoir, and the more complex the formed fracture system after fracturing will be.

The key indexes of shale gas targets have been set up based on the relationship between main geological and engineering parameters and tested gas production (Table 1), which provide the bases for choosing and optimizing horizontal well targets.

Table 1 Threshold values of shale gas target selection

Geological parameter	Limit value	Engineering parameter	Limit value
Shale gas content/(m^3/t)	≥3.5	Brittleness index/%	≥50
Thickness/m	≥3.0	Difference of horizontal stresses/MPa	≤10
Pressure coefficient/f	≥1.5	Development degree of fractures	Developed

4.2 Classification indexes of shale gas horizontal well evaluation

In order to simplify the shale gas evaluation index system from numerous parameters in the process of development, and to accurately reflect the production capacity of gas wells, five main indexes, thickness, gas content, pressure coefficient, sand dosage and flow-back rate, are selected through correlation analysis. These geological and engineering indexes are relatively independent to each other, and can reflect the production of shale gas well collectively. Comprehensive coefficient of these indexes is used to evaluate shale gas wells. The comprehensive coefficient can be expressed as:

$$\psi = \ln \frac{(h/h_a)(C/C_a)(\beta/\beta_a)(S/S_a)}{\eta/\eta_a}$$

where, C for gas content, m^3/t; h for thickness of high quality shale, m; S for sand dosage in fracturing, t; Ψ for comprehensive coefficient; β for pressure coefficient, f; η for flow-back rate, %; subscript a represents regional average.

Specific approach of establishing evaluation standards: the internal rate of return (IRR) of 12% and 8% are taken as threshold values to classify shale gas wells into Ⅰ, Ⅱ, Ⅲ classes (Table 2). According to the gas price 1.335 Yuan/m^3 (excluding taxes, including subsidies of 0.3 Yuan/m^3) and single well total investment of 55 million Yuan, the cumulative production is determined when IRR is 12% and 8%. According to Figure 1 (a), threshold values of comprehensive classification coefficient of well class Ⅰ, Ⅱ and Ⅲ can be speculated; based on Figure 1(b), we can obtain the limits of testing production of each class. A comprehensive classification standard has been established consequently by integrating comprehensive classification coefficient, testing production, cumulative production and internal rate of return (Table 2).

Table 2 Classification of "comprehensive classification coefficient"

Well class	Comprehensive classification coefficient	Testing production /(10^4m^3/d)	Cumulative production /10^4m^3	Internal rate of return /%
Ⅰ	>1.603	>9.62	>8380	>12
Ⅱ	1.456 ~ 1.603	8.64 ~ 9.62	7651 ~ 8380	8 ~ 12
Ⅲ	<1.456	<8.64	<7651	<8

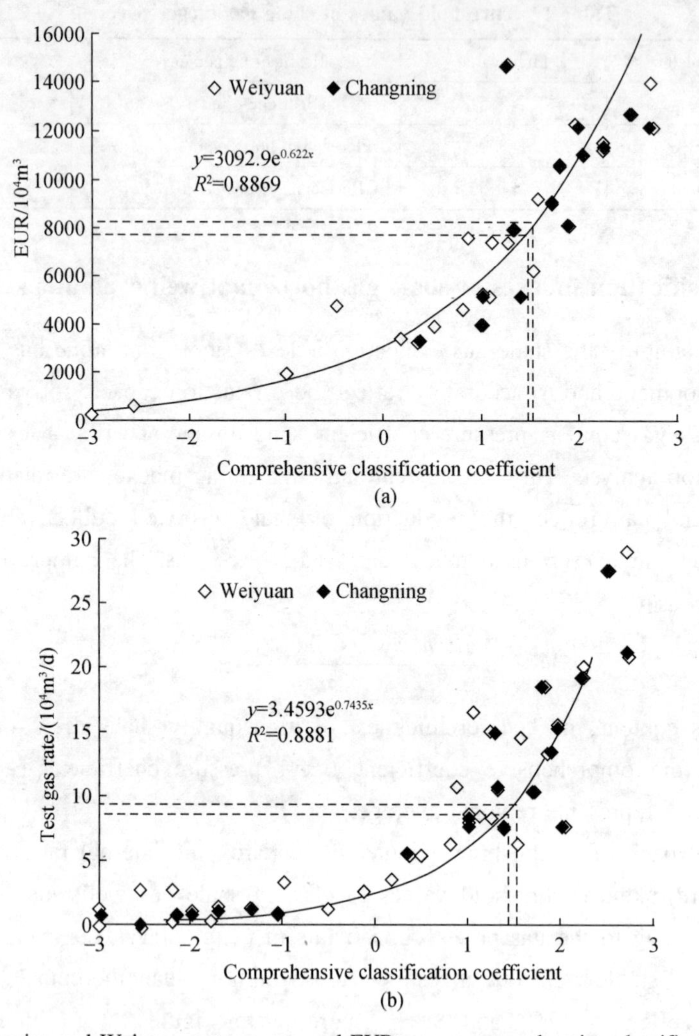

Figure 1 Changning and Weiyuan test gas rate and EUR versus comprehensive classification coefficient

5 Productivity evaluation technologies of shale gas

The inherent pore fracture system of nanometer-sized pores and micrometer-sized fractures of the shale reservoir together with the millimeter-sized crakes introduced by fracturing form the complex fracture network system of the shale gas reservoir. Because the reservoir space in shale differ widely in scale, shale gas flows in two ways, seepage and desorption diffusion, which is different essentially from that in conventional gas reservoirs. Therefore, the seepage flow model based on the Darcy flow can not describe the complex shale gas flow.

5.1 Modeling

The mechanism of micro-scale desorption and seepage have been firstly studied based on the characteristics of shale gas flow, and the principle of gas desorption, multi-scale flow mechanism and diffusion process from matrix to the fracture system subsequently been investigated to figure out the physical essence of shale gas flow. Multiple fracture model has been built by compound linear flow on the basis of simulated pressure control equation which takes desorption and non-Darcy flow into consideration. Meanwhile, continuous model was used to describe matrix and natural fractures, discrete model was used to describe the major fractures of large scale, the fractal model[5-8] was used to describe the complexity of flow space in the fracture network system (Figure 2). All these models are combined into a complete mathematical seepage model of multistage fracturing horizontal well.

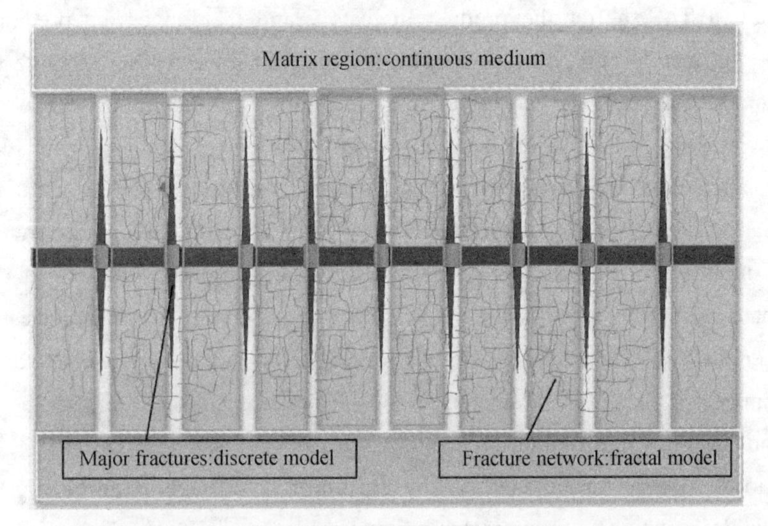

Figure 2 Physical model of multi-stage fractured horizontal well

The conditions for forming complex fracture network system by fracturing include difference of horizontal stress of less than 5MPa, developed natural fractures, and high pumping rate of fracturing fluid etc. Field practices in the past decade in North America have proved that shale gas reservoirs are more difficult to form fracture network than conventional reservoirs. Most of the fractures introduced by horizontal well fracturing are major fractures[9]. Differences of horizontal stresses of China's shale reservoirs are larger, so it's more difficult to form truly effective fracture network. Therefore, the physical model built in this paper is the major fracture oriented volume transformation model.

The fractal and continuum geometry theories have been used to describe the scale-span of fractures aiming at the coupling system composed of artificial and natural fractures, characterize spatial distribution pattern of fracture system visually, and build more accurate seepage production model to simulate the gas well production performance. As a result, more precise production indexes and dynamic prediction can be obtained.

5.2 Probabilistic productivity evaluation

In order to obtain reasonable development parameters, analysis methods for production performance and productivity evaluation were presented with core of unstable seepage mathematic model of multistage fractured horizontal wells, combined with shale gas production data. It takes a long time to reach quasi-steady state of shale gas production after stimulation[10]. For example, Barnett shale gas field takes 2.3a before its permeability reached $0.1 \times 10^{-3} \mu m^2$ after stimulation. It will take 230a for its permeability to reach $0.001 \times 10^{-3} \mu m^2$[11]. Therefore, quasi-steady data are difficult to be obtained. Meanwhile, there is no evident regularity of early production data, and many flow status could not be clearly reflected, which leads to the multiplicity of solution as the number of unknown parameters (i. e., formation, fracture) is more than the constraint equations (i. e. characteristic flow section). Different solutions of the equations make large impact on the recovery prediction. The shorter the production history and the longer the prediction time, the more influence the factors endure. According to the their characteristics, the value ranges of key parameters such as fracture interval, effective permeability and radius of major fracture were confirmed with auxiliary information including fracturing design, micro-seismic monitoring and laboratory measurements. An associated probabilistic analysis model was constructed, based on the Monte Carlo simulation to determine the credible region under different distribution probability of input parameter. In this paper, probabilistic approaches provide a distribution of reserves estimates with three confidence levels (P_{10}, P_{50} and P_{90}) and a corresponding 80% confidence interval ($P_{10} \sim P_{90}$). Meanwhile, we regard the confidence level of P_{50} as the reference result. Value ranges of the final parameters could be determined according to the mathematical expectation and the given upper and lower bounds of probabilistic results.

The method is illustrated as below, taking the linear flow stage as an example (Figure 3): ①Establishreliable equations of parameter combinations by dynamic data analysis and linear regression method ($x_f K^{0.5}$ is a constant). ② Determine the probability distribution model of fundamental parameters such as formation thickness and permeability, etc, by statistical analysis [Figure 4(a)], it is worth noting that the lower limit of permeability is from core testing and the upper limit is determined by the formula of radius probing[12]. ③Sample randomly from the per-

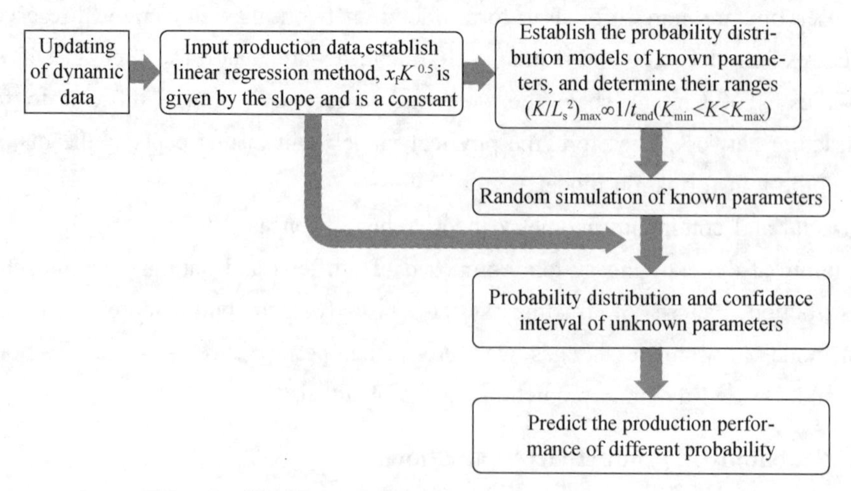

Figure 3 Workflow of probabilistic productivity evaluation

K. permeability, $10^{-3} \mu m^2$; L_s. fracture spacing, m; t_{end}. time of linear flow, d; x_f. fracture half-length, m

meability values based on stochastic simulation to calculate the unknown parameters combined with constraint equations. ④ Rearrange the results from small to large, to get the probability distribution and corresponding confidence range of unknown parameters[Figure 4(b)].

The production performance of shale gas wells was predicted with the linear flow model, based on the permeability and controlled reserves of single well at the probability of P_{10}, P_{50} and P_{90}, furthermore, quantitative risk assessment of production dynamic and cumulative production were performed(Figure 5). Average cumulative production of single wells at P_{50} is $(0.6 \sim 1.0) \times 10^8 m^3$ according to the production data of 270 stimulated horizontal wells in China.

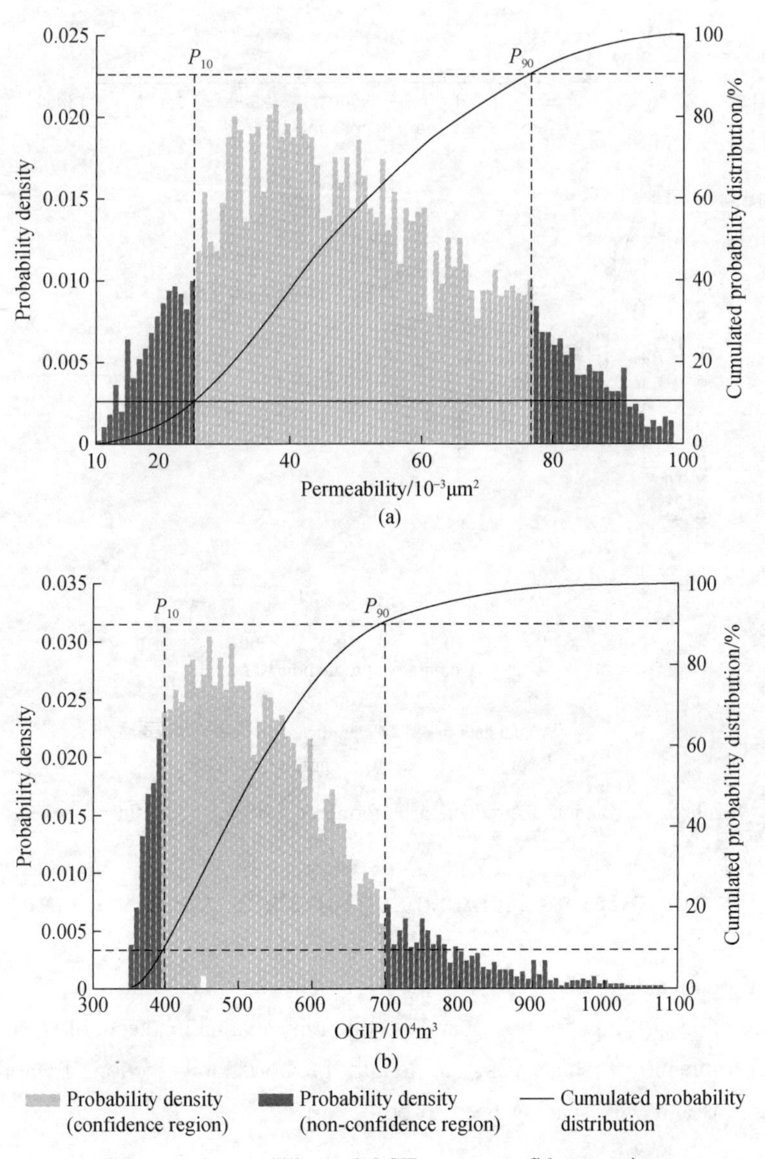

Figure 4 Permeability and OGIP versus confidence region

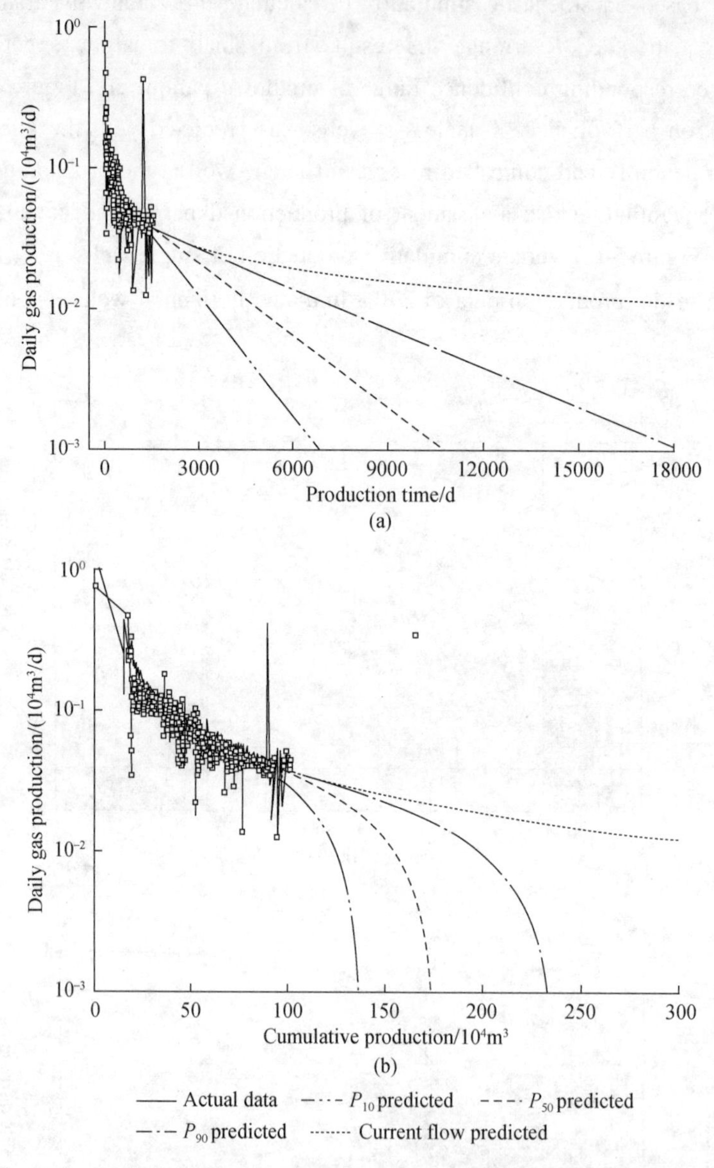

Figure 5　Prediction of production performance at different confidence levels

6　Parameter optimization technologies for shale gas development

Technology demands of shale gas well development can be basically satisfied by combining geological evaluation, drilling, fracturing and productivity evaluation technologies, but to realize reasonable development of shale gas, the length of horizontal section, fracture placement, production systems and well spacing need to be optimized.

6.1　Optimization of horizontal section length and fracture parameters

Shale reservoirs are very tight, large scale staged fracturing in horizontal wells can greatly increase the gas drainage area and single well production. Under the constraint of horizontal length, fracture parameters such as stage spacing, fracturing stage number and length of major fractures can directly influence well production. As the horizontal well parameters and fracture parameters are all connected, analysis of the influence of one factor on production while keeping others constant is not reasonable. Therefore, optimizing the fracture parameters of horizontal wells requires an integrated optimization of several parameters simultaneously[13,14]. Gas well productivity is strongly influenced by the horizontal section length of the fractured well, number of fractures, flow conductivity and fracture length. The optimization idea is: ① enlarging the contact area between the fracture network system and formation by increasing the fracture number and length; ② balancing the relationship between inflows and outflows of fracture network through adjusting the limited fracture conductivity; ③ reducing the interference between fractures by regulating the fracture spacing and relative location with the closed boundary.

In addition, shale gas development is marginal in benefit now, so the costs of various engineering links are stringent, and parameters of horizontal well and fracturing should be comprehensively demonstrated by combining the theoretical research and actual operation conditions.

In Haynesville shale gas field, with geological condition is similar to that of southern China, shale gas production increased year by year in the initial development stage, the rate of initial production increase, however, slacked gradually with the increase of horizontal section length (Figure 6). Based on the practical experience in Haynesville, in consideration of the actual operation ability, engineering risks and economic benefits, and the principle of "one trip" drilling of the horizontal section[15], it is recommended the horizontal section length be between 1500 ~ 1600m in the period of productivity construction, and the recommended length should be adjusted according to the actual geological and engineering conditions.

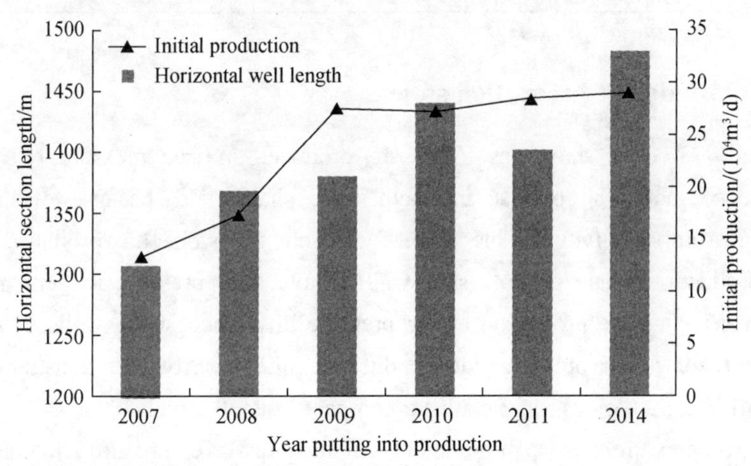

Figure 6 Variation of horizontal length and initial production with time in Haynesville shale play (no data in 2012 and 2013)

6.2 Optimization of well spacing

At present, horizontal well deployment from wellpad, factory-like drilling and fracturing, and large-scale continuous operation are widely adopted in drilling shale gas horizontal wells. If the well spacing is too large, some remaining gas reserve will be left behind forever; on contrary, if the spacing is too small, the cumulative production of single well will decrease, which will lower the development economic benefit. The production history of China's shale gas is short, and the well spacing is determined mainly based on micro-seismic monitoring results in the early stage of development. But the monitoring of test production performance to date shows that the well spacing determined from micro-seismic monitoring is too large[16]. It can be seen from Table 3, compared with the four major shale gas fields in the Unite States, the average proppant dosage of single stage in shale gas developing areas of south Sichuan Basin is much lower. Besides that, the length of fracture is shorter due to the bedding restriction. Therefore, reasonable well spacing should be smaller. Development well spacing should be optimized through theoretical study, interference well-testing and dynamic data analysis, in combination with field test.

Table 3　Comparison of well spacing between typical areas in Southern Sichuan and the United States

Area	Horizontal section length/m	Well controlled area/km^2	Average well controlled area/km^2	Average well spacing/m	Proppant dosage of single stage/t
Barnett	1219	0. 24 ~ 0. 65	0. 45	280	129. 7
Haynesville	1402	0. 16 ~ 2. 27	0. 50	260	162. 3
Marcellus	1128	0. 16 ~ 0. 65	0. 42	260	181. 2
Eagle Ford	1494	0. 32 ~ 2. 59	0. 60	300	112. 6
Southern Sichuan Basin	1448	0. 36 ~ 1. 10	0. 65	400 ~ 500	97. 5

6. 3　Optimization of production system

Barnett and Marcellus shale gas fields are produced in free pressure release and large pressure difference mode. In contrast, in Haynesville shale field, because of high formation pressure and pressure-sensitivity of the shale formation, the wells are produced at controlled pressure and production. Shale gas plays of CNPC have high pressure in general, so wells in Changning and Weiyuan are producing at big pressure difference, while wells in Zhaotong area are producing at controlled pressure and production. Since the two demonstration areas take different fracturing technologies, they can't be compared quantitatively.

Theoretically, pressure release production will lead fast pressure drop in major fractures, consequently, pressure sensitive effect will cause rapid permeability decline in areas around the major fractures, thus the fast formation of reservoir damage area, which would forbids peripheral gas getting into the major fractures and cause production decrease of well (Figure 7). Numerical simulation based on the data of production wells shows that different production systems differ widely in cumulative production (Figure 7); and pressure controlling production gets 28% higher cumulative production than pressure release production. Therefore, comparison test between nearby wellpads or wells in the same wellpad could be conducted to select the better production system.

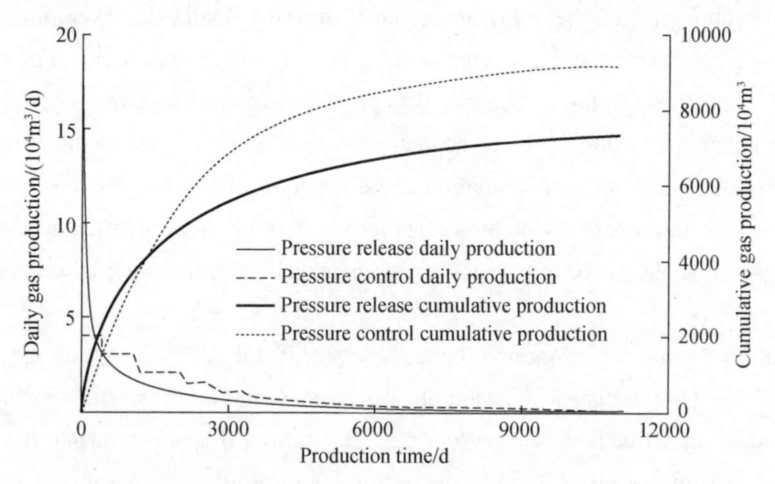

Figure 7 Comparison of predicted production between pressure release and pressure control production system considering stress-sensitivity

7 Development prospects and technical direction of shale gas

As a new source of energy, shale gas has huge exploration and development potential. According to the latest evaluation of CNPC shale gas project team of nation project on oil & gas, recoverable resources of marine shale gas in southern China are 8.82×10^{12} m^3, in which 2.10×10^{12} m^3 are in Fuling, eastern Sichuan Basin, and 4.37×10^{12} m^3 in Changning, Weiyuan, Fushun-Yongchuan, Zhaotong, in southern Sichuan Basin. With the promotion of geologic theories and progress of technologies in China, the marine shale gas production is expected to reach 300×10^8 m^3 by 2020[2]; the shale gas is likely to become the most productive type of natural gas by 2030. But due to the complex geological conditions in China and the current low prices of oil & gas in the world, large-scale efficient development of China's shale gas is facing lots of problems.

Theories of China's shale gas are still not complete, and comparative studies between sub-layers and regions need to be deepened further. Although existent studies have revealed the basic laws and features of shale gas accumulation, in such studies, shale formations are often taken as homogenous reservoirs, differences between different layers and areas haven't been examined carefully, therefore, calculation of reserves and selection of production relay block lack effective guidance.

Percolation mechanism and shale gas productivity evaluation methods based on the artificial fracture network and nanometer pores should be further improved. Pores in shale are mostly nano-scale, which lead to gas flow state different from conventional reservoirs, especially in complex fracture network system created by volume fracturing, the combined effect of multi-scale flow space and nano-effect makes the flow space description and seepage pattern characterization very complex, and brings about huge challenge to productivity evaluation.

The key technologies of shale gas development haven't finalized. Development policies are essential issues for scientific and efficient development of oil or gas fields. For example, well spacing influences the ultimate recovery of gas reservoir, production system affects the cumulative production of single wells. The determination of these key technology polices only depend on theoretical analysis and simulation so far, because of the deficiency of practical data or big error caused by improper test methods. Therefore, it is of great significance to demonstrate key technologies of shale gas development comprehensively by combining theories with and field data.

From the viewpoint of economic benefits, even if the costs of shale gas drilling and fracturing in China have reduced considerably in recent years, they are still too high. The comprehensive cost of signal well should be less than 50 million to achieve an internal rate of return of 12%, at the current gas price. To transfer resource into production as soon as possible, further improvement and innovation should be encouraged in both technology and management.

8 Conclusions

Compared with the United States, shale gas in China has deeper buried depth, more complicate structures and surface conditions, which add difficulties to the shale gas development. At present, shale gas reservoirs less than 3500m deep have been put into preliminary scale development. Shale gas exploration is heading toward deeper formations, and shale gas development has wider prospects.

An evaluation criterion of shale gas well classification considering dynamic and static parameters and economic indicators has been basically established, but needs further improvement. Unstable mathematic seepage flow model of horizontal wells and probabilistic evaluation method of productivity based on multi-scale space and various flow states is nearly perfect.

Optimization of horizontal section length, fracture placement, production system and well spacing are the keys to ensure the overall proper development of shale gas reservoirs. The theoretical research is basically mature, and field testing is urgently needed.

Establishment of shale gas development theory, optimization of evaluation methods and control of development cost are the core tasks in large scale shale gas development.

References

[1] Energy Information Administration. Shale in the United States[EB/OL]. https://www.eia.gov/energy_in_brief/article/shale_in_the_united_states.cfm[2016-07-26].

[2] China Geological Survey. China shale gas resources survey report [R]. Beijing: China Geological Survey, 2014.

[3] Jia A L, Guo J L, He D B. Perspective of development in detailed reservoir description[J]. Petroleum Exploration and Development, 2007, 34(6): 691-695.

[4] Liang X, Zhang T S, Yang Y, *et al.* Microscopic pore structure and its controlling factors of overmature shale in the lower Cambrian Qiongzhusi Fm, northern Yunnan and Guizhou provinces of China[J]. Natural Gas Industry, 2014, 34(2): 18-26.

[5] Chang J C, Yortsos Y C. Pressure-transient analysis of fractal reservoirs[J]. SPE Reservoir Evaluation & Engineering, 1990, 5(1): 31-38.

[6] Tong D K, Chen Q L, Liao X W, *et al.* Non-linear fluid mechanics in porous media[M]. Beijing: Petroleum Industry Press, 2002.

[7] Raghavan R, Chen C. Fractional diffusion in rocks produced by horizontal wells with multiple, transverse fractures of finite conductivity[J]. Journal of Petroleum Science and Engineering, 2013, 109(9): 133-143.

[8] Raghavan R, Chen C. Fractured-well performance under anomalous diffusion[J]. SPE Reservoir Evaluation & Engineering, 2013, 16(3): 237-245.

[9] Wong S W. Geomechanics of multiples fractures in horizontal wells[R]. Brisbane: Society of Petroleum Engineers, 2015.

[10] Anderson D M, Nobakht M, Moghadam S, *et al.* Analysis of production data from fractured shale gas wells [R]. SPE 131787, 2010.

[11] Yu W, Sepehrnoori K. Optimization of multiple hydraulic fractured horizontal wells in unconventional gas reservoirs[R]. SPE 164509, 2013.

[12] Wang J L, Jia A L, Ning B, *et al.* Analysis of the unsteady production data of a gas well based on pseudo-time function[J]. Natural Gas Industry, 2014, 34(10): 1-7.

[13] Valko P P, Economides M J. Heavy crude production from shallow formations: Long horizontal wells versus horizontal fractures[R]. SPE 50421, 1998.

[14] Wang J L, Jia A L, Wei Y S, *et al.* Pseudo steady productivity evaluation and optimization for horizontal well with multiple finite conductivity fractures in gas reservoirs[J]. Journal of China University of Petroleum (Edition of Natural Science), 2016, 40(1): 100-107.

[15] Zhang D X, Yang T Y. Environmental impacts of hydraulic fracturing in shale gas development in the United States[J]. Petroleum Exploration and Development, 2015, 42(6): 801-807.

[16] Zhu W Y, Qi Q, Ma Q, *et al.* Unstable seepage modeling and pressure propagation of shale gas reservoirs [J]. Petroleum Exploration and Development, 2016, 43(2): 261-267.

页岩气与致密气开发特征与开发技术对比分析

位云生　贾爱林　何东博　王军磊　韩品龙　金亦秋

（中国石油勘探开发研究院）

摘要：页岩气与致密气同属于非常规气藏，是我国目前天然气产量增长的主体。两种非常规气藏宏观上均具有储层致密且大面积连续分布的特点，均需要通过压裂改造技术才能获得高产，但在基本地质特征、开发特征及开发策略与技术等方面仍存在诸多差异。地质沉积背景决定了储层的岩性、物性、含气性及展布形态，从而也决定了相适应的钻完井及压裂改造工艺技术；原始储层与工艺措施影响多孔介质格架，气体原始赋存形式影响流体流动状态，两者共同决定不同类型气藏的开发策略和开发特征。两种气藏详细的对比分析，为深刻认识与合理开发两种非常规资源提供相互借鉴的依据。

关键词：海相页岩气；陆相致密气；地质特征；生产特征；开发策略与技术；对比分析

我国致密气与页岩气资源规模巨大，是我国天然气持续增长的两个重要支柱。根据2015年国土资源部资源评价最新结果，全国致密气技术可采资源量$12\times10^{12}m^3$[1]；页岩气技术可采资源量$21.8\times10^{12}m^3$[2]。两者均属于非常规资源类型，既具有相似性，又具有明显差异。致密气开发历程早于页岩气，形成的一些地质认识、开发技术和经验，得到了向页岩气延伸的应用；页岩气开发难度比致密气更大，技术要求更高，地质认识的深化和开发技术的进一步发展，可以反哺致密气的深入开发。两种气藏类型的对比分析以及开发特征认识和技术对策的相互借鉴，对推动两个领域的发展具有现实意义。

1　我国页岩气与致密气基本地质特征

根据沉积背景，我国页岩气分为海相、过渡相、陆相三大类，其中海相页岩气可采资源量占主体[3]，特别是四川盆地及其周缘的海相页岩气已成功初步规模开发。我国致密气主要包括陆相砂岩、海相碳酸盐岩、火山岩等多种类型，其中陆相砂岩可采资源量占主体，特别是鄂尔多斯盆地的陆相砂岩已成为我国天然气产量增长的主体[1]。因此，本文重点以四川盆地及其周缘海相页岩气与鄂尔多斯盆地陆相砂岩致密气进行对比分析。

1.1　四川盆地海相页岩气基本地质特征

四川盆地及其周缘主要发育下志留统龙马溪组和下寒武统筇竹寺组两套厚度较大的页岩储层，其中龙马溪组埋深相对较浅，是最有利的勘探开发层系。龙马溪组下部龙一段龙$一_1$亚段的黑色页岩是最有利的开发目的层，埋深2300～4500m，主要受控于海相深水陆棚沉积环境（图1），黑色碳质页岩、硅质页岩以及含钙黑色页岩发育，具有特低孔渗和储源同岩的特征，部分地区发育天然微裂缝，由于有机质的存在，气体以游离态和吸附态两种状态共存，埋深4000m以浅的有利区面积为$2.1\times10^4km^2$，资源量达$9.6\times10^{12}m^3$，平均资源丰度为

$4.6\times10^{8}m^{3}/km^{2}$[2]。

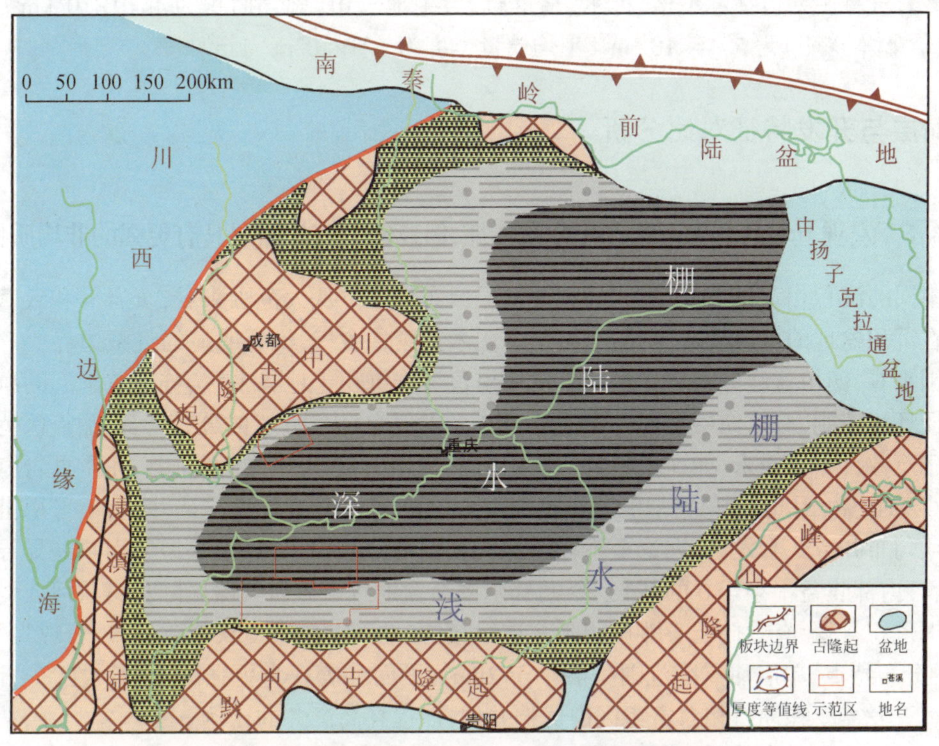

图1 四川盆地及其周缘龙马溪组页岩沉积环境

1.2 鄂尔多斯盆地致密砂岩气基本地质特征

鄂尔多斯盆地主要发育下、中二叠统的太原组、山西组和下石盒子组等多套致密砂岩储层,其中山西组山$_1$段和下石盒子组盒$_8$段是最有利的勘探开发层系。山$_1$段和盒$_8$段埋深3200~3500m,主要受控于陆相辫状河沉积环境(图2),心滩和河道底部充填的中粗砂岩是

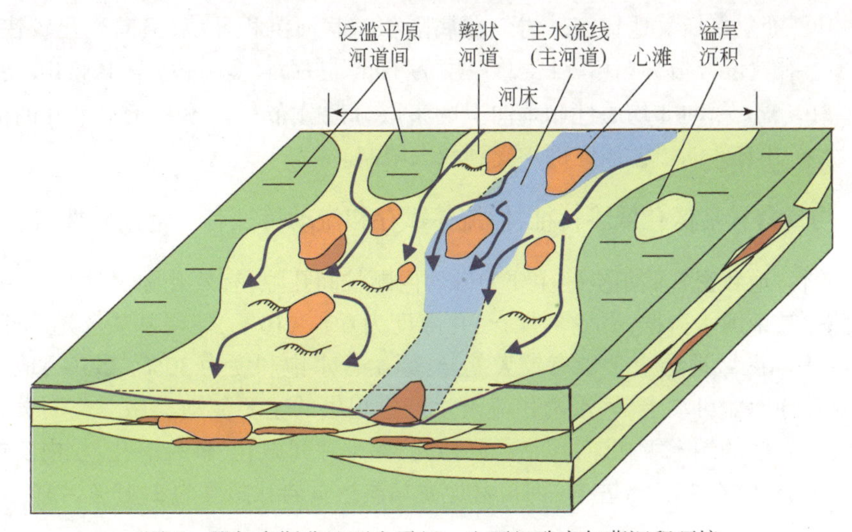

图2 鄂尔多斯盆地石盒子组—山西组致密气藏沉积环境

主要的有效砂体,整体具有低孔低渗特征,气体以干气或气水共存状态存在,鄂尔多斯盆地致密砂岩气藏主要包括苏里格、神木、乌审旗、大牛地气田,含气面积为 $4.19\times10^4km^2$,探明储量(含基本探明)为 $5.44\times10^{12}m^3$,平均储量丰度为 $1.3\times10^8m^3/km^2$[4]。

2 储层与开发特征对比分析

2.1 宏观上均具有大面积连片分布特征,但致密气有效储层具有更强的非均质性

不同的沉积相背景,决定不同的有效储层展布特征。从大范围来看,页岩气和致密气均分布在沉积盆地的负向构造单元,上万平方千米连续性含气,没有明显的气藏边界[5]。从较小范围来看,四川盆地海相页岩由富有机质黑色页岩沉积于地势平缓、分布范围广的海相深水陆棚环境控制,有效储层平面分布连续、稳定,表现为"大甜点"的分布特征,甜点区范围可达数十至数百平方千米,如在长宁-昭通区块内,横向跨度 40km 的剖面上,优质页岩厚度变化率小于 10%,因此,四川盆地海相页岩气在一定范围内不需要地质选井,利于采用批量平台化均匀布井方式开发,且理论上不存在地质Ⅲ类井(地质Ⅲ类井就是地质因素差异导致产量低,且内部收益率小于 0 的井)[2]。而鄂尔多斯盆地致密砂岩气藏属陆相辫状河或辫状河三角洲沉积背景,有效储层心滩微相为非连续相,在平面上呈透镜状零散分布,表现为"小甜点"的分布特点,单层甜点区范围一般小于 $2km^2$,平面上表现为更强的非均质性[6,7],因此,致密气开发需要在储量相对富集区内优选井位,以提高高产井的比例。

2.2 页岩气优质储层段集中分布,致密气有效储层多层分散分布

四川盆地五峰组—龙马溪组海相页岩优质储层段是指总有机碳含量(TOC)大于 2%、含气量大于 $2m^3/t$,硅质、碳酸盐岩、黄铁矿等脆性矿物含量大于 45% 的储层[8]。长宁、威远、昭通区块优质储层段厚度在 35m 左右,集中分布五峰组—龙$一_1$亚段底部(图 3),从下至上包括五峰组,龙$一_1^1$、龙$一_1^2$、龙$一_1^3$三个小层及龙$一_1^4$小层的底部。页岩气优质储层段集中分布主要也是由海相深水陆棚沉积环境稳定、变化慢的特点决定的,利于平台水平井组的部署和实施。相反,鄂尔多斯盆地致密砂岩气藏属陆相辫状河沉积环境,横向变化较快,垂向上有效储层多层叠置或分散分布(图 4),总厚度为 10m 左右,且盒$_{8上}$、盒$_{8下}$、山$_1$、山$_2$均有分布,单层厚度为 1~5m,横向非均质性和垂向上的多层分散分布[6,7],不利于水平井的部署和实施,直井和水平井有效储层钻遇比例均受到较大限制。

2.3 均具有异常低孔低渗特征,但流体赋存和流动特征有一定差异性

鄂尔多斯盆地致密气储集空间主要为砂岩内颗粒间孔隙,有效孔隙尺寸主要是微米级,纳米级孔隙多被束缚水占据,裂缝不发育,孔隙度为 6%~10%,储层基质渗透率小于 $0.1\times10^{-3}\mu m^2$;在不同区域地层水含量有较大差异,原始含水饱和度为 30%~50%,地层水含量较高的区域存在原始可动水,气水两相流是生产过程中的主要流动类型[9]。而页岩气储集空间主要为页岩内的纳米级孔隙,包括粒间微孔、黏土片间微孔、颗粒溶孔、粒内溶蚀孔及有机质孔等,平均孔隙在 100nm 左右,四川盆地龙马溪组页岩孔隙度为 2.43%~5%[8];基质

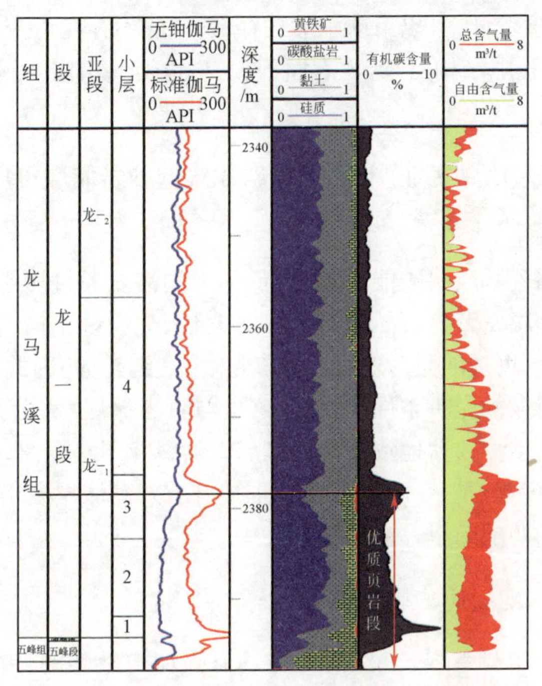

图3 四川盆地龙马溪组页岩优质储层段纵向分布

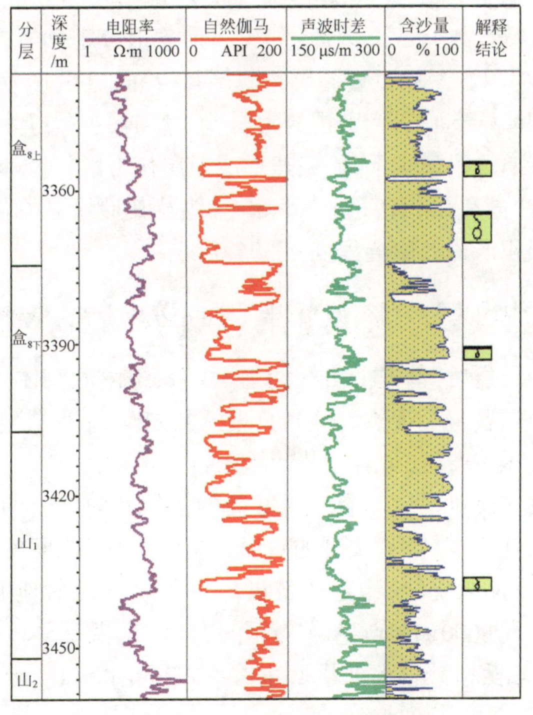

图4 鄂尔多斯盆地致密气藏有效储层纵向分布

渗透率极低,为 $10^{-5}\sim10^{-3}$mD①[10,11],微裂缝发育区,渗透率可达 0.1mD 左右[12];由于页岩气自生自储,含水饱和度较低,且以束缚水形式存在,一般不存在可流动的地层水,含气饱和度可高达 80% 以上,但孔隙结构复杂,比表面积大,与致密气相比,吸附气含量占比大[13-15],游离气与吸附气共存。

2.4 均需要改造来达到工业产量,但页岩和致密砂岩储层可压裂改造特征具有明显差异

页岩与致密砂岩储层基质渗流能力均较低,两者均需要通过压裂改造实现气井工业产量。但四川盆地页岩储层发育有天然裂缝和页理,且天然裂缝走向与目前最大水平主应力方向一般有一定的夹角,页理面是薄弱面,可诱导人工裂缝转向,因此,页岩储层一般具备人工压裂改造形成复杂缝网的先决条件。同时,这也与页岩储层“大液量、大砂量、大排量”的压裂技术相吻合。而鄂尔多斯盆地致密砂岩储层天然裂缝不发育,有效砂体零散分布,且规模较小,因此,适合“适度规模”人工压裂改造,形成与有效砂体规模相匹配的人工裂缝尺寸,以及与基质渗流能力相匹配的单一主裂缝系统。

2.5 开发过程中,地层应力敏感性不同

页岩和致密砂岩在实验室岩心尺度下,均具有强应力敏感的特征,即随有效应力的增加,储层渗透性逐渐降低[16],且渗透率越低,下降比例越大。但在气藏尺度下,致密砂岩中有效砂体为非连续分布,非有效储层为连续相,有效砂体内压力下降后,周围非有效储层支撑以及透镜状有效砂体的受力分解,有效砂体受应力作用较小,压敏效应对储层的伤害较小,对气井产量和累积产量影响较小,因此,鄂尔多斯盆地致密砂岩应力敏感效应对单井累积产量的影响幅度在 5% 以内。而页岩储层呈薄层状分布,且页理发育,应力变化极易造成页岩储层变形,影响气井生产动态,特别是放大压差生产时,主裂缝短期内迅速泄压,压敏效应导致近主裂缝区储层渗透率急剧下降,快速形成储层伤害区,过早阻挡外围气体进入主裂缝系统,严重影响气井产量和累积产量[17],四川盆地及其周缘五峰组—龙马溪组海相页岩应力敏感效应对单井累积产量的影响程度在 30% 以上。

2.6 气井产量曲线均为“L”形,但页岩气早期递减更快、低产期的生产时间更长

页岩气与致密气井压裂后,首先采出的均为裂缝系统内的流体。页岩气采用“千方砂、万方液”的大规模体积压裂,且受天然裂缝和页理的共同影响,形成复杂的裂缝网络,与基质的接触面积大,因此,放压生产时,页岩气井初始产量高于致密气井,但裂缝的储集能力是有限的,随着气体的采出,基质对裂缝与裂缝对井筒的供给能力之间的矛盾将会凸显,具体体现是气井产量迅速递减;同时页岩基质更加致密,基质与裂缝之间的导流能力差异更明显,因此,页岩气井早期产量递减更快、稳定产量更低[18-21]。四川盆地五峰组—龙马溪组海相页岩水平井单井累积产气 $8000\times10^4\text{m}^3$ 左右,首年累积产量占总累积产量的 30% 左右;鄂尔多斯盆地盒$_8$—山$_1$段陆相致密砂岩水平井单井累积产气 $6500\times10^4\text{m}^3$ 左右,首年累积产量占

① 达西(D),1D=$0.986923\times10^{-12}\text{m}^2$。

总累积产量的25%左右(图5)。

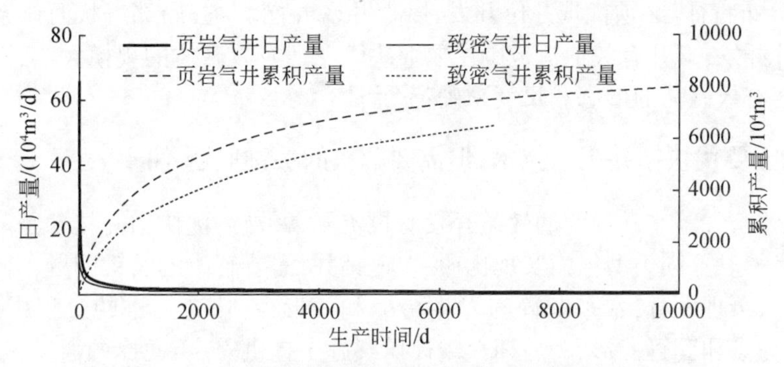

图5 页岩气井与致密气井日产量与累积产量对比

随着气井产量的下降,基质与裂缝的供给能力逐渐趋于平衡,页岩气与致密气井产量递减幅度均明显减小;同样页岩气井人工裂缝与基质的接触面积较大,即泄气面积较大,且页岩储层具有"大甜点"连续分布的特点,因此,页岩气井低产期时间更长,四川盆地五峰组—龙马溪组海相页岩气井生产周期长达30年左右。相反,致密气井"适度规模"压裂后人工裂缝与基质的接触面积相对较小,且有效砂体规模也有限,因此,致密气井低产期时间相对较短,鄂尔多斯盆地盒$_8$—山$_1$段致密砂岩气井生产周期一般小于20年。

3 开发策略与技术对比分析

3.1 均需要优选相对富集区,但考虑的因素各有侧重

致密砂岩气藏相对富集区优选侧重于地质因素。在鄂尔多斯盆地致密砂岩储层中,辫状河沉积体系叠置带是心滩等有效砂体多期叠置发育的有利相带,为流体提供良好的储集空间,是气藏形成相对富集区的基础;致密砂岩气属二次运移成藏,成藏时气体对孔隙中地层水的驱替程度有所差异,因此,区域性含水特征的变化也是影响相对富集区的重要因素。致密砂岩气藏储层横向非均质性强,纵向分散分布,往往需要完钻一定数量的评价井才能确定相对富集区。

页岩气相对富集区优选地质与工程因素并重。海相页岩气优选相对富集区时,除考虑有机质发育的深水陆棚沉积体系的分布外,保存条件、地层压力系数、储层可压性、埋藏深度也是不可或缺的条件,保存条件和地层压力系数反映储层的物质基础,储层可压性保障压裂改造效果,埋藏深度考验目前的工程技术水平。海相页岩储层纵横向分布相对稳定,在研究区构造特征分析的基础上,仅需完钻少量评价井即可确定相对富集区。

3.2 均采用小井距密井网开发,但井型井网有一定差异

由于储层基质致密,页岩气与致密气单井控制的有效泄流范围均较小,需要密井网开发,但井型井网不同。页岩气有利层段集中,且横向分布稳定,具有"大甜点"的分布特征,长水平井+大规模体积压裂是效益开发的必然选择,且利于采用平台化均匀布井方式开发,由于原始基质渗透率达到纳达西级别,气井有效泄流范围主要取决于改造规模大小,开发井距

主要依据人工裂缝的横向延伸长度来确定。而致密气多层分散分布，且横向连续性差，具有“小甜点”的分布特征，必须优选井位开发，以直井密井网为主，局部辫状河体系叠置区选择性采用水平井开发，气井有效泄流范围主要取决于钻遇有效砂体的规模大小，因此井网井距主要依据钻遇有效砂体的形态和尺度来确定[6,22]。

3.3 均需要优快钻井工艺技术，但需要解决的关键问题不同

页岩气、致密气等非常规气的效益开发对投资异常敏感，优快钻井技术是降低开发投资的主要手段之一。不同类型气藏实现优质、快速钻井，需要解决的关键问题不同。页岩储层呈薄层状分布，页理发育，水平井筒易发生垮塌和变形，在快速钻进的过程中，保持井眼光滑，防止井壁垮塌和套管变形，为后期大规模压裂施工提供安全、便捷的通道，是页岩气钻井工艺需要解决的关键问题。目前主要通过选用旋转导向工具和油基泥浆实现优快钻井的需求，且随着技术的进步，机械钻速不断提高，已由初期的4m/h提升至目前的7m/h。致密砂岩储层砂泥岩交互分布，且有效砂体规模较小，水平井需要钻穿多个有效砂体才能达到高产，在泥岩中的安全钻进及前方砂体的准确导向，是致密气钻井工艺需要攻克的关键问题。随着钻井工艺技术的不断进步，致密气水平井有效砂体钻遇比例已由2006年的50%提高至目前的70%以上。

3.4 核心开发工艺均是压裂投产，但压裂改造方式和压裂规模具有较大差异

致密气和页岩气基质储层均无工业产能，均必须进行压裂改造才能获得工业产量。鄂尔多斯盆地致密砂岩气藏，为适应有效砂体规模尺度，水平井采用水力喷射或裸眼封隔器压裂工具进行多段“适度规模”压裂，即采用以胍胶为主的压裂液体系、3～5m^3/min小排量注入、单段注入液量100m^3左右、加砂量35t左右，从趾端到根端逐段实施，每段形成单一主裂缝，构建多条致密储层“高速公路”，单井压裂周期3～5d，压裂后的单井产量3×10^4～$5\times10^4m^3$，达到直井的3～4倍。

在四川盆地海相页岩气开发方面，“十二五”期间压裂改造工艺技术的长足进步，直接实现了页岩气从无效资源向有效工业产量的技术跨越，特别是大型分段多簇体积压裂工艺技术和“工厂化”作业模式直接推动了页岩气的效益开发[23,24]。实践证明，在气井压裂改造范围内，形成复杂的缝网系统是页岩气储层改造追求的目标，也是页岩气井获得高产的必要条件。这也就决定了页岩气的压裂工艺技术不同于致密气：全程大排量低黏滑溜水，可形成复杂裂缝；粉砂+低密度支撑剂、段塞式加注，可保障施工顺利完成；“千方砂、万方液”，可满足改造强度；大通径桥塞或可溶桥塞，既可满足分段的需求，又利于压后快速投产。目前四川盆地五峰组—龙马溪组海相页岩气水平井主体泵注排量12～14m^3/min、100目粉砂+40/70目低密度陶粒、60～80m段长、单段注入液量1800m^3左右、单段加砂量100t左右，压后平均测试产量达到$18\times10^4m^3/d$以上。同时，“拉链式”压裂作业模式[25,26]（图6）大大提高了作业效率、降低了压裂成本，如一个6口井的平台，先进行半支3口井拉链压裂，这3口井的压裂返排液供另半支2口井拉链压裂重复使用，这2口井的压裂返排液供最后1口井单独压裂重复使用，该作业模式使平台半支平均压裂周期缩短至30d。

3.5 均采用单井自然递减、井间接替的气田生产方式，但单井配产方式有一定差异

页岩气和致密气井均采用自然递减、后期增压的方式生产，由于早期产量递减较快，均

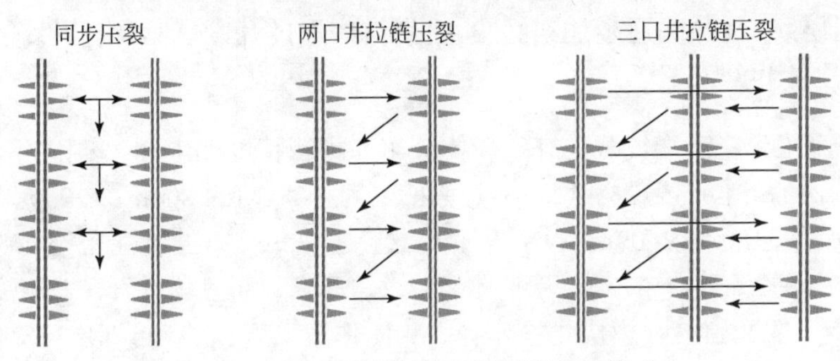

图6 同步压裂与“拉链式”压裂作业模式

需要不断地钻井投产进行产量接替，才能保持区块的上产或稳产，但两种气藏单井配产方式存在差异。基于薄层状页岩储层以及压裂后复杂缝网强应力敏感的特点，建议采用“控压限产”的生产方式，在满足早期井筒携液和经济开发的情况下，尽可能降低初期配产，使作用于SRV区的有效应力稳定变化，可延缓主裂缝周围储层被伤害的时间，延长周边次级裂缝有效导流能力的持续时间。建议四川盆地及其周缘五峰组—龙马溪组海相页岩气早期配产$6\times10^4\sim8\times10^4m^3/d$，单井累积产量具备在放压生产单井累积产量的基础上提高30%以上的潜力。致密砂岩储层在气藏尺度下应力敏感效应对单井累积产量的影响很小，因此，致密砂岩气井可放大压差生产，但考虑生产压差放大到一定程度后，气井出砂风险明显增加，故气体流速以小于冲蚀速度为宜。

4 页岩气与致密气开发技术的相互借鉴

4.1 致密气以成本为目标导向的开发技术优化是页岩气降低开发成本的主要方向

鄂尔多斯盆地致密砂岩气藏开发初期，制约其规模开发的核心问题是开发成本高、没有经济效益。解决途径是首先根据单井生产能力倒算单井综合开发成本最高界限，以苏里格致密气田为例，立足苏里格平均单井累积产量$2200\times10^4m^3$的客观条件，在当时的气价条件下，测算单井综合投资800万元是气田效益开发的临界点(内部收益率8%)；然后根据倒逼技术进步和管理创新，通过快速钻井、压裂优化、丛式井部署等技术以及引入“5+1”合作开发的管理模式，实现了单井综合开发成本由1400万元降低到800万元以内的目标，推动了苏里格大气田的规模建产和效益开发。目前，我国南方海相页岩气处于早期开发阶段，实现效益开发面临同样的问题，借鉴致密气效益开发的思路，立足页岩气平均单井累积产量$8000\times10^4m^3$的认识，提出单井综合投资4000万元、单位操作成本0.2元/m^3的成本目标和各环节的成本构成，同时提出页岩气降低单井成本的三点具体建议：优化井场与井距、适当增加水平井长度和单个平台钻井数量；加强市场化力度，提高关键设备和工具的国产化水平，优化液体和支撑剂体系及用量，力争钻完井成本至少降低40%；推广数字化、信息化管理，减少用工人数，降低操作成本。

4.2 对比致密气井网逐步加密经验，页岩气应通过开发试验优化井距，实施一次性井网部署

鄂尔多斯盆地致密砂岩气藏受有效砂体分布和预测精度的限制，采用井网逐步加密的方式进行多轮次的井位部署，持续提高气藏采收率，从早期的800m×1200m井网、采收率17%，第一轮加密至800m×600m井网、采收率提高至32%，第二轮加密至400m×600m井网、预计采收率可提高至47%左右，且整体保持效益开发。而海相页岩气普遍采用批量化布井、平台化设计、工厂化作业、模块化操作等大兵团作战，尽可能降低单井开发成本，才能做到效益开发，因此，单井作业或分散加密是不可取的，井网逐步加密的思路不是最优的开发部署方式。要提高页岩气采收率，应结合页岩气效益开发的施工特点，首先开展小范围变井距先导试验，优化开发井距，然后按最优井距，根据地面地形条件，实施一次性井网部署，同一平台，双钻机或多钻机同步钻井，拉链式压裂，平台同一半支井统一排采投产。

4.3 如何降低低效井比例，致密气更注重储层预测和井位优选，页岩气更注重工艺技术

在单井开发技术和不同类型井最终可采储量（EUR）一定的前提下，如何降低低效井比例是提高区块累积产量和储量采出程度的关键。针对不同类型气藏的地质特征，降低低效井比例的具体措施也存在较大差异。致密气藏有效砂体空间上分布具有强非均质性，进行储层精细描述和井位优选，是降低致密气低效井比例的关键。相反，页岩气优质储层纵横向分布稳定，注重工艺技术与地层的适应性以及工艺技术本身的成功率，是降低页岩气低效井比例的关键。

4.4 页岩气工厂化作业及新型压裂工艺技术，有助于更低品位资源的有效开发动用

受页岩气井体积压裂后开发效果的启发，我国致密气也开展了大型分段压裂改造试验，初步取得了一定的效果。优化致密气大型分段多簇压裂改造工艺、实施“工厂化”压裂作业模式，是未来致密气进一步提高单井产量和开发效益最现实的途径。

目前，针对页岩气效益开发发展的低成本开发技术，如滑溜水+小粒径石英砂、长水平段钻井+趾端阀辅助压裂、后勺子井型钻井和压裂等技术在页岩气规模效益开发中的成功应用，也将成为我国其他类型更低品位资源有效开发动用的关键技术。

5 结论与认识

海相页岩气与陆相致密砂岩气同属于非常规气藏，大面积连片分布，储量基础和开发潜力巨大。通过对比分析两者的异同，得出以下几点结论和认识：

（1）地质特征差异是本质特点。海相页岩气相比陆相致密砂岩气而言，储层非均质性弱，是批量平台化布井的基础。极低的储层渗透率是工程技术特别是压裂方式选择的关键影响因素，基质渗透率越低，气井达到工业气流需要的泄流面积越大，要求裂缝对基质储层的切割程度就越高。另外，页岩气自生自储，储层中有机质的存在影响了气体在地下的赋存

状态,吸附气与游离气共存,也是与致密气流动状态差异的主要原因之一。

(2)工程技术差异是适应手段。页岩气开发评价阶段,直井压裂、水平井多段压裂初步反映了页岩气的开发潜力,但无法获得高产井,主要原因是裂缝间距大、供给范围小。为适应页岩储层的地质特点,长水平井钻井与大型体积压裂改造技术成为页岩气井获得工业产量及高产的关键措施,也是适应比致密气藏更复杂储层开发的核心技术。

(3)生产特征差异是必然结果。致密气藏储层压裂后,形成了多条主裂缝,基质内气体通过自身的渗流能力直接流向主裂缝,从而进入井筒采出,单位面积的基质供给能力强,气体流动路径简单,因此,基质对裂缝的供给与裂缝对井筒的供给差异较小。而页岩储层通过体积压裂后,形成了尺度差异大且分布错综复杂的缝网系统,气体流动路径为基质—微裂缝—次裂缝—主裂缝—井筒,流动路径复杂,且单位面积的基质供给能力弱,因此,基质—微裂缝—次裂缝对主裂缝的供给能力与主裂缝对井筒的供给能力差别较大,这是页岩气早期产量高、短时间内产量快速下降的根本原因。

参考文献

[1] 马新华,贾爱林,谭健,等. 中国致密砂岩气开发工程技术与实践[J]. 石油勘探与开发,2012,39(5):572-579.

[2] 贾爱林,位云生,金亦秋. 中国海相页岩气开发评价关键技术进展[J]. 石油勘探与开发,2016,43(6):1-8.

[3] 董大忠,邹才能,杨桦,等. 中国页岩气勘探开发进展及发展前景[J]. 石油学报,2012,33(S.1):107-114.

[4] 杨华,刘新社,杨勇. 鄂尔多斯盆地致密气勘探开发形势与未来发展展望[J],中国工程科学,2012,14(6):40-48.

[5] 邹才能,董大忠,王玉满,等. 中国页岩气特征、挑战及前景(一)[J]. 石油勘探与开发,2015,42(6):689-701.

[6] 何东博,贾爱林,冀光,等. 苏里格大型致密砂岩气田开发井型井网技术[J]. 石油勘探与开发,2013,40(1):79-89.

[7] 贾爱林. 中国储层地质模型20年[J]. 石油学报,2011,32(1):181-188.

[8] 邹才能,等. 非常规油气地质(第二版)[M]. 北京:地质出版社,2013.

[9] 王国亭,何东博,王少飞,等. 苏里格致密砂岩气田储层岩石孔隙结构及储集性能特征[J]. 石油学报,2013,34(4):660-666.

[10] Sakhaee-Pour A, Steven L B. Gas permeability of shale[J]. SPE Reservoir Evaluation & Engineering,2012,15(4):401-409.

[11] Michael D B, Xia W W, Shelton J. Shale Gas Play Screening and Evaluation Criteria[J]. China Petroleum Exploration,2009,14(3):51-64.

[12] 糜利栋,姜汉桥,李俊键,等. 页岩储层渗透率数学表征[J]. 石油学报,2014,35(5):928-934.

[13] 王敬,罗海山,刘慧卿,等. 页岩气吸附解吸效应对基质物性影响特征[J]. 石油勘探与开发,2016,43(1):1-8.

[14] Wei Y, Sepehrnoori K, Tadeusz W P. Modeling gas adsorption in Marcellus Shale with Langmuir and BET isotherms[J]. Society of Petroleum Engineering Journal,2016,21(2):589-600.

[15] Wei Y, Sepehrnoori K, Tadeusz W P. Evaluation of gas adsorption in Marcellus Shale[C]. Paper SPE 170801presented at the SPE Annual Technical Conference and Exhibition held in Amsterdam, The

Netherlands, 27-29 October 2014.

[16] 张睿,宁正福,杨峰,等. 页岩应力敏感实验与机理[J]. 石油学报,2015,36(2):224-231.

[17] 朱维耀,马东旭,朱华银,等. 页岩储层应力敏感性及其对产能影响[J]. 天然气地球科学,2016, 27(5):892-897.

[18] Wei Y S, He D B, Wang J L, *et al.* A coupled model for fractured shale reservoirs with characteristics of continuum media and fractal geometry [C]. Paper SPE 176843 presented at the SPE Asia Pacific Unconventional Resources Conference and Exhibition held in Brisbane, Austrilia, 9-11 November 2015.

[19] 齐亚东,王军磊,庞正炼,等. 非常规油气井产量递减规律分析新模型[J]. 中国矿业大学学报,2016, 45(4):772-778.

[20] Wang J L, Yan C Z, Jia A L, *et al.* Rate decline analysis of multiple fractured horizontal well in shale reservoir with triple continuum[J]. J. Cent. South Univ, 2014, 21:4320-4329.

[21] 王军磊,贾爱林,何东博,等. 致密气藏分段压裂水平井产量递减规律及影响因素[J]. 天然气地球科学,2014,25(2):278-285.

[22] 何东博,王丽娟,冀光,等. 苏里格致密砂岩气田开发井距优化[J]. 石油勘探与开发,2012,39(4): 458-464.

[23] Waters G, Dean B, Downie R, *et al.* Simultaneous hydraulic fracturing of adjacent horizontal wells in the Woodford Shale [C]. Paper SPE 119635presented at the 2009 SPE Hydraulic Fracturing Technology Conference held in The Woodlands, Texas, USA, 19-21 January 2009.

[24] Rafiee M, Soliman M Y, Pirayesh E. Hydraulic fracturing design and optimization: a modification to zipper frac[C]. Paper SPE 159786 presented at the SPE Eastern Regional Meeting held in Lexington, Kentucky, USA, 3-5 October 2012.

[25] Qiu F D, Xu J, Pope T L. Simulation study of zipper fracturing using an unconventional fracture model [C]. Paper SPE 175980 presented at the SPE/CSUR Unconventional Resources Conference held in Calgary, Alberta, Canada, 20-22 October 2015.

[26] Sierra L, Mayerhofer M. Evaluating the benefits of zipper fracs in unconventional reservoirs[C]. Paper SPE 168977 presented at the SPE Unconventional Resources Conference-USA held in The Woodlands, Texas, USA, 1-3 April 2014.

二、开发地质类

四川盆地志留系龙马溪组优质页岩储层特征与开发评价

贾成业[1] 贾爱林[1] 韩品龙[1] 王建君[2] 袁 贺[1] 乔 辉[1]

(1. 中国石油勘探开发研究院;2. 中国石油浙江油田公司勘探开发研究院)

摘要:随着页岩气国家级示范区的建设,中国页岩气进入规模开发阶段;到2016年12月,长宁—威远、昭通和涪陵国家级页岩气示范区实现年产气量$79\times10^8m^3$,到2030年页岩气规划年产量达到$800\times10^8\sim1000\times10^8m^3$,展现出良好的发展前景。四川盆地龙马溪组地层厚达300m左右,进一步识别出优质页岩段,评价优质页岩储层特征是开发阶段的首要任务。依据笔石带特征和沉积相演化,将龙马溪组划分为龙一、龙二2个层段;志留纪早期深水陆棚沉积环境控制了优质页岩的空间分布,岩心观察识别出龙一段底部黑色碳质页岩和硅质页岩是优质页岩发育的有利岩相,优质页岩具有高有机碳含量、高脆性矿物含量、高含气量的特征;优质页岩段地质储量丰度集中,平均储量丰度达到$1.85\times10^8m^3/km^2$,是页岩气规模化开发的有利层位。

关键词:四川盆地;优质页岩;储层特征;碳质页岩;硅质页岩;含气量

0 引言

页岩气是非常规天然气的主要类型之一,是指赋存于富有机质泥页岩及其夹层中,以吸附或游离状态为主要存在方式的非常规天然气,成分以甲烷为主,是一种清洁、高效的能源资源[1]。据美国能源信息署(EIA)公开数据[2],2015年美国页岩气产量达到15086.64Bcf(折合4272.05×10^8 m^3),超过致密气成为开发规模最大的非常规天然气。中国是继美国和加拿大之后第三个实现页岩气商业化开发的国家。2009年12月18日,中国第一口页岩气评价井威201井开钻并成功获得商业气流;2013年1月,国家发展和改革委员会(以下简称国家发改委)和国家能源局批准设立首批页岩气国家级示范区,有效推动了页岩气勘探开发进程;到2016年12月,四川长宁—威远、滇黔北昭通和重庆涪陵3个国家级海相页岩气示范区已实现年产量79×10^8 m^3;到2030年,全国页岩气规划产量达到$800\times10^8\sim1000\times10^8$ m^3,页岩气展现出良好的发展前景。

四川盆地及其周缘志留系龙马溪组富有机质海相页岩分布面积广、厚度大,是最有利的页岩气开发层系。前人对志留系龙马溪组的研究主要集中在古生物地层学、岩相古地理、沉积环境和有机地球化学等方面。有些学者对志留系笔石类型和发育特征的研究为志留系地层划分建立了统一的标准[3-7],有些学者对志留系古地理和沉积环境开展了研究工作[8-10],还有学者对志留系龙马溪组地球化学特征进行了分析[11-13]。前人的研究成果为龙马溪组有利区评价和建产区优选提供技术支持,奠定了龙马溪组页岩规模开发的基础。但随着页岩气开发评价工作的深入,根据页岩气开发生产的需要,在龙马溪组厚达300m左右的地层中进一步识别优质页岩段,明确优质页岩段特征并开展相关评价,是四川盆地页岩气进入规模开发阶段亟待解决的任务。

1　龙马溪组地层结构与沉积环境

1.1　龙马溪组地层结构

尹赞勋[14]创建了龙马溪页岩一名，命名剖面位于湖北秭归新滩东南1km的龙马溪。全国地层委员会根据龙马溪组垂向沉积序列将龙马溪组分为上段、下段两部分(表1)，下段以黑色、灰黑色泥页岩为主，底部为黑色碳质/硅质页岩，产丰富的笔石化石；上段以灰绿色、黄绿色泥页岩为主。在川东南地区与下伏奥陶系观音桥组整合接触，在黔北地区与下伏奥陶系临湘组地层呈假整合接触；与上覆石牛栏组呈整合接触(图1)。上覆石牛栏组岩性以灰色粉砂岩、深灰色灰岩为主，底部岩性为灰岩与龙马溪组顶部灰色/深灰色泥岩界线明显[图1(a)]；下伏观音桥组为薄层状介壳灰岩，富含生物介壳，与龙马溪组底部碳质页岩界线明显[图1(b)]。

表1　四川盆地龙马溪组地层结构与特征

年代地层				重要化石带		长宁—威远—昭通地区
系	统	组	段	笔石带	牙形刺带	
志留系	中统	上覆地层				石牛栏组
	下统	龙马溪组	上段	*Sedgwickii communis triangulates*	*D. staurogn-athoides*	灰黑色、黑色泥岩，顶部见薄层状或透镜状粉砂岩
			下段	*Leei cyphus vesiculosus acuminatus persculptus*	*D. kentuck-yensis*	黑色页岩、硅质页岩、碳质页岩，底部页理发育、富含笔石及黄铁矿结核/条带
奥陶系	上统	下伏地层				观音桥组

注：据文献[10]修改

(a)与上覆石牛栏组呈整合接触

(b)与下伏观音桥组呈整合接触

图1　志留系龙马溪组地层接触关系

1.2　龙马溪期沉积环境

安静、稳定的半深水-深水还原环境是海洋富有机质沉积物生成、发育和保存的最佳环境[15]。龙马溪期早期继承了五峰期的沉积特点，沉积主体为半局限浅海相的深水陆棚，沉积速率慢，沉积时间上超过龙马溪期的一半，但沉积厚度为100～135m，仅占整个龙马溪组地层厚度的30%～45%，发育一套黑色碳质、硅质页岩和黑色页岩组合，龙马溪组早期沉积中心位于泸州—永川、石柱—彭水地区，这一时期半局限深水陆棚水体较深，安静、稳定，沉积厚度大、有机质丰富（笔石发育）、有机质类型以Ⅰ型为主、热演化程度高，是龙马溪组有效烃源岩发育的主要时期（图2）。龙马溪组晚期，主要发育深灰色泥页岩、灰色泥岩和粉砂质泥岩组合，沉积主体为浅水陆棚，沉积速率明显大于早期，沉积厚度为150～165m，沉积中心主要在泸州—永川、石柱地区以及张家界地区；该时期水体较浅，有机质相对不发育，笔石含量明显降低。

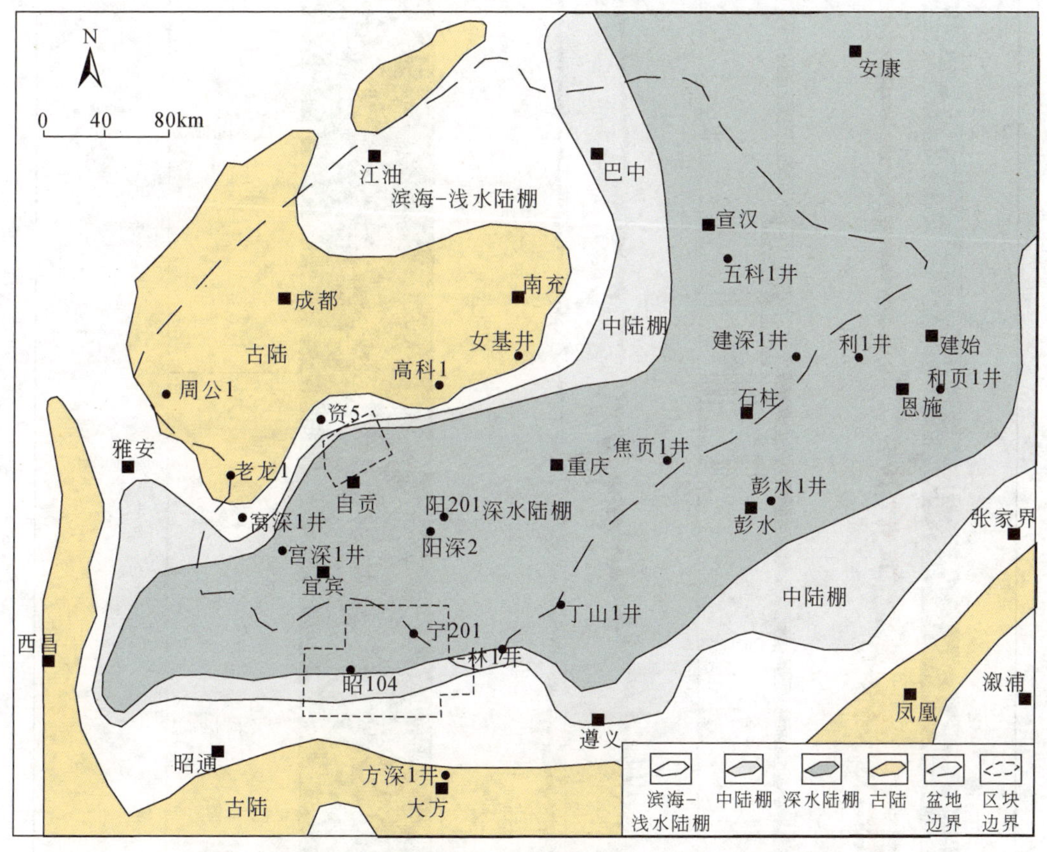

图2　志留系龙马溪组早期沉积环境图（据文献[10]和[15]修改）

1.3　龙马溪组小层结构

参照全国地层委员会将龙马溪组分为上段、下段两部分，结合石油天然气行业标准（SY/T 5363—1997）《含油气层系划分标准》[16]，将志留系含油气层系龙马溪组含油气层组自下而上划分并命名为龙一段、龙二段。

以宁203井为例（图3），龙二段为浅水陆棚沉积，岩性为底部黑色/灰黑色页岩，中部深灰色泥页岩，顶部灰色泥岩，偶见粉砂质条带，该段黏土含量相对较高平均为72%，硅质和碳酸盐岩含量相对较低，有机碳含量低于1%；龙一段顶部为深水陆棚向浅水陆棚过渡相，底部为深

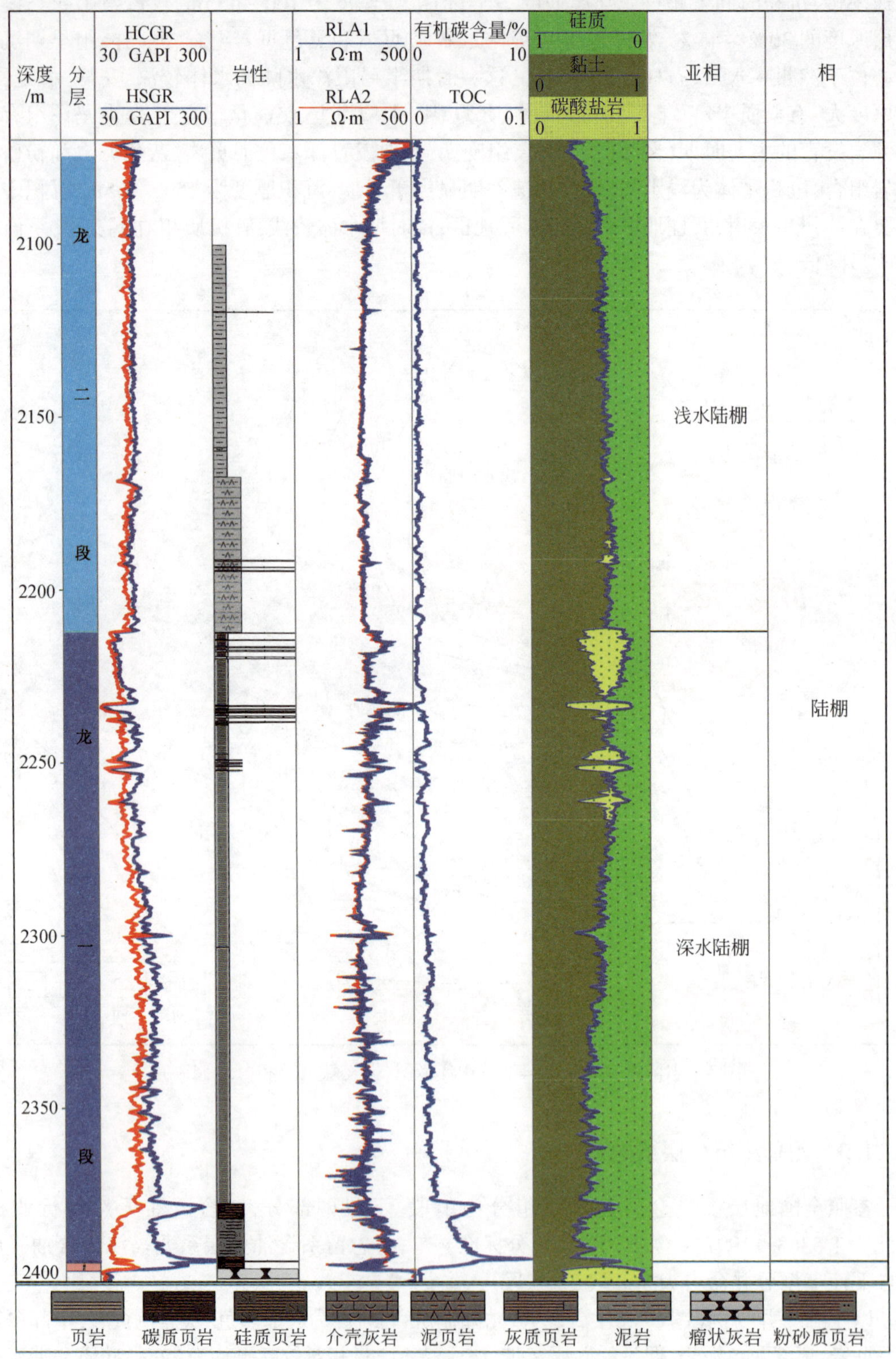

图3　四川盆地典型页岩气井龙马溪组地层综合柱状图

HCGR. 无铀伽马；HSGR. 标准伽马；RLA1. 浅电阻率；RLA2. 深电阻率；TOC. 总有机碳含量

水陆棚相，该段底部为有机质含量高的黑色硅质页岩和黑色碳质页岩，中部为黑色页岩，向上逐步发育深灰色页岩，有机碳含量从底部到顶部呈逐渐降低的趋势，底部富有机质页岩段平均有机碳含量达4%，是优质页岩发育的主要部位；中部和上部黑色、深灰色页岩段有机碳含量低于2%。

2 龙马溪组优质页岩特征

2.1 岩相类型及特征

四川盆地龙马溪组厚达300m左右，通过对探区内23口井近2000m岩心观察，共识别出7种主要岩相，即黑色碳质页岩、黑色硅质页岩、黑色页岩、深灰色泥页岩、灰色泥岩、灰色泥灰岩和泥质粉砂岩[17-19]。其中，黑色碳质页岩和黑色硅质页岩是优质页岩段发育的有利岩相。

黑色碳质页岩：页理发育，含大量碳化有机质，有机碳含量为4%～15%，笔石化石非常发育，层面上笔石覆盖面积可达70%以上，呈黑色、强染手的特征；主要矿物有黏土矿物、石英、长石，含黄铁矿和方解石，石英含量≤40%［图4(a)，(i)］。

黑色硅质页岩：纹层发育，表面较光滑，有机碳含量为3%～4%，笔石覆盖层面面积一般<30%，呈黑色；矿物成分中石英含量>50%，黏土含量<40%［图4(b)，(i)］。

黑色页岩：纹层发育，有机碳含量为2%～3%，笔石覆盖层面面积一般<10%，矿物成分中石英含量为40%～50%，矿物成分中石英含量为40%～50%，黏土含量约为50%［图4(c)］。

黑色碳质页岩、黑色硅质页岩和黑色页岩分布于龙马溪组下部，反映出水体较深的还原环境，为深水陆棚相特征，利于形成有利的烃源岩，TOC均大于2%，是页岩气开发有利岩相；深灰色泥页岩、灰色泥岩、灰色泥灰岩和泥质粉砂岩主要发育在龙马溪组上部，反映出水体变浅，由缺氧的还原环境逐渐过渡为富氧的氧化环境，有机质保存条件较差，有机质较少，有机碳含量<2%，为浅水陆棚沉积环境。

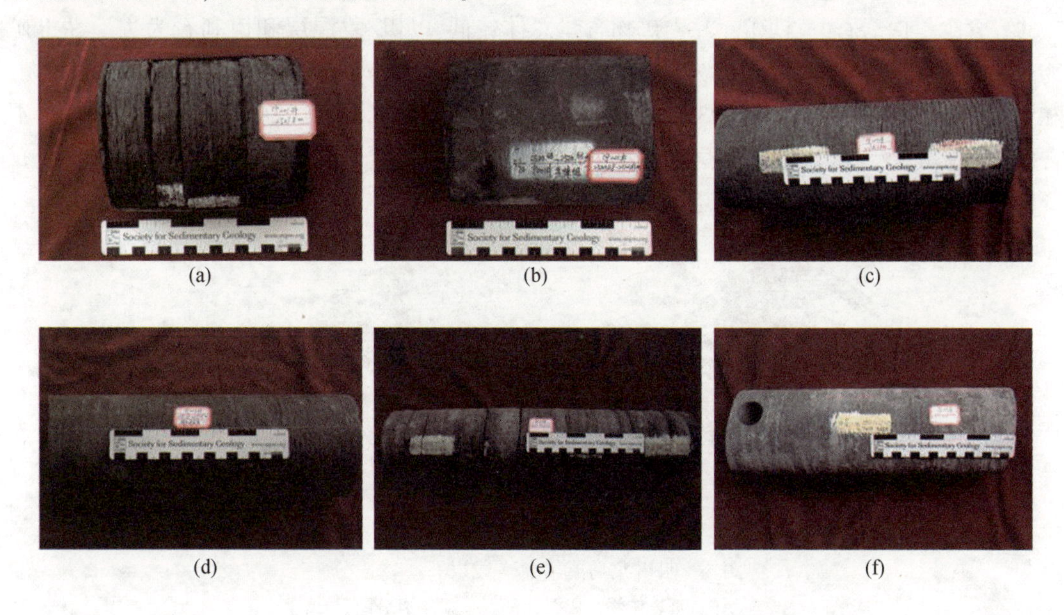

(a) (b) (c)

(d) (e) (f)

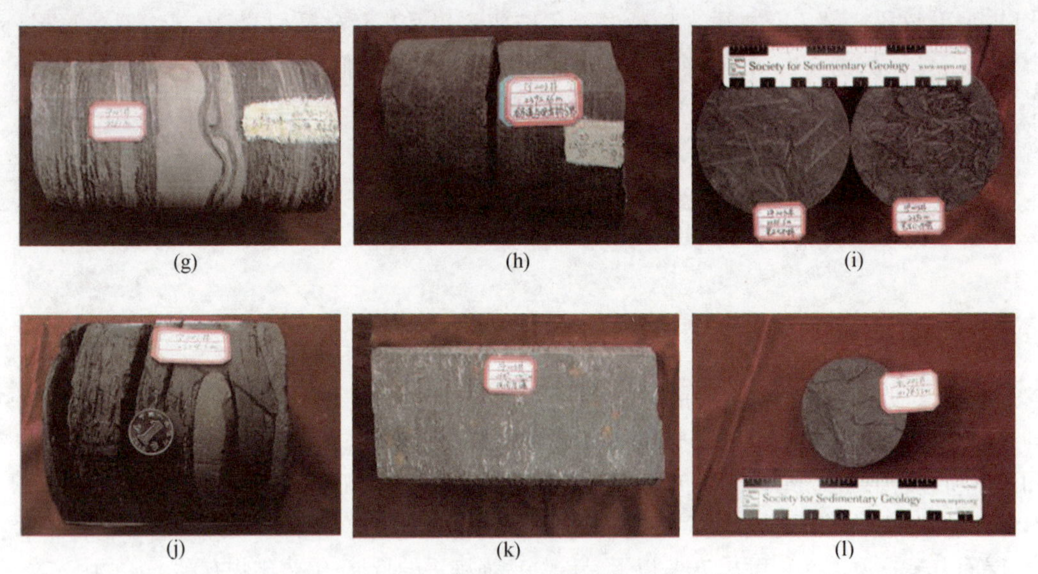

图4 四川盆地龙马溪组岩相类型与标志性特征

(a)宁201井,2507.8m,黑色碳质页岩;(b)宁201井,2520.48m,黑色硅质页岩;(c)宁203井,2378.38m,黑色页岩;(d)宁203井,2279.58m,深灰色页岩;(e)宁203井,2232.5m,深灰色泥页岩;(f)宁203井,2116.34m,灰色泥灰岩;(g)宁203井,2161m,粉砂岩;(h)龙马溪组与观音桥组分界面岩性变化(介壳化石);(i)宁203井,2386.6m(黑色硅质页岩)与2392m(黑色碳质页岩)处笔石含量对比;(j)宁201井,2504.1m,黄铁矿结核;(k)宁203井,2347.1m,高角度张开缝;(l)威203井,3178.5m,方解石充填构造缝

2.2 矿物组分

岩样X射线衍射分析表明,页岩含矿物类型包括:石英、长石、方解石、白云石、黄铁矿和黏土等矿物,其中黏土矿物主要为伊利石、伊/蒙混层和绿泥石。龙马溪组龙一段优质页岩矿物成分中石英含量最高,平均含量占37%;其次是方解石和斜长石,平均含量分别为28%和7.6%,含少量白云石、黄铁矿。黏土矿物含量总体较低,以伊/蒙混层和伊利石为主。黏土矿物含量为5.1%~58.2%,平均为23.8%;黏土矿物均以伊利石(4%~95%,平均为60.8%)和绿泥石(0%~73%,平均为23.9%)为主,次为伊/蒙混层(0%~92%,平均为15.3%),不含蒙脱石(图5)。以石英、方解石、斜长石等脆性矿物为主要组分,同时不含高岭石、蒙脱石等膨胀矿物,有利于压裂施工作业。

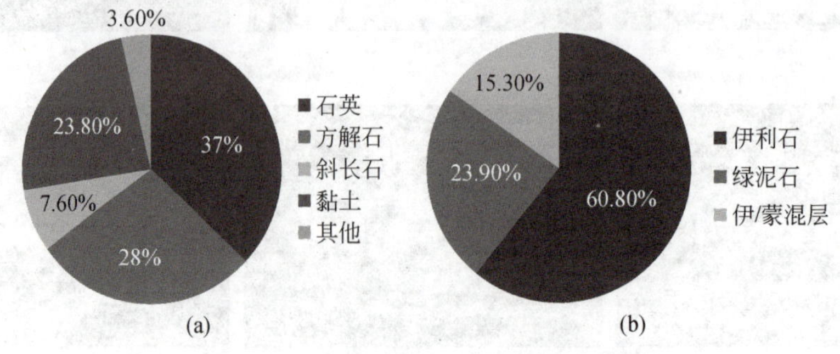

图5 四川龙马溪组龙一段岩石矿物成分

(a)页岩矿物组成图;(b)黏土矿物成分图

2.3 储集空间类型与特征

孔隙是页岩的主要储集空间[20]。采用 HITACH IM400 型抛光仪对页岩样品进行氩离子抛光操作;然后将抛光后的样品固定在样品台上,利用 eIGMA 型热场发射扫描电镜观察孔隙形状和分布规律。扫描电镜照片显示(图 6),龙马溪组页岩孔隙类型包括:有机孔、晶间孔、晶内溶孔和粒间孔。

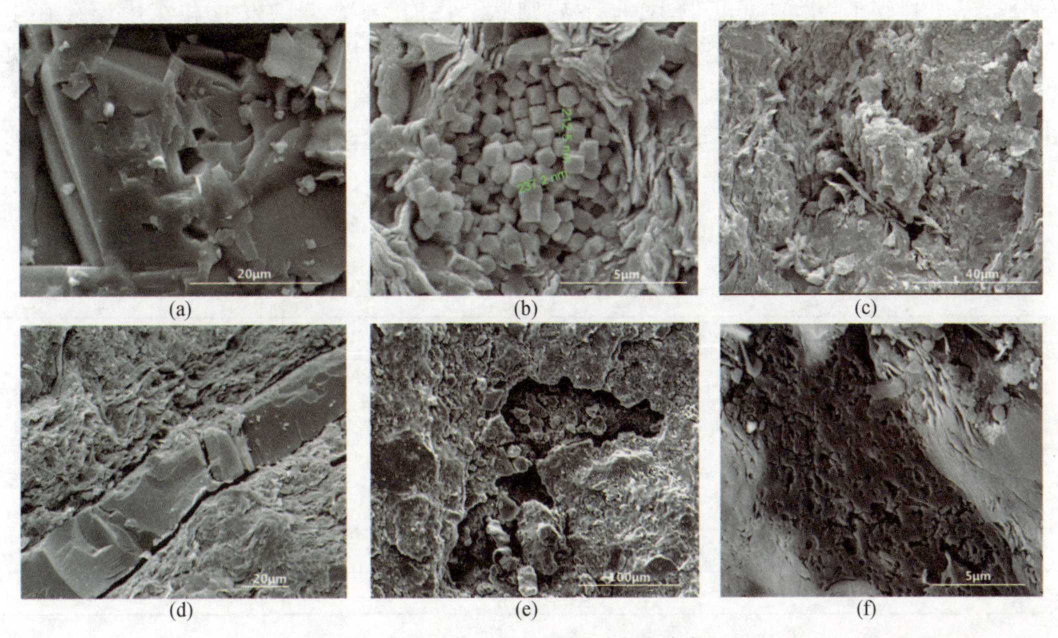

图 6　四川盆地龙马溪组页岩孔隙类型

(a) YS112 井,2402.76m,方解石晶内孔;(b) YS108 井,2389.8m,黄铁矿晶间孔;(c)宁 203 井,2336.3m,粒间孔;(d)成岩缝,威 201 井,1517.21m,成岩收缩缝;(e) YS112 井,2382.3m,晶间溶孔;(f)宁 203 井,2378.0m,有机质孔

有机质孔在页岩中广泛发育,形成于后期有机质热演化阶段。当烃源岩达到生烃门限时,有机质开始向烃类物质转化,泥页岩中开始产生生烃、排烃作用形成的有机质孔隙。有机质孔隙一般呈片麻状、凹坑状、椭圆形、多角形及不规则形状,孔径范围变化较大,为 5 ~ 1500nm,属于中–大孔范围,不仅发育纳米级有机质孔隙,甚至发育微米级有机质孔隙,常伴生发育黄铁矿晶间孔、黏土矿物片间孔、矿物颗粒孔等孔隙类型。

晶间孔、晶内溶孔同属于粒内孔隙。粒内孔隙多为成岩作用后期改造形成,原生较少。黄铁矿晶间孔在页岩中发育较广泛。黄铁矿团块被局部溶解后,充填有机质及矿物雏晶,有机质常与黄铁矿颗粒呈包裹与被包裹关系,颗粒内中孔和有机质大孔发育,孔隙连通性较好。方解石和白云石晶内溶孔相对较少,且孔径较小,溶孔仍保存完整的形态,没有挤压变形的迹象,表明晶内溶孔于埋藏期溶蚀作用下形成。

2.4 裂缝发育特征

构造缝、页理缝和成岩收缩缝是页岩中最常见的裂缝类型[21,22]。构造缝通常分为充填型和未充填型 2 种[图 4(k),(l)]。在龙马溪组龙一段可见高角度构造缝。在构造裂缝面上通常可见明显的镜面和擦痕现象,裂缝宽度为 0.5 ~ 1mm,多数井段裂缝密度为 0.1 ~ 4 条/m,裂缝多被方解石充填(表 2)。构造缝尺寸通常在毫米级,是天然气运移和渗流的优势通道。页理一般呈水平、片状,无论是露头还是岩心都可以观察到,是页岩的独特特征。安静稳定的深水环境下,沉积速率低,沉积物呈纹层状堆积,但由于有机质含量高,在成岩压实作用下,有机质和黏土矿物强烈收缩,在薄层状的页理间形成页理缝,页理缝基本在微米级。成岩收缩缝是成岩过程中由于水分的蒸发在脱水作用下形成的微裂缝,常贴矿物颗粒分布,部分可以切割碎屑,缝的宽度很小,呈不规则状。目前,对裂缝的发育和作用还存在争议[23]。由于页岩系统渗透率极低,通常在纳达西级别,一定规模的构造裂缝有利于提高页岩储层的渗透性,形成优势渗流通道,是页岩气井高产的关键因素。但如果构造裂缝影响了压裂或者破坏了盖层就会起到负面的作用。同样,页理缝也有利于改善页岩储层的导流能力,但有研究表明在压裂过程中页理缝是薄弱面,流体和压力容易沿着页理面延伸,可能不利于压裂复杂缝网的形成。

表 2　长宁—威远区块宁 203 井取心井段裂缝发育统计表

取心井段/m	裂缝条数/条	裂缝宽度/mm	裂缝长度/cm	充填类型
2298.5 ~ 2299.7	1	0.1 ~ 1.0	44	方解石充填
2301.6 ~ 2303.8	1	0.5 ~ 2.0	32	方解石充填
2328.3 ~ 2329.1	1	0.1 ~ 2.0	20	方解石充填
2330.2 ~ 2335.6	2	0.1 ~ 0.5	13	方解石充填
2347.1 ~ 2347.3	1	0.5 ~ 3.0	20	未充填、张开缝
2348.2 ~ 2348.7	1	0.5 ~ 2.0	50	未充填、张开缝

2.5　含气特征

在页岩储层中,天然气的赋存形式有 3 种状态:一是以吸附气的形式吸附在有机质和黏土颗粒表面;二是以自由气的形式存在于岩石基质孔隙和裂缝中;此外还有很少一部分气体以溶解状态存在于干酪根、沥青等物质中[18,22]。目前,对于页岩含气量的评价有 2 种方法。一种是现场解吸附法,即钻井过程中从产层段取心,岩心取出后放入解析罐中,样品装罐后采用量管进行计量吸附气量,自然解吸附后,通过粉碎岩样解析岩样中的残余气,由于取样过程中从井底到解析罐环节中有部分气体损失,通过解吸附曲线外推,求取该部分气量即损失气,解吸附气、残余气和损失气相加即为总含气量。目前,该方法广泛应用于矿场评价页岩含气量,但损失气量采用曲线外推法,难以准确测定,因此该方法存在较大误差。另外一种是实验室测定法(等温吸附实验),将钻井过程中取心样品用保鲜膜密封,在实验室内对样品进行破碎,破碎后颗粒粒径为 0.2 ~ 0.3mm;将粉碎后的页岩样品称重后加入样品缸,密封后将其放到恒温箱中,实验通过精确计量参照缸和样品缸的压力与温度来计算气体吸附量。目前实验室等温吸附实验通常采用甲烷、氦气或 CO_2 等单一气体成分进行实验室测定,与气藏内实际多组分气体相差较大,因此,该方法也存在一定的误差。

现场解吸附法和实验室测定均表明:龙马溪组页岩具有良好的含气性。采用现场解吸附法测得 2491.8 ~ 2522.8m 处岩样总含气量为 1.52 ~ 2.09m^3/t(表 3);而采用实验室等温吸附实验,2516.7m 处岩样甲烷作为吸附气体,$T=79.6$℃时测定不同压力下岩样总含气量,并拟合曲线外推至气藏压力下,预计 $P=49$MPa 状态下岩样总含气量约为 2.72m^3/t(图7)。鉴于以上2种技术方法的局限性,现场解吸附法无法准确测量损失气量,造成总含气量评价结果偏小;实验室等温吸附法,采用甲烷单一气体组分作为吸附气体,而与多组分气藏流体组分不同,甲烷分子体积小,同等比表面积下,势必造成吸附气量偏大,引起总含气量偏高。因此,2种测试方法结果均有一定的偏差。更精确评价页岩含气量仍然是页岩气开发评价的一项重大技术挑战。

表3 宁201井现场解吸附法含气量测定数据表

样品号	井深/m	损失气时间/h	损失气量/(m^3/t)	解吸附气量/(m^3/t)	残余气量/(m^3/t)	总含气量/(m^3/t)
1	2491.8	7.28	0.729	0.23	0.564	1.523
2	2505.0	6.15	1.064	0.192	0.72	1.976
3	2516.4	6.33	0.998	0.171	0.92	2.088
4	2522.8	6.07	0.999	0.215	0.594	1.809

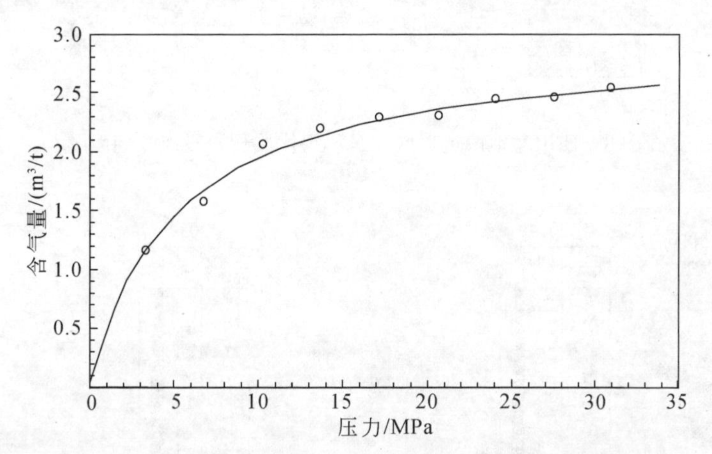

图7 宁201井2523.1m岩样甲烷等温吸附曲线(TOC=2.51%,$T=79.6$℃)

3 龙马溪组优质页岩开发评价

3.1 优质页岩发育模式

沉积环境对高有机质丰度的页岩发育具有非常重要的作用,决定了海相沉积物的类型及分布特点,是提供有机质的生存环境,是决定有机质丰度和页岩品质的重要因素。有机质富集与生物的产率有关,同时还与沉积物的堆积速度密切有关。研究表明[10,24]:龙马溪组早期,海平面快速上升,四川盆地处于大面积缺氧的深水陆棚环境,沉积速率为1.5~9.3m/Ma,藻类、放射虫、笔石等浮游生物大繁盛,表层水体浮游生物产率大,TOC值为2.1%~8.4%,发育富笔石碳质页岩、硅质页岩等优质页岩相;龙马溪组晚期,沉降中心向川中和川北迁移,海平面大幅度下降，逐步转变为浅水-半深水陆棚，海水封闭性进一步增强，沉积速率上升至9.5～384.4m/Ma，水体由缺氧还原环境演变为弱还原-氧化环境，浮游生物大规模减少，同时有机质保存条件逐步变差。因此，优质页岩的发育和分布与沉积相带的分布紧密相关。

志留纪早期经历了两期海进—海退的海平面变化，笔石、藻类等浮游生物经历了繁盛—稀疏—繁盛—灭绝的生长发育轮回，岩心和测井上均显示，纵向上龙马溪组龙一段间隔发育两套富笔石、高有机碳含量的碳质页岩，中间夹一套中笔石密度、中有机碳含量的硅质页岩，顶部为贫笔石、低有机碳含量的黑色页岩。底部优质页岩段含两套碳质页岩和一套硅质页岩段（图8，图9）。

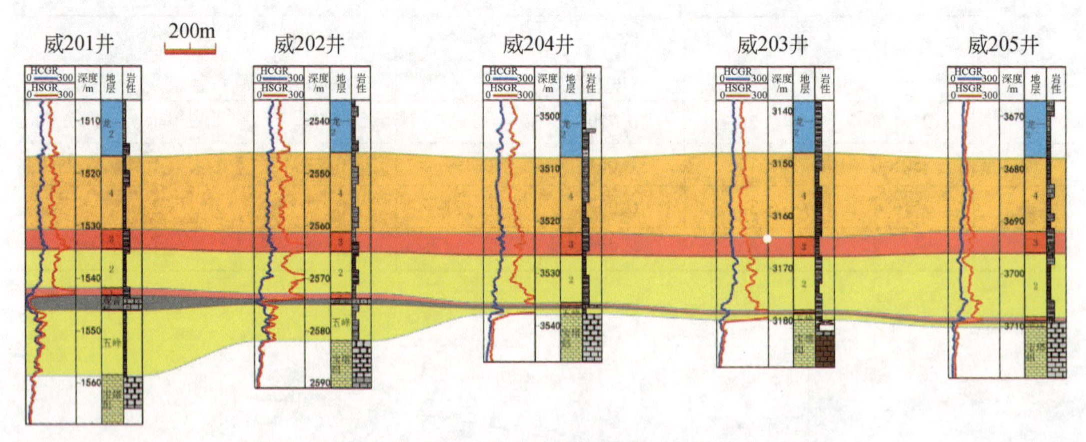

图8　四川盆地威远地区龙马溪组优质页岩段连井剖面图

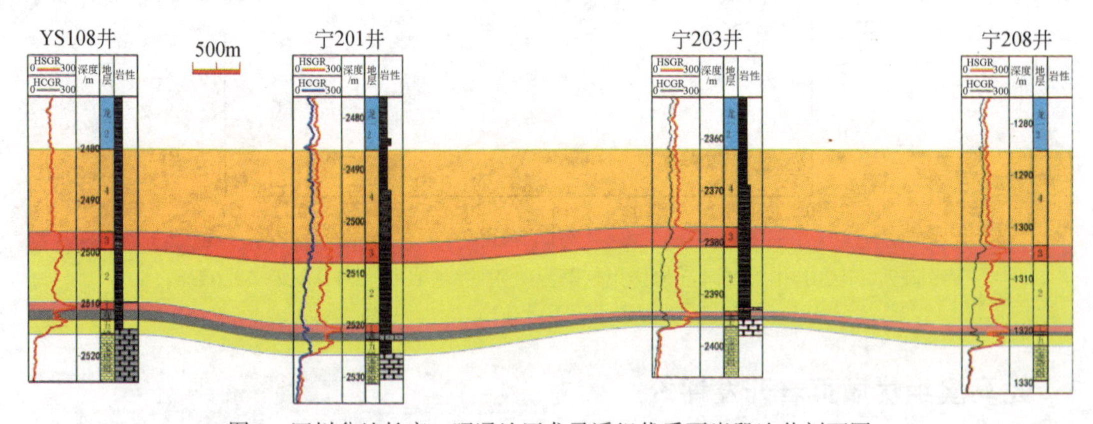

图9　四川盆地长宁—昭通地区龙马溪组优质页岩段连井剖面图

3.2　优质页岩空间分布

沉积相带分布具有大面积、连片分布的特征，这决定了页岩的空间分布也具有平面上大面积、连片分布的特征。龙马溪组优质页岩厚度受深水陆棚沉积相的整体控制，整体上厚度变化幅度不大，为10～20m。空间上越靠近沉积中心，厚度相对略大，如长宁区块较靠近泸州—永川沉积中心，龙马溪组优质页岩厚度为15～20m；威远、昭通区块位于深水陆棚沉积区域边部，厚度相对较小，平均厚度为10～15m。但沉积环境对优质页岩储层参数影响相对较大(表4)。浮游生物受水体深度的影响更明显,越靠近沉积中心,水体更安静、稳定,为浮游生物的生长和保持提供了更好的缺氧-还原环境;同时,越靠近沉积中心,受陆源物质作用的影响更小,沉积物中黏土含量越低。因此,越靠近沉积中心,有机碳含量更高,随之优质页岩的含气量、脆性矿物含量也越高,优质页岩储层品质更好。

表4 四川盆地龙马溪组优质页岩段储层参数统计

区块	优质页岩厚度/m	有机碳含量/%	脆性矿物含量/%	含气量/(m^3/t)	储量丰度/($10^8m^3/km^2$)
长宁	17.6	5.08	72.9	5.39	2.41
威远	13.5	4.33	67.2	4.91	1.70
昭通	11.2	3.17	69.0	4.85	1.52

3.3 优质页岩地质储量评价

国土资源部油气资源战略研究中心制定了页岩气有利区标准,主要指标包括:有机碳含量高于2%,页岩厚度稳定、单层厚度不小于30m,含气量不小于$1m^3/t$。该标准主要针对页岩气勘探评价和目标区优选,而对于开发阶段,评价的重点在目的层位的确定和评价,四川盆地两口评价井脉冲中子俘获测井监测显示,目前压裂规模下五峰组—龙一段压裂缝高为12.0~13.4m[25]。因此,进入规模开发阶段之后,龙一段底部的优质页岩段是页岩气开发评价的重点目标。

龙马溪组优质页岩段纵向上两套碳质页岩段TOC、含气量等储层参数优于硅质页岩段,同时由于富含有机质,密度相对较低。根据长宁—威远—昭通页岩气示范区现有评价井资料分析,目前矿场评价中含气量(单位岩石体积的含气体积)与优质页岩厚度采用体积法计算地质储量丰度,评价结果显示优质页岩段地质储量丰度平均达到$1.85\times10^8m^3/km^2$(表4),具备良好的资源条件,是页岩气规模开发的有利层位。

4 结论

四川盆地龙马溪组经历了由深水陆棚到浅水陆棚的沉积环境转化。早志留世为深水陆棚沉积,浮游生物繁盛,沉积速率低,水体为还原环境,是优质页岩发育的有利相带;晚志留世为半深水-浅水陆棚沉积,浮游生物凋零,沉积速率高,水体为半还原-氧化环境,不利于有机质生长和保存。

根据化石带、岩性等区域地层发育特征,龙马溪组进一步细分为龙一段和龙二段。龙马溪组底部龙一段,优质页岩发育,从岩相和测井响应特征识别出龙一段深水陆棚环境下发育的两套富笔石、高有机碳含量的碳质页岩,中间夹一套中笔石密度、中有机碳含量的硅质页岩是优质页岩段。优质页岩段具有高脆性矿物含量、高含气量、高地质储量丰度的特征,有利于压裂改造,是页岩气规模化开发的有利层位。

参考文献

[1] 国家能源局. 页岩气发展规划(2011-2015年)[EB/OL]. www.ndrc.gov.cn/zcfb/zcfbtz/201203/t20120316_46.7518.html[2012-03-16].

[2] U. S. Energy Information Administration. Shale in the United States[R]. Washington DC,2016.
[3] 戎嘉余,陈旭. 中国志留纪年代地层学述评[J]. 地层学杂志,2000,24(1):27-35.
[4] Chen X, Fan J X, Melchin M J, *et al*. Hirnnantian (latest Ordovician) graptolites from the upper Yangtze region, China[J]. Palaeontology,2005,48(2):1-47.
[5] 陈旭,戎嘉余,周志毅,等. 上扬子区奥陶-志留纪之交的黔中隆起和宜昌上升[J]. 科学通报,2001,46(12):1052-1056.
[6] Fan J X, Peng P A, Melchin M J, *et al*. Carbon isotopes and event stratigraphy near the Ordovician-Silurian boundary, Yichang, south China[J]. Palaeoecology,2009,276(1-4):160-169.
[7] 樊隽轩,吴磊,陈中阳,等. 四川兴文麒麟乡五峰组—龙马溪组黑色页岩的生物地层序列[J]. 地层学杂志,2013,37(4):513-520.
[8] 王鸿祯. 中国古地理图集[M]. 北京:地图出版社 1985.
[9] 万方,许效松. 川滇黔桂地区志留纪构造-岩相古地理[J]. 古地理学报,2003,5(2):180-185.
[10] 张春明,张维生,郭英海. 川东南-黔北地区龙马溪组沉积环境及对烃源岩的影响[J]. 地学前缘,2012,19(1):136-145.
[11] 梁狄刚,郭彤楼,陈建平,等. 南方四套区域性海相烃源岩的分布[J]. 海相油气地质,2008,13(2):1-16.
[12] 王清晨,严德天,李双建. 中国南方志留系底部优质烃源岩发育的构造-环境模式[J]. 地质学报,2008,3:289-299.
[13] 白志强,刘树根,孙玮,等. 四川盆地西南雷波地区五峰组—龙马溪组页岩储层特征[J]. 成都理工大学学报:自然科学版,2013,40(5):521-531.
[14] 尹赞勋. 关于龙马溪页岩[J]. 地质论评,1943,8(z1):1-8.
[15] 金之钧,胡宗全,高波,等. 川东南地区五峰组—龙马溪组页岩气富集与高产控制因素[J]. 地学前缘,2016,23(1):1-10.
[16] 国家能源局. 中华人民共和国石油和天然气行业标准(SY/T 5363—1997)[S]. 北京:石油工业出版社,1997.
[17] 冉波,刘树根,孙玮,等. 四川盆地及周缘下古生界五峰组—龙马溪组页岩岩相分类[J]. 地学前缘,2016,23(2):96-107.
[18] 邹才能,董大忠,王玉满,等. 中国页岩气特征、挑战及前景(一)[J]. 石油勘探与开发,2015,42(6):689-701.
[19] 邹才能,董大忠,王玉满,等. 中国页岩气特征、挑战及前景(二)[J]. 石油勘探与开发,2016,36(2):166-178.
[20] 刘树根,冉波,郭彤楼,等. 四川盆地及周缘下古生界富有机质黑色页岩:从优质烃源岩到页岩气产层[M]. 北京:科学出版社,2014.
[21] 胡宗全,杜伟,彭勇民,等. 页岩微观孔隙特征及源-储关系——以川东南地区五峰组-龙马溪组为例[J]. 石油与天然气地质,2015,36(6):1001-1008.
[22] 王香增,张丽霞,李宗田,等. 鄂尔多斯盆地延长组陆相页岩孔隙类型划分及其油气地质意义[J]. 石油与天然气地质,2016,37(1):1-7.
[23] 郭彤楼. 中国式页岩气关键地质问题与成藏富集主控因素[J]. 石油勘探与开发,2016,43(3):317-326.
[24] 邹才能,董大忠,王社教,等. 中国页岩气形成机理、地质特征及资源潜力[J]. 石油勘探与开发,2010,37(6):641-653.
[25] 贾成业,贾爱林,何东博,等. 页岩气水平井产量影响因素分析[J]. 天然气工业,2017,37(4):80-88.

基于物质基础的页岩气储层分类与评价

金亦秋　韩品龙　位云生　贾成业　袁　贺

（中国石油勘探开发研究院）

摘要：储层分类与评价是对储层的综合认识和评判，并为进一步勘探和高效开发提供指导。本文在页岩储层综合研究的基础上，总结主要参数的特征和控制因素，提出含气量的大小反映了孔隙度、含水饱和度及总有机碳含量等主要参数的性质，是页岩储层开发潜力的直观表征。根据含气量的大小将储层划分为Ⅰ、Ⅱ、Ⅲ、Ⅳ4类，其中Ⅰ类储层含气量大于$4m^3/t$，为优质储层；Ⅱ类储层含气量为$2\sim4m^3/t$；Ⅲ类储层含气量为$1\sim2m^3/t$；Ⅳ类储层含气量小于$1m^3/t$。归纳了不同类型储层的判别公式并总结各自的岩性、物性和电性特征。结合地层划分与测井解释结果，对威远—长宁—昭通地区储层连井剖面进行了储层划分和评价。

关键词：页岩气；储层分类；含气量；吸附气；游离气

我国页岩气资源分布面积广，有利层系众多，海陆相兼备，可采资源量极为可观。目前已在四川盆地及其周缘下古生界海相龙马溪组和筇竹寺组实现勘探开发的重大突破，截至2016年8月，于四川长宁—威远、云南昭通、重庆涪陵三个海相页岩气示范区完钻水平井400余口，压裂投产300余口。2015年累积产气44.7亿m^3，2016年产气超过90亿m^3，初步进入大规模开发阶段。

页岩气储层综合评价以地质、钻井、测井和分析化验资料为基础，对储层矿物成分、物性特征、孔隙类型和裂缝分布等因素进行深入研究，以实现对页岩气可采资源量的精确估算。在此基础之上，对储层进行合理分类，为不同级别储层的有序开发提供科学指导。页岩地层孔隙以纳米级微孔隙为主，地层流体并非简单的达西运动，利用常规的测井手段进行有效储层界限的划分和气水界面的识别存在困难。此外，吸附和游离两种不同的气体赋存方式共同决定着资源量的大小，二者含量共同影响着储层潜力的大小，在储层评价中也应予以全面的考量。

前人通过大量的研究，对不同类型气藏的评价和分类标准进行了有益探讨。张龙海和刘忠华[1]通过对比发现，在运用储集层品质指数和地层流动带指数两种岩石物理分类方法对低孔低渗储集层进行分类时，前者具有更高的符合程度。周晓峰等[2]研究了川东北地区须家河组五段（T_3x_5）低渗致密砂岩储层的孔隙结构，利用孔隙度、渗透率、压汞曲线特征、孔喉半径等进行储层的评价。涂乙等[3]采用灰色关联分析法定量判断各参数的权重，利用获得的综合评价因子对页岩气储层进行分类。张金川等[4]通过调研国内外页岩气田的开发现状，总结了页岩气的8条选区评价标准：①有机碳含量>0.3%；②成熟度（R_o）≥0.4%；③埋藏深度<4500m；④富有机质泥页岩集中发育，有效厚度>9m；⑤含气量>0.5 m^3/t；⑥黏土矿物含量>30%，脆性矿物含量>45%；⑦孔隙度>1%；⑧渗透率>$0.001\times10^{-3}\mu m^2$。

本文在深入分析中国海相页岩气储层参数的基础上，优选页岩气储层评价参数，划分Ⅰ类储层、Ⅱ类储层、Ⅲ类储层和Ⅳ类储层，并确定对应的界限值。根据页岩的独特性，总结不同类型储层的岩性、电性与物性规律，形成储层类型识别的综合方法。在此基础上，利用此分类标准对威远、长宁和昭通三个示范区页岩储层进行对比评价。需要说明的是，地层两向应力差、天然裂缝发育程度和脆性指数等工程因素同样影响着页岩气的开发效果，对它们的测量和评价方法明显有别于传统的地质因素，有待进一步的研究。

1　储层分类的标准

页岩气储层评价需要考虑诸多参数，这些参数有着不同的地质意义，但彼此间并非完全独立。总有机碳含量（TOC）和有机质成熟度（R_o）反映了烃源岩的产气条件，有机质在生烃的过程中会形成大量的微孔隙，是游离气重要的储集空间，同时有机质和黏土矿物的表面也是吸附气的聚集场所，因此 TOC 在一定程度上控制着含气量的大小，尤其是吸附气的含量与其数值密切相关[5,6]。孔隙度（Φ）和渗透率（k）是储层物性的反映，页岩储层致密，孔隙度值为 1% ~8%，水平渗透率通常小于 $1\times10^{-3}\mu m^2$，垂向渗透率小于 $0.001\times10^{-3}\mu m^2$。孔隙作为游离气的储存空间，决定了游离气含量的高低。由于页岩水平井加水力压裂的开发方式，水力裂缝大大提高了储层的排烃能力，渗透率对储层的影响相对较小。针对目前开发区块内的页岩储层进行评价与分类，其埋藏深度大多符合当前的工艺条件，而垂向上不同类型储层的叠置情况可以替代富有机质页岩的有效厚度用以预测开发的效益。

页岩储层含气量是指每吨岩石所含天然气折算到标准温度和压力条件下（25℃、1atm①）的天然气总量，包括孔隙和裂缝中的游离气、吸附在有机质和黏土表面的吸附气、溶解于液态烃中的烃相溶解气和溶解于地层水当中的水相溶解气，我国南方海相页岩由于经过长期的埋藏演化，有机质成熟度高（R_o>3%），烃类以甲烷为主，因此烃相溶解气和水相溶解气含量可忽略。含气量参数是储层开发能力的直观体现，同时也是总有机碳含量、有机质成熟度、地层压力系数、储集空间大小和储层吸附能力的综合反映。含气量的数值虽然可以由密闭取心测试直接获取，但目的层段连续的含气量数值往往来自测井解释。本文采用斯伦贝谢公司的含气量解释结果，通过元素俘获测井（ECS）和常规测井解释计算得到 TOC 含量，利用 Langmuir 气体体积与 TOC 的关系，求取地层温度和压力下的吸附气体积。在精确得到黏土矿物及其类型和地层孔隙度的基础上，利用双水模型，采用 ELANplus 优化解释程序，得到游离气饱和度，结合孔隙度、地层压力系数等，计算地层的游离气含量[7]。

本文以含气量作为主要的评价与分类指标，参考国外页岩气田的开发实践[8,9]，结合蜀南示范区各个层位含气量的差异及不同层位钻遇比例下的气井开发效果，将储层划分为Ⅰ、Ⅱ、Ⅲ、Ⅳ4 种类型（表 1），其中Ⅰ类储层含气量大于 $4m^3/t$，为优质储层；Ⅱ类储层含气量为 $2\sim4m^3/t$；Ⅲ类储层含气量为 $1\sim2m^3/t$；Ⅳ类储层含气量小于 $1m^3/t$，一般来讲，这类储层开发效益较低，不能单独构成有效开发层段。为了突出优质页岩的开发潜力，将Ⅰ类储层进一步划分为两个类型，Ⅰ-a 型储层含气量大于 $6m^3/t$，Ⅰ-b 型储层含气量为 $4\sim6m^3/t$。Ⅰ-a 型储层的含气量可以高达 $12m^3/t$，是最优质的页岩储层类型。

① 1atm=101325Pa

表1 页岩气储层分类指标

参数	I		Ⅱ	Ⅲ	Ⅳ
	I-a	I-b			
含气量/(m^3/t)	>6	4~6	2~4	1~2	<1

计算含气量时,游离气含量主要取决于孔隙度与含水饱和度,吸附气含量与总有机碳含量直接相关,可以利用孔隙度、含气饱和度、总有机碳含量三个参数来进行储层类型的判别(图1)。利用交会图,得出不同类型储层的判别公式,Ⅰ类储层 $\Phi S_g+1.4\mathrm{TOC}>4$;Ⅱ类储层 $2.3<\Phi S_g+1.4\mathrm{TOC}<4$;Ⅲ类储层 $1.3<\Phi S_g+1.4\mathrm{TOC}<2.3$;Ⅳ类储层 $\Phi S_g+1.4\mathrm{TOC}<1.3$。其中,$\Phi$ 为孔隙度,%;S_g 为含气饱和度,%;TOC为总有机碳含量。

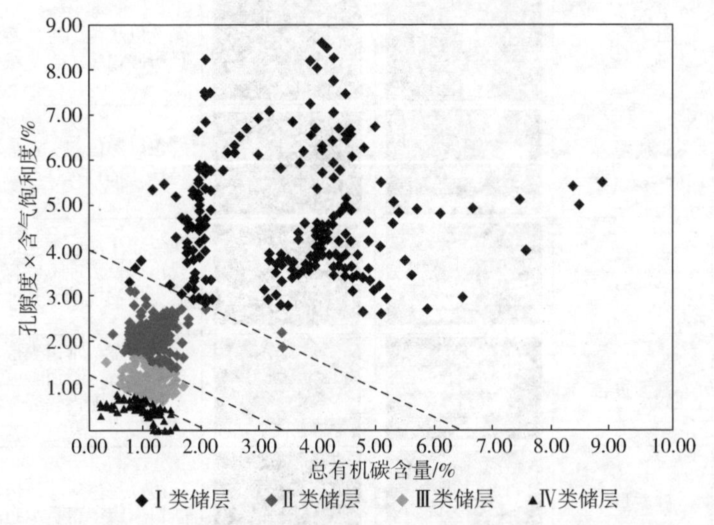

图1 长宁地区不同类型储层的孔隙度、含气饱和度与总有机碳含量的交会图
(数据来自岩心样品测试)

2 储层岩性、物性和电性特征

2.1 不同类型储层岩性特征

通过对蜀南威远、长宁和昭通示范区多口页岩气取心井的精细观察和描述,结合其镜下微观特征和剖面位置的储层含气量特征,将不同类型储层的岩性特征进行归纳总结(图2)。本文采用的岩性划分方案是根据岩石中碳质、硅质、钙质和碎屑矿物含量的多少来确定的。Ⅰ类储层为黑色碳质页岩,主要发育在龙一$_1^1$小层;Ⅱ类储层包括含碳硅质页岩和含碳钙质页岩,主要发育于龙一$_1^2$、龙一$_1^3$小层;Ⅲ类储层主要为含碳粉砂质页岩、碳质粉砂质页岩,可见于龙一$_2$小层;Ⅳ类储层主要为粉砂质页岩、钙质页岩和硅质页岩,这些岩性常见于龙马溪组上部层位。

储层类型	岩性	岩心照片	镜下照片	特征描述
Ⅰ类储层	碳质页岩			颜色以黑色为主，染手程度高，含有大量已碳化的有机质；页理发育；笔石化石丰富，以雕笔石、栅笔石为主
Ⅱ类储层	含碳硅质页岩	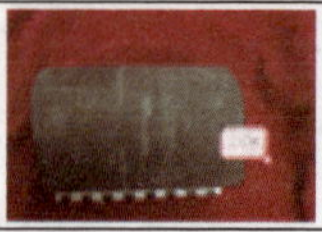		颜色以黑色为主，岩心表面平整，页理较发育；笔石化石丰富；参差状断口；常见黄铁矿结核或条带；岩性致密，硬度大，脆性高
	含碳钙质页岩			颜色以黑色、深灰色为主；笔石发育；矿物以方解石为主，微裂缝发育
Ⅲ类储层	含碳粉砂质页岩			颜色以灰色为主，断面不平整，可见粉砂颗粒；含笔石化石
	碳质粉砂质页岩			颜色以黑色为主，页理发育，断面平整；笔石发育
Ⅳ类储层	粉砂质页岩		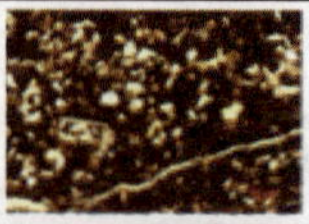	灰色、深灰色为主，通常以夹层形式出现，页理不发育，参差状断口，常见生物扰动或其他水动力结构
	钙质页岩			灰色为主，常以夹层形式出现，页理不发育，断口参差，且岩心表面不平整，脆性较好，常见黄铁矿结核
	硅质页岩		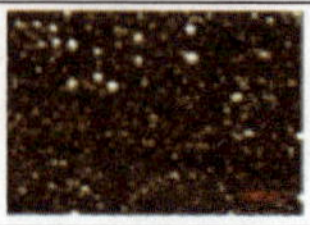	颜色以灰色、深灰色为主；贝壳状断口；镜下可见大量的硅质结核

图2　不同类型储层的典型岩性特征

页岩储层各具典型的岩性，需要说明的是，由于烃类的运移作用（尽管在页岩中这种作用非常缓慢），储层含气量在纵向上通常是渐变过渡的，垂向岩性的变化并不一定导致储层类型的变化。所以，页岩岩性可作为储层评价的参考，而非严格的划分标准。

2.2　不同类型储层的物性特征

页岩天然储层（基质）具有极低的渗透率，喉道的大小和结构不足以有效沟通孔隙空间，难以使游离或解吸的甲烷顺利运移至井筒；通过水力压裂，在页岩储层中形成人工裂缝与天然裂缝耦合的复杂裂缝网络，缩短气体流入裂缝系统的距离，改善地层传导气体的能力。基于以上原因，有些学者认为基质渗透率的评价并不重要。但笔者发现，地层压裂之后，赋存于裂缝内的气体容量本身有限，更多的天然气从基质块由内向外（裂缝面）排出，而基质内孔喉的大小和结构直接决定了岩石破碎单元中能够有效排烃的动用范围，因此，渗透率是影响页岩气最终可采储量的主控因素，对页岩气的开发评价至关重要。受气体滑脱效应的影响，不同压力下测得的岩心渗透率有所差异，由于真实地下压力较高，采用校正后的克氏渗透率来作为页岩渗透能力的反映。图 3 表明,页岩孔隙度与渗透率没有明显的相关性。从微观孔隙结构上来说,孔隙空间主要为有机质中的纳米级孔隙,而渗透率的主要贡献者却是页岩内的微小裂缝。在页岩地层埋深、压实的过程中,温度和压力持续升高,自由水被逐渐排出,残留部分难以运移的束缚水占据了部分孔隙空间,剩余的孔隙空间才能填充甲烷等烃类气体,因此,含水饱和度(S_w)是影响气体含量的重要因素。孔隙度与含水饱和度的关系较复杂,在优质页岩段,含水饱和度随着孔隙度的增加有先增后减的趋势(图 4)。

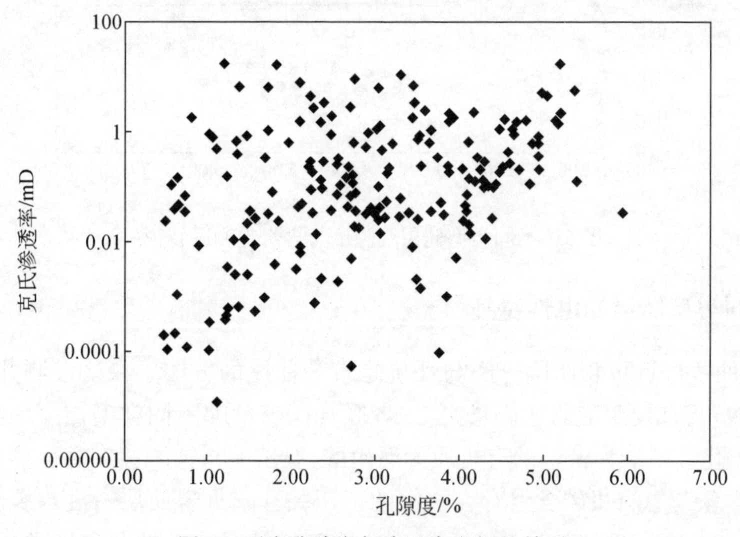

图 3 页岩孔隙度与渗透率之间的关系

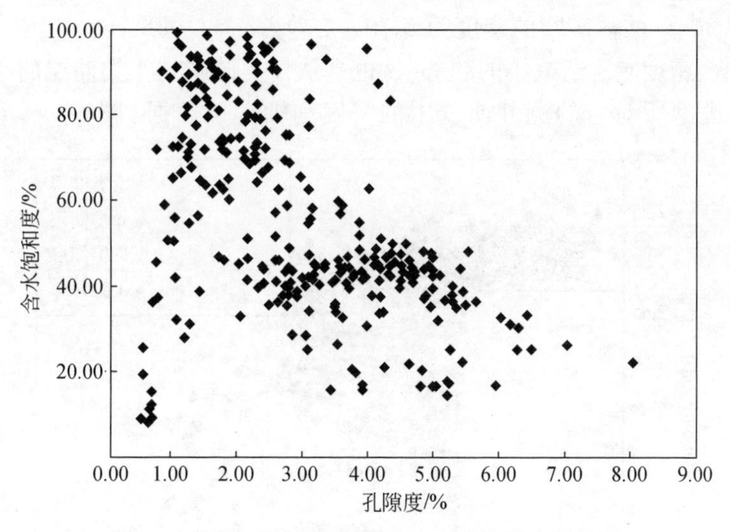

图 4 不同类型储层的孔隙度和含水饱和度的关系

由图 5 可知, Ⅰ类储层 $\Phi>2.8\%$,$S_w<35\%$; Ⅱ类储层 $3.2\%<\Phi<6.0\%$,$35\%<S_w<70\%$; Ⅲ类储层 $2.8\%<\Phi<5.2\%$,$35\%<S_w<70\%$; Ⅳ类储层 $\Phi<2.8\%$ 。

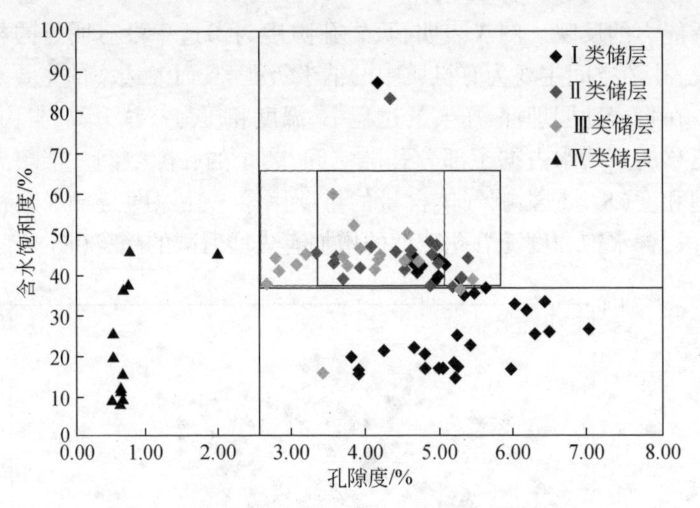

图5　不同类型储层孔隙度与含水饱和度特征

2.3　不同类型储层的电性特征

通过测井曲线进行页岩储层分类与评价是较为便捷的方法。在诸多测井系列中，铀含量与密度（DEN）具有较明确的地质意义，二者联合能够反映不同类型储层的特征。前人研究表明，与铀矿化关系最为密切的有机质为腐殖酸，腐殖酸与铀酰存在着强烈的吸附与络合关系[10,11]，铀含量与总有机碳含量存在一定的正相关性。页岩储层密度与孔隙大小和各类矿物的含量密切相关，研究表明[12]，填充孔隙空间的流体（地层水和烃类）密度为0.1～1.3g/cm^3，有机质的密度约为1.3g/cm^3，黏土矿物密度为2.6～2.9g/cm^3，长石和石英的密度约为2.65g/cm^3，碳酸盐矿物的密度为2.70～2.71g/cm^3。可见，随着有机质含量的增加和孔隙度的增大，密度有逐渐减小的趋势，这也是从Ⅳ类储层到Ⅰ类储层的明显变化规律。根据不同储层类型所对应的测井曲线中的密度和铀含量数据（图6），Ⅰ类储层DEN≤

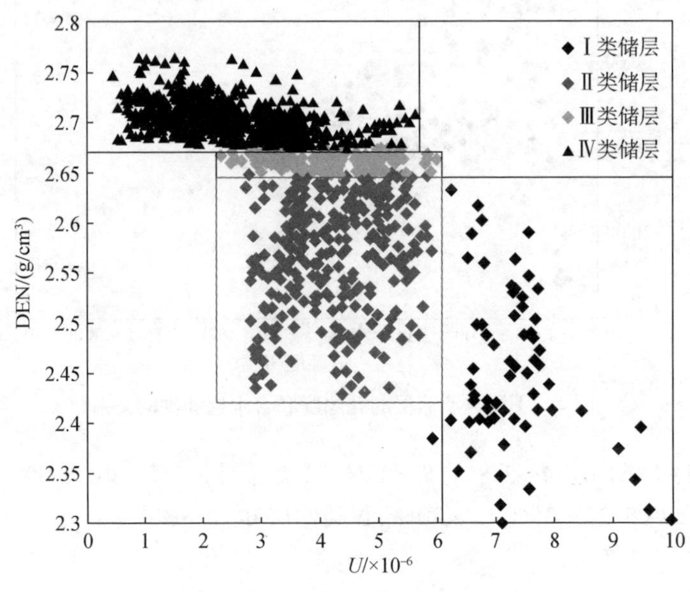

图6　不同类型储层铀含量与密度的关系

2.64g/cm³,$U \geq 6\times10^{-6}$;Ⅱ类储层 2.42g/cm³<DEN<2.64g/cm³,$2.3\times10^{-6}<U<6\times10^{-6}$;Ⅲ类储层 2.64g/cm³<DEN<2.67g/cm³,$2.3\times10^{-6}<U<6\times10^{-6}$;Ⅳ类储层 DEN≥2.67g/cm³,$U<5.8\times10^{-6}$。

3 蜀南地区页岩储层分类与评价

当前蜀南地区主要开发的页岩层系为上奥陶统—下志留统龙马溪组富有机质页岩。五峰-龙马溪组自下而上可以划分为凯迪、赫南特、鲁丹、埃隆和特列奇5阶13个笔石带,以埃隆阶沉积厚度最大,凯迪阶和赫南特阶厚度较小。根据伽马能谱、电阻、密度等曲线,可将五峰-龙马溪组划分为3个三级层序:SQ_1为五峰组期一套海进-海退旋回,下部海进体系域岩性为黑色含笔石碳质页岩和硅质页岩,高位体系域岩性为介壳灰岩,厚度较薄,一般在20cm左右。SQ_2为龙马溪组早期为海侵-海退沉积响应下的深水陆棚沉积,笔石化石丰富,岩性以黑色碳质页岩、硅质页岩为主,有机质丰度高。从下向上显示明显差异,龙马溪组底部发育大量笔石,页理发育,碳酸盐岩含量高,地层脆性较强。向上碳酸盐岩含量迅速减少,黏土矿物增加,笔石含量减少,其类型由雕笔石和栅笔石逐渐变为螺旋笔石、半耙笔石,预示着水体环境由深变浅;SQ_3为龙马溪组中晚期的一套半深水-浅水的沉积建造,有机质含量明显减少,钙质条带与粉砂质泥岩条带频繁发育,为一套浅水陆棚相沉积。结合沉积旋回特征并参考全国地层委员会对龙马溪组的划分方案,将龙马溪组自上而下划分为龙一段和龙二段,分别与SQ_2和SQ_3相对应,其中龙一段自下而上又可划分为龙一$_1$、龙一$_2$两个亚段,为了精细开发的需要,将龙一$_1$亚段进一步划分为龙一$_1^1$、龙一$_1^2$、龙一$_1^3$、龙一$_1^4$四个小层。

对威远—长宁—昭通三个示范区的典型井区内的典型井进行储层的连井划分(图7),威远区块与长宁和昭通相距较远,且不同位置的井真实深度相差较大,为了更好突出储层细节,剖面中的井间距非等比例显示,井身深度比例尺和井间海拔比例尺也是不一致的。整体来看,页岩储层横向分布稳定,龙一$_1$的四个小层厚度约为35m(除了威204井区,因其更靠近盆地中心,厚度略有增加),龙一$_1^1$、龙一$_1^2$、龙一$_1^3$、龙一$_1^4$四个小层,储层质量向下逐渐变好,Ⅰ类储层所占比例持续增大。储层的埋深对页岩气储层质量具有关键的影响,威204井优质页岩段埋深大约3500m,几乎已达到当前开发的极限深度,其页岩储层的含气量性极高,且纵向分布深度大,甚至在龙一$_1^4$小层也有厚层的Ⅰ-a类储层。深度较浅的宁201井、宁203井、威201井、YS112井Ⅰ类储层所占比例依次降低,至最浅的YS111井,只有龙一$_1^1$小层为Ⅰ-a类。尽管有学者提出了页岩储层深度富集带L-W模型[13],即储层含气性随着深度的增加会先增大后减小,但实际表明,在3500m浅的区域,深度对含气性有积极的影响。究其原因,一方面,深度的增大使孔隙压力随之增压,游离气与吸附气含量同步增高;另一方面,由于页岩气的异常高压和低级的渗透率,在深度增加的同时孔隙体积得以保存,或者压力的增大弥补了孔隙减小的部分,含气量整体呈增大的趋势。

五峰组的储层性质在不同地区存在较大差异,在长宁和昭通地区,包括观音桥段在内的五峰组厚度较薄(2.63~6.36m),但储层含气性较好,多为Ⅰ类、Ⅱ类储层。威远地区厚度由盆地内部向边缘逐渐增大,至威201井达到最厚的8.48m,但储层含气性差,多为Ⅳ类储层。

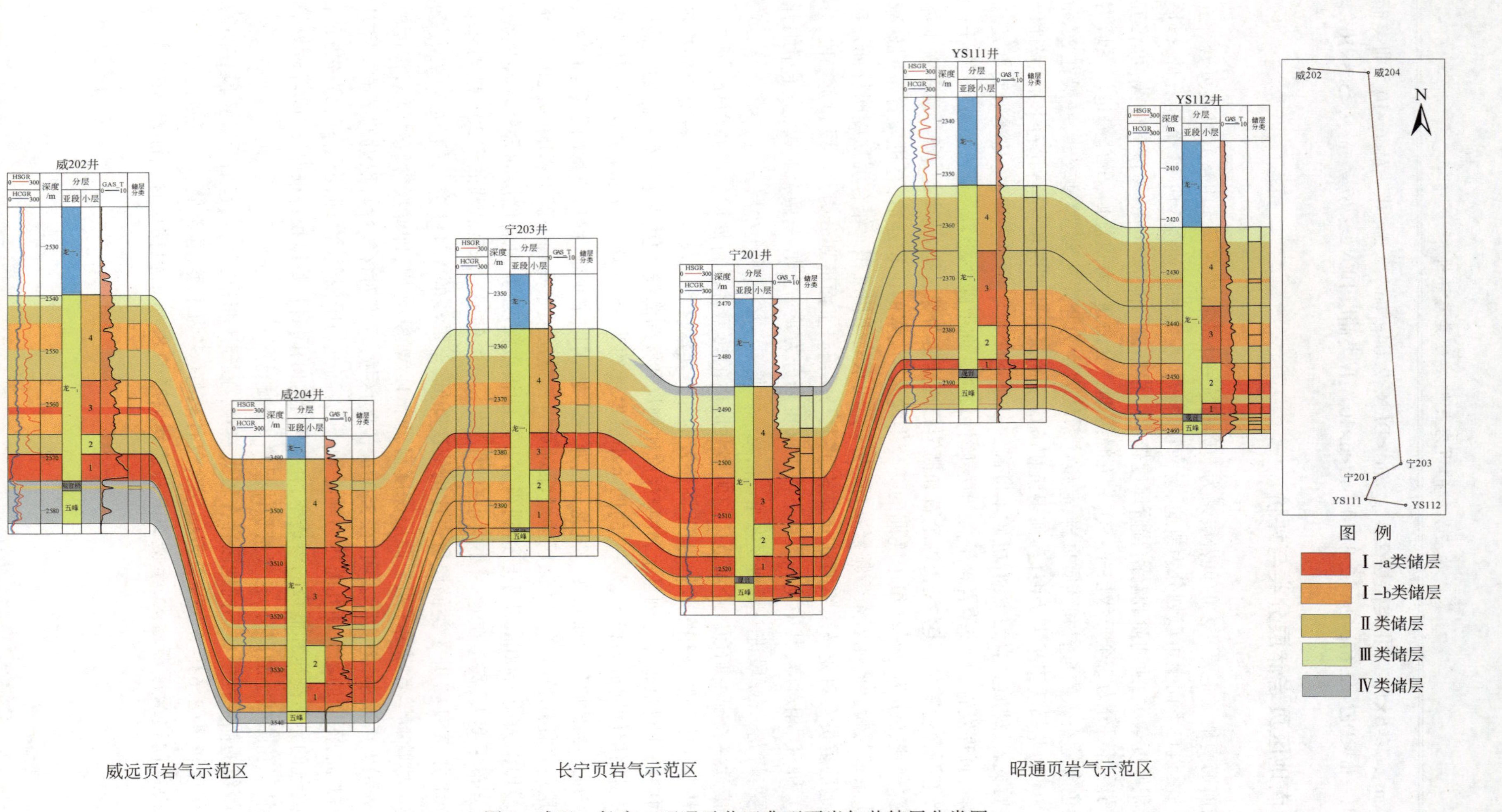

图7 威远—长宁—昭通示范区典型页岩气井储层分类图

4 结论

(1)在页岩储层各参数中,含气量可以表征孔隙度、含水饱和度和总有机碳含量等参数,可以作为储层划分的依据,根据含气量的不同,将储层划分为Ⅰ、Ⅱ、Ⅲ、Ⅳ4种类型,Ⅰ类储层含气量大于$4m^3/t$;Ⅱ类储层含气量为$2\sim4m^3/t$;Ⅲ类储层含气量为$1\sim2m^3/t$;Ⅳ类储层含气量小于$1m^3/t$。由于含气量多为测井解释的结果,可以间接利用孔隙度、含水饱和度和总有机碳含量进行联合划分:Ⅰ类储层$\Phi S_g+1.4TOC>4$;Ⅱ类储层$2.3<\Phi S_g+1.4TOC<4$;Ⅲ类储层$1.3<\Phi S_g+1.4TOC<2.3$;Ⅳ类储层$\Phi S_g+1.4TOC<1.3$。

(2)根据测井曲线和岩心分析测试数据,总结了不同类型储层的岩性、物性和电性特征,可以辅助储层类型的判断。Ⅰ类储层为黑色碳质页岩,$\Phi>2.8\%$,$S_w<35\%$,$DEN\leqslant2.64g/cm^3$,$U\geqslant6\times10^{-6}$;Ⅱ类储层包括含碳硅质页岩和含碳钙质页岩,$3.2\%<\Phi<6.0\%$,$35\%<S_w<70\%$,$2.42g/cm^3<DEN<2.64g/cm^3$;Ⅲ类储层主要为含碳粉砂质页岩、碳质粉砂质页岩,$2.8\%<\Phi<5.2\%$,$35\%<S_w<70\%$,$2.64g/cm^3<DEN<2.67g/cm^3$,$2.3\times10^{-6}<U<6\times10^{-6}$;Ⅳ类储层主要为粉砂质页岩、钙质页岩和硅质页岩,$\Phi<2.8\%$,$DEN\geqslant2.67g/cm^3$,$U<5.8\times10^{-6}$。

(3)对威远—长宁—昭通地区进行储层划分,在同一的沉积背景下,现今的埋藏深度是影响储层性质的重要因素。从各小层来看,优质储层主要分布于龙一$_1^1$和龙一$_1^3$小层。五峰组储层性质差异大,在威远地区向盆地边缘逐渐增厚,但整体含气量极低,在长宁和昭通地区储层厚度相对稳定,且开发潜力较好。

参考文献

[1] 张龙海,刘忠华. 低孔低渗储集层岩石物理分类方法的讨论[J]. 石油勘探与开发,2008,35(6):763-767.

[2] 周晓峰,陈波,凡睿,等. 川东北地区须家河组四段特低渗致密砂岩储层孔隙结构特征及分类评价[J]. 石油天然气学报,2015,37(5):10-15.

[3] 涂乙,邹海燕,孟海平,等. 页岩气评价标准与储层分类[J]. 石油与天然气地质,2014,35(1):153-158.

[4] 张金川,林腊梅,姜生玲,等. 页岩气资源潜力评价方法与有利区优选标准操作手册[M]. 北京:中国地质大学出版社,2011.

[5] 马行陟,柳少波,姜林,等. 页岩吸附气含量测定的影响因素定量分析[J]. 天然气地球科学,2016,27(3):488-493.

[6] 李武广,杨胜来. 考虑地层温度和压力的页岩吸附气含量计算新模型[J]. 天然气地球科学,2012,23(4):791-796.

[7] 吴庆红,李晓波. 页岩气测井解释和岩心测试技术——以四川盆地页岩气勘探开发为例[J]. 石油学报,2011,32(3):484-488.

[8] Kaiser M J, Yu Y. Louisiana Haynesville Shale—Characteristics, production potentialof Haynesville Shale Wells Described[J]. Oil & Gas Journal,2011,109(19):68-79.

[9] Loucks R G, Reed R M, Ruppel S C, *et al.* Morphology, genesis, and distribution of nanometer-scale pores in siliceous mudstones of the Mississippian Barnett shale[J]. Journal of Sedimentary Research, 2009, 79: 848-861.

[10] 郭庆银,李子颖,王文广. 内蒙古西胡里吐盆地有机质特征及其与铀矿化的关系[J]. 铀矿质,2005,21(1):16-22.

[11] 薛伟,薛春纪,涂其军,等. 鄂尔多斯盆地东北缘侏罗系铀矿化与有机质的某些关联[J]. 地质论评,2009,(3):361-369.

[12] Mavlo G,Mukerji T,Dvorkin J. 岩石物理手册:孔隙介质中地震分析工具[M]. 合肥:中国科学技术大学出版社,2008.

[13] 刘人和,郭伟. 页岩气深度富集带理论研究及应用[J]. 科技创新导报,2015,(1):10-14.

页岩气储层微观孔隙结构特征及发育控制因素
——以川南—黔北××地区龙马溪组为例

魏祥峰[1,2]　刘若冰[1]　张廷山[2]　梁　兴[3]

（1. 中国石化勘探南方分公司；2. 西南石油大学油气藏地质及开发工程国家重点实验室；
3. 中国石油浙江油田分公司）

摘要：利用扫描电镜以及比表面积分析仪产生的试验数据、吸附脱附曲线对页岩气储层储集空间类型、微观孔隙结构的系统研究表明，川南—黔北××地区龙马溪组页岩气储层储集空间多样，包括了残余原生粒间孔、晶间孔、矿物铸模孔、次生溶蚀孔、黏土矿物间微孔、有机质孔以及构造裂缝、成岩收缩微裂缝、层间页理缝、超压破裂缝等基质孔隙和裂缝类型。发现研究区龙马溪组泥页岩比表面积和孔体积都较大且具有良好的正相关性，并认为微孔隙越发育，泥页岩的比表面积和孔体积越大，越有利于泥页岩对页岩气的吸附储集。建立了泥页岩的孔隙模型，并利用吸附脱附曲线分析了研究区龙马溪组泥页岩的微观孔隙结构特征，指出研究区龙马溪组泥页岩以极为发育的微孔为主，其中为泥页岩提供最大量孔体积和表面积的孔隙主要为Ⅲ类细颈瓶状（墨水瓶状）孔和Ⅰ类开放透气性孔。认为总有机碳含量、伊/蒙间层矿物含量以及有机质成熟度是控制研究区龙马溪组页岩气储层微观孔隙结构的主要因素。

关键词：页岩气储层；储集空间类型；微观孔隙结构；控制因素；龙马溪组；川南—黔北××地区

富有机质泥页岩作为特殊的储层，具多微孔性、低渗透率特点[1-4]，其储集方式不同于常规的油气藏，页岩气以吸附气和游离气为主，吸附态主要赋存于泥页岩中有机质和黏土矿物表面，游离态则存在于孔隙和裂隙中，还有少量溶解于液态烃和水中[3-7]。泥页岩主要由黏土矿物和有机质等成分组成，它们形成了不同类型、大小及形态的孔隙，而孔隙的性质（类型、大小及形态等）影响着泥页岩的比表面积和微孔隙的相对丰度，这直接关系到页岩气的吸附性、解吸性及其在泥页岩中的流动性。随着近年来人们对页岩气资源的开发与利用的重视，研究泥页岩的微观孔隙特征已成为一项重要的基础性工作，并指出泥页岩储层岩石的孔隙结构是影响其储集能力和页岩气开采的主要因素[8,9]，研究表明，大多数页岩的孔隙率主要依赖于小于10μm孔隙的孔隙体积的发育程度[10,11]，其中10nm左右的纳米孔隙含量丰富[2,4,11]，因此本文以川南—黔北××地区重点探井为依托，对下志留统龙马溪组泥页岩进行了系统采样，利用扫描电镜、比表面积分析仪产生的吸附脱附曲线对页岩气储层储集空间类型、微观孔隙结构进行了系统研究，并探讨了控制储层微观孔隙结构的主要因素，以期对正确评价泥页岩储气性能、揭示页岩气富集规律乃至对南方页岩气勘探和开发具有重要的意义。

1　地质背景

川南—黔北××地区位于四川、云南、贵州三省交界边缘，区域构造上位于上扬子板块的西南部，北接四川盆地，南毗滇东—黔中隆起，东部与武陵拗陷相邻。该地区龙马溪组为一套深灰色至灰黑色粉砂质泥页岩、碳质泥页岩、硅质泥页岩夹泥质类砂岩，岩性在纵向上具有一定的渐变性，总体上具有向上颜色逐渐变浅、笔石含量、碳质含量、有机质丰度逐渐减少以及粉砂质和灰质含量逐渐增多的特征。研究区龙马溪组分布于中北部，总体上呈由南薄北厚的特点，厚度主要在190～260m，南部受滇东—黔中隆起影响而向南逐渐减薄并出现区域性薄层灰岩和粉砂岩夹层，颜色变浅，并大致在彝良龙街—赫章花坭—镇雄盐源—芒部—毛坝—毕节核桃树一线附近发生尖灭(图1)。

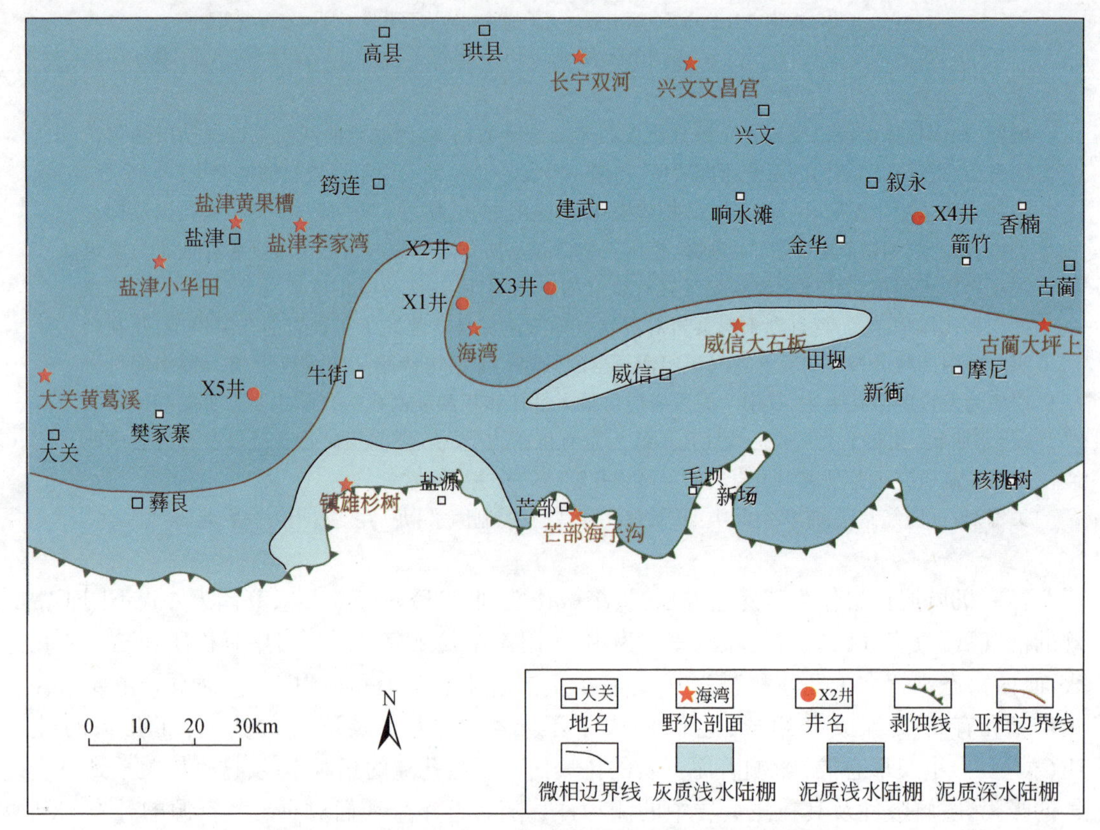

图1 川南—黔北××地区龙马溪组早期岩相古地理图

研究表明，区内龙马溪组沉积期沉积水体总体处于缺氧环境，缺氧程度、沉积水体深度和古生产力由龙马溪组沉积早期到晚期具有相似的变化趋势，即有水体含氧量增大、水体变浅以及古生产力降低的特点，尤其在龙马溪组沉积初期沉积水体具有能形成优质泥页岩的厌氧、水体较深以及高古生产力的条件；而在平面上龙马溪组沉积初期富含有机质的暗色泥页岩分布与深水泥质陆棚展布范围基本一致，主要分布于研究区的中北部(图1)。

2 储集空间类型及特征

2.1 孔隙

基质孔隙是泥页岩的基质块体单元中未被固态物质充填的空间。泥页岩中基质孔隙发育,引用霍多特分类,按孔径大小可将其划分为微孔(<10nm)、小孔(10~100nm)、中孔(孔径1000~100nm)和大孔(孔径>1000nm),其中微孔构成泥页岩主要的吸附空间;小孔为泥页岩毛细凝结和扩散的主要区域;中大孔则为渗流和层流的主要区域。按成因可将基质孔隙区分为残余原生粒间孔、晶间孔、矿物铸模孔、次生溶蚀孔、黏土矿物间微孔以及有机质孔(图2)。

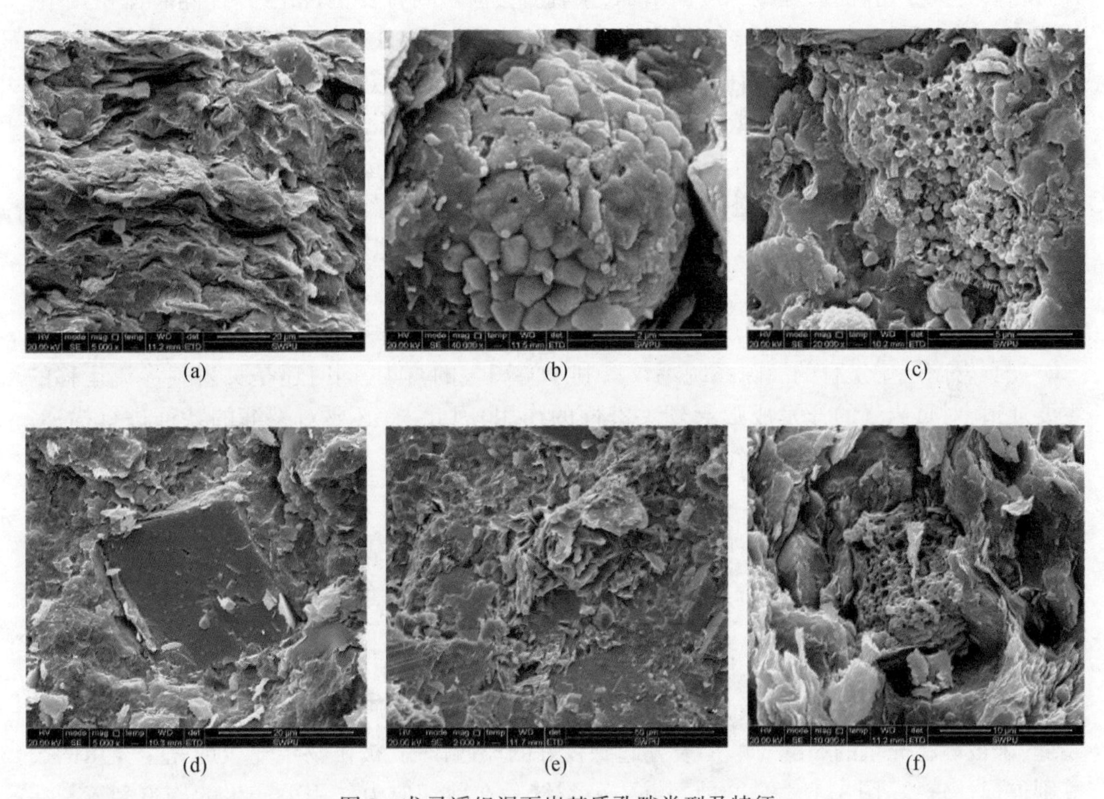

图2　龙马溪组泥页岩基质孔隙类型及特征

(a)残余原生粒间孔,X3井,龙马溪组,1397.91m;(b)晶间孔,X3井,龙马溪组,1409.60m;(c)矿物铸模孔,X3井,龙马溪组,1422.54m;(d)次生溶蚀孔,X3井,龙马溪组,1414.59m;(e)黏土矿物间微孔,X3井,龙马溪组,1422.54m;(f)有机质孔,X3井,龙马溪组,1409.60m

(1)残余原生粒间孔。残余原生粒间孔是原生粒间孔经过成岩作用中的压实、失水改造后残留的粒间孔隙空间[图2(a)]。这种孔隙与常规储层的残余原生粒间孔相似,通常随埋藏深度的增加而缩小。

(2)晶间孔。晶间孔是环境稳定和介质条件适当情况下,矿物结晶形成的晶间微孔隙,其孔径多分布在10~500nm。区内龙马溪组泥页岩中最常见的晶间孔为缺氧环境下形成的草莓状黄铁矿晶粒间的孔隙[图2(b)]。

(3)矿物铸模孔。泥页岩形成初期,其混杂的矿物晶体(如黄铁矿)在成岩阶段压实作用下,因晶体坚固,其几何形态不易发生形变,而在一定水动力或酸性流体介质条件下,矿物晶体遭受这些流体的冲击或溶蚀而发生脱落,留下了大量与晶形大体相仿的印坑,扫描电镜下观察到的矿物铸模孔孔径多在100~500nm[图2(c)]。

(4)次生溶蚀孔。泥页岩中常含有长石及碳酸盐等易溶矿物,在空气、地下水或有机质脱羧后产生的酸性水作用下溶蚀而产生的次生孔隙,这类孔隙又可分为粒内溶孔[图2(d)]和粒间溶孔。粒内溶孔孔径相对较小,主要分布在0.05~2μm;粒间溶孔孔径相对较大,主要分布在1~20μm。

(5)黏土矿物间微孔。主要为黏土矿物伊利石之间的微孔隙。当泥页岩孔隙水偏碱性并且富含钾离子时,随着埋藏深度的增加,蒙脱石会向伊利石发生转化,并伴随着体积减小,从而产生微裂(孔)隙,这种微裂(孔)隙孔径相对较小,主要分布在0.02~2μm[图2(e)]。

(6)有机质孔。泥页岩中有机质孔隙是泥页岩中有机质在热裂解生烃过程中形成的孔隙。据Jarvie等[12]研究表明,有机碳含量为7%的泥页岩在生烃演化过程中,消耗35%的有机碳可使泥页岩的孔隙度增加4.9%。产生的有机质孔隙孔径主要分布在2~1000nm[图2(f)],其中微孔和小孔所占比例较大,其对泥页岩的比表面积和孔体积贡献较大,对泥页岩的吸附性起着巨大的积极作用。

2.2　裂缝

泥页岩储层中发育的裂隙系统不仅有利于游离气的富集,同时还是页岩气渗流运移的主要通道,对页岩气的开发成败起到关键性的作用,因此,有必要对裂缝的特征进行描述。根据裂缝的成因,可将裂缝区分为构造裂缝(张裂缝和剪裂缝)、成岩收缩微裂缝、层间页理缝和超压破裂缝。

(1)构造裂缝。构造裂缝是泥页岩经一次或多次构造应力破坏而形成的,是裂缝中最主要的类型,可出现在泥页岩层的任何部位。根据力学性质的不同,又可分为张裂缝和剪裂缝[图3(a)~(c)]。①张裂缝。张裂缝是在张应力作用下产生的构造裂缝。在岩心上观察到的宏观张性裂缝缝宽和长度变化较大,通常裂缝面粗糙不平,多数已被矿物半充填或完全充填[图3(a)]。在扫描电镜下最常观察到的微观张裂缝通常近于垂直于层面切穿顺层裂缝,未被矿物充填的裂缝对顺层裂缝起到良好的连通作用,被矿物半充填或完全充填的裂缝则连通性较差[图3(b)]。②剪裂缝。剪裂缝是在剪切应力作用下产生的构造裂缝。在岩心上观察到的宏观剪裂缝较张裂缝少,其产状变化也较大,但多为低角度缝。其裂缝面通常平直光滑,在裂缝面上具有擦痕、阶步或微错动现象[图3(c)]。

(2)成岩收缩微裂缝。成岩收缩微裂缝为成岩过程中在上覆地层压力下泥页岩岩层失水、均匀收缩、干裂以及重结晶等作用产生内应力形成的裂缝。在扫描电镜下常见,其连通性较好,开度变化较大,部分被次生矿物充填[图3(d)]。

(3)层间页理缝。层间页理缝主要是指页岩中页理间平行于层理纹层面间的孔缝,为沉积作用过程中的产物。通常形成于强水动力条件,由一系列薄层页岩组成,层间页理缝通常为页岩间力学性质较薄弱的界面,常易于剥离。层间页理缝在区内泥页岩中极为常见,其开度一般较小,有时被其他矿物半充填或完全充填[图3(e)]。

(4)超压破裂缝。超压破裂缝是指在封闭状态下,由泥页岩中的黏土矿物转化脱水、生烃或水热增压等综合作用形成高异常流体压力或有机质在演化过程中产生局部异常压力,造成岩石发生破裂而形成的裂缝。这种裂缝一般不呈组系出现[图3(f)]。

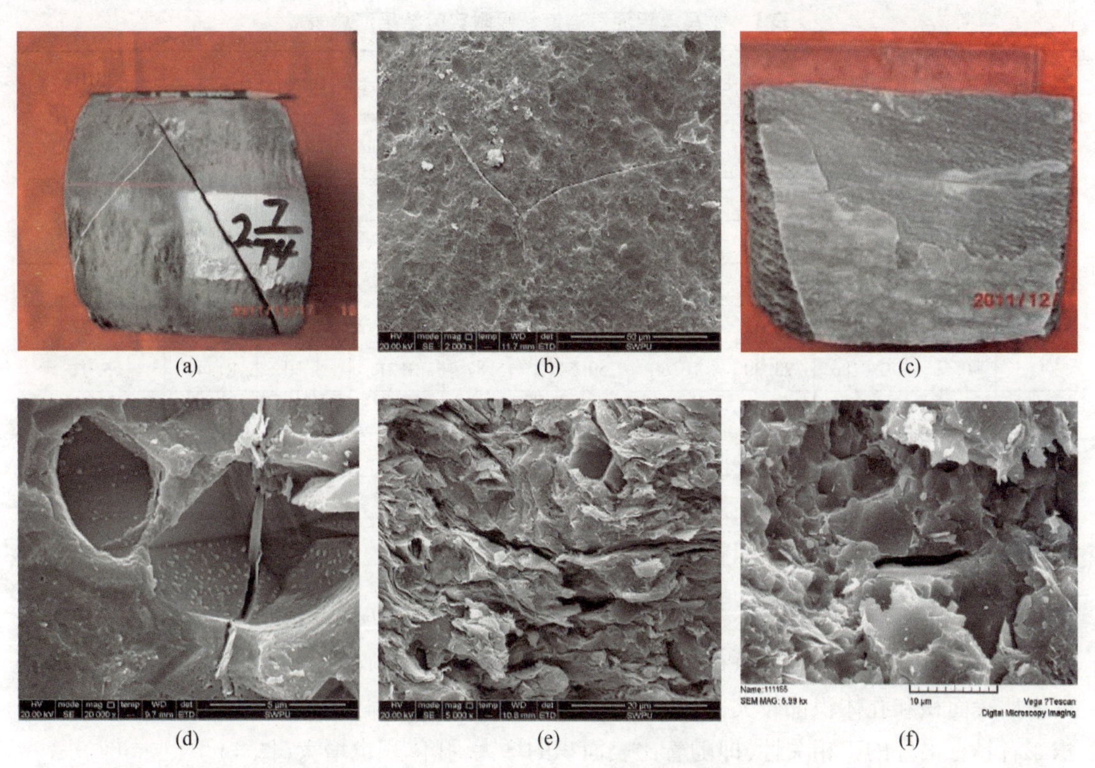

图3 龙马溪组泥页岩裂缝类型及特征

(a)张裂缝,被矿物全充填,X3井,龙马溪组,1197.12～1197.21m;(b)张裂缝,被矿物半充填,X3井,龙马溪组,1419.02m;(c)剪裂缝,见擦痕,X3井,龙马溪组,1309.18～1309.24m;(d)成岩收缩微裂缝,被矿物半充填,X3井,龙马溪组,1439.44m;(e)层间页理缝,X3井,龙马溪组,1397.91m;(f)成岩收缩微裂缝,X2井,龙马溪组,1875.21m

3 孔隙比表面积和孔体积特征

页岩气在泥页岩层中的储集量依赖于基质孔隙的比表面积和孔体积的大小。测定比表面积和孔体积的方法较多,但目前公认的最好的方法为液氮吸附法,主要是因为这种方法能对微–中孔的发育情况进行详细描述。

3.1 样品采集及实验测试

采集研究区X1井共计8个样品,见表1。实验由油气藏地质及开发工程国家重点实验室利用ASAP2020比表面测定仪进行,样品经150℃真空充分脱气4h,除去杂质气体后,放在盛有液氮的杜瓦瓶中与仪器分析系统相连,在77.3K进行等温物理吸附–脱附测定,孔径测量范围为1.5～300nm,吸附–脱附相对压力(P/P_0)范围为0.001～0.998;得到样品的等温吸脱附曲线数据和平均孔径数据;比表面积选用多点BET模型线性回归得到;孔体积和孔径分布则利用毛细凝聚模型BJH法计算得到。

表 1　龙马溪组泥页岩液氮吸附实验结果汇总表

样品号	TOC/%	BJH 孔体积 /(mL/g)	各孔径段体积比/%			BET 比表面积 /(m^2/g)	各孔径段比表面积比/%			BJH 平均孔径 /nm
			>100nm	10 ~ 100nm	<10nm		>100nm	10 ~ 100nm	<10nm	
X1-1	0. 13	0. 0098	30. 41	18. 47	51. 12	6. 226	4. 82	6. 66	88. 53	6. 56
X1-2	0. 21	0. 0181	19. 36	20. 30	60. 34	12. 598	2. 22	6. 2	91. 58	5. 87
X1-3	0. 08	0. 01104	19. 66	23. 10	57. 25	8. 691	2. 34	7. 41	90. 25	5. 41
X1-4	0. 82	0. 0133	26. 83	21. 25	51. 92	9. 363	3. 69	7. 52	88. 80	5. 97
X1-5	0. 66	0. 0186	23. 80	16. 66	59. 54	15. 87	3. 21	5. 30	81. 49	5. 07
X1-6	1. 26	0. 022	10. 01	12. 6	77. 39	20. 357	0. 94	3. 41	96. 65	4. 62
X1-7	2. 4	0. 0216	11. 4	14. 09	74. 5	20. 463	1. 11	3. 67	95. 22	4. 6
X1-8	2. 46	0. 0164	16. 41	11. 25	72. 34	14. 225	0. 23	5. 55	94. 22	5. 06

3. 2　比表面积和孔体积特征

测试结果表明,研究区泥页岩储集层的 BET 比表面积为 6. 226 ~ 20. 463m^2/g,平均为 13. 474m^2/g;BJH 孔体积为 0. 0098 ~ 0. 0220mL/g,平均为 0. 01635mL/g。这说明了研究区泥页岩比表面积和孔体积都较大,有利于页岩气的吸附(表 1);且泥页岩的孔比表面积和孔体积二者具有很好的正相关性,即随着比表面积的增大,孔体积也增大(图 4)。

在不同类型孔的中,微孔的孔体积和比表面积所占比例最大,分别可占到 51. 12% ~ 77. 39% 和 81. 49% ~96. 65%,而小孔及中孔所占比例较小,且二者相差不大(表 1)。同样可以发现,研究区泥页岩样品的平均孔径在 4. 60 ~ 6. 56nm(表 1),所有样品几乎都具有孔径(r)在 3 ~ 7nm 的微孔隙对孔体积值贡献最大的特点(图 5),且从孔体积累积曲线图看出(图 5),样品的孔径在 r<10nm 时,累积曲线很陡;而在 $r\geqslant$10nm 时,累积曲线逐渐得平缓,这也说明了泥页岩的微孔占据了最大的孔体积。因此本文认为微孔隙越发育,泥页岩的孔体积、比表面积越大,越有利于泥页岩对页岩气的吸附储集。

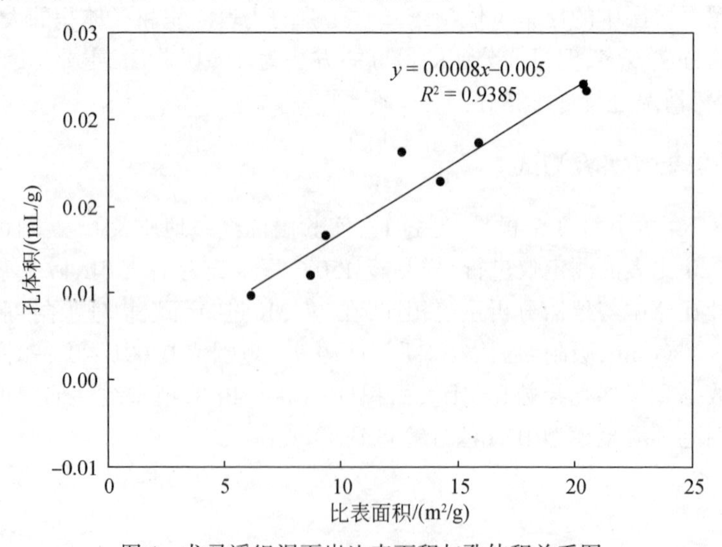

图 4　龙马溪组泥页岩比表面积与孔体积关系图

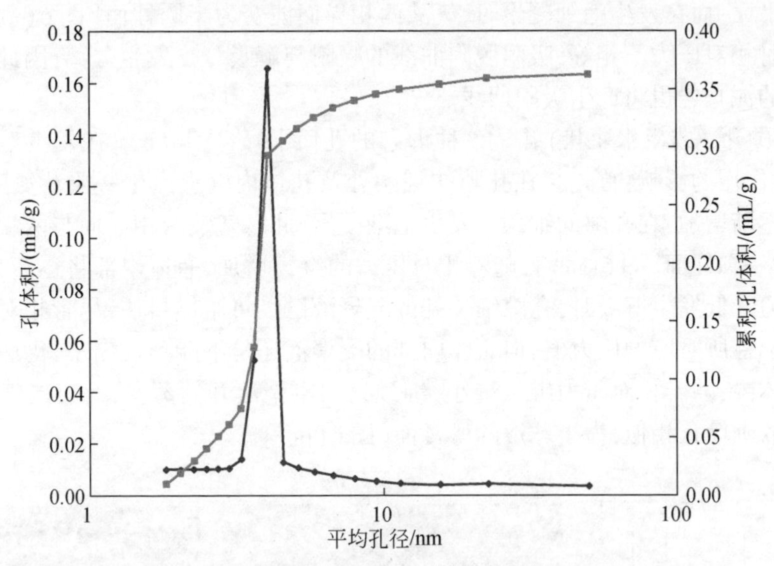

图5 X1井190.04m样品的孔体积分布曲线图

3.3 吸附脱附曲线与泥页岩的孔隙形态特征

泥页岩的等温吸附曲线在一定的压力范围内常与脱附曲线发生分离,形成所谓的吸附回线。由于泥页岩孔隙的具体形态的不同,同一孔隙在发生毛细凝聚和蒸发时的相对压力可能相同,也可能不相同,在相同时,吸附脱附曲线的吸附分支与解吸分支重叠;反之,吸附脱附曲线的两个分支便会分离,形成吸附回线。因此,可以利用吸附回线来进行孔隙形态特征的分析。

3.3.1 泥页岩中不同形态的孔及其对吸附回线的贡献

泥页岩中的孔隙形态各异,只有非常少的孔隙与某种典型的几何形状相符合。但为了更方便地讨论不同形态的孔隙对吸附回线的贡献,常把它们理想化为数种几何模型,这样便可以根据吸附脱附曲线的特征分析泥页岩中的孔隙结构组成,具体假设的孔隙模型如下:

(1)一端封闭的圆筒状孔。该种形态的孔[图6(a)~(c)]在毛细凝聚和蒸发时,处于同等的相对压力条件,气液两相界面都为相同种类半球形的弯月面[图6(a)];因此,具有这种孔隙形态的泥页岩其吸附脱附曲线的吸附与解吸分支发生重叠,不会产生吸附回线。

(2)两端开口的圆筒状孔。该种形态的孔[图6(d)~(f)]在发生毛细凝聚和蒸发时,气液两相界面分别为一个圆柱面和两个类半球形的弯月面[图6(d)];在发生毛细凝聚时的相对压力要大于毛细蒸发时的,因此具此种孔隙形态的泥页岩其吸附脱附曲线的吸附与解吸分支发生分离,形成吸附回线。

(3)一端封闭的平行板状孔或尖劈形孔。具有这两种形态的孔[图6(g)~(i)]在毛细凝聚和蒸发时,同样处于同等的相对压力条件,气液两相界面都是相同的半圆柱面[图6(g)];因此,具有这两种孔隙形态结构的泥页岩其吸附脱附曲线的吸附与解吸分支会发生重叠,不会产生吸附回线。

(4)四边都开口的平行板状孔。该种形态的孔[图6(j)~(l)]在发生毛细凝聚时,气液两相界面为平面；而在发生毛细凝聚时，气液两相界面演变为半圆柱面[图6(j)]，其在发生凝聚与蒸发时相对压力不相等，吸附脱附曲线的吸附与解吸分支发生分离，因此具有此种孔隙形态结构的泥页岩也将产生吸附回线。

(5)细颈瓶状(或墨水瓶状)孔。该种形态的孔[图6(m)~(o)]在发生毛细凝聚初始阶段，因吸附作用的影响，首先会在孔隙的细颈处及孔体内壁上产生一层液氮吸附层；随着相对压力的逐渐增加，在孔隙的细颈处发生毛细凝聚，此时气液两相界面与两端都开口的圆筒孔相似，为一圆柱面[图6(m)]；随着相对压力的继续增加，瓶体内部将逐渐充满凝聚液。而当相对压力降低发生解吸时，孔隙细颈处的凝聚液已经将瓶体封住，尽管相对压力接近或达到瓶体内半径所对应的压力值，但瓶体内部的凝聚液还是不能蒸发出来，仍然以液态的形式保留在瓶体内；随着相对压力继续减小，细颈处的凝聚液开始蒸发，此时的气液两相界面与一端封闭的圆筒孔相似，为半球形的弯月面[图6(m)]。

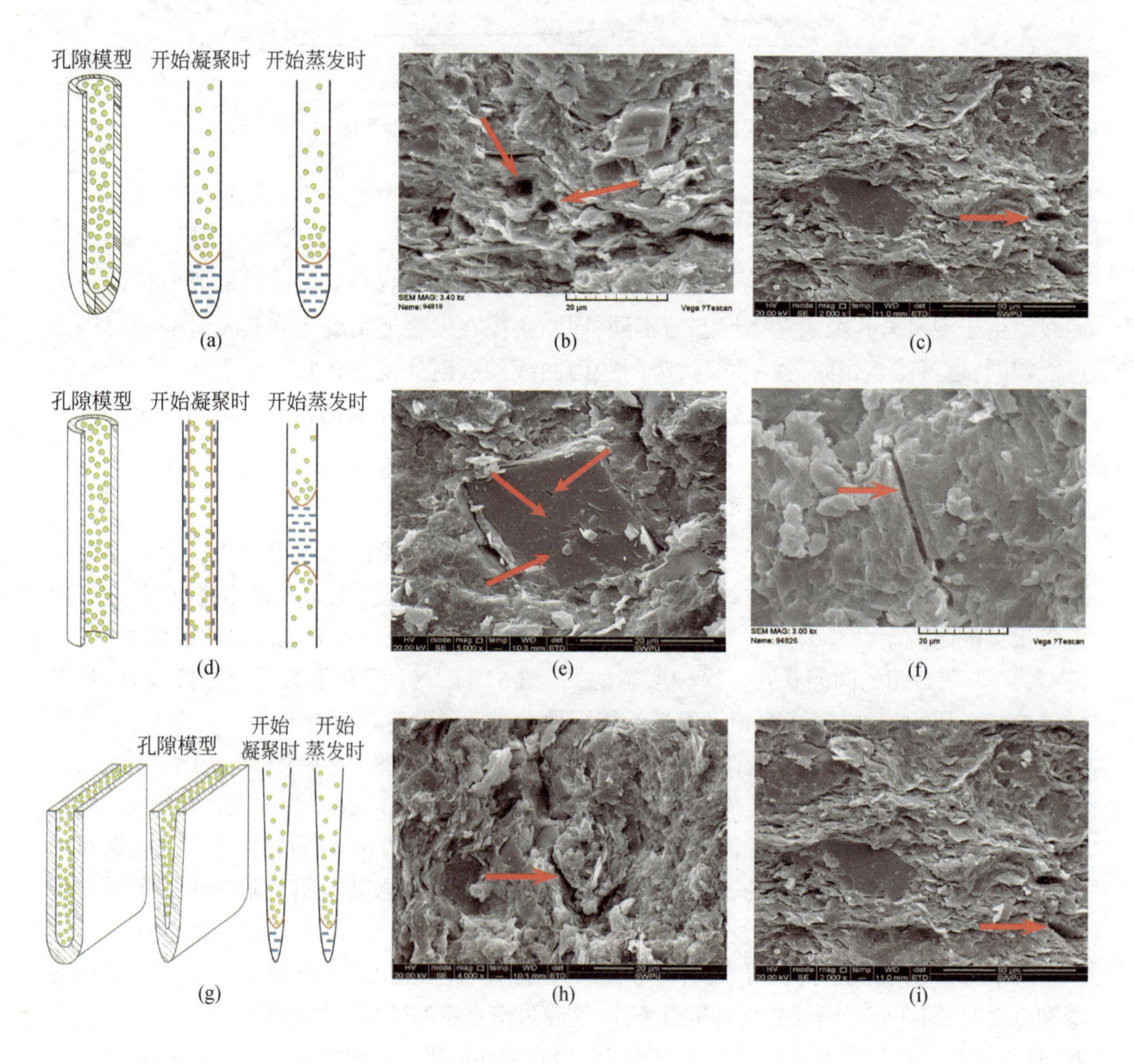

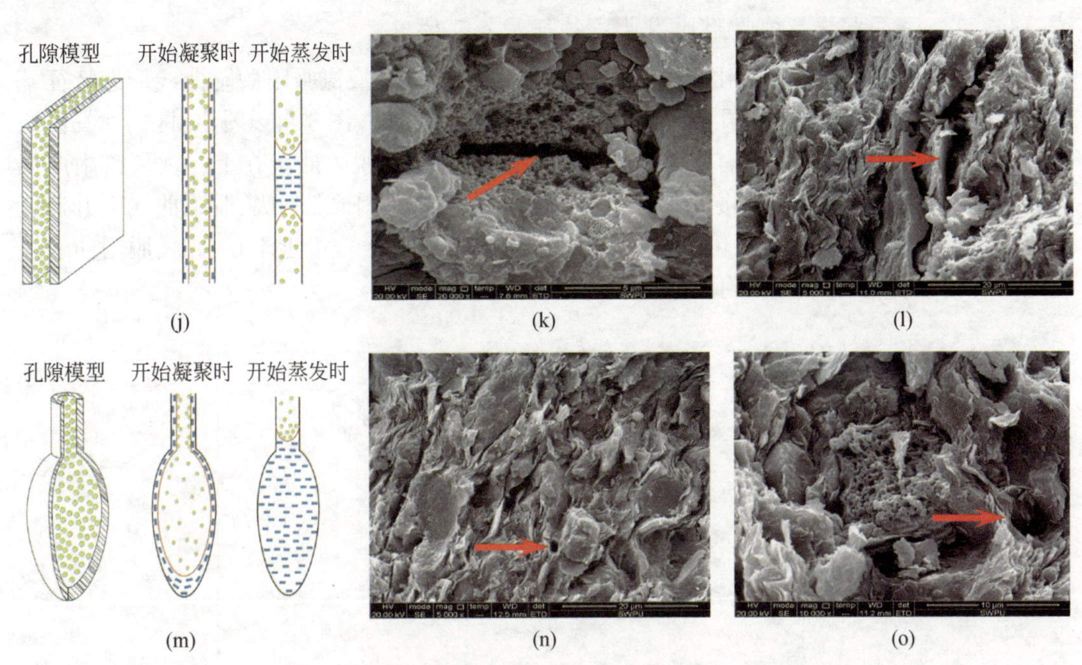

图6 龙马溪组泥页岩孔隙模型及孔隙内气液界面特征

(a)一端封闭的圆筒状孔模型以及开始凝聚和蒸发时的气-液界面;(b)一端封闭的圆筒状孔(横切面),X1井,龙马溪组,203.43m;(c)一端封闭的圆筒状孔(纵切面),X3井,龙马溪组,1397.91m;(d)两端开口的圆筒状孔模型以及开始凝聚和蒸发时的气-液界面;(e)两端开口的圆筒状孔(横切面),X3井,龙马溪组,1414.59m;(f)两端开口的圆筒状孔(纵切面),X1井,龙马溪组,203.51m;(g)一端封闭的平行板状孔或尖劈形孔模型以及开始凝聚和蒸发时的气-液界面;(h)一端封闭的平行板状孔或尖劈形孔(横切面),X3井,龙马溪组,1439.44 m;(i)一端封闭的平行板状孔或尖劈形孔(纵切面),X3井,龙马溪组,1397.91m;(j)四边都开口的平行板状孔模型以及开始凝聚和蒸发时的气-液界面;(k)四边都开口的平行板状孔(横切面),X3井,龙马溪组,1434.97m;(l)四边都开口的平行板状孔(纵切面),X3井,龙马溪组,1409.60m;(m)细颈瓶状(或墨水瓶状)孔模型以及开始凝聚和蒸发时的气-液界面;(n)细颈瓶状(或墨水瓶状)孔(横切面),X3井,龙马溪组,1404.99m;(o)细颈瓶状(或墨水瓶状)孔(纵切面),X3井,龙马溪组,1409.60m

在毛细凝聚时与毛细蒸发时相对压力并不相同,在细颈处吸附脱附曲线的吸附与解吸分支发生分离,产生吸附回线;等到细颈处凝聚液蒸发完毕,相对压力已经远低于瓶体内吸附质蒸发时所需要的相对压力值,此时瓶体内部的全部凝聚液会被骤然蒸发出,因此肯定会产生吸附回线,且在脱附曲线上具有一个急剧下降的拐点(图7,图8)。

通过以上孔隙模型的建立,根据孔隙的形态结构及其能否产生吸附回线,又可将泥页岩中的孔隙区分为3类,各类孔隙特征见表2。

表2 根据孔隙的形态结构及其能否产生吸附回线的孔隙类型划分

孔隙类型	孔隙形态	能否产生吸附回线	孔隙模型类型	其他特点
Ⅰ类	开放透气性孔隙	能	两端开口圆筒状孔、四边都开口的平行板状孔	
Ⅱ类	一端封闭的不透气性孔隙	否	一端封闭的圆筒状孔、一端封闭的平行板状孔或尖劈形孔	
Ⅲ类	细颈瓶状(墨水瓶状)孔隙	能	细颈瓶状(墨水瓶状)孔	在脱附曲线上有一个急剧下降的拐点

3.3.2　泥页岩样品的吸附脱附曲线特征及其孔的意义

通过测试实验所得的研究区龙马溪组 8 个泥页岩的低温液氮吸附脱附曲线形态特征基本相同，这反映区内泥页岩中孔隙形态特征基本相同。从图 7、图 8 可以看出，区内龙马溪组泥页岩样品的吸/脱附曲线具有以下特点：①吸附曲线在下，脱附曲线在上；②吸、脱附曲线都随相对压力的增大，处于缓慢上升状态；③在相对压力接近于 1 时，吸、脱附曲线上升速度加快；④吸附回线出现在相对压力在 0.4 ~ 1.0 时；⑤在相对压力接近于 0.5 时，脱附曲线上出现了明显的拐点 G，致使脱附曲线近乎陡直下降。

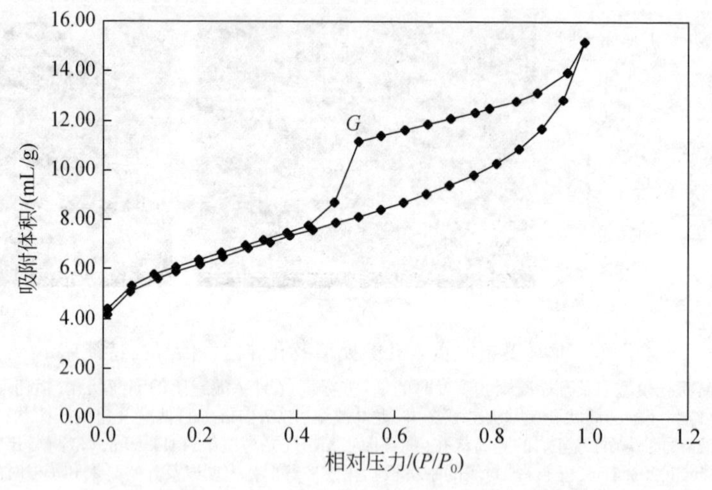

图 7　X1 井 190.04m 样品的吸附脱附曲线图

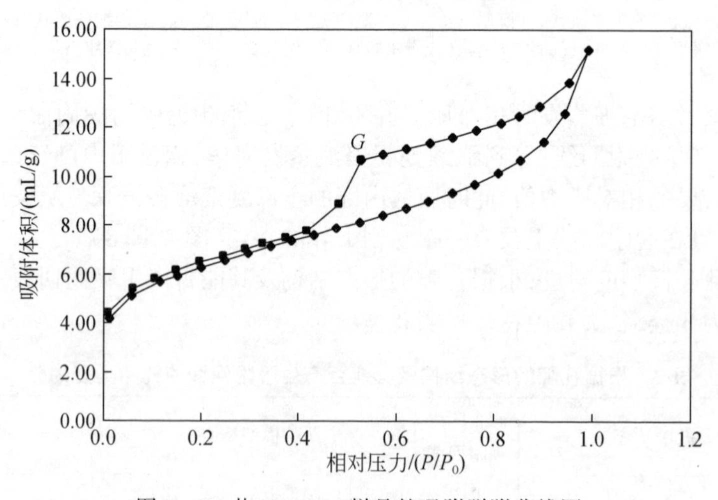

图 8　X1 井 203.43m 样品的吸附脱附曲线图

吸附曲线在下，脱附曲线在上（图 7，图 8），说明退氮速度比进氮速度慢，反映了随着压力的升高，泥页岩样品中并没有产生了新的裂隙和孔隙使退氮速度加快，仍保留原来的孔隙系统。在吸附时，吸附曲线的前半段上升比较平稳并且呈向上微凸起的形状，表明该时期为由单分子层逐渐向多分子层吸附过渡的阶段；而在后半段，特别是相对压力接近 1 时，曲线上升速度加快，表明在泥页岩中较大孔径的孔隙里发生了毛细凝聚现象，从而造成了气体吸附量的快速增大(图7，图8)。

吸附脱附曲线在相对压力处于0.4～1.0时出现吸附回线，且在脱附曲线分支上具有明显的拐点G(图7，图8)，这反映了泥页岩样品的孔隙系统比较复杂。在相对压力较低处(0～0.4)，吸附曲线与脱附曲线基本重叠，不产生吸附回线，这说明在较小孔径范围内的孔隙的形态多为一端封闭的半不透气性孔，即Ⅱ类孔；在相对压力较高处(0.4～1.0)，出现了吸附回线且具有明显的拐点G，说明具有较大孔径的孔隙，其形态必然存在着Ⅲ类孔，即细颈瓶状(墨水瓶状)孔；同时也可能存在Ⅰ类开放透气性孔和Ⅱ类一端封闭不透气性孔，这是因为Ⅰ类开放透气性孔虽然没有拐点G，但也可以产生吸附回线，而Ⅱ类一端封闭不透气性孔对回线没有贡献，因此这两种类型孔在曲线上产生的效应有可能被Ⅲ类细颈瓶状(墨水瓶状)孔所掩盖。

区内龙马溪组泥页岩样品的吸附回线(图7，图8)可以解释为：在解吸退氮的初始过程中，随着相对压力的降低，由于开放透气性孔和Ⅲ类细颈瓶状(墨水瓶状)孔在毛细凝聚与蒸发时气液两相界面形状的不同，会产生吸附回线。而样品中存在各级孔径范围的孔隙，随着相对压力的降低，较大孔的凝聚液首先开始蒸发，造成吸附量逐渐减少，脱附曲线随之逐渐下降。当相对压力降低到脱附曲线上拐点G所对应的值时，意味着最小一个孔径的细颈瓶状(墨水瓶状)孔隙里的凝聚液即将蒸发出来，相对压力稍一降低，孔隙里的全部凝聚液会一涌而出，在脱附曲线上表现出急剧下降的特征，在开放性孔隙或细颈瓶状(墨水瓶状)孔隙瓶体内部凝聚液蒸发完毕后，仅一端封闭不透气性Ⅱ类孔隙内的凝聚液仍未解吸，随着相对压力的继续降低，这类孔隙内的凝聚液也逐渐被蒸发出来，但吸附曲线和脱附曲线基本重合。

根据BJH法求半径r公式[13]，即$r=-2\gamma V_m/[RT\ln(P/P_0)]+0.354[-5/\ln(P/P_0)]^{1/3}$可得，在相对压力为0.4时，对应的孔隙直径为3.3nm；在相对压力(拐点G所对应的相对压力)为0.5时，对应的孔隙直径为4.0nm；在相对压力为0.8时，对应的孔隙直径为10nm。由此可知，区内龙马溪组泥页岩存在各级孔径的孔隙，但以极为发育的微孔为主。其中$r<3.3$nm的微孔隙主要为Ⅱ类不透气性孔；$3.3\text{nm}\leq r<4.0$nm的微孔隙主要为Ⅰ类开放透气性孔，也不排除Ⅱ类不透气性孔的存在；$r\geq 4.0$nm的微孔和小孔主要为能产生拐点G的Ⅲ类细颈瓶状(墨水瓶状)孔和另外一种能产生吸附回线的Ⅰ类开放透气性孔，以及少量不能产生吸附回线的Ⅱ类不透气性孔。

前述研究表明，研究区龙马溪组泥页岩孔隙平均孔径处在4.60～6.56nm，孔径处在该范围内的孔隙主要为Ⅲ类细颈瓶状(墨水瓶状)孔和Ⅰ类开放透气性孔，其中细颈瓶状(墨水瓶状)孔的存在虽然有利于页岩气的吸附，但透气性较差，不利于页岩气的解吸与扩散。但当相对压力降低到拐点G所对应的压力值以下时，瓶体内的页岩气会在瞬间快速解吸转变为游离气，页岩气会在短时间内涌出，因此在相对压力降低到临界压力值附近，要防止页岩气的突出。

4 页岩气储层微观孔隙结构的控制因素

4.1 总有机碳含量(TOC)

泥页岩总有机碳含量是衡量烃源岩生烃潜力的重要参数，富有机质页岩中有机质孔的平均孔径远小于无机质的平均孔径[14]。研究区龙马溪组泥页岩 TOC 与孔体积、孔比表面积关系图(图9)表明,TOC 与孔体积和比表面积间呈现较为显著的对数关系(R^2分别为0.6511和0.606),即随着 TOC 增大,泥页岩孔体积、孔比表面积随之增大,这反映了 TOC 是控制龙马溪组泥页岩中微观孔隙孔体积及其比表面积的主要控制因素之一。

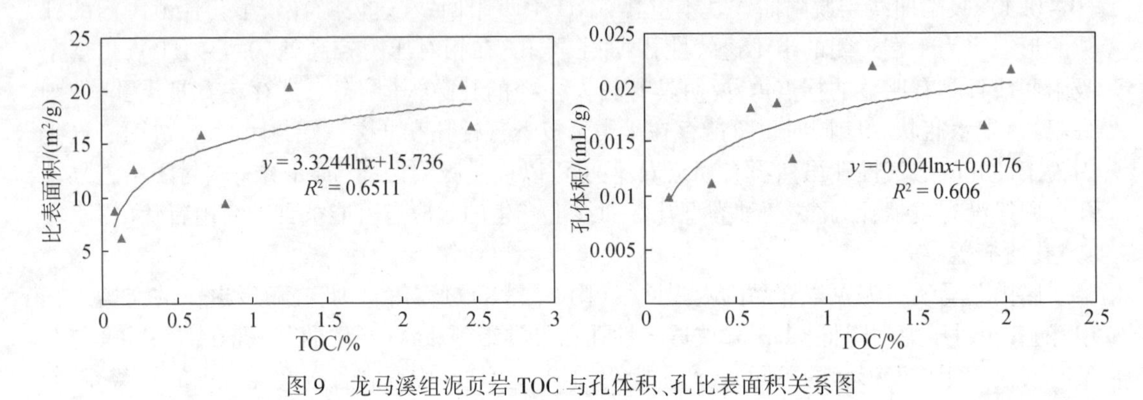

图9 龙马溪组泥页岩 TOC 与孔体积、孔比表面积关系图

4.2 黏土矿物类型及含量

页岩气储层中黏土矿物具有较高的微孔隙体积和较大的比表面积[15]。但不同黏土矿物晶层及孔隙结构不同,比表面积也存在很大的差别[11,16]。研究区龙马溪组泥页岩中黏土矿物主要为伊利石、伊/蒙间层和绿泥石,见极少量的高岭石(表3),其中伊/蒙间层是伊利石和蒙脱石的间层矿物。研究发现,研究区龙马溪组泥页岩伊/蒙间层矿物含量与孔体积、孔比表面积关系图(图10)表明,伊/蒙间层矿物含量与孔体积和比表面积间呈现良好的对数关系(R^2分别为0.7435 和0.8541),即随着伊/蒙间层矿物含量增大,泥页岩孔体积、孔比表面积随之增大,这反映了伊/蒙间层矿物含量同样是控制龙马溪组泥页岩中微观孔隙孔体积及其比表面积的主要控制因素之一。究其原因,这主要是在研究区龙马溪组泥页岩中黏土矿物类型中,伊/蒙间层是伊利石和蒙脱石的间层矿物,而蒙脱石具有的比表面积(800cm²/g)明显高于其他类型的黏土矿物(伊利石 30cm²/g、绿泥石 15cm²/g、高岭石15cm²/g),从而为伊/蒙间层矿物提供了大量的比表面积,蒙脱石的高比表面积是由于蒙脱石除提供了外表面积,还存在由层间结构提供的内表面积[16,17]。

表3 X1 井龙马溪组泥页岩矿物成分分析表

样品号	黏土矿物相对含量/%								全岩定量分析/%							I/S/%	BJH 孔体积/(mL/g)	BET 比表面积/(m²/g)
	K	C	I	S	I/S	%S	C/S	%S	黏土	石英	钾长石	斜长石	方解石	白云石	黄铁矿			
X1-2	2	18	58		22	10			43	32	2	6	14	2	1	9.46	0.0181	12.598
X1-3		29	54		17	10			16	39	3	12	19	11		2.72	0.01104	8.691
X1-5		20	62		18	10			40	30	6	8	10	4	2	7.2	0.0186	15.87
X1-6	6	14	45		35	10			49	35		3	9	3	1	17.15	0.022	20.357
X1-7	1	10	52		37	10			55	23	2	4	6	6	4	20.35	0.216	20.463
X1-8		14	69		17	10			20	37		1	21	20	1	3.4	0.0164	14.225

注:K 表示高岭石;C 表示绿泥石;I 表示伊利石;S 表示蒙皂石;I/S 表示伊/蒙间层;C/S 表示绿/蒙间层;%S 表示间层比

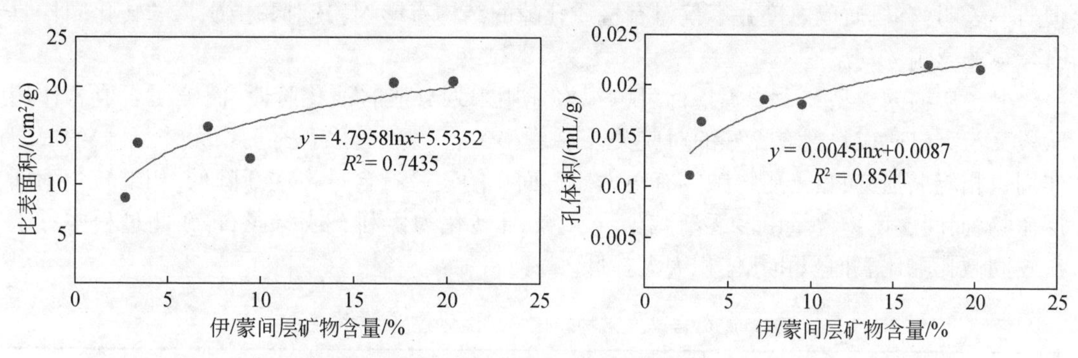

图 10 龙马溪组泥页岩伊/蒙间层矿物含量与孔体积、孔比表面积关系图

4.3 有机质成熟度(R_o)

页岩气微观孔隙结构随有机质成熟度的变化关系复杂。这是因为有机质成熟度不仅会造成有机质中纳米孔隙的孔隙结构变化，同时还会引起黏土矿物之间的转化，造成了黏土矿物之间微孔隙比表面积的改变。

4.3.1 有机质成熟度对有机质孔隙结构的影响

据统计[5]，页岩气以吸附态赋存于泥页岩中有机质和黏土矿物表面的含量占页岩气总含量的 20% ~85%。因此，页岩气的极限吸附量在一定程度上能反映泥页岩比表面积和孔体积的大小。研究发现，在 TOC 和伊/蒙间层矿物含量非常相近的情况下，有机质越大，研究区具有高有机质成熟度泥页岩样品的吸附量随 R_o 的增大而增大(图 11)。这说明了在 TOC 和伊/蒙间层矿物含量非常相近的情况下，对页岩气吸附量引起变化的主要为提供大量吸附空间有机质孔，即随 R_o 的增大，泥页岩中有机质孔的孔隙结构会发生变化，微孔和小孔逐渐增加，在有机质演化过程中大量增加的小孔和微孔为页岩气的吸附提供了更大量的孔体积和比表面积，以致泥页岩对页岩气的吸附能力也大大提高。

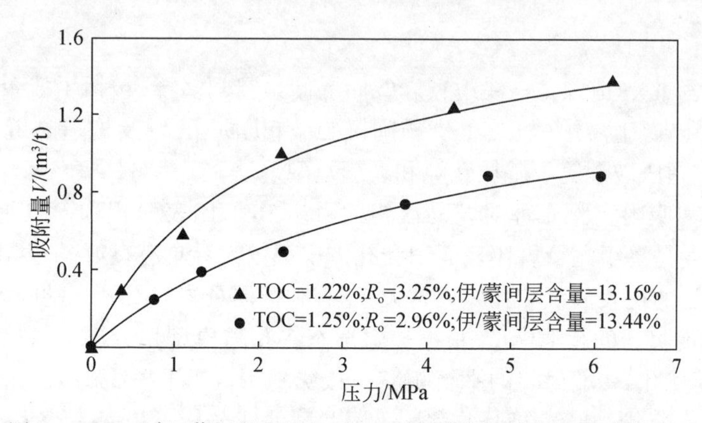

图 11 TOC 和伊/蒙间层含量相近、R_o 不同的泥页岩等温吸附曲线图

4.3.2 有机质成熟度对黏土矿物类型和含量的影响

以上的分析是在研究区龙马溪组泥页岩中伊/蒙间层矿物含量相近的前提下进行的。

但研究表明,有机质成熟度并不仅对有机质孔隙的结构有影响,其对黏土矿物的转化同样具有较大的影响。

黏土矿物的类型和含量等指标可以很好地用来划分成岩演化阶段,而碎屑岩成岩作用阶段又与有机质成熟度有良好的对应关系[18-24](表4)。通常随着 R_o 的增大,泥页岩的成岩作用也相应加强,而黏土矿物中具有很大比表面积的蒙脱石含量将逐渐降低,相继转化为间层矿物,而间层矿物含量由多逐渐减少,最终全部转化为伊利石或绿泥石,在此过程中黏土矿物间微孔隙比表面积和孔体积大大降低。

表4　泥页岩黏土矿物转化与热演化程度的对应关系

温度/℃	有机质成熟度 R_o/%	I/S 矿物中 I 含量/%	伊利石结晶度 IC($\Delta 2\theta$)	黏土矿物组合	油气阶段
<100	<0.6	<60	—	伊/蒙间层,高岭石,蒙脱石	未成熟,低成熟生物气,低熟油
100~180	0.6~1.6	60~85	>0.42	伊/蒙间层,高岭石,伊利石,绿泥石	成熟,油和湿气
180~212	1.6~2.6	>85	—	伊/蒙间层,伊利石,绿泥石,绿/蒙间层	高成熟,干气
>212	>2.6	无 I/S 混层	0.42~0.25	伊利石,绿泥石	过成熟

注:I/S 表示伊/蒙间层;I 表示伊利石;S 表示蒙脱石

由此可见,有机质成熟度确实控制着泥页岩的孔隙结构。但其对泥页岩孔隙结构的控制又处在一个“矛盾体”中,即泥页岩中有机质孔隙的比表面积和孔体积随有机质成熟度的增高而增大,而黏土矿物间微孔隙的比表面积和孔体积随有机质成熟度的增高而降低。因此,需要寻找有机质成熟度(R_o值)的一个“临界点”,使泥页岩的比表面积和孔体积最大,从而使其对页岩气的吸附能力最大。

5　结论

(1)川南—黔北××地区龙马溪组页岩气储层储集空间多样,包括了残余原生粒间孔、晶间孔、矿物铸模孔、次生溶蚀孔、黏土矿物间微孔、有机质孔,以及构造裂缝、成岩收缩微裂缝、层间页理缝、超压破裂缝等基质孔隙和裂缝类型。

(2)利用液氮吸附实验原理对龙马溪组泥页岩 BET 比表面积和 BJH 孔体积进行了研究,发现滇黔北龙马溪组泥页岩比表面积和孔体积都较大且具有良好的正相关性;认为微孔隙越发育,泥页岩的比表面积和孔体积越大,越有利于泥页岩对页岩气的吸附储集。

(3)建立了泥页岩的孔隙模型,根据孔隙形态将孔隙模型划分出一端封闭的圆筒状孔、两端开口的圆筒状孔、一端封闭的平行板状孔或尖劈形孔、四边都开口的平行板状孔,以及细颈瓶状(或墨水瓶状)孔等类型;并利用吸附脱附曲线分析了川南—黔北××地区龙马溪组泥页岩的微孔隙结构特征,指出区内龙马溪组泥页岩以极为发育的微孔为主,其中为泥页岩提供最大量孔体积和表面积的孔隙主要为Ⅲ类细颈瓶状(墨水瓶状)孔和Ⅰ类开放透气性孔,其中细颈瓶状(墨水瓶状)孔有利于页岩气的吸附,但透气性较差,不利于页岩气的解吸与扩散。

(4)总有机碳含量、伊/蒙间层矿物含量以及有机质成熟度是控制川南—黔北××地区龙马溪组页岩气储层孔隙结构的主要因素。页岩气储层孔隙的比表面积和孔体积随有机碳或伊/蒙间层矿物含量增大而增大;而有机质成熟度对页岩气储层孔隙比表面积和孔体积的控制处在一个"矛盾体"中。

参考文献

[1] Schettler P D, Parmely C R. Contributions to total storage capacity in Devonian shales[J]. SPE 23422, 1991: 77-88.

[2] Bowker K A. Recent development of the Barnett Shale play, Fort Worth Basin: West Texas[J]. Geological Society Bulletin, 2003, 42(6): 1-11.

[3] Montgomery S L, Jarvie D M, Bowker K A, *et al.* Missiissippian Barnett Shale, Fort Worth Basin, North-central Texas: Gas-shale play with multitrillion cubic foot potential[J]. AAPG Bulletin, 2005, 89(2): 155-175.

[4] Bowker K A. Recent developments of the Barnett Shale play, Fort Worth Basin[J]. West Texas Geological Society Bulletin, 2005, 42(6): 4-11.

[5] Curtis J B. Fractured shale-gas systems[J]. AAPG Bulletion, 2002, 86(11): 1921-1938.

[6] 王祥,刘玉华,张敏,等. 页岩气形成条件及成藏影响因素研究[J]. 天然气地球科学,2010,21(2):350-356.

[7] 王飞宇,贺志勇,孟晓辉,等. 页岩气赋存形式和初始原地气量(OGIP)预测技术[J]. 天然气地球科学,2011,22(3):501-510.

[8] Ambrose R J, Hartman R C, Campos M D, *et al.* New pore-scale considerations for shale gas in place calculations[J]. SPE131772, 2010: 1-17.

[9] 陈尚斌,朱炎铭,王红岩,等. 川南龙马溪组页岩气储层纳米孔隙结构特征及其成藏意义[J]. 煤炭学报,2012,37(3):438-444.

[10] Bustin R M, Bustin A M M, Cui X, *et al.* Impact of shale properties on pore structure and storage characteristics[R]. SPE 119892, 2008.

[11] 吉利明,邱军利,夏燕青,等. 常见黏土矿物电镜扫描微孔隙特征与甲烷吸附性[J]. 石油学报,2012,33(2):249-256.

[12] Jarvie D M, Hill R J, Ruble T E, *et al.* Unconventional shale-gas systems: The Mississippian Barnett shale of north-central Texas as one model for thermogenic shale-gas assessment[J]. AAPG Bulletin, 2007, 91(4): 475-499.

[13] 杜玉娥. 煤的孔隙特征对煤层气解吸的影响[D]. 西安科技大学硕士学位论文,2010.

[14] Kang S M, Fathi E, Ambrose R J, *et al.* Carbon dioxide storage capacity of organic-rich shales[J]. SPE 134583, 2010: 1-17.

[15] Ross D J K, Bustin R M. Characterizing the shale gas resource potential of Devonian Mississippian strata in the Western Canada sedimentary basin: application of an integrated formation evalution[J]. AAPG Bulletin, 2008, 92(1): 87-125.

[16] 赵杏媛,张有瑜. 黏土矿物与黏土矿物分析[M]. 北京:海洋出版社,1990.

[17] Passey Q R, Bohacs K M, Esch W L, *et al.* From oil-prone source rock to gas-producing shale reservoir: geologic and petrophysical characterization of uniconventional shale-gas reservoirs[R]. SPE 131350, 2010.

[18] Hoffman J, Hower J. Clay mineral assemblages as low grade metamorphic geothermometers application to the thrust faulted disturbed belt of Montana in Scholle PA and Schluger PS. eds. aspects of diagenesis[J]. Society of Economic Paleontologists and Mineralogists Special Publication 1979, 26: 55-80.

[19] Nadeau P H. Burial and contact metamorphism in the Mancos shale[J]. Clays and Clay Minerals,1981,29:249-259.

[20] Arkai P. Chlorite crystallinity:An empirical approach and correlation with illite crystallinity, coal rank andmineral facies as exemplified by Palaeozoic and Mesozoic rocks of north east Hungary[J]. Jour Metamorphic Geol. 1991,9:23-734.

[21] 赵孟为. 划分成岩作用与埋藏变质作用的指标及其界线[J]. 地质论评,1995,41(3):238-244.

[22] Ji J F,Browne P R L. Relationship between illite crystallinity and temperature in active geothermal system of New Zealand[J]. Clays and Clay Minerals,2000,48(1):139-144.

[23] 肖丽华,孟元林,牛嘉玉,等. 歧口凹陷沙河街组成岩史分析和成岩阶段预测[J]. 地质科学,2005,40(3):346-362.

[24] 刘伟新,王延斌,秦建中. 川北阿坝地区三叠系黏土矿物特征及地质意义[J]. 地质科学,2007,42(3):469-482.

页岩气吸附解吸效应对基质物性影响特征

王 敬[1] 罗海山[2] 刘慧卿[1] 林 杰[3] 李立文[3] 林文鑫[4]

(1. 中国石油大学(北京)石油工程教育部重点实验室;2. Department of Petroleum & Geosystems Engineering, The University of Texas at Austin;3. 中国石油华北油田公司;4. 西南石油大学地球科学与技术学院)

摘要:为了研究页岩气降压开采过程中吸附气解吸作用对基质表观物性(如有效孔隙半径、有效孔隙度、表观渗透率)及气体流动机制的影响,推导了吸附解吸作用下页岩基质孔隙有效半径和表观渗透率动态模型,建立了考虑吸附解吸影响基质表观物性和气体传输机制的页岩气渗流数学模型。采用有限体积法对模型进行求解,利用实验及矿场数据验证了模型的可靠性,最后应用该模型研究了页岩气开采过程中基质物性参数、气体流动机制变化特征,以及吸附效应对页岩气开发的影响规律。研究结果表明,页岩气开采过程中基质孔隙有效半径、有效孔隙度和表观渗透率逐渐变大,体积压裂改造区域流动机制由滑脱流转变为过渡流;忽略吸附层影响将导致地质储量和产气量被严重高估;随着吸附层厚度增加,累积产气量变化不大,但采收率逐渐降低。

关键词:页岩气;吸附作用;解吸作用;基质孔隙;表观物性;气体流动机制;渗流模型

页岩气的大规模开发影响世界天然气供给格局,预计到2020年,全球页岩气产量将达到$4000\times10^8\mathrm{m}^3$[1]。但是,页岩气开发难度巨大,因为页岩气藏渗透率极低,一般为$10^{-9}\sim10^{-6}\mu\mathrm{m}^2$,孔隙度低于10%[2]。页岩气孔隙尺寸非常小,据统计,半径小于10nm的孔隙体积占总孔隙体积的42%,部分孔隙和流动通道半径甚至小于2nm,只有少量的孔隙半径大于50nm[3,4]。页岩气藏为典型的自生自储气藏,气体主要以自由气和吸附气的形式储存于气藏中,吸附气含量达20%~80%,吸附气必须首先发生解吸才能开采出来[5,6]。由于页岩孔隙尺寸为纳米级,且大量的气体吸附在孔隙表面,气体分子流动和吸附气解吸过程同时进行,因此,页岩气在基质孔隙中的传输机理复杂,既要受到孔隙尺寸、孔隙压力的影响,又要受到吸附、解吸过程影响,从而最终影响页岩气开采。目前,大量研究中没有考虑吸附气对页岩孔隙尺寸的影响,一方面导致地质储量被高估[7,8],另一方面无法考虑页岩气藏开采过程中吸附、解吸效应对页岩表观物性和气体流动机制的影响。

本文首先推导了吸附、解吸作用下的页岩基质孔隙有效半径和表观渗透率动态模型,据此建立了考虑吸附、解吸效应对基质表观物性和气体传输机制影响的页岩气渗流数学模型,然后采用有限体积法对耦合模型进行求解、验证,最后应用该模型研究吸附、解吸过程中基质物性参数变化特征、气体流动机制变化特征,以及吸附效应对页岩气开发的影响。

1 基质孔隙中页岩气分布特征及流动机理

如图1所示,基质孔隙中气体主要包括自由气、吸附气两部分。大量研究表明[9,10],页岩气吸附规律符合描述单层吸附的Langmuir模型,饱和吸附状态下有效半径R_e为孔隙绝对半径R_0与甲烷分子直径d_{CH_4}之差,随着气体采出,孔隙压力降低,孔隙表面的吸附气发生解吸并参与流动,吸附层厚度变薄,有效孔隙半径变大。页岩气基质孔隙尺寸较小,所以气体分子表面吸附、解吸所导致的有效半径变化对基质表观物性造成的影响不能忽略。

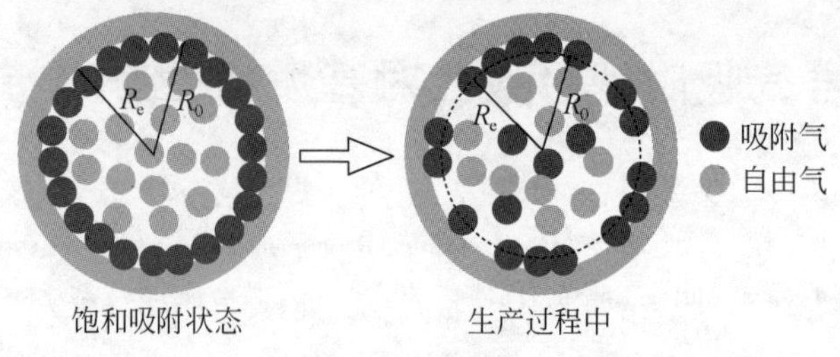

图1　页岩气在基质孔隙中分布示意图

图2为基于克努森数 Kn 划分的气体流动机制，克努森数定义为气体分子平均自由程 λ 与孔隙平均直径 D 之比[11,12]：

$$Kn=\frac{\lambda}{D} \tag{1}$$

式中，Kn 为克努森数；λ 为分子平均自由程，m；D 为孔隙平均直径，m。

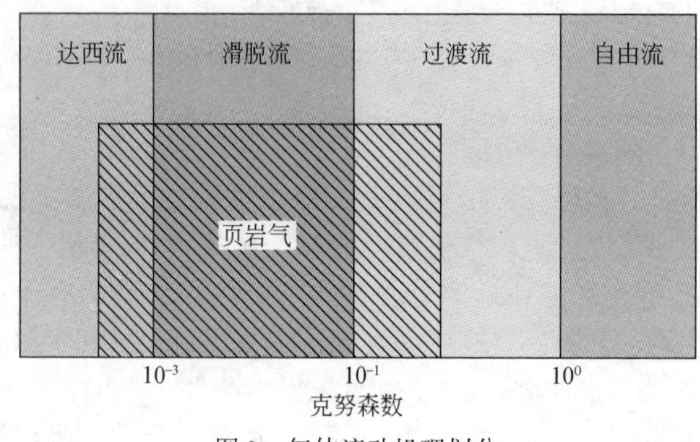

图2　气体流动机理划分

当 $Kn\leqslant10^{-3}$ 时，气体流动为达西流；当 $10^{-3}<Kn\leqslant10^{-1}$ 时，气体流动为滑脱流；当 $10^{-1}<Kn\leqslant10$ 时，气体流动为过渡流；当 $Kn>10$ 时，气体流动为自由流。从前面分析可知，吸附气导致孔隙有效半径低于真实半径，并且随着吸附气解吸，有效半径逐渐增大，进而影响气体流动机制。因此，实际流动过程中，Kn 为气体分子平均自由程 λ 与孔隙平均有效直径 D_e 之比：

$$Kn=\frac{\lambda}{D_e} \tag{2}$$

式中，D_e 为孔隙平均有效直径，m。

2　页岩气渗流数学模型

2.1　物质守恒方程

为了建立页岩气渗流数学模型，提出以下假设：①页岩基质孔隙为球形；②页岩气藏不含水，或为束缚水条件且束缚水对吸附气没有影响；③页岩气吸附符合 Langmuir 模型；④气

藏为等温系统，不考虑能量交换；⑤页岩气吸附解吸瞬间达到平衡。则页岩气在纳米级孔隙中渗流的质量守恒方程可表示为

$$\nabla\cdot\left(\frac{K_a\rho_g}{\mu_g}\nabla\Phi_g\right)+\rho_g q_g=\frac{\partial}{\partial t}(\rho_g\phi_e)+\frac{\partial\dot{m}}{\partial t} \tag{3}$$

式中，K_a 为页岩孔隙中气体流动表观渗透率，m^2；ρ_g 为页岩气密度，kg/m^3；μ_g 为页岩气黏度，$Pa\cdot s$；Φ_g 为气体势函数，Pa；q_g 为气相注入/采出速度，$m^3/(m^3\cdot s)$；t 为时间，s；ϕ_e 为有效孔隙度，f；$\dot{m}$ 为单位体积岩石吸附气质量，kg/m^3。

2.2 页岩气渗流表观渗透率模型

气体在纳米孔中的渗流机制包括达西流、滑脱流、过渡流和自由流，Shi 等[13]推导了涵盖这几种流动机制的表观渗透率表达式：

$$K_a=K_0\left[\frac{410(1+5Kn)}{410+Kn^4}+\frac{128\tau_m\mu_g\delta^2}{3k_B\phi_e}\sqrt{\frac{1000\pi R}{MT}}\frac{Kn^5}{410+Kn^4}\right] \tag{4}$$

式中，K_0 为无吸附气时基质绝对渗透率，m^2；τ_m 为孔隙迂曲度；δ 为气体分子碰撞直径，m，通常取甲烷分子直径；k_B 为 Boltzmann 常数，1.3806×10^{-23} J/K；R 为气体常数，$8.314Pa\cdot m^3/(mol\cdot K)$；$M$ 为相对分子质量；T 为温度，K。

但是，气体分子吸附在孔隙表面导致基质孔隙绝对渗透率降低，因此，基质绝对渗透率是吸附量的函数，则式(4)应变形为

$$K_a=K_{0c}\left[\frac{410(1+5Kn)}{410+Kn^4}+\frac{128\tau_m\mu_g\delta^2}{3\kappa_B\phi_e}\sqrt{\frac{1000\pi R}{MT}}\frac{Kn^5}{410+Kn^4}\right] \tag{5}$$

式中，K_{0c}为考虑吸附气影响的校正的基质绝对渗透率，m^2。

同时，在页岩气解吸过程中，孔隙有效半径不断变化，进而影响克努森数。页岩储集层微观孔隙结构研究发现，有机质孔隙为近球形或椭球形[14,15]，因此近似将页岩基质孔隙看作球形，则无吸附气状态下单个孔隙的体积为

$$V_0=\frac{4}{3}\pi R_0^3 \tag{6}$$

式中，V_0 为无吸附气时孔隙体积，m^3；R_0 为无吸附气时孔隙半径，m。

饱和吸附状态下，单个孔隙的有效体积为

$$V_{sa}=\frac{4}{3}\pi R_{sa}^3 \tag{7}$$

式中，V_{sa}为饱和吸附状态下孔隙体积，m^3；R_{sa}为饱和吸附状态下孔隙半径，m。

由式(6)和式(7)可得

$$V_0-V_{sa}=\frac{4}{3}\pi(R_0^3-R_{sa}^3)=mV_L \tag{8}$$

式中，m 为单个孔隙体岩石质量，kg；V_L 为 Langmuir 体积，m^3/kg。

由于页岩气在基质孔隙中服从 Langmuir 吸附定律，则

$$V_0-V_e=\frac{4}{3}\pi(R_0^3-R_e^3)=mV_L\frac{p_g}{p_g+p_L} \tag{9}$$

式中，V_e 为某压力下孔隙有效体积，m^3；R_e 为有效孔隙半径，m；p_g 为气相压力，Pa；p_L 为

Langmuir 压力,Pa。

联立式(8)和式(9)可得

$$\frac{R_0^3 - R_{sa}^3}{R_0^3 - R_e^3} = \frac{p_g + p_L}{p_g} \tag{10}$$

因此,任意压力下的有效孔隙半径为

$$R_e = \sqrt[3]{\frac{R_0^3 p_L + R_{sa}^3 p_g}{p_g + p_L}} \tag{11}$$

气体分子平均自由程为[11,12]

$$\lambda = \frac{\kappa_B T}{\sqrt{2}\,\pi\delta^2 p_g} \tag{12}$$

联立式(11)、式(12)可得克努森数表达式为

$$Kn = \frac{\kappa_B T \sqrt[3]{p_g + p_L}}{2\sqrt{2}\,\pi d_{CH_4}^2 p_g \sqrt[3]{R_0^3 p_L + R_{sa}^3 p_g}} \tag{13}$$

式中,d_{CH_4} 为甲烷分子直径,m。

无吸附气时孔隙半径可由 Carman-Kozeny 方程求得[16,17]

$$R_0 = 2\sqrt{2\tau_m}\sqrt{\frac{K_0}{\phi_0}} \tag{14}$$

式中,ϕ_0 为无吸附气时基质孔隙度。

大量研究表明,页岩孔隙的吸附能力受到众多因素的影响,如有机质含量、黏土矿物含量、孔隙迂曲度以及比表面积等[18,19],因此,吸附气含量可能低于或高于理想状态下的单层吸附气量,为此,引入系数 α 校正饱和状态下吸附层厚度,则饱和吸附状态下的孔隙有效半径为

$$R_{sa} = R_0 - \alpha d_{CH_4} \tag{15}$$

式中,α 为页岩孔隙单层饱和吸附校正系数。

式(15)反映了实际单层饱和吸附量与理想单层饱和吸附量的差别,理想饱和单层吸附时 $\alpha=1$;迂曲度大或比表面积大等因素可能导致实际单层吸附量高于理想光滑孔隙表面吸附量,此时 $\alpha>1$;有机质含量低或黏土矿物少等因素可能导致实际单层吸附量低于理想光滑孔隙表面吸附量,此时 $\alpha<1$;总体来说,α 接近于 1,其值通过拟合吸附实验参数获得。

同时,基质有效孔隙度可根据式(9)求得

$$\phi_e = \phi_0 - \rho_r V_L \frac{p_g}{p_g + p_L} \tag{16}$$

式中,ρ_r 为岩石密度,kg/m³。

因此,联立式(5)、式(11)、式(14)、式(16)可得到页岩气生产过程中基质表观渗透率的表达式:

$$K_a = \frac{1}{8\tau_m}\left(\phi_0 - \rho_r V_L \frac{p_g}{p_g + p_L}\right)\left(\frac{R_0^3 p_L + R_{sa}^3 p_g}{p_g + p_L}\right)^{\frac{2}{3}} \times \left[\frac{410(1+5Kn)}{410+Kn^4} + \frac{128\tau_m \mu_g \delta^2}{3\kappa_B \phi_e}\sqrt{\frac{1000\pi R}{MT}}\frac{Kn^5}{410+Kn^4}\right] \tag{17}$$

2.3 吸附解吸效应

Ambrose 等[6]的分子动力学实验研究表明,孔隙表面吸附气处于超临界状态,其密度为自由气密度的1.8~2.5倍,因此单位体积岩石吸附的气体质量表达式为

$$\dot{m}=\rho_r\alpha_d\rho_g G_a \tag{18}$$

$$G_a=V_L\frac{p_g}{p_g+p_L} \tag{19}$$

式中,α_d 为吸附气与自由气密度之比;G_a 为页岩孔隙吸附能力,m^3/kg。

3 数学模型求解及验证

3.1 数学模型求解

页岩气渗流数学模型为复杂的耦合方程系统,为了提高计算效率、精度和稳定性,本文采用有限体积隐式方法求解数学模型,式(3)可变形为

$$-\sum_j(f_{ij,n+1}\boldsymbol{l}_{ij})(A_{ij}\boldsymbol{n}_{ij})\Delta t_{n+1}+s_{i,n+1}\Delta V_i\Delta t_{n+1}-(m_{i,n+1}-m_{i,n})\Delta V_i=0 \tag{20}$$

式中

$$m_i=(\rho_g\phi_e+\rho_r\alpha_d\rho_g G_a)_i \tag{21}$$

$$s_j=(\rho_g q_g)_i \tag{22}$$

$$f_{ij}=T_{ij}[(p_g+\rho_g gD)_i-(p_g+\rho_g gD)_j] \tag{23}$$

式中,f_{ij}为网格 i 与 j 间的质量流量,$kg/(m^2\cdot s)$;l_{ij}为沿 i 到 j 的单位向量;A_{ij}为网格 i 与 j 的接触面积,m^2;n_{ij}为网格 i 和 j 界面上的单位法向向量;Δt 为时间步长,s;i 为网格序号;j 为与 i 网格相邻的任意网格序号;s_i 为第 i 个网格源汇项,$kg/(m^3\cdot s)$;ΔV_i 为第 i 个网格体积,m^3;m_i 为 i 网格的总质量密度,kg/m^3;n,n+1 分别为本时刻和下一时刻;T_{ij}为网格 i 与 j 间的传导率,$kg/(m^2\cdot Pa\cdot s)$;g 为重力加速度,$g=9.8m/s^2$。

式(20)~式(23)中大多数参数或变量均为压力的非线性函数,所以采用 N-R(Newton-Raphson)方法进行求解,式(20)可写为

$$-\sum_j F_{ij,n+1}(p_{i,n+1},p_{j,n+1})+S_{i,n+1}(p_{i,n+1})-[M_{i,n+1}(p_{i,n+1})-M_{i,n}(p_{i,n})]=0 \tag{24}$$

式中,$F_{ij}=(f_{ij}\boldsymbol{l}_{ij})\cdot(A_{ij}\boldsymbol{n}_{ij})$,$S_i=s_i\Delta V_i\Delta t$,$M_i=m_i\Delta V_i$,$p_i$、$p_j$ 分别为 i、j 网格气相压力。

应用 N-R 迭代并给定残余向量 $\boldsymbol{R}$,则

$$R_i=-\sum_j F_{ij,n+1}(p_{i,n+1},p_{j,n+1})+S_{i,n+1}(p_{i,n+1})-[M_{i,n+1}(p_{i,n+1})-M_{i,n}(p_{i,n})] \tag{25}$$

式中,R_i 为质量余量,kg。

则稀疏雅可比矩阵 $\boldsymbol{J}$ 可以通过下面方法求得

$$J_{ij'}=\begin{cases}-\sum_j\frac{\partial F_{ij,n+1}}{\partial p_{i,n+1}}+\frac{\partial S_{i,n+1}}{\partial p_{i,n+1},}-\frac{\partial M_{i,n+1}}{\partial p_{i,n+1}} & j'=i\\ -\frac{\partial F_{ij',n+1}}{\partial p_{j',n+1}} & j'=i\pm1,j'=i\pm n_x,j'=i\pm n_xn_y\\ 0 & \text{其他}\end{cases} \tag{26}$$

式中，$J_{ij'}$ 为雅可比矩阵中第 i 行第 j' 个元素；n_x 为 x 方向网格数；n_y 为 y 方向网格数。

计算得到 $\boldsymbol{R}$ 和 $\boldsymbol{J}$ 后，便可以求解线性代数方程组：

$$\boldsymbol{J}\delta \boldsymbol{p}_{k,n+1} = -\boldsymbol{R} \tag{27}$$

式中，$P_{k,n+1}$ 为 $n+1$ 时间步第 k 步牛顿迭代的压力，Pa；$\boldsymbol{R}$ 为残余向量。

求解后，$\boldsymbol{p}_{k,n+1}$ 可通过式(28)求得

$$\boldsymbol{p}_{k,n+1} = \boldsymbol{p}_{k-1,n+1} + \delta \boldsymbol{p}_{k,n+1} \tag{28}$$

3.2　模型验证

本文分别采用物理实验数据和矿场生产数据验证了模型的可靠性。首先应用 Roy 等[20]和 Civan 等[21]验证气体渗流模型所采用的实验数据[22]验证模型的可靠性。气体微尺度流道流动实验在 5 个注入压力下(135kPa、170kPa、205kPa、240kPa 和 275kPa)注入气体，出口压力设定为 100.8kPa，测定微尺度流道沿程压力分布，模型参数参考相关文献[20]。根据上述参数采用建立的数学模型进行模拟计算并拟合，结果如图 3 所示，可见，计算值与实验值拟合效果较好。然后，应用 Barnett 页岩气藏一口典型井的实际数据进行了模型验证，生产数据及模型参数参考相关文献[23]。根据模型参数建模后，采用体积法计算地质储量，并与以往模型[24](不考虑吸附层影响)计算结果及采用 Ambrose 等[6]公式计算的理论地质储量进行对比(表 1)，然后对该井开发过程进行数值模拟研究，拟合其产气速度(图 4)，可见，新模型具有较好的可靠性。

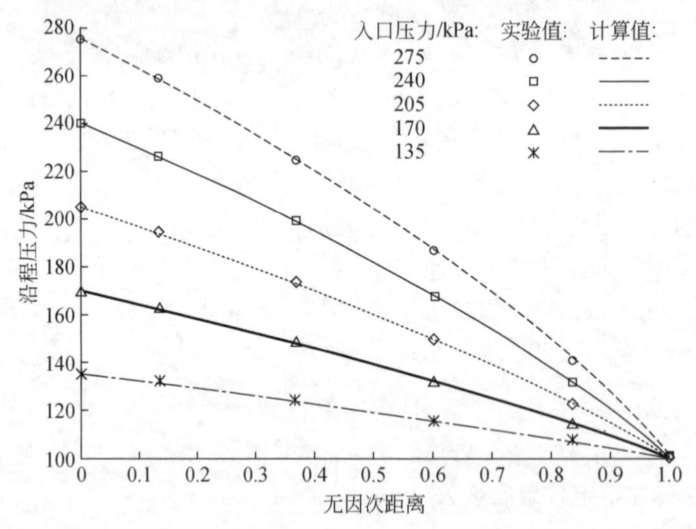

图 3　本文模型计算值与毛细管实验实测值对比(无因次距离为距注入端距离与毛细管长度之比)

表 1　页岩气储量计算结果对比

页岩气类型	理论值	以往模型计算结果		本文模型计算结果	
		储量/$10^8 m^3$	误差/%	储量/$10^8 m^3$	误差/%
吸附气	1.32	1.265	4.17	1.265	4.17
自由气	1.33	1.997	50.20	1.295	2.63

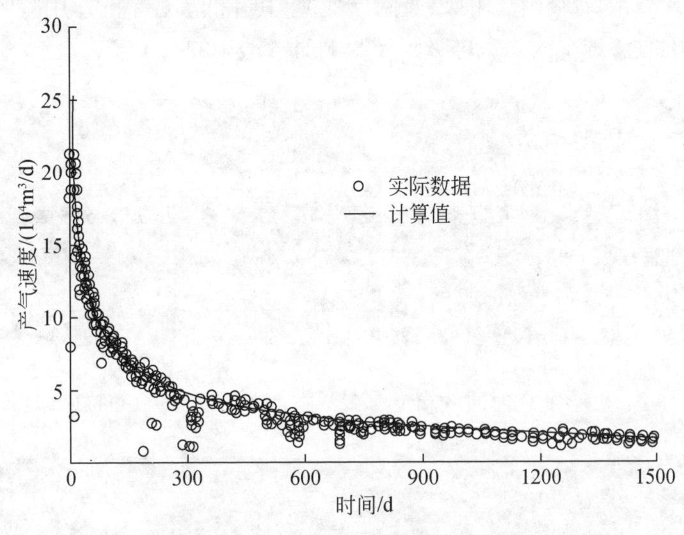

图4 Barnett 页岩生产数据拟合结果

4 页岩气开发过程中基质表观物性参数变化规律

为了认识页岩气开发过程中页岩气吸附解吸效应对基质表观物性参数影响规律，使用上述拟合 Barnett 页岩生产数据得到的模型进行模拟研究，模型具体参数如下：模型尺寸为 890m×436m×90m，埋深为 1600m，地层压力为 21.0MPa，气藏温度为 340K，井底流压为 3.5MPa，绝对孔隙度为 0.06，束缚水饱和度为 0.7，岩石密度为 2580kg/m^3，孔隙迂曲度为 1.1，基质渗透率为 $0.00014\times10^{-3}\mu m^2$，微裂缝渗透率为 $2\times10^{-3}\mu m^2$，微裂缝间距为 1m；水力裂缝渗透率为 $1000\times10^{-3}\mu m^2$，裂缝开度为 0.003m，裂缝间距为 30m，裂缝半长为 40m，裂缝高度为 90m，水平井水平段长度为 890m，页岩孔隙单层饱和吸附校正系数为 1.3，吸附气与自由气密度之比为 1.8，页岩气黏度为 0.014mPa · s，气体体积系数为 0.0046，甲烷分子直径为 0.38nm。

4.1 基质表观物性变化特征

部分页岩气以吸附气的形式储存在岩石表面，气体有效流动通道减小，有效孔隙度降低，生产过程中，随着压力降低，吸附在孔隙表面的气体发生脱附，有效流动通道和有效孔隙度逐渐升高[7,25]。图 5 ~ 图 7 分别为不同时间 Barnett 页岩水平井筒附近基质孔隙有效半径、有效孔隙度和表观渗透率分布。从图中可以出，初始条件下，页岩气藏基质有效孔隙度为 0.045 左右，仅为绝对孔隙度的 75% 左右；随着页岩气开采，吸附气从孔隙表面解吸，大量吸附气变为自由气，基质孔隙有效半径逐渐增大，有效孔隙度增加，表观渗透率变大，并且近井地带压力下降快，解吸量大，因此井筒周围表观物性变化明显。生产 30a 后，基质孔隙有效半径增加 5% 左右，有效孔隙度大约增加 10%，而表观渗透率增加 1 倍左右，由于页岩气的特殊流动机制，井筒周围表观渗透率远高于绝对渗透率。图 8 为不考虑吸附层对基质渗透率影响时渗透率分布，对比图 7、图 8 可以看出，开采初期，不考虑吸附层影响时渗透率约为考虑吸附层影响时渗透率的 1.5 倍，随着吸附气解吸，吸附层变薄，其降低渗透率的作用减弱，开采至 30a 时，不考虑吸附层影响时渗透率约为考虑吸附层影响时渗透率的 1.3 倍。

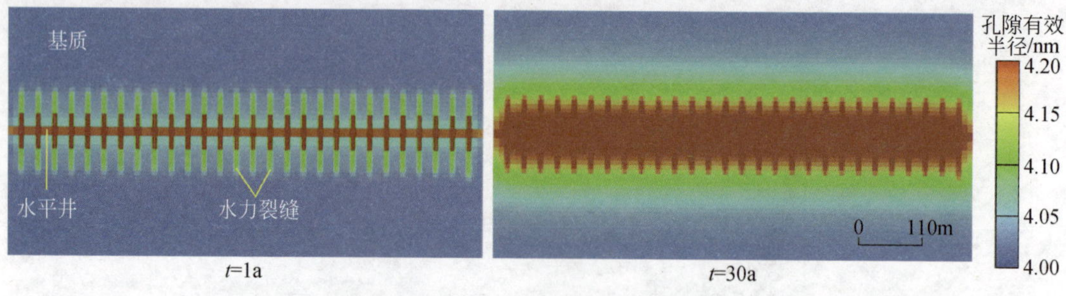

图 5　不同时间 Barnett 页岩基质孔隙有效半径分布特征

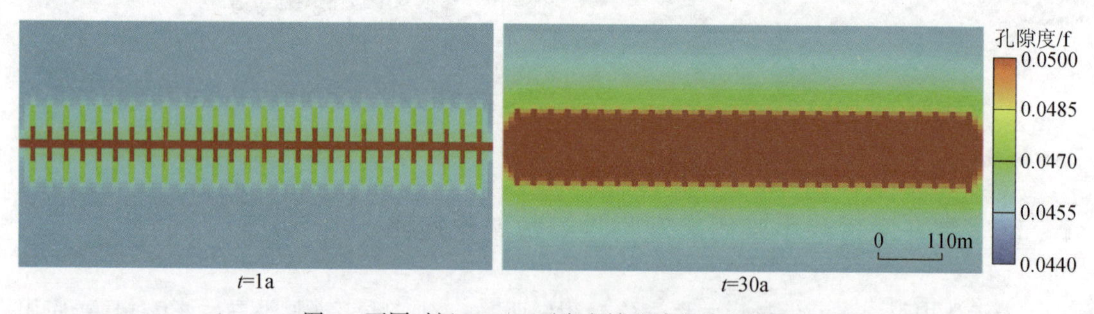

图 6　不同时间 Barnett 页岩有效孔隙度分布特征

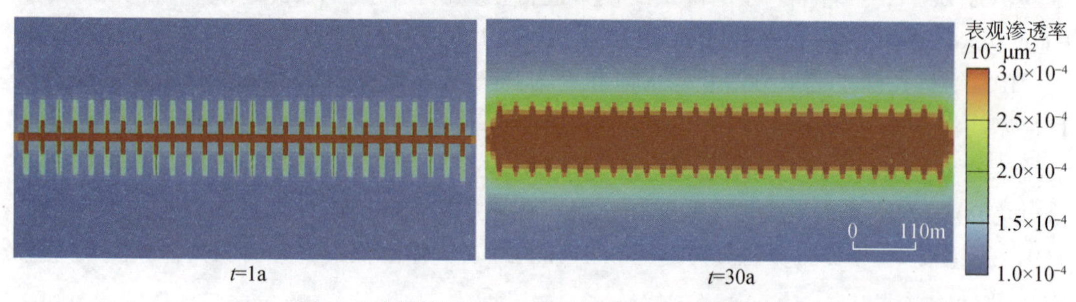

图 7　不同时间 Barnett 页岩表观渗透率分布特征

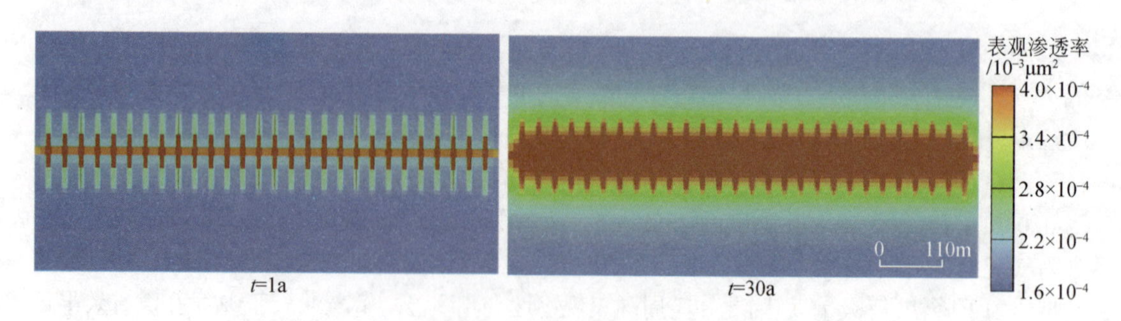

图 8　不考虑吸附气影响时 Barnett 页岩表观渗透率分布特征

4.2 基质孔隙中流动机制变化特征

图9为开采30a时气藏克努森数分布,从图中可以看出,Kn为0.05~0.45,根据Kn对流动机制的判别标准,气藏中存在滑脱流和过渡流两种流动机制。图10为气藏开发过程中过渡流区域变化特征,从图中可以看出,页岩气开采过程中,Kn逐渐变大,过渡流区域逐渐变大,即部分滑脱流区域转变为过渡流。原因在于,Kn与压力、孔隙平均有效半径呈反比,孔隙压力降低会导致Kn增加,孔隙平均半径增加会导致Kn降低,但是由于体积压裂区域压力大幅降低,压力起主导作用,随着页岩气采出,Kn增加。

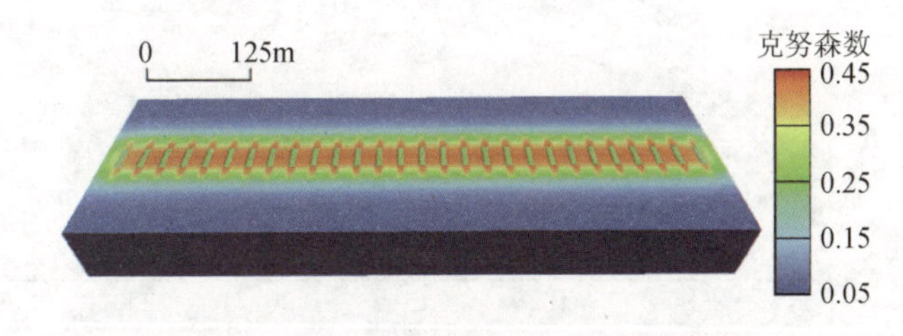

图9 $t=30$a时气藏克努森数分布特征

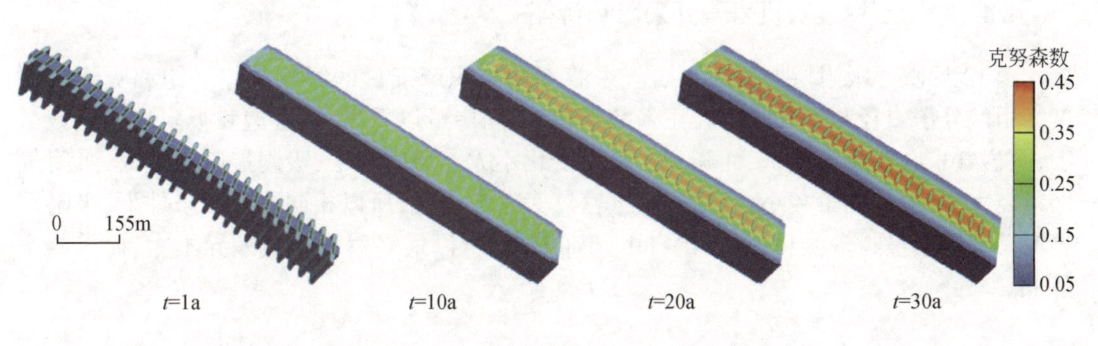

图10 Barnett页岩基质过渡流区域变化特征

4.3 吸附气对页岩气开采影响特征

吸附层的存在影响基质孔隙度、渗透率以及气体流动机制,研究者往往忽略这一点,因此笔者研究了考虑吸附层影响和不考虑吸附层影响等多种情况下的产气量特征。根据地质参数建立地质模型并进行开发过程模拟计算,得到图11所示4种情况下的地质储量和累积产气量:忽略吸附气对孔隙度和渗透率的影响(即将吸附层看作虚拟体积,现行商业软件均采用该处理方法)时,地质储量为$3.26\times10^8m^3$,累积产气量约为$1.35\times10^8m^3$;仅考虑吸附层降低孔隙度,而忽略其对表观渗透率的影响,地质储量为$2.56\times10^8m^3$,累积产气量约为$1.13\times10^8m^3$;考虑吸附气对孔隙度和渗透率的影响,即实际页岩气藏的情况,地质储量为$2.56\times10^8m^3$,累积产气量约为$1.08\times10^8m^3$,采收率42.2%;如果气藏中气体仅以自由气形式存在,即致密气藏的情况,地质储量仅为$2.00\times10^8m^3$,累积产气量约为$0.95\times10^8m^3$,采收率47.5%。由此可见,将吸附层当作虚拟体积处理,会使地质储量被严重高估,导致计算产气量偏高;仅考虑吸附层对孔隙度的影响,虽然地质储量与实际页岩气藏情况相符,但是由于渗透率被高估,最终计算的产气量仍偏高;对比实际页岩气藏和致密气藏的情况可以发现,页岩气藏储量较高,但由于吸附气较自由气难以采出,最终采收率致密气藏高于页岩气藏。

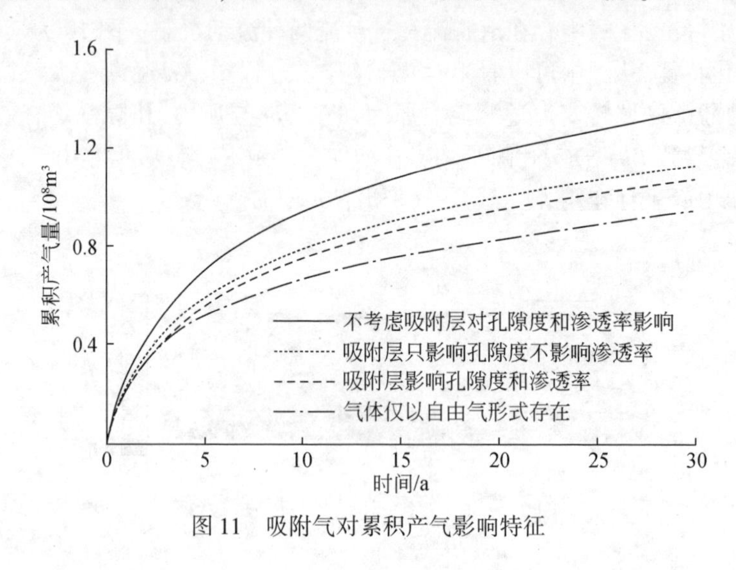

图 11　吸附气对累积产气影响特征

4.4　吸附层厚度对页岩气开采影响特征

图 12 反映了吸附层厚度对页岩气采收率和累积产气量的影响规律。饱和吸附层厚度为甲烷分子直径与饱和吸附校正系数乘积,从图中可以看出,随着饱和吸附校正系数增大,页岩气采收率逐渐降低,而累积产气量变化不大。原因在于,吸附层厚度越大,吸附气储量越大、自由气储量越小,但是吸附气密度大于自由气,所以相同孔隙度条件下,吸附层越厚总地质储量越多,但吸附气比自由气难采出,所以最终累积产气差异不大,而采收率降低。

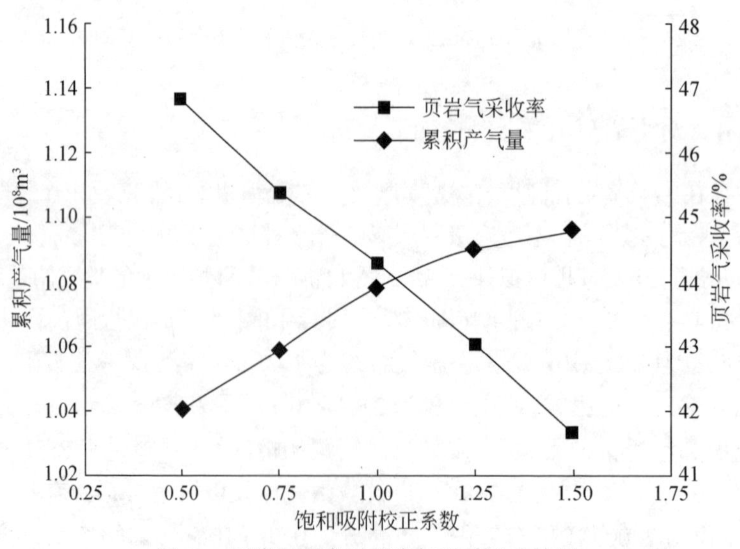

图 12　吸附层厚度对页岩气开采影响特征

5 结论

基于页岩气在基质孔隙中的分布特征和吸附解吸规律推导了页岩吸附、解吸作用下的有效孔隙半径和表观渗透率动态模型,然后根据质量守恒定律建立了页岩气渗流数学模型,采用有限体积隐式法求解渗流模型,并利用实验及矿场数据验证了模型的可靠性。应用该模型研究表明:页岩气开采过程中基质孔隙有效半径、有效孔隙度和表观渗透率逐渐变大,体积压裂改造区域流动机制由滑脱流转变为过渡流;忽略吸附层对基质物性影响将导致地质储量和产气量被严重高估;随着吸附层厚度增加,累积产气量变化不大,但页岩气采收率逐渐降低。

参考文献

[1] 邹才能,董大忠,王玉满,等. 中国页岩气进展、挑战及前景(一)[J]. 石油勘探与开发,2015,42(6):689-701.

[2] Ghanizadeh A, Aquino S, Clarkson C R, *et al.* Petrophysical and geomechanical characteristics of Canadian tight oil and liquid-rich gas reservoirs[R]. SPE 171633,2014.

[3] Akkutlu I Y, Fathi E. Multiscale gas transport in shales with local kerogen heterogeneities[R]. SPE 146422,2011.

[4] Adesida A G, Akkutlu I Y, Resasco D E, *et al.* Kerogen pore size distribution of Barnett shale using DFT analysis and Monte Carlo simulations[R]. SPE 147397,2011.

[5] Curtis J B. Fractured shale-gas systems[J]. AAPG Bulletin,2002,86(11):1921-1938.

[6] Ambrose R J, Hartman R C, Diaz -Campos M, *et al.* New pore-scale considerations for shale gas in place calculations[R]. SPE 131772,2010.

[7] Hartman R C, Wearherford L, Ambrose R J, *et al.* Shale gas-in-place calculations part II: Multi-component gas adsorption effects[R]. SPE 144097,2011.

[8] Sigal R F. The effect of gas adsorption on storage and transport of methane in organic shales[R]. New Orleans, Louisiana: SPWLA 54th Annual Logging Symposium,2013.

[9] 王香增,高胜利,高潮. 鄂尔多斯盆地南部中生界陆相页岩气地质特征[J]. 石油勘探与开发,2014,41(3):294-304.

[10] Ross D J K, Bustin R M. Impact of mass balance calculations on adsorption capacities in microporous shale gas reservoirs[J]. Fuel,2007,86:2696-2706.

[11] Javadpour F, Fisher D, Unsworth M. Nanoscale gas flow in shale gas sediments[J]. Journal of Canadian Petroleum Technology,2007,46(10):55-61.

[12] Bird R B, Stewart W E, Lightfoot E N. Transport phenomena[M]. New York: John Wiley & Sons Inc. ,2002.

[13] Shi J T, Zhang L, Li Y S, *et al.* Diffusion and flow mechanisms of shale gas through matrix pores and gas production forecasting[R]. SPE167226,2013.

[14] 杨峰,宁正福,胡昌蓬,等. 页岩储层微观孔隙结构特征[J]. 石油学报,2013,34(2):301-311.

[15] 侯宇光,何生,易积正,等. 页岩孔隙结构对甲烷吸附能力的影响[J]. 石油勘探与开发,2014,41(2):248-256.

[16] Carman P C. Flow of gases through porous media[M]. London: Butterworths Scientific Publications,1956.

[17] Civan F. Effective correlation of apparent gas permeability in tight porous media[J]. Transport in Porous Media,2010,82:375-384.

[18] Rououerol J, Rououerol F, Sing K S W. Adsorption by powers and porous solids: Principles, methodology and

applications[M]. London:Academic Press,1999.

[19] Solar C, Blanco A G, Vallone A, *et al.* Adsorption of methane in porous materials as the basis for the storage of natural gas[M]. Rijeka:InTech Europe,2010.

[20] Roy S, Raju R, Chuang H F, *et al.* Modeling gas flow through microchannels and nanopores[J]. Journal of Applied Physics,2003,93(8):4870-4879.

[21] Civan F, Rai C S, Sondergeld C H. Shale-gas permeability and diffusivity inferred by improved formulation of relevant retention and transport mechanisms[J]. Transport in Porous Media,2011,86:925-944.

[22] Pong K C, Ho C M, Liu J Q, *et al.* Non-linear pressure distribution in uniform micro-channels: Applications of microfabrication to fluid mechanics[J]. ASME Fluids Engineering Division,1994,197:51-56.

[23] Yu W, Sepehrnoori K. Simulation of gas desorption and geomechanics effects for unconventional gas reservoirs [R]. SPE 165377,2013.

[24] CMG. GEM user guide[M]. Calgary:Computer Modeling Group Ltd.,2011.

[25] Xiong X, Devegowda D, Michel G G, *et al.* A fully-coupled free and adsorptive phase transport model for shale gas reservoirs including non-Darcy flow effects[R]. SPE 159758,2012.

考虑井壁稳定及增产效果页岩气水平井段方位优化方法

范翔宇[1,2] 吴 昊[2] 殷 晟[3] 夏宏泉[1] 陈 平[1] 雷 梦[1] 覃 勇[4]

(1. 西南石油大学"油气藏地质及开发工程"国家重点实验室;2. 西南石油大学石油与天然气工程学院;
3. 中国石油塔里木油田公司;4. 中国石油川庆钻探工程公司钻井液技术服务公司)

摘要:针对页岩气水平井钻井过程中常见的井壁失稳垮塌问题,为提高井壁稳定性和为后期压裂增产提供良好的条件,分析了层理面产状和地应力类型对井壁稳定性的影响,并对压裂缝与水平井段夹角,以及与水平气井产量之间的关系进行了研究,进而提出了通过对水平井段方位的优化来提高井壁稳定性和改善压裂增产效果的技术思路;利用川南地区页岩气井的测井、地应力及岩石力学资料,通过页岩层理面破坏模型及本体破坏模型分析计算出不同方位角下的井壁坍塌压力,找出井壁稳定性最好时井眼与层理面产状和地应力类型之间的关系;通过对压裂缝延伸方向的分析,得出了压裂缝的一般延伸规律;利用产能公式计算裂缝与水平井段呈不同夹角时的产量,以此获得了裂缝与水平井段夹角影响压裂效果的规律。还以川南地区的 W201-H3 井为例,充分考虑页岩气井壁稳定及增产效果,提出了有针对性的水平井段方位优化设计,使该井能在保持良好井壁稳定性的前提下获得良好的增产效果。

关键词:页岩气;水平井段;方位;优化;井壁失稳;增产效果;四川盆地南

四川地区的威远、长宁及富顺—永川是中国石油页岩气勘探开发的主要区块。据中国石油与美国新田石油勘探公司共同开展的四川盆地页岩气资源潜力评价,结果显示四川盆地寒武系页岩气资源量为 $7.14\times10^{12} \sim 14.6\times10^{12}\ m^3$,志留系页岩资源量为 $2\times10^{12} \sim 4\times10^{12}\ m^3$,仅这两套页岩资源量就相当于四川盆地常规天然气资源总量的 1.5~2.5 倍[1]。

地层失稳垮塌等工程问题对钻井安全、进度的制约不容忽视,处理井下地层垮塌造成的工程问题直接使钻井成本上升,影响钻井的经济效益[2]。不同于常规天然气储层,页岩的储层非均质性极强,这决定了其独特的开发方式。而水平井技术是开发页岩气的核心技术之一[3]。页岩气水平井需要解决的关键技术是:采用最优化设计方法,确定出页岩气水平井钻井的最优化轨迹,以此来提高钻井时的井壁稳定性和为后期增产压裂提供良好的条件。其中,水平井段的方位优化设计是整个页岩气水平井轨迹优化设计的重点。这直接关系到钻井过程中的井壁稳定性和后期增产压裂的效果。

1 考虑水平井段井壁失稳的方位优化设计

页岩气钻井工程中出现的复杂情况以井壁失稳垮塌为主。引起井壁失稳的因素主要有力学和化学两方面。通过对川南地区的威远、长宁工区的页岩气水平井钻井过程中出现的井壁失稳现象进行分析,发现某井在采用油组分在 90% 以上、水组分不到 10% 的油基防塌钻井液时,也会发生井壁垮塌。在对工区页岩岩心进行水化膨胀率测试时发现,纯水导致的页岩膨胀率小于 4%,而现场使用的油基泥浆导致的岩心膨胀率为零。这表明了造成页岩气水平井井壁失稳的主要原因是力学因素而非化学因素。

1.1 地应力与水平段井眼方位的关系特征

不同地应力分布条件下水平井水平段的井眼方位对井壁稳定影响的规律存在较大的差异。地应力大小和方向的确定对于井壁稳定性与井眼方位相关性分析具有至关重要的作用[4]。地层岩石所处的地应力场通常由垂向应力 S_V、最大水平地应力 S_{Hmax} 和最小水平地应力 S_{Hmin} 组成。根据垂直主应力 S_V 与两个水平主应力之间的关系,可将地应力分为三种类型,即正常地应力类型($S_V>S_{Hmax}>S_{Hmin}$)、走滑地应力类型($S_{Hmax}>S_V>S_{Hmin}$)和反转地应力类型($S_{Hmax}>S_{Hmin}>S_V$)。

通过常用的莫尔-库仑破坏准则,对四川地区三种不同类型的水平井进行计算分析,得出了不同地应力类型下井眼方位影响水平段井壁稳定的规律[5]:在正常及反转地应力下,随着井斜角的增加,井壁稳定性逐渐变差,即直井比水平井井壁稳定性好;在走滑地应力类型下,随着井斜角的增加,井壁稳定性逐渐变强,即水平井井壁稳定性最好。单从井壁稳定性方面考虑,钻井或完井过程中,为保持良好的井壁稳定性,在正常地应力、走滑地应力、反转地应力类型下水平井井眼轨迹最优方位和最大水平主应力方向之间的夹角分别为90°、45°和0°。

本文以川南地区筇竹寺组页岩储层的 W201-H3 井为例进行水平段方位优化,表1为该井储层井段井壁稳定性计算参数。可以看出,该地区地应力符合 $S_{Hmax}>S_V>S_{Hmin}$,属于走滑地应力类型。

表1 筇竹寺组井壁稳定计算参数

构造	井深/m	S_1/MPa	S_{Hmax}/MPa	S_V/MPa	S_{Hmin}/MPa	地层压力/MPa	泊松比
威远	2627.51	243.605	66.476	60.433	49.193	28.87	0.233
	2686.59	366.6	67.97	61.792	46.309	29.52	0.162
	2723.85	309.95	68.913	62.649	46.850	29.93	0.206

1.2 水平段井眼方位优化的影响因素

本文在考虑了地应力类型及页岩层理面产状对水平段井壁稳定性的影响和后期压裂产生的裂缝走向和其与水平段之间的夹角关系对增产效果的影响两方面的基础上对水平段井眼方位进行了优化设计。

1.2.1 基于地应力类型的水平段方位优化设计

通过常用的莫尔-库仑破坏准则,对走滑地应力类型的水平井进行计算分析,得出了该地应力类型下水平段井眼方位影响井壁稳定的规律(图1)。

通过图1可以看出,在走滑地应力类型下,单从井壁稳定性方面考虑,钻井或完井过程中,为保持良好的井壁稳定性,水平井段的最优方位与最大水平主应力方向之间的夹角应为45°。

1.2.2 基于层理面产状的水平段方位优化设计

页岩地层的破坏与井壁围岩的受力情况、层理面的强度、层理面的产状及井眼轨迹有着

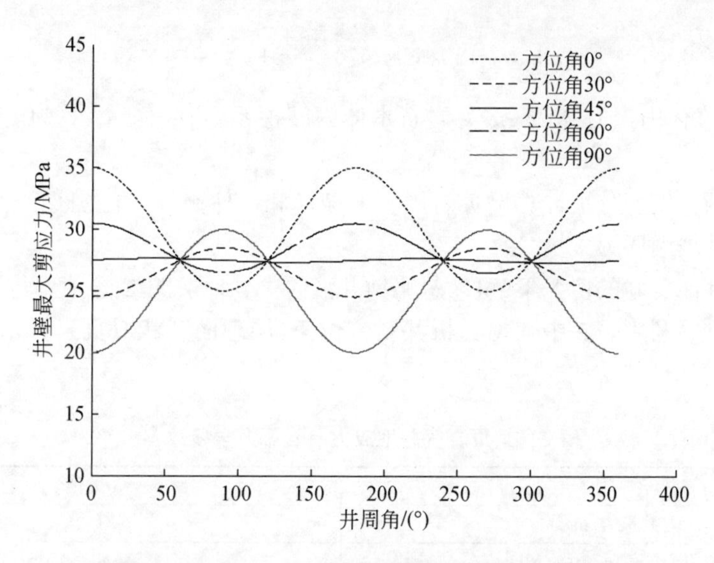

图1 走滑地应力类型井眼方位影响井壁坍塌的规律图

密切的关系。在水平井开发页岩气过程中,钻遇大斜度井段或水平段页岩地层时,地层中会存在一些低强度的薄弱层理面,在较小的钻井液柱压力下先于岩石本体发生破坏。计算分析并比较页岩发生坍塌时层理面的坍塌压力和本体破坏坍塌压力可得到页岩气水平段的坍塌压力。

根据斜井井壁破坏力学模型对井壁进行应力分析和坐标转换[6],可计算出水平井井壁上的最大、最小有效主应力 σ_1'、σ_3'。

(1)层理面破坏准则[7]如下。岩石沿层理面破坏时正应力 σ 与剪应力 τ 应满足:

$$\tau = C_w + \sigma \tan\varphi_w \tag{1}$$

式中,C_w 为层理面内聚力,MPa;φ_w 为层理面内摩擦角,(°)。

通过正应力与剪应力的计算模型及层理面内摩擦角与内摩擦系数的关系可得

$$\sigma_1' - \sigma_3' = \frac{2(C_w + \mu_w \sigma_3')}{2\cos\beta(\sin\beta - \mu_w \cos\beta)} \tag{2}$$

式中,σ_1'为最大有效主应力,MPa;σ_3'为最小有效主应力,MPa;C_w 为层理面内聚力,MPa;μ_w 为层理面的内摩擦系数,$\mu_w = \tan\varphi_w$;φ_w 为层理面的内摩擦角,rad;β 为层理面的法向与σ_1'夹角,rad。

由式(2)可知,当 $\beta = \varphi_w$ 或 $\beta = \frac{\pi}{2}$ 时,$\sigma_1' - \sigma_3' \to \infty$,此时层理面不产生滑动;而当 $\varphi_w < \beta < \frac{\pi}{2}$ 时,层理面才有可能破坏。所以,层理面产生滑动的条件是

$$\varphi_w < \beta < \frac{\pi}{2} \tag{3}$$

且 $\sigma_1' - \sigma_3'$的大小必须满足式(2)的关系。

(2)岩石本体发生破坏时,根据莫尔-库仑破坏准则:

$$\tau = C + \sigma \tan\varphi \tag{4}$$

通过内摩擦角的正切与余弦的关系并令 $\mu = \tan\varphi$,有[8]

$$\sigma_1'-\sigma_3'=2\mu(C+\mu\sigma_3')\left[\left(\frac{1}{\mu^2}+1\right)^{\frac{1}{2}}+1\right] \tag{5}$$

式中，C 为岩石本体内聚力，MPa；μ 为岩石本体内摩擦系数，$\mu=\tan\varphi$；φ 为岩石本体内摩擦角，(°)。

通过上述两种模型分别计算出页岩层理面破坏和本体破坏时的坍塌压力，取较大者即为水平段井壁的坍塌压力。

在钻井液性能良好的情况下，可以忽略钻井液向页岩地层的渗透，把页岩井壁近似当作不渗透井壁处理。利用表 2 中数据分析井斜方位角、层理面产状对页岩气水平井井壁稳定性的影响。

表 2　某页岩气井地应力及岩石力学参数[9]

参数	数值
井深 H/m	1525
水平最大主应力 σ_H/MPa	47.69
上覆岩石压力 σ_v/MPa	35.075
内摩擦系数 μ	0.183
岩石本体内摩擦角 φ/(°)	40
岩石层理面内摩擦角 φ_w/(°)	23
地层压力 P_p/MPa	16
水平最小主应力 σ_h/MPa	28.975
有效应力系数 η	0.9
地层孔隙度 φ/%	5
岩石本体内聚力 C/MPa	17
岩石层理面内聚力 C_w/MPa	5

利用表 2 中的数据，运用 MATLAB 计算了当页岩层理面倾角为 15°、45°、75°，走向分别为 0°、15°、45°、75°和 90°时，井壁坍塌压力随井眼方位的变化规律，通过曲线拟合得到的结果如图 2～图 4 所示。

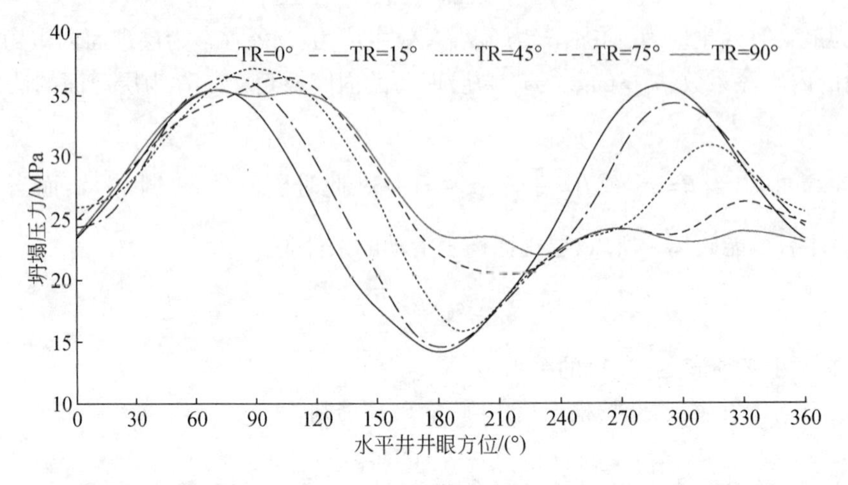

图 2　层理面倾角 15°时坍塌压力随层理面走向及井眼方位变化规律曲线

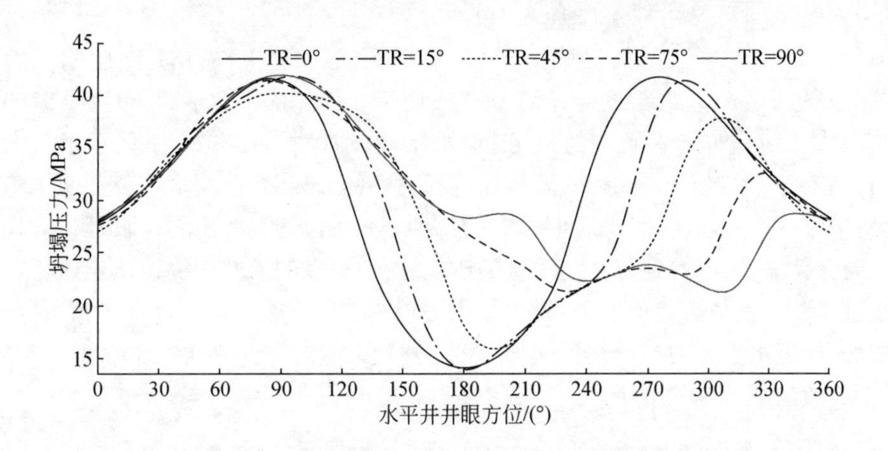

图3 层理面倾角45°时坍塌压力随层理面走向及井眼方位变化规律曲线

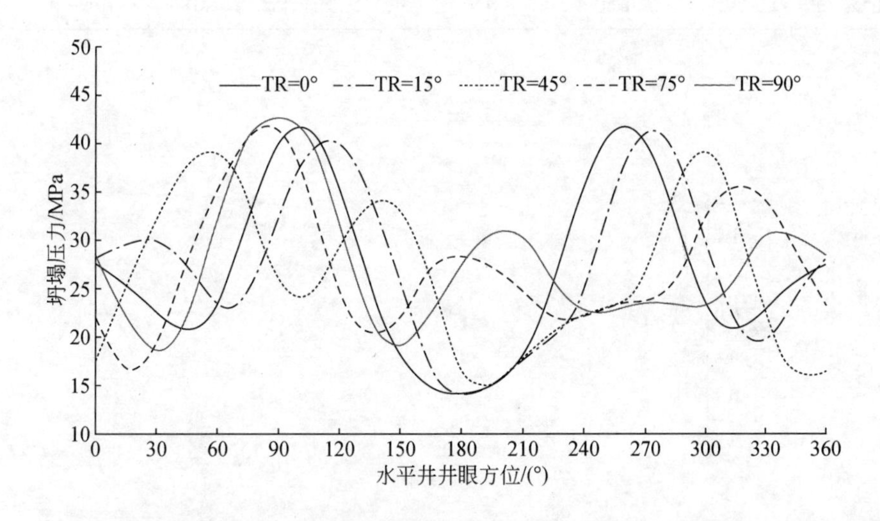

图4 层理面倾角75°时坍塌压力随层理面走向及井眼方位变化规律曲线

从图2~图4中分析得出以下规律。

(1)页岩层理面的倾角对页岩井壁剪切破坏的影响：在页岩层理面倾角小于45°时，页岩井壁的坍塌主要是层理面的剪切破坏，页岩本体几乎不发生垮塌；页岩层理面倾角大于45°，在特定的井眼方位处页岩井壁会发生本体剪切破坏；当页岩本体发生剪切破坏时，坍塌压力随页岩层理面的走向化不明显，即表现出不同的层理面走向对应的井壁坍塌压力在一定的井眼方位处会重合。

(2)页岩层理面的走向影响井壁剪切破坏形式：当页岩层理面倾角大于等于45°，页岩井壁岩石会发生本体的破坏，且破坏范围存在一定的规律，即在水平井井眼方位与页岩层理面的走向呈180°夹角左右的范围内，井壁岩石会发生本体破坏。此时，页岩井壁比较稳定，有利于钻井施工的顺利进行。

1.2.3　考虑增产效果的水平段方位优化设计

大部分页岩气是以游离状态赋存于页岩储层天然裂缝中，为了提高后期压裂增产的效果，应尽量使压裂产生的裂缝与井筒垂直或交错。而压裂时所产生的诱导缝的延伸方向都是与最大主应力方向一致。通过分析，水平段设计时应尽量选择垂直于最大水平主应力方向，有利于压裂缝延伸得更远，从而连通更多的天然裂缝，增大与储层页岩气接触的表面，提高采收率。所以，仅考虑后期增产效果时，理论上应使水平段井眼延伸方向与最小主应力方向保持一致，以得到与井筒垂直的压裂缝，使更多的天然裂缝彼此连通。

但通过气井产量与裂缝角度关系公式[10]可以得出，裂缝数量一定时，随着裂缝与井筒夹角的增加，压裂水平井的产量逐渐增加(表 3)。当裂缝与井筒夹角大于 60°时产量增幅减小较快(图 5)。

表 3　裂缝数量一定时不同裂缝与水平段夹角下水平井的产量

夹角/(°)	10	20	30	40	50	60	70	80	90
产量/($10^4m^3/d$)	0.795	0.920	1.009	1.076	1.126	1.162	1.186	1.199	1.204

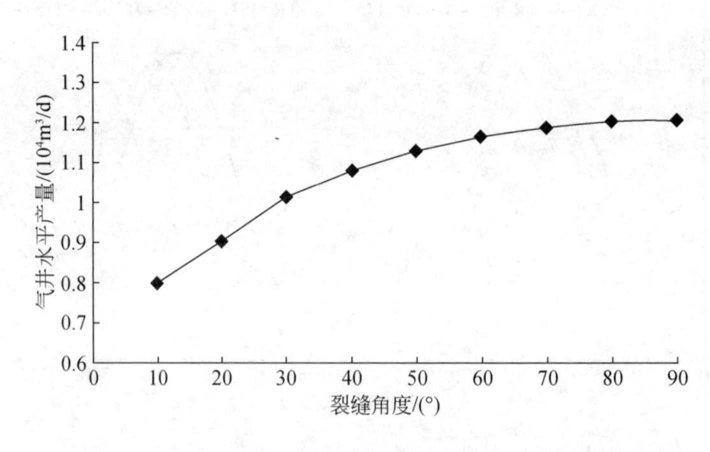

图 5　裂缝角度与产量关系图

2　优化结果的对比分析

通过上述分析，以川南地区筇竹寺组页岩储层的 W201-H3 井为例进行水平段方位优化。

结合地区地质资料，得到筇竹寺组的地层倾向为 SE20°，所以该层位的层理面走向为 70°，层理面倾角取 45°[11]。通过前边考虑层理面对井壁稳定性的影响的分析，结合表 1 中的数据进行水平段方位优化。

通过计算水平段处于各个方位时泥浆的安全密度窗口来确定井壁稳定性，以此指导井眼方位的选取，如图 6 ~ 图 8 所示。

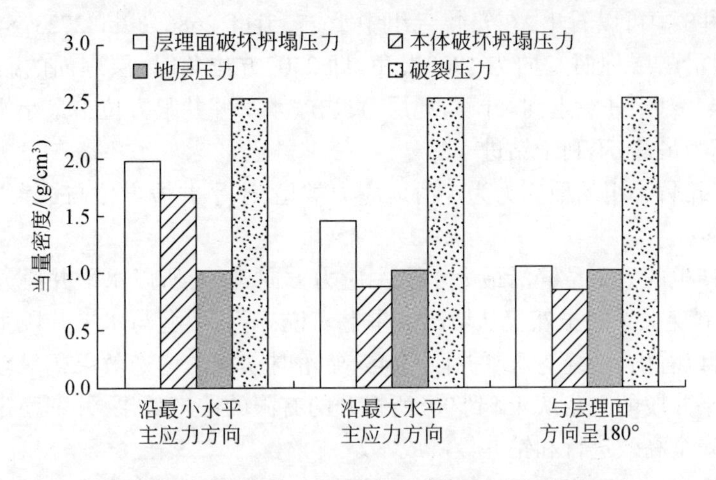

图6 井深2627.51m处各方位坍塌压力对比

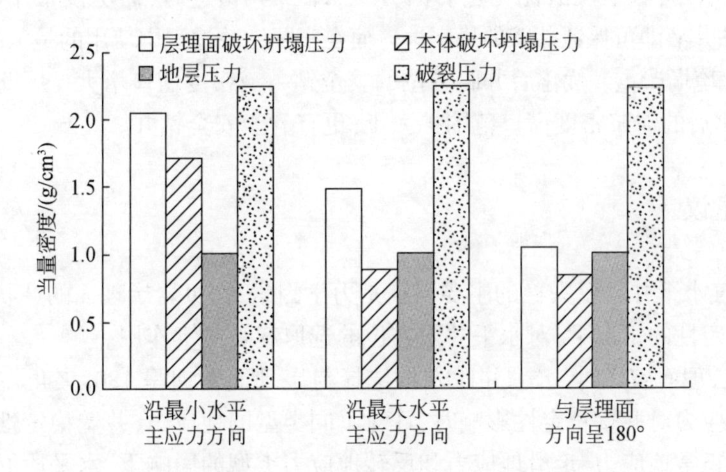

图7 井深2686.59m处各方位坍塌压力对比

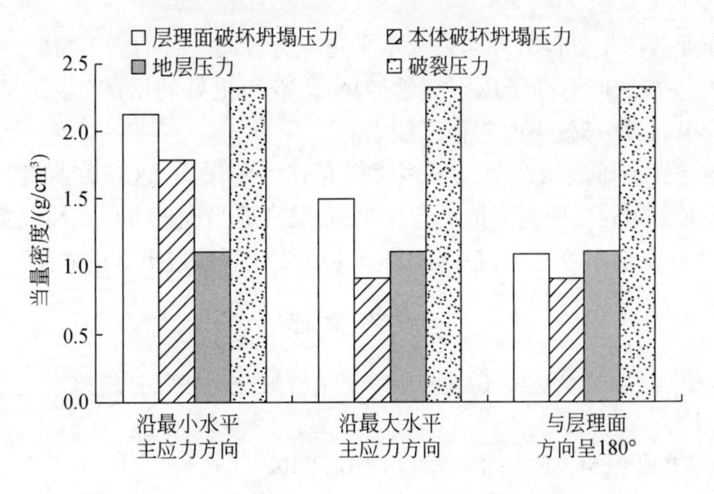

图8 井深2723.85m处各方位坍塌压力对比

从图 6 ~ 图 8 中可以看出,在筇竹寺组中 2627.51m、2686.59m、2723.85m 三个深度位置,当水平井段沿与层理面走向为 180°夹角,即 250°方位角时,层理面的破坏坍塌压力最小,井壁最稳定。对于任意走向和倾角地层,页岩气水平井井眼方位沿最小水平主应力方向时,井壁坍塌压力最高,不利于钻进[12]。

研究发现,筇竹寺组的地应力为走滑地应力类型,水平井段方位与最大主应力呈 45°夹角时井壁最稳定。

压裂增产产生的人工裂缝沿最大水平主应力方向延伸,所以水平井段方位与裂缝夹角同样为 45°,呈斜交关系。虽然仅从增产效果上来说,不如裂缝与水平井段垂直的增产效果好,但从图 5 可以看出,裂缝与水平井段呈 45°夹角时获得的增产效果仍然较为明显。相对于压裂缝和水平井段的夹角从 45°到 90°所提升的有限的增产效果,井壁稳定性直接关系到钻井作业能否安全高效进行和钻井经济成本。

通过筇竹寺组井壁稳定力学计算可知,沿最大、最小水平主应力方位钻进时的安全泥浆密度窗口都较窄,当水平段沿优化过后的方位,即 254°钻进时,储层段采用密度为 1.42 ~ 2.01g/cm^{3}的钻井液即可保证井壁的稳定性。而在实际钻井过程中使用的是 1.41 ~ 1.82g/cm^{3}的钻井液,井壁垮塌严重。优化后井眼方位的安全钻井液密度窗口增大。因此相比实际钻井泥浆密度,优化后的泥浆密度窗口范围扩大了,更有利于安全钻井。

3　结论与建议

(1)页岩气水平井井壁失稳的主要因素是力学因素,而页岩层理面的产状和地应力类型在考虑井壁稳定性的情况下,对水平井段方位的选取起着关键作用。

(2)考虑层理面破坏时,水平井段方位选择与层理面走向呈 180°夹角左右井壁稳定性最好。考虑地应力对井壁稳定性影响时,针对不同类型的地应力,井壁稳定性最好的方位不尽相同。即在正常地应力、走滑地应力和反转地应力类型的影响下,水平段的最优方位分别应选择与最大水平主应力呈 90°方向、与最大水平主应力呈 45°方向和沿最大水平主应力方向。

(3)压裂作业会产生延伸方向与最大水平主应力方向一致的人工裂缝,在水平井眼方向垂直于最大水平主应力时有助于压裂缝延伸更远,获得更好的增产效果。但当裂缝与水平井段夹角大于 60°时,增产效果改善幅度较小。

(4)井壁稳定性与压裂增产效果很多时候存在一定矛盾,这导致两者很难同时达到最优,这就需要根据实际的现场施工情况进行权衡,通常应当优先考虑获得良好的井壁稳定性以保证钻井作业的安全高效进行,在此前提下尽量使压裂获得更好的效果。

参 考 文 献

[1] 闫存章,黄玉珍,葛春梅,等. 页岩气是潜力巨大的非常规天然气资源[J]. 天然气工业,2009,29(5):1-6.

[2] 范翔宇. 油气钻井地质[M]. 重庆:重庆大学出版社,2010.

[3] 陈安明,张辉,宋占伟. 页岩气水平井钻完井关键技术分析[J]. 石油天然气学报,2012,34(11):98-103.

[4] 刘向君,罗平亚,孟英峰. 地应力场对井眼轨迹及稳定性的影响研究[J]. 天然气工业,2004,24(9):

59-57.
[5] 李玉飞,付永强,唐庚,等. 地应力类型影响定向井井壁稳定的规律[J]. 天然气工业,2012,32(3):78-80.
[6] Zhang J G. The Important of Shale Properties on Wellbore Stability[D]. The University of Texas at Austin,2005.
[7] Aadnoy B S. Stability of highly inclined boreholes[J]. SPE 16052,1987.
[8] Lee H, Ong S H, Azeemuddin M, *et al*. A wellbore stability model for formation with anisotropic rockstrengths[J]. Journal of Petroleum Science and Engineering,2012:108-119.
[9] 金衍,陈勉. 井壁稳定力学[M]. 北京:科学出版社,2012.
[10] 黄利平. 低渗透气藏压裂水平井产能预测与裂缝参数优化研究[D]. 西南石油大学硕士学位论文,2011.
[11] 何涛,李茂森,杨兰平,等. 油剂钻井液在威远地区页岩气水平井中的应用[J]. 钻井液与完井液,2012,29(3):1-5.
[12] 范翔宇. 复杂钻井地质环境描述[M]. 北京:石油工业出版社,2012.

三、气藏工程类

页岩气不稳定渗流压力传播规律和数学模型

朱维耀[1]　亓　倩[1]　马　千[1]　邓　佳[1]　岳　明[1]　刘玉章[2]
（1. 北京科技大学土木与环境工程学院；2. 中国石油勘探开发研究院）

摘要：利用稳定状态依次替换法，研究了页岩基质储集层内压力扰动的传播规律，得到动边界随时间变化的关系，考虑解吸、扩散、滑移作用及动边界的影响，建立页岩气不稳定渗流数学模型。采用拉普拉斯变换，求解了内边界定产、外边界为动边界条件下的不稳定渗流压力特征方程。结合中国南方某海相页岩气藏储集层参数，应用 MATLAB 编程，计算分析了页岩气不稳定渗流压力特征及其影响因素。研究表明，页岩气开采过程中，压力传播具有动边界效应，动边界随时间延续向外传播，且传播速度逐渐减慢；动边界使压力传播速度变慢，储集层压力下降减缓；页岩气解吸使压力传播速度减慢，地层压力下降减缓；扩散系数越大，地层压力下降越慢，且扩散系数影响逐渐减小。在气藏开采过程中，扩散、滑移对产气量贡献逐渐增加，占主要地位；渗流及解吸对产气量贡献逐渐减小后趋于平稳。

关键词：页岩气；不稳定渗流；压力传播规律；数学模型；动边界

页岩气具有多尺度流动的特征，并以吸附、游离或溶解状态赋存于纳米-微米级页岩孔隙及裂缝中，气体产出机理主要为解吸、扩散、滑移等，为低速强非线性渗流[1-5]。Pascal[6]、刘慈群[7]、李凡华和刘慈群[8]认为：低渗透油藏渗流具有启动压力梯度，压力扰动的传播并非瞬时到达无穷远，其渗流规律就是一个动边界问题；页岩储集层具有与低渗油藏类似的动边界压力传播特性；压力方程的建立和求解较难，考虑动边界问题就更难，至今尚未见这方面的研究报道。虽然目前国内外对页岩气开发的研究已进入快速发展阶段，然而由于页岩气储集层条件复杂，对页岩气渗流和产能递减规律的研究，大多数研究成果给出的仅是页岩气渗流规律及其影响因素，尽管也有部分文献提出了具体的渗流方程，但考虑的因素较少[9-12]，方程过于简化，难以更好地反映页岩气的低速强非线性渗流规律。为此，有必要揭示其页岩气的非线性流动规律，研究页岩气不稳定渗流特征，以便选取合理有效的开发方式和增产手段，为页岩气开发提供理论依据。

本文基于压力传播的稳定状态依次替换法对页岩气渗流动边界问题进行研究，进而建立考虑解吸、扩散、滑移及动边界影响的页岩气不稳定渗流数学模型，并推导和求解。结合中国南方某海相页岩储集层参数，分析页岩气不稳定渗流压力特征及其影响因素。

1　页岩气压力传播规律

在解决不稳定渗流压力动态的问题时，可以把不稳定渗流过程的每一瞬间状态看作稳定的，这种方法称为稳定状态依次替换法[13]。

当页岩气投入开发、页岩储集层被打开后，形成的压力降将逐渐向外传播，设某时刻 t，压力降传到 $R(t)$ 处，在 $R(t)$ 范围内形成压降漏斗，$R(t)$ 即为渗流过程中的扰动边界。$R(t)$ 随时间逐渐增大，压力降所波及的边缘为条件影响边缘，在该边缘上压力等于原始地层压力。

1.1　非线性渗流对动边界的影响

页岩储集层非常致密,主要为纳米-微米级孔隙,管壁和流体之间的微观作用力使气体在纳米-微米孔隙中流动时出现类似油藏启动压力梯度现象。页岩气的流动不仅有渗流过程,还存在扩散、滑移、解吸流动,气体流动总体表现为非线性流动,流动阻力比常规天然气大。滑脱效应附加了一种滑脱动力,但在驱动力小于气固间吸附作用所产生的阻力后,气体同样不能流动,即压力传播具有一定的动用范围。因此在不稳定渗流过程中压力扰动随时间延续逐渐向外传播,其边界条件也是一个动边界问题。

对于纳米-微米孔隙页岩储集层,气体在其中流动时,由于储集层渗透率极低,流动已偏离达西定律,扩散、滑移作用对储集层内气体流动影响增加。朱维耀等[14]建立了考虑扩散、滑移的纳米-微米孔隙气体流动方程:

$$v=-\frac{K_0}{\mu}\left(1+\frac{3\pi a}{16K_0}\frac{\mu D_K}{p}\right)\frac{dp}{dx} \tag{1}$$

式中,a 为克努森数 Kn 有关的修正系数(当 $0\leqslant Kn<0.001$, $a=0$; $0.001\leqslant Kn<0.1$, $a=1.2$; $0.1\leqslant Kn<10$, $a=1.34$);D_K 为扩散系数,m^2/s;K_0 为储集层绝对渗透率,m^2;p 为储集层压力,Pa;μ 为气体黏度,Pa·s;v 为气体渗流速度,m/s;x 为渗流距离,m。

则任一瞬间,地层压力为

$$\left(p+\frac{3\pi a\mu D_K}{16K_0}\right)^2=\left(p_e+\frac{3\pi a\mu D_K}{16K_0}\right)^2-\frac{\left(p_e+\frac{3\pi a\mu D_K}{16K_0}\right)^2-\left(p_w+\frac{3\pi a\mu D_K}{16K_0}\right)^2}{\ln\frac{R_1(t)}{r_w}}\ln\frac{R_1(t)}{r} \tag{2}$$

式中,p_e 为外边界压力,Pa;p_w 为内边界压力,Pa;$R_1(t)$ 为微尺度效应影响动边界,m;r 为距井筒距离,m;r_w 为井筒半径,m。

如图 1 所示,在地层中半径为 r 处取出厚度为 h,宽度为 dr 的微小圆环体,其体积为 $2\pi rh dr$,此单元体中游离态气体的原始质量为 $2\pi rh\varphi_i\rho_i dr$。在给定时刻 t,该单元体中残留气体质量为 $2\pi rh\varphi\rho dr$,因此,从单元体孔隙中采出的游离态气体质量为 $2\pi rh(\varphi_i\rho_i-\varphi\rho)dr$。

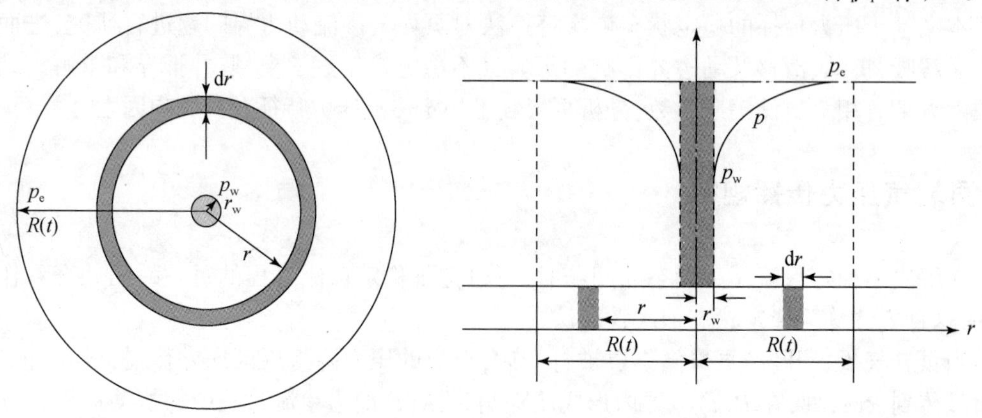

图 1　页岩气平面径向流示意图

考虑吸附态气体的解吸,采出的总气体量为

$$Q_{sc}=\int_{r_w}^{R_1(t)}2\pi rh[(\varphi_i\rho_i-\varphi\rho)+q_dt]\mathrm{d}r+\pi r_w^2h(\varphi_i\rho_i-\varphi_w\rho_w) \tag{3}$$

式中,Q_{sc}为地层中采出的总气体量,m^3;h为气层厚度,m;φ为储集层孔隙度,f;ρ为气体密度,kg/m^3;q_d为单位体积页岩单位时间的解吸量,$kg/(m^3\cdot s)$;t为生产时间,s;下标i为原始地层条件;下标w为内边界井底处。

代入气体状态方程,联立式(2)、式(3)进一步求得

$$Q_{sc}=\frac{\pi h\varphi CT_{sc}Z_{sc}\rho_{gsc}}{p_{sc}TZ}\left[\left(p_e+\frac{3\pi a\mu D_K}{16K_0}\right)^2-\left(p_w+\frac{3\pi a\mu D_K}{16K_0}\right)^2\right]\times\frac{R_1^2(t)-r_w^2}{2\ln\dfrac{R_1(t)}{r_w}}+\pi hq_dt[R_1{}^2(t)-r_w^2] \tag{4}$$

式中,T为地层温度,K;T_{sc}为标准状态下温度,K;Z为气体压缩因子,无因次;Z_{sc}为标准状态下气体压缩因子,无因次;p_{sc}为标准压力,Pa;ρ_{gsc}为标准状态下气体密度,kg/m^3;C为气体等温压缩系数,Pa^{-1}。

气井流量按稳定渗流公式可写为

$$q_{sc1}=\frac{\pi K_0hZ_{sc}T_{sc}\rho_{gsc}}{p_{sc}T\mu Z\ln\dfrac{R_1(t)}{r_w}}\left[\left(p_e+\frac{3\pi a\mu D_K}{16K_0}\right)^2-\left(p_w+\frac{3\pi a\mu D_K}{16K_0}\right)^2\right] \tag{5}$$

式中,q_{sc1}为标准条件下气井流量,m^3/s。

假设页岩气井采用定产量生产,即q_{sc1}为常数,则$Q_{sc}=q_{sc1}t$,联立式(4)、式(5)可得:

$$Q_{sc}=\frac{\varphi\mu Cq_{sc1}}{2K_0}[R_1^2(t)-r_w^2]+\pi hq_dt[R_1^2(t)-r_w^2] \tag{6}$$

由此得出页岩储集层压力扰动传播动边界随时间变化的关系为

$$R_1(t)=\sqrt{\frac{q_{sc1}t}{\dfrac{\varphi\mu Cq_{sc1}}{2K_0}+\pi hq_dt}} \tag{7}$$

中国南方下志留统龙马溪组海相页岩气藏C的孔隙度为0.07,绝对渗透率为$0.0005\times10^{-3}\mu m^2$;地层温度为396.15K,标准状态下气体压缩因子为1,真实气体压缩因子为0.89;气体黏度为0.027mPa·s;外边界压力为24MPa;井筒半径0.1m;气藏厚度为30.5m;平均解吸量为3.3707×10^{-10} $kg/(m^3\cdot s)$;扩散系数为$8.4067\times10^{-7}cm^2/s$,定产量为$200m^3/d$。

利用气藏C基本参数,采用式(7)绘制不同渗透率条件下页岩气储集层动边界随时间变化关系图。由图2可见,同一时刻,渗透率越大,压力扰动传播动边界越远。在生产初期,压力扰动传播边界扩展较快,随着生产时间的延续,传播速度逐渐减慢。当渗透率大于$0.5\times10^{-3}\mu m^2$时,压力快速传播到动边界;当渗透率小于等于$0.5\times10^{-3}\mu m^2$时,压力随着时间的推进逐渐向外传播,因此页岩储集层中压力传播需考虑动边界的影响。

1.2 页岩储集层压力传播动边界问题

部分学者认为在低渗透气藏渗流问题中,由于启动压力梯度的影响,压力并不是瞬间传播到无穷远,而是随着时间的推进逐渐向外传播,压力的传播边界称为动边界。分析启动压力梯度和非线性渗流对压力传播边界的影响,本文认为只有在外加压力梯度大于启动压力

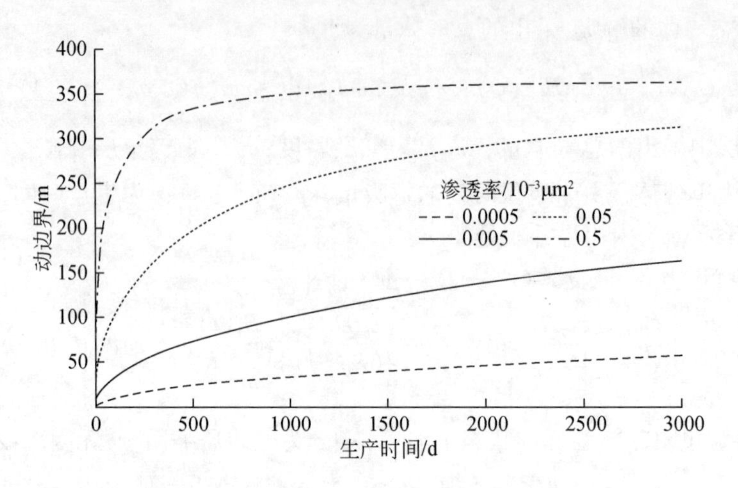

图 2　不同渗透率条件下动边界随时间变化关系

梯度时，气体才发生流动。在页岩气开采过程中，由于启动压力梯度的存在，压力传播存在一定的范围。建立考虑启动压力梯度的稳态径向流常微分方程及其内边界定产、外边界定压的边界条件：

$$\begin{cases} \dfrac{\mathrm{d}^2 m}{\mathrm{d}r^2}+\dfrac{1}{r}\dfrac{\mathrm{d}m}{\mathrm{d}r}-G_t\dfrac{\mathrm{d}m}{\mathrm{d}r}=0 \\ r=r_w,\ r\dfrac{\partial m}{\partial r}=\dfrac{q_{sc2}Tp_{sc}}{\pi K_0 hZ_{sc}T_{sc}} \\ r=r,\ m=m_{R_2(t)} \end{cases} \tag{8}$$

式中，$G_t=CG$；G 为启动压力梯度，Pa/m；$m=2\int_{p_a}^{p}\dfrac{p}{\mu Z}\mathrm{d}p$；$p_a$ 为某一已知压力，Pa；q_{sc2} 为启动压力梯度影响标准条件下气井流量，m^3/s；$R_2(t)$ 为启动压力梯度影响动边界，m；m 为拟压力函数，Pa/s。

求解得到启动压力梯度下页岩储集层中直井稳态渗流表达式：

$$q_{sc2}=\pi h\frac{K_0T_{sc}Z_{sc}\rho_{gsc}}{p_{sc}T\mu Z\mathrm{e}^{-G_t r_w}}\left[\frac{p_e^2-p_w^2}{\mathrm{Ei}(1,-G_t r_e)-\mathrm{Ei}(1,-G_t r_w)}\right] \tag{9}$$

式中，r_e 为气井供给半径，m；Ei 为幂积分函数。

地层中采出的总气体量：

$$Q_{sc}=\pi h\frac{\varphi CT_{sc}Z_{sc}\rho_{gsc}}{p_{sc}TZ}(p_e^2-p_w^2)\times\frac{\mathrm{e}^{G_tR_2(t)}[G_tR_2(t)-1]-\mathrm{e}^{G_tr_w}(G_tr_w-1)}{G_t^2\{\mathrm{Ei}[1,-G_tR_2(t)]-\mathrm{Ei}(1,-G_tr_w)\}}+\pi hq_dt[R_2^2(t)-r_w^2] \tag{10}$$

将式(10)代入考虑启动压力梯度的页岩储集层直井稳态渗流表达式[式(9)]，得页岩气储集层压力扰动传播动边界随时间变化的关系为

$$t=\frac{\varphi\mu C}{K_0}\frac{\mathrm{e}^{G_tR_2(t)}[G_tR_2(t)-1]+1}{G_t^2}+\frac{\pi hq_dt}{q_{sc2}}R_2^2(t) \tag{11}$$

当 $G_t\to 0$ 时，即得到不考虑启动压力梯度的页岩储集层压力扰动传播动边界与时间的变化关系：

$$t=\frac{\varphi\mu C}{2K_0}R_2^2(t)+\frac{\pi hq_{\mathrm{d}}t}{q_{\mathrm{sc2}}}R_2^2(t) \tag{12}$$

图3为不同启动压力梯度下动边界随时间变化关系。可见,在同一时刻,启动压力梯度越大,压力扰动传播动边界越小。

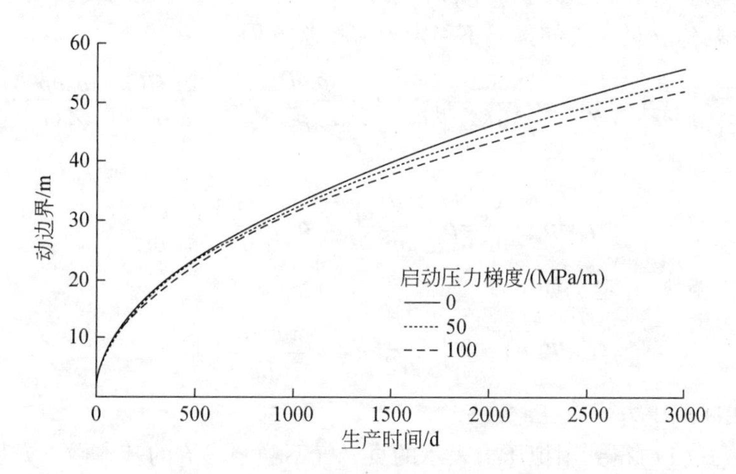

图3　不同启动压力梯度下动边界随时间变化关系

2　页岩气不稳定渗流模型及求解

页岩气在储集层中的流动包括3个过程:甲烷从页岩基质表面解吸;甲烷通过页岩基质和微孔隙流动;甲烷在裂缝系统流动。本文考虑页岩储集层气体解吸、扩散、滑移等非线性渗流特征在不稳定渗流过程中对压力扰动传播动边界的影响,利用压力传播的稳定状态依次替换法,建立页岩气不稳定渗流模型,并进行推导和求解。

2.1　页岩气吸附-解吸模型

通过吸附-解吸实验得到不同平衡压力下页岩吸附-解吸过程的含气量变化。将得到的实验结果与吸附-解吸模型对比,结果表明,Langmuir 模型拟合程度很高且形式简单,适于描述页岩气的吸附过程[15]。

Langmuir 吸附模型[15]的数学表达式如下:

$$V_{\mathrm{E}}=V_{\mathrm{L}}\frac{p}{p+p_{\mathrm{L}}} \tag{13}$$

式中,V_{E} 为气体吸附量,$\mathrm{m^3/m^3}$;V_{L} 为 Langmuir 体积,$\mathrm{m^3/m^3}$;p_{L} 为 Langmuir 压力,表示吸附量为最大吸附量一半时的压力,Pa。

考虑瞬时平衡条件,解吸量可以表示为

$$V_{\mathrm{d}}=V_{\mathrm{L}}\left(\frac{p_{\mathrm{i}}}{p_{\mathrm{i}}+p_{\mathrm{L}}}-\frac{p}{p+p_{\mathrm{L}}}\right) \tag{14}$$

式中,V_{d} 为单位体积基质累积解吸量,$\mathrm{m^3/m^3}$。

2.2 页岩气不稳定渗流模型

在考虑解吸、扩散、滑移作用的页岩气非线性渗流方程基础上[16,17]，基于天然气渗流的连续性方程、运动方程和状态方程，考虑不稳定渗流过程中压力扰动传播动边界的影响，引入动边界的模型，建立页岩气储集层不稳定渗流控制方程：

$$\frac{1}{r}\frac{\partial}{\partial r}\left[r\frac{1}{\mu Z}\left(p+\frac{3\pi a\mu D_{\mathrm{K}}}{16K_0}\right)\frac{\partial p}{\partial r}\right]+\frac{p_{\mathrm{sc}}T}{T_{\mathrm{sc}}Z_{\mathrm{sc}}\rho_{\mathrm{gsc}}K_0}q_{\mathrm{d}}=\frac{\varphi\mu C(p)}{K_0}\frac{p}{\mu Z}\frac{\partial p}{\partial t} \tag{15}$$

式中

$$q_{\mathrm{d}}=\rho_{\mathrm{gsc}}\frac{\partial V_{\mathrm{d}}}{\partial t}=\rho_{\mathrm{gsc}}\frac{\partial V_{\mathrm{d}}}{\partial p}\frac{\partial p}{\partial t}=-\rho_{\mathrm{gsc}}\frac{p_{\mathrm{L}}V_{\mathrm{L}}}{(p+p_{\mathrm{L}})^2}\frac{\partial p}{\partial t} \tag{16}$$

引入拟压力函数，并定义如下：

$$\Psi(p)=2\int_{p_{\mathrm{a}}}^{p}\frac{1}{\mu Z}\left(p+\frac{3\pi a\mu D_{\mathrm{K}}}{16K_0}\right)\mathrm{d}p \tag{17}$$

式中，Ψ 为非线性拟压力函数，Pa/s。

由式(15)、式(17)得到用拟压力表示的页岩气不稳定渗流的基本微分方程：

$$\frac{1}{r}\left[\frac{\partial}{\partial r}\left(r\frac{\partial\Psi}{\partial r}\right)\right]=\frac{\varphi\mu C_{\mathrm{t}}^*}{K_0}\frac{\partial\Psi}{\partial t} \tag{18}$$

式中

$$C_{\mathrm{t}}^*=C_{\mathrm{k}}+C_{\mathrm{d}} \tag{19}$$

$$C_{\mathrm{k}}=C(p)\frac{p}{p+\dfrac{3\pi a\mu D_{\mathrm{K}}}{16K_0}} \tag{20}$$

$$C_{\mathrm{d}}=\frac{p_{\mathrm{sc}}TZ}{T_{\mathrm{sc}}Z_{\mathrm{sc}}\varphi}\frac{p_{\mathrm{L}}V_{\mathrm{L}}}{(p+p_{\mathrm{L}})^2}\frac{1}{p+\dfrac{3\pi a\mu D_{\mathrm{K}}}{16K_0}} \tag{21}$$

式中，C_{t}^* 为综合压缩系数，Pa^{-1}；C_{k} 为扩散压缩系数，Pa^{-1}；C_{d} 为解吸压缩系数，Pa^{-1}。

由于 μC_{t}^* 是压力的函数，式(18)为非线性方程，为将其线性化，Agarwal 等[18]针对气井给出了拟时间 $t_{\mathrm{a}}{}^*$ 的定义：

$$t_{\mathrm{a}}^*(p)=\mu_{\mathrm{i}}C_{\mathrm{ti}}^*\int_0^t\frac{\mathrm{d}t}{\mu C_{\mathrm{t}}^*} \tag{22}$$

式中，t_{a}^* 为拟时间，s。

则式(18)可转换为

$$\frac{1}{r}\left[\frac{\partial}{\partial r}\left(r\frac{\partial\Psi}{\partial r}\right)\right]=\frac{\varphi\mu_{\mathrm{i}}C_{\mathrm{ti}}^*}{K_0}\frac{\partial\Psi}{\partial t_{\mathrm{a}}^*} \tag{23}$$

2.3 页岩气不稳定渗流模型求解

当页岩气直井以某一恒定产量生产时，内边界定产，外边界定压，受动边界影响，泄压半径为当前时刻压力扰动传播到的距离，则其定解条件如下：

$$\Psi\big|_{t=0}=\Psi_{\mathrm{i}}(p=p_{\mathrm{i}})$$

$$r\frac{\partial\Psi}{\partial r}\bigg|_{r=r_{\mathrm{w}}}=\frac{q_{\mathrm{sc}}p_{\mathrm{sc}}T}{\pi K_0 hZ_{\mathrm{sc}}T_{\mathrm{sc}}}$$

$$\Psi\big|_{r=r_{\mathrm{e}}=R(t)}=\Psi_{\mathrm{i}}(p=p_{\mathrm{i}})$$

为方便求解，将式(23)用无因次量表示，无因次距离：

$$r_{\mathrm{D}}=\frac{r}{r_{\mathrm{w}}} \tag{24}$$

无因次拟时间：

$$t_{\mathrm{aD}}^{*}=\frac{K_0 t_{\mathrm{a}}^{*}}{\varphi\mu_{\mathrm{i}}C_{\mathrm{ti}}^{*}r_{\mathrm{w}}^{2}} \tag{25}$$

无因次拟压力：

$$\Psi_{\mathrm{D}}=\frac{\pi K_0 hZ_{\mathrm{sc}}T_{\mathrm{sc}}}{p_{\mathrm{sc}}q_{\mathrm{sc}}T}(\Psi_{\mathrm{i}}-\Psi) \tag{26}$$

无因次流量：

$$q_{\mathrm{D}}=\frac{p_{\mathrm{sc}}T}{\pi K_0 hZ_{\mathrm{sc}}T_{\mathrm{sc}}(\Psi_{\mathrm{i}}-\Psi_{\mathrm{w}})}q_{\mathrm{sc}} \tag{27}$$

式中，Ψ_{w} 为井底非线性拟压力，Pa/s。

则式(23)可化为

$$\frac{\partial^2\Psi_{\mathrm{D}}}{\partial r_{\mathrm{D}}{}^2}+\frac{1}{r_{\mathrm{D}}}\frac{\partial\Psi_{\mathrm{D}}}{\partial r_{\mathrm{D}}}=\frac{\partial\Psi_{\mathrm{D}}}{\partial t_{\mathrm{aD}}^{*}} \tag{28}$$

引入拉普拉斯变换函数：

$$\bar{\Psi}_{\mathrm{D}}(s)=\int_0^{\infty}\Psi_{\mathrm{D}}(r_{\mathrm{D}},t_{\mathrm{aD}}^{*})\mathrm{e}^{-st_{\mathrm{aD}}^{*}}\mathrm{d}t_{\mathrm{aD}}^{*} \tag{29}$$

式中，$\bar{\Psi}_{\mathrm{D}}$ 为拉普拉斯变换非线性拟压力函数，Pa/s。

则式(28)变换为

$$\frac{\mathrm{d}^2\bar{\Psi}_{\mathrm{D}}}{\mathrm{d}r_{\mathrm{D}}{}^2}+\frac{1}{r_{\mathrm{D}}}\frac{\mathrm{d}\bar{\Psi}_{\mathrm{D}}}{\mathrm{d}r_{\mathrm{D}}}=s\bar{\Psi}_{\mathrm{D}}(s) \tag{30}$$

式中，s 为拉普拉斯算子。

代入对应的定解条件：

$$\frac{\mathrm{d}\bar{\Psi}_{\mathrm{D}}}{\mathrm{d}r_{\mathrm{D}}}\bigg|_{r_{\mathrm{D}}=1}=-\frac{1}{s};\quad \bar{\Psi}_{\mathrm{D}}\big|_{r_{\mathrm{D}}=R_{\mathrm{D}}(s)}=0$$

求得

$$\bar{\Psi}_{\mathrm{D}}(r_{\mathrm{D}},s)=\frac{\mathrm{I}_0(R_{\mathrm{D}}\sqrt{s})\mathrm{K}_0(r_{\mathrm{D}}\sqrt{s})-\mathrm{K}_0(R_{\mathrm{D}}\sqrt{s})\mathrm{I}_0(r_{\mathrm{D}}\sqrt{s})}{s^{3/2}[\mathrm{I}_1(\sqrt{s})\mathrm{K}_0(R_{\mathrm{D}}\sqrt{s})+\mathrm{K}_1(\sqrt{s})\mathrm{I}_1(R_{\mathrm{D}}\sqrt{s})]} \tag{31}$$

式中，I_0 为零阶第一类虚变量贝塞尔函数；I_1 为一阶第一类虚变量贝塞尔函数；K_0 为零阶第二类虚变量贝塞尔函数；K_1 为一阶第二类虚变量贝塞尔函数。

经过逆变换，求得地层任意一点压力变化规律为

$$\Psi(r,t_a^*)=\Psi_i-\frac{q_{sc}p_{sc}T}{\pi K_0 h Z_{sc}T_{sc}}\ln\frac{R(t_a^*)}{r}+\frac{q_{sc}p_{sc}T}{\pi K_0 h Z_{sc}T_{sc}}$$

$$\times\sum_{n=1}^{\infty}\frac{e^{-\beta_n^2\frac{K_0 t_a^*}{\varphi\mu_i C_{ti}^* r_w^2}}J_0^2(r_{eD}\beta_n)[Y_1(\beta_n)J_0(r_D\beta_n)-J_1(\beta_n)Y_0(r_D\beta_n)]}{\beta_n[J_0^2(r_{eD}\beta_n)-J_1^2(\beta_n)]} \tag{32}$$

式中，β_n 为式(33)第 n 个正根：

$$J_1(\beta_n)Y_0(\beta_n r_{eD})-Y_1(\beta_n)J_0(\beta_n r_{eD})=0 \tag{33}$$

式中，J_0 为零阶第一类贝塞尔函数；J_1 为一阶第一类贝塞尔函数；Y_0 为零阶第二类贝塞尔函数；Y_1 为一阶第二类贝塞尔函数。

当 $r=r_w$ 时，得井底压力变化规律：

$$\Psi_w=\Psi_i-\frac{q_{sc}p_{sc}T}{\pi K_0 h Z_{sc}T_{sc}}\times\left\{\ln\frac{R(t)}{r_w}-2\sum_{n=1}^{\infty}\frac{e^{-\beta_n^2 t_D}J_0^2[\beta_n R_D(t_D)]}{\beta_n^2\{J_1^2(\beta_n)-J_0^2[\beta_n R_D(t_D)]\}}\right\} \tag{34}$$

3　页岩气不稳定渗流储集层压力分布特征

根据前面推导出的考虑解吸、扩散的页岩气不稳定渗流压力分布规律，结合气藏 C 参数，应用 MATLAB 编程计算，对页岩气不稳定渗流压力分布及其影响因素进行分析。

图 4 为不同时间地层压力分布曲线，由图可见，随着生产时间的延长，地层压力逐渐向外传播。在动边界影响范围内，页岩气储集层及气体释放弹性能，形成一个压降漏斗。动边界影响范围以外的地区，由于没有压力扰动，气体并不流动，且动边界的传播速度逐渐减慢。

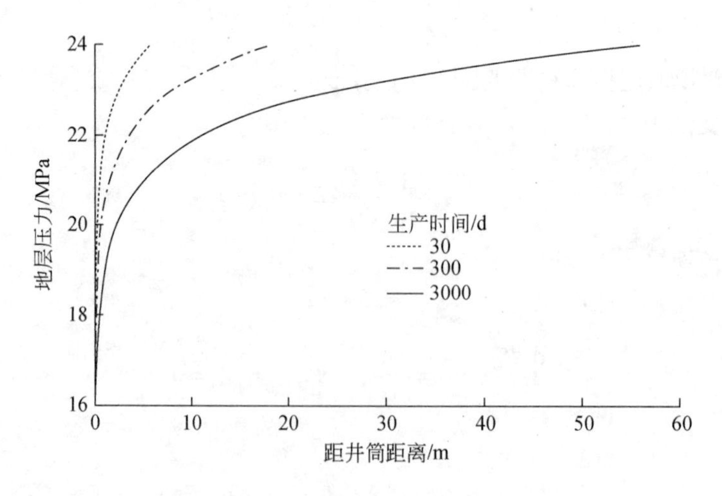

图 4　不同生产时间地层压力分布曲线

图 5 为不同产量条件下井底压力随时间变化曲线，由图可见，井底压力随时间增加而降低，且减小趋势逐渐减缓，在前 50d，井底流压下降较快。同一时刻，产量越大，地层压力下降越多。

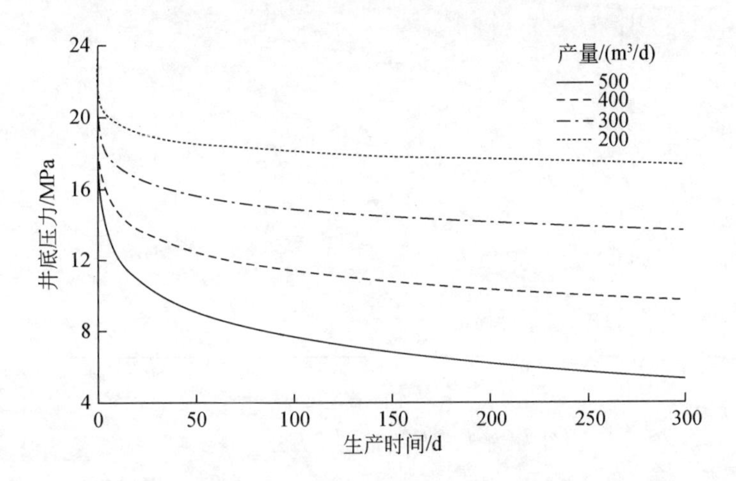

图5 不同产气量条件下井底压力随生产时间变化曲线

图6为动边界对地层压力分布的影响,由图可见,考虑动边界影响时,压力扰动范围减小,地层压力分布下降减缓。对于超致密的纳微米孔隙页岩储集层,压力扰动随时间逐渐向外传播,且速度较慢,因此,考虑动边界影响的压力分布更贴近实际,更能准确地指导页岩气的生产。

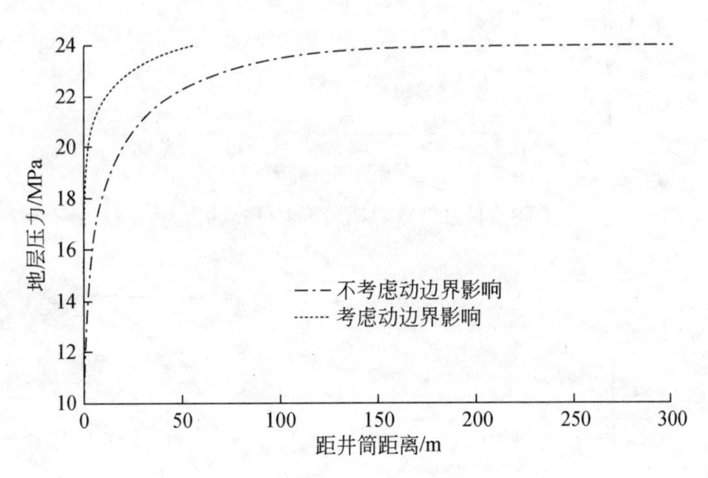

图6 动边界对地层压力分布的影响

图7为解吸量对地层压力分布的影响,由图可见,考虑解吸量时地层压力较高,吸附气体的解吸使地层压力传播速度减慢,地层压力下降减缓。

图8为扩散系数对地层压力分布的影响,由图可见,定产条件下,在动边界影响范围内,扩散系数越大,地层压力下降越慢,且扩散系数的影响逐渐减小。

图9为解吸、渗流、扩散、滑移及对产气量贡献率,由图可见,扩散、滑移在页岩气藏开采过程中占主要地位。在气藏开采过程中,随着开采的进行,渗流、解吸对产气量的贡献逐渐减小后趋于平稳,而扩散、滑移的贡献逐渐增加。生产初期,页岩气藏压降较小,基质中的游离气滑移、扩散,对产气量的贡献逐渐增加;随着生产的进行,解吸气不断释放,贡献逐渐趋于平稳。

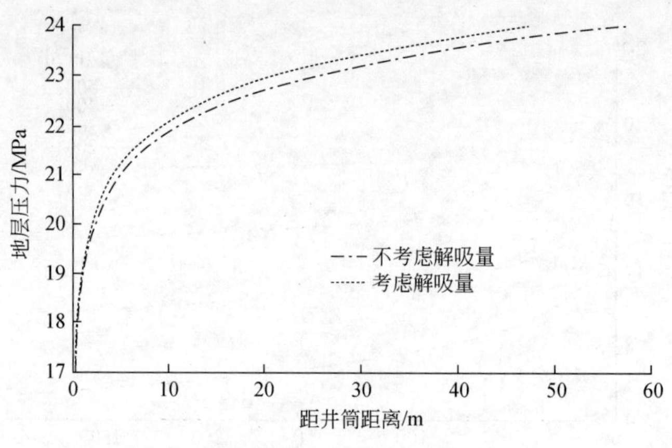

图 7　解吸气体对地层压力分布的影响

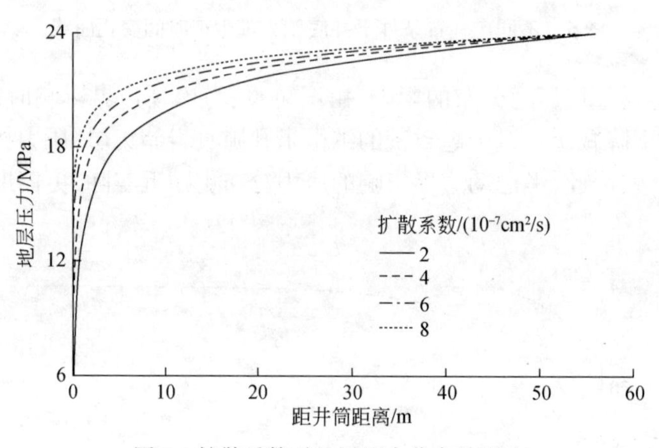

图 8　扩散系数对地层压力分布的影响

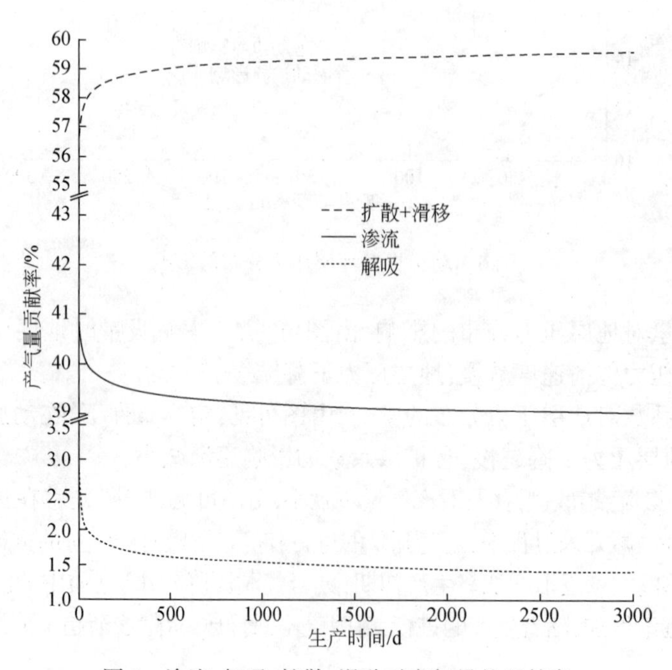

图 9　渗流、解吸、扩散、滑移对产气量的贡献率

4 结论

基于页岩气吸附-解吸模型及纳米-微米孔隙气体流动模型,建立了考虑解吸、扩散、滑移的页岩气不稳定渗流数学模型。引入拟压力、拟时间函数及天然气扩散、解吸压缩系数,得到了页岩气不稳定渗流控制方程。

利用稳定状态依次替换法推导了考虑解吸、扩散、滑移的页岩气储集层压力扰动传播动边界随时间变化的关系式。研究表明,动边界随时间增加逐渐增大,渗透率越大,动边界传播越远。

通过拉普拉斯变换,推导求解了内边界定产、外边界为动边界的页岩气储集层不稳定渗流压力特征方程,得出了井底压力变化规律。结合中国南方某海相页岩气藏参数,计算分析了页岩气不稳定渗流储集层压力分布及其影响因素。研究结果表明,地层压力分布在动边界影响范围内形成压降漏斗,在动边界影响范围以外,没有压力扰动,气体并不流动;动边界随时间向外传播,且传播速度逐渐减慢。井底压力随时间增加而降低,且降低趋势逐渐减缓。与不考虑动边界的压降曲线对比,考虑动边界影响的地层压力传播慢,压力下降减缓。气体解吸扩散影响压力的传播速度,解吸使压力传播速度减慢,地层压力下降减缓;扩散系数越大,地层压力下降越慢,且扩散系数的影响逐渐减小。在气藏开采过程中,扩散、滑移对产气量的贡献逐渐增加,且占主要地位;渗流及解吸对产气量的贡献逐渐减小后趋于平稳。用本模型方法描述页岩气储集层的压力分布特征更符合实际。

参考文献

[1] 宁正福,王波,杨峰,等. 页岩储集层微观渗流的微尺度效应[J]. 石油勘探与开发,2014,41(4):445-452.
[2] 邹才能,董大忠,王玉满,等. 中国页岩气特征、挑战及前景(一)[J]. 石油勘探与开发,2015,42(6):689-701.
[3] 杨峰,宁正福,胡昌蓬,等. 页岩储层微观孔隙结构特征[J]. 石油学报,2013,34(2):301-311.
[4] 钟太贤. 中国南方海相页岩孔隙结构特征[J]. 天然气工业,2012,32(9):1-4.
[5] 李治平,李智锋. 页岩气纳米级孔隙渗流动态特征[J]. 天然气工业,2012,32(4):50-53.
[6] Pascal H. Non- steady flow through porous media in the presence of a threshold gradient[J]. Acta Mechanica,1981,39(3-4):207-224.
[7] 刘慈群. 有起始比降固结问题的近似解[J]. 岩土工程学报,1982,4(3):107-109.
[8] 李凡华,刘慈群. 含启动压力梯度的不定常渗流的压力动态分析[J]. 油气井测试,1997(1):1-4.
[9] 尹虎,王新海,姜永,等. 页岩气藏渗流数值模拟及井底压力动态分析[J]. 长江大学学报(自然科学版),2012,9(8):68-71.
[10] 于荣泽,张晓伟,卞亚南,等. 页岩气藏流动机理与产能影响因素分析[J]. 天然气工业,2012,32(9):10-15.
[11] 程远方,董丙响,时贤,等. 页岩气藏三孔双渗模型的渗流机理[J]. 天然气工业,2012,32(9):44-47.
[12] 李亚洲,李勇明,罗攀,等. 页岩气渗流机理与产能研究[J]. 断块油气田,2013,20(2):186-190.
[13] 葛家理. 油气层渗流力学[M]. 北京:石油工业出版社,1982.
[14] 朱维耀,马千,邓佳,等. 纳微米级孔隙气体流动数学模型及应用[J]. 北京科技大学学报,2014,36(6):709-715.

[15] 郭为,熊伟,高树生,等. 页岩气等温吸附/解吸特征[J]. 中南大学学报(自然科学版),2013,44(7):2836-2840.

[16] Gerami S, Pooladi- Darvish M, Morad K, *et al*. Type curves for dry CBM reservoirs with equilibrium desorption[J]. Journal of Canadian Petroleum Technology,2008,47(7):48-56.

[17] Deng J, Zhu W Y, Ma Q. A new seepage model for shale gas reservoir and productivity analysis of fractured well[J]. Fuel,2014,124(15):232-240.

[18] Agarwal R G, Gardner D C, Fussell D D. Analyzing well production data using combined-type-curve and decline-curve analysis concepts[J]. SPE Reservoir Evaluation & Engineering,1999,2(5):478-486.

考虑岩石变形效应的页岩气渗流模型

代　宇[1]　贾爱林[1]　尚福华[2]

(1. 中国石油勘探开发研究院;2. 东北石油大学)

摘要:页岩气藏孔渗结构具有强烈的多尺度性,渗流机理复杂,纳米级基质孔隙克努森扩散效应,裂缝应力敏感效应,以及气体解吸收缩效应等多重机制对页岩气多尺度流动特征及页岩气产能模型都有一定影响。文中建立考虑纳米级基质孔隙克努森扩散流、裂缝应力敏感变形、基质解吸收缩效应协同作用的非线性渗流数学模型,应用全隐式有限差分和牛顿迭代法进行数值求解。对相关因素分析得到,裂缝变形负相关于中前期气体产能;而基质解吸收缩正相关于中后期气体产能。实际生产过程中,应当结合不同生产阶段,合理调整页岩气生产条件,协同考虑裂缝变形和基质解吸收缩耦合效应,最终优化页岩气生产制度,提高页岩气采收率。

关键词:页岩气;压裂水平井;克努森扩散;应力敏感;解吸收缩;产能

页岩气藏孔渗结构具有强烈的多尺度性,其微观结构特殊、渗流机理复杂,气体解吸、克努森扩散效应、应力敏感效应、非达西效应以及滑脱效应等多重机制对页岩气多尺度流动特征及页岩气产能模型都具有一定的影响[1-5]。Abdassah 和 Ershaghi 首次建立考虑微可压缩流体径向流动三重孔隙模型[6]。El-Banbi 提出页岩气裂缝储层线性三孔模型,并推导一定内外边界条件下的拉普拉斯空间解[7]。赵玉龙等在考虑页岩气解吸扩散的基础上,依据点源函数和势的叠加原理,建立页岩气产能方程[8]。石军太等从微观渗流机理着手,同时考虑页岩气解吸扩散,依据双孔板状模型建立页岩气能预测模型[9]。王海涛综合考虑微裂缝应力敏感效应,采用点源函数和势的叠加原理建立页岩气不稳定压力分析模型[10]。可以发现,以上页岩气产能模型,并没有考虑纳米级孔隙中的克努森扩散效应[11,12],同时气体解吸会导致基质收缩[13],气体流动性增强,其与微裂缝应力敏感效应的协同效应,以及其对产能和生产制度的影响亟待研究。

本文建立考虑基质解吸收缩效应和天然裂缝应力敏感协同作用的渗流数学模型,应用全隐式有限差分和牛顿迭代法进行数值求解,绘制了页岩气产量递减曲线,并分析了相关因素对气井产量变化规律的影响。

1　渗流模型建立

页岩气多级压裂地层简化为块状双重介质模型,考虑矩形封闭地层中心一口水平井定压力生产,井底拟流动压力为 ψ_{wf},人工裂缝垂直于井筒,关于井筒对称且均匀分布,取其中一条裂缝等效单元,考虑地层和人工裂缝双区复合,其他条件做如下假设:

储层具有双孔介质特征,在初始条件下,地层各处的拟压力为 ψ_i;考虑天然裂缝应力敏感效应以及基质解吸收缩效应,基质中为克努森扩散;页岩气微可压缩,压缩系数恒定;页岩气解吸满足朗缪尔(Langmuir)等温吸附方程;忽略重力和毛细管力影响。

1.1　应力敏感效应

考虑气体拟压力形式下的应力敏感对天然裂缝渗透率影响可以写为

$$k_f = k_{fi} e^{-\beta(\psi_i - \psi_f)} \tag{1}$$

式中，k_f为考虑应力敏感的天然裂缝渗透率，μm^2；k_{fi}为原始天然裂缝渗透率，μm^2；ψ_i为原始地层拟压力，$MPa^2/(mPa \cdot s)$；ψ_f为地层中天然裂缝拟压力，$MPa^2/(mPa \cdot s)$；β为应力敏感系数，$mPa \cdot s/MPa^2$。

1.2　基质变形效应

随着地层压力下降，页岩储层中的吸附气体开始解吸，页岩基质收缩引起渗流通道的增大对渗透率有重要影响。采用 Bangham 固体变形理论描述压力下降时吸附气解吸对页岩气渗透率影响。

$$\Delta\varepsilon = \rho_s \frac{RT}{EV_b} \int_{\psi_i}^{\psi_m} \frac{V}{\psi} d\psi \tag{2}$$

式中，$\Delta\varepsilon$为有效应力下的页岩收缩程度，无量纲；ρ_s为岩石密度，kg/m^3；R为气体常数，$MPa \cdot L/(mol \cdot K)$；$T$为绝对温度，K；$E$为杨氏模量，MPa；$V_b$为气体摩尔体积，$10^{-3} m^3/mol$；$\psi$为地层拟压力，$MPa^2/(mPa \cdot s)$；$\psi_m$为基质孔隙拟压力，$MPa^2/(mPa \cdot s)$；$V$为吸附气含量，$m^3/t$。

结合朗缪尔方程，将式(2)积分得到基质收缩量百分数为

$$\Delta\varepsilon = \rho_s \frac{RTV_m}{EV_b} \ln\left(\frac{\psi_L + \psi_m}{\psi_L + \psi_i}\right) \tag{3}$$

式中，V_m为饱和吸附气含量，m^3/t；ψ_L为朗缪尔拟压力，$MPa^2/(mPa \cdot s)$。

随着储层压力的降低，吸附气体开始解吸导致基质收缩，同时裂隙内的有效应力增加，岩体也产生膨胀变形，则总变形量为

$$\Delta\varepsilon = \frac{\rho_s RTV_m}{EV_b} \ln \frac{\psi_L + \psi_m}{\psi_L + \psi_0} + \frac{\mu Z}{P_i} C_g (\psi_i - \psi_m) \tag{4}$$

式中，μ为气体黏度，cP①；Z为气体偏差因子，无量纲；C_g为气体压缩系数，MPa^{-1}；P_i为原始地层拟压力，MPa。

对于页岩气开发过程中，气体解吸基质内部收缩孔隙通道变大，得出基质孔隙度和储层形变间的关系：

$$\frac{\phi_m}{\phi_{mi}} = 1 + \left(1 + \frac{2}{\phi_{mi}}\right) \Delta\varepsilon \tag{5}$$

式中，ϕ_m为基质孔隙度；ϕ_{mi}为基质初始孔隙度。

进一步根据理想毛管束模型，得到基质直径为

$$D = D_i \sqrt{1 + \left(1 + \frac{2}{\phi_{mi}}\right) \left[\frac{\rho_s RTV_m}{EV_b} \ln \frac{\psi_L + \psi_m}{\psi_L + \psi_0} + \frac{\mu Z}{P_i} C_g (\psi_i - \psi_m)\right]} \tag{6}$$

---

① 厘泊(cP)，$1cP = 10^{-3} Pa \cdot s$。

1.3 纳米孔表观渗透率模型

根据 Javadpour 等[11]相关研究结果,结合气体通量守恒原理,纳米级基质孔隙表观渗透率表达式为

$$K_{\mathrm{m}}=(1-\alpha)K_{\infty}+\alpha D_{\mathrm{k}}\mu c_{\mathrm{g}} \tag{7}$$

式中,K_{∞}为基质固有渗透率;α 为气体分子自由程大于基质孔隙直径(D)的分子所占总的分子量的比例;D_{k}为克努森扩散系数,其三者表达式分别可以表示为

$$\alpha=\mathrm{e}^{-D/\lambda},\lambda=\frac{K_{\mathrm{B}}T}{\sqrt{2}\,\pi\delta^{2}P}\times10^{-6}$$

$$D_{\mathrm{k}}=\frac{D}{3}\sqrt{\frac{8RT}{\pi M}} \tag{8}$$

$$K_{\infty}=\frac{D^{2}}{8}$$

式中,λ 为分子自由程,m;K_{B}为玻尔兹曼常数,1.38×10^{-23} J/K;δ 为分子碰撞直径,m;P 为地层压力,MPa;M 为分子量,g/mol。

将式(8)代入式(7)得到,基质孔隙中的气体表观渗透率为

$$k_{\mathrm{m}}=\left[(1-\mathrm{e}^{-D/\lambda})\frac{D^{2}}{32}+\mathrm{e}^{-D/\lambda}\frac{D_{\mathrm{K}}\mu C_{\mathrm{g}}}{1000}\right]\times10^{12} \tag{9}$$

1.4 渗流模型

考虑解吸的页岩气从基质到天然裂缝属于克努森扩散的拟稳态窜流,天然裂缝到人工裂缝具有不稳定线性流特征,依据质量守恒原理可以得到基质、天然裂缝和人工裂缝中流动方程为

人工裂缝:
$$\frac{\partial^{2}\psi_{\mathrm{F}}}{\partial y^{2}}+\frac{2k_{\mathrm{fi}}}{k_{\mathrm{F}}w_{\mathrm{F}}}\mathrm{e}^{-\beta(\psi_{\mathrm{i}}-\psi_{\mathrm{f}})}\left.\frac{\partial\psi_{\mathrm{f}}}{\partial x}\right|_{x=w_{\mathrm{F}}/2}=\frac{\phi_{\mathrm{F}}\mu_{\mathrm{g}}c_{\mathrm{tF}}}{3.6k_{\mathrm{F}}}\frac{\partial\psi_{\mathrm{F}}}{\partial t} \tag{10}$$

天然裂缝:
$$\frac{\partial^{2}\psi_{\mathrm{f}}}{\partial x^{2}}+\beta\left(\frac{\partial\psi_{\mathrm{f}}}{\partial x}\right)^{2}+\alpha\frac{k_{\mathrm{m}}}{k_{\mathrm{fi}}}\mathrm{e}^{\beta(\psi_{\mathrm{i}}-\psi_{\mathrm{f}})}(\psi_{\mathrm{m}}-\psi_{\mathrm{f}})=\frac{\phi_{\mathrm{f}}\mu_{\mathrm{gi}}c_{\mathrm{tf}}}{k_{\mathrm{fi}}}\mathrm{e}^{\beta(\psi_{\mathrm{i}}-\psi_{\mathrm{f}})}\frac{\partial\psi_{\mathrm{f}}}{3.6\partial t} \tag{11}$$

基质:
$$-\alpha k_{\mathrm{m}}(\psi_{\mathrm{m}}-\psi_{\mathrm{f}})=\frac{\mathrm{d}\phi_{\mathrm{m}}}{\mathrm{d}\psi_{\mathrm{m}}}\frac{P_{\mathrm{i}}}{Z}\frac{\partial\psi_{\mathrm{m}}}{3.6\partial t}+\phi_{\mathrm{m}}C_{\mathrm{g}}\mu\frac{\partial\psi_{\mathrm{m}}}{3.6\partial t}+\rho_{\mathrm{s}}\frac{P_{\mathrm{sc}}T}{T_{\mathrm{sc}}}\frac{V_{\mathrm{L}}\psi_{\mathrm{L}}}{(\psi_{\mathrm{L}}+\psi_{\mathrm{i}})^{2}}\frac{\partial\psi_{\mathrm{m}}}{3.6\partial t} \tag{12}$$

式中,C_{tj}(j=F,f,m)分别为人工裂缝、天然裂缝和基质系统的综合压缩系数,MPa^{-1};k_{F},k_{fi},k_{m}分别为人工裂缝渗透率、天然裂缝渗透率和基质表观渗透率,D;t 为时间,h;P_{sc}为地面标准状况下的压力,MPa;T_{sc}为地面标准状况下的温度,K;α 为基质岩块形状因子,m^{-2};w_{F}为人工裂缝宽度,m。

考虑基质收缩变形效应和吸附,引入新的基质压缩系数:

$$\begin{aligned}C_{\mathrm{tm}}=&\frac{P_{\mathrm{i}}}{\mu Z}\left(1+\frac{2}{\phi_{\mathrm{mi}}}\right)\left(\frac{\rho_{s}RTV_{\mathrm{m}}}{EV_{0}}\frac{1}{\psi_{\mathrm{L}}+\psi_{\mathrm{m}}}-\frac{\mu Z}{P_{\mathrm{i}}}C_{\mathrm{p}}\right)+2\rho_{\mathrm{s}}\frac{P_{\mathrm{sc}}T}{\mu\phi_{\mathrm{mi}}T_{\mathrm{sc}}}\frac{V_{\mathrm{m}}\psi_{\mathrm{L}}}{(\psi_{\mathrm{L}}+\psi_{\mathrm{i}})^{2}}\\&+C_{\mathrm{g}}\left\{1+\left(1+\frac{2}{\phi_{\mathrm{mi}}}\right)\left[\frac{\rho_{\mathrm{s}}RTV_{\mathrm{m}}}{EV_{0}}\ln\frac{\psi_{\mathrm{L}}+\psi_{\mathrm{m}}}{\psi_{\mathrm{L}}+\psi_{0}}+\frac{\mu_{Z}}{P_{\mathrm{i}}}C_{\mathrm{p}}(\psi_{\mathrm{i}}-\psi_{\mathrm{m}})\right]\right\}\end{aligned} \tag{13}$$

进一步,通过无因次化控制方程可得

人工裂缝为

$$\frac{\partial^2\psi_{FD}}{\partial y_D^2}+\frac{2}{F_{CD}}e^{-\beta_D\psi_{fD}}\frac{\partial\psi_{fD}}{\partial x_D}\bigg|_{x_D=w_{FD}}=\frac{1}{\eta_{FD}}\frac{\partial\psi_{FD}}{\partial t_D} \tag{14}$$

初始条件：
$$\psi_{FD}(y_D,0)=0 \tag{15}$$

内边界条件：
$$\psi_{FD}(0,t_D)=1 \tag{16}$$

外边界条件：
$$\frac{\partial\psi_{FD}}{\partial y_D}\bigg|_{y_D=1}=0 \tag{17}$$

天然裂缝为

$$\frac{\partial^2\psi_{fD}}{\partial x_D^2}-\beta_D\left(\frac{\partial\psi_{fD}}{\partial x_D}\right)^2+\delta e^{\beta_D\psi_{Df}}(\psi_{mD}-\psi_{fD})=\frac{e^{\beta_D\psi_{fD}}}{\eta_{fD}}\frac{\partial\psi_{fD}}{\partial t_D} \tag{18}$$

初始条件：
$$\psi_{fD}(x_D,0)=0 \tag{19}$$

内边界条件：
$$\psi_{fD}\left(\frac{w_{FD}}{2},t_D\right)=\psi_{FD}\left(\frac{w_{FD}}{2},t_D\right) \tag{20}$$

外边界条件：
$$\frac{\partial\psi_{fD}}{\partial x_D}\bigg|_{x_D=x_{eD}}=0 \tag{21}$$

基质为

$$-\delta(\psi_{mD}-\psi_{fD})=\frac{\gamma}{\eta_{mD}}\frac{\partial\psi_{mD}}{\partial t_D} \tag{22}$$

初始条件：
$$\psi_{mD}(t_D=0)=0 \tag{23}$$

式中,定义的无因次量为

无因次拟压力：
$$\psi_{jD}=\frac{\psi_i-\psi_j}{\psi_i-\psi_{wf}},j=F,f,m \tag{24}$$

无因次时间：
$$t_D=\frac{3.6k_{fi}t}{\phi_f\mu_{gt}c_{tf}y_F^2} \tag{25}$$

无因次应力敏感系数：
$$\beta_D=\beta(\psi_i-\psi_{wf}) \tag{26}$$

窜流系数：
$$\delta=\alpha\frac{k_m}{k_{fi}}y_F^2 \tag{27}$$

无因次导压系数：
$$\eta_{FD}=\frac{k_F\phi_f C_{tf}}{\phi_F C_{tF}k_{fi}},\eta_{fD}=1,\eta_{mD}=\frac{k_m\phi_f C_{tf}}{\phi_{mi}C_{tm}k_{fi}} \tag{28}$$

无因次裂缝导流能力：
$$F_{CD}=\frac{k_F w_F}{y_F k_{fi}} \tag{29}$$

无因次窜流系数：
$$\delta=\alpha\frac{k_m}{k_{fi}}y_F^2 \tag{30}$$

无因次距离：
$$y_D=\frac{y}{y_F},x_{eD}=\frac{x_e}{y_F},w_{FD}=\frac{w_F}{y_F} \tag{31}$$

渗透率极差：
$$\gamma=\frac{k_m}{k_{fi}} \tag{32}$$

根据达西定律得到单条裂缝产量与无因次拟力的关系：

$$q_{sc}=\frac{4.98\times10^{-3}k_F w_F h(\psi_i-\psi_{wf})T_{sc}}{P_{sc}Ty_F}\frac{\partial\psi_{DF}}{\partial y_D}\bigg|_{y_D=0} \tag{33}$$

式中，y_F为人工裂缝半长，m；x_e为裂缝半间距，m。

2 模型求解

式(14)、式(18)、式(22)是关于拟压力的强非线性偏微分方程，难以求出其解析解，因此采用数值解法，利用全隐式有差分法对其离散求解。将式(14)～式(22)运用全隐式有限差分法离散后的方程为

人工裂缝为

$$\frac{\psi_{DF(j+1)}^{n+1}-2\psi_{DF(j)}^{n+1}+\psi_{DF(j-1)}^{n+1}}{\Delta y_D^2}+\frac{2}{F_{CD}}e^{-\beta_D\psi_{fD(j)}^{n+1}}\frac{\psi_{Df(i+1)}^{n+1}-\psi_{Df(i-1)}^{n+1}}{2\Delta x_D}=\frac{1}{\eta_{FD}}\frac{\psi_{DF(j)}^{n+1}-\psi_{DF(j)}^{n}}{\Delta t_D} \tag{34}$$

天然裂缝为

$$\frac{\psi_{Df(i+1,j)}^{n+1}-2\psi_{Df(i,j)}^{n+1}+\psi_{Df(i-1,j)}^{n+1}}{\Delta x^2}-\beta_D\left(\frac{\psi_{Df(i+1,j)}^{n+1}-\psi_{Df(i-1,j)}^{n+1}}{2\Delta x}\right)^2+\delta e^{\beta_D\psi_{Df(i,j)}^{n+1}}(\psi_{Dm(i,j)}^{n+1}-\psi_{Df(i,j)}^{n+1})$$
$$=\frac{e^{\beta_D\psi_{Df(i,j)}^{n+1}}(\psi_{Df(i,j)}^{n+1}-\psi_{Df(i,j)}^{n})}{\eta_{fD}\Delta t_D} \tag{35}$$

基质为

$$-\delta(\psi_{Dm(i,j)}^{n+1}-\psi_{Df(i,j)}^{n+1})=\frac{\gamma_{(i,j)}^{n+1}}{\eta_{mD(i,j)}^{n+1}}\frac{\psi_{Dm(i,j)}^{n+1}-\psi_{Dm(i,j)}^{n}}{\Delta t_D} \tag{36}$$

边始条件：

$$\psi_{DF(j)}^0=0,\psi_{Df(i,j)}^0=0,\psi_{Dm(i,j)}^0=0,$$
$$\psi_{DF(1)}^n=1,\psi_{DF(k)}^n-\psi_{DF(k-1)}^n=0,$$
$$\psi_{Df(kk)}^n-\psi_{Df(kk-1)}^n=0$$

对于式(34)～式(36)，利用牛顿迭代法进行数值求解。

3 影响因素分析

通过有限差分和牛顿迭代法获得无因次产量 q_D随无因次时间 t_D的变化关系，并通过无因次关系，做出各种岩石变形条件下的产量变化规律。因素分析中选取页岩储层基本物性参数，计算不同岩石变形效应下的页岩气产能并分析其微观渗流特征。其中参数包含：$P_i=10MPa$，$P_{wf}=2MPa$，$T=350K$，$h=20m$，$y_F=100m$，$x_e=100m$，$w_F=0.002m$，$P_{ormi}=0.08$，$P_{orf}=0.005$，$P_{orF}=0.2$，$C_t=0.0002MPa^{-1}$，$K_F=0.5D$，$K_{fi}=10^{-5}D$，$P_L=5MPa$，$\rho_s=2.56t/m^3$。

图1反映的是不同裂缝变形程度下的单缝产能变化规律，裂缝应力敏感系数 β 分别取 0.0001mPa·s/MPa^2、0.0003mPa·s/MPa^2、0.0005mPa·s/MPa^2。应力敏感系数越大，生产中前期产量越大，后期裂缝渗透率降低到一定程度，应力敏感系数影响不大。

图2反映的是分别考虑和忽略基质解吸收缩效应下的单缝产能变化规律。由图可以明显看出，考虑基质解吸收缩效应，随生产基质压力降低，气体解吸导致基质变形收缩，孔隙度和渗透率变大，生产中期产量明显增大。

图3反映的是不同基质岩石模量下的单缝产能变化规律，基质杨氏模量 E 分别取 20GPa、30GPa、40GPa。杨氏模量越大，气体解吸导致基质变形收缩越大，基质随压力降低，

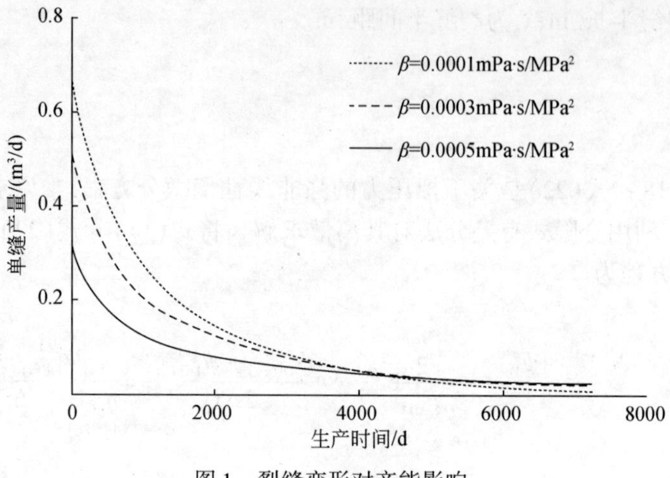

图 1　裂缝变形对产能影响

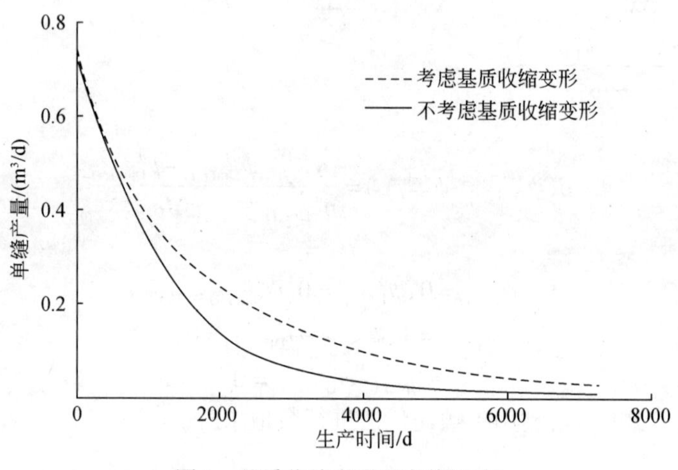

图 2　基质收缩变形对产能影响

孔隙度和渗透率变大,生产中期产量越大,后期页岩气产能主要受控于吸附气体供给,杨氏模量引起的基质变形对产能影响不大。

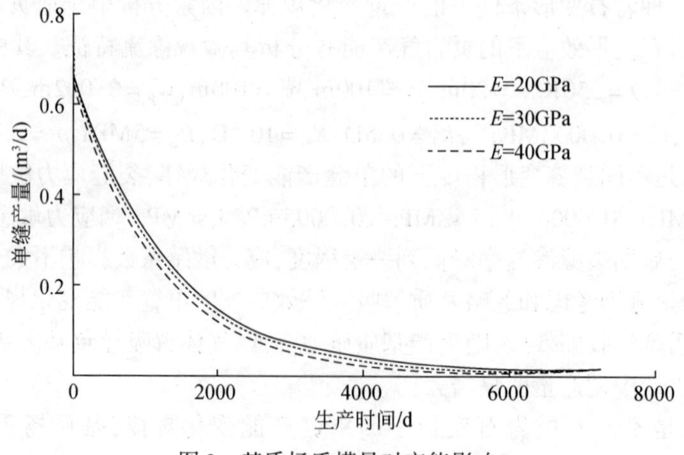

图 3　基质杨氏模量对产能影响

图4反映的是4种不同基质收缩与裂缝变形组合条件产能变化规律。明显可以看出，裂缝变形影响生产中前期而基质解吸收缩影响生产中后期。基质解吸收缩正相关于产能而裂缝变形负相关于气体产能。因此，实际生产过程中，在一定的基质变形参数条件下，考虑不同生产阶段，合理控制页岩气生产压差，协同考虑裂缝变形和基质解吸收缩耦合效应。

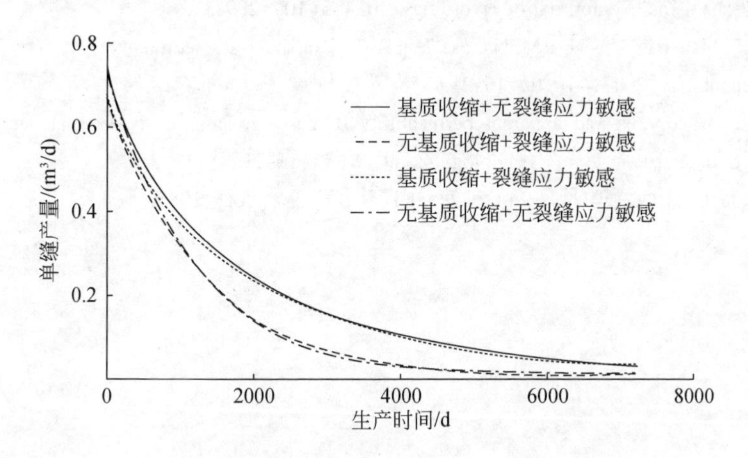

图4 不同基质收缩与裂缝变形组合对产能影响

4 结论

文中建立考虑纳米级基质孔隙克努森扩散流、裂缝应力敏感变形、基质解吸收缩效应协同作用的非线性渗流数学模型，应用全隐式有限差分和牛顿迭代法进行数值求解。对相关因素分析得到，裂缝变形影响生产中前期，而基质解吸收缩效应影响生产中后期，并且，基质解吸收缩正相关于产能而裂缝变形负相关于气体产能。因此，实际生产过程中，在一定的地层参数条件下，应当结合不同生产阶段，合理调整页岩气生产条件，协同考虑裂缝变形和基质解吸收缩耦合效应，最终优化页岩气生产制度，提高页岩气采收率。

参考文献

[1] 王世谦. 中国页岩气勘探评价若干问题评述[J]. 天然气工业，2013，33(12)：13-29.

[2] 姜呈馥，程玉群，范柏江，等. 陆相页岩气的地质研究进展及亟待解决的问题——以延长探区上三叠统延长组长7段页岩为例[J]. 天然气工业，2014，34(2)：27-33.

[3] 张小涛，吴建发，冯曦，等. 页岩气藏水平井分段压裂渗流特征数值模拟[J]. 天然气工业，2013，33(3)：47-52.

[4] 张士诚，牟松茹，崔勇. 页岩气压裂数值模型分析. 天然气工业，2011，31(12)：81-84.

[5] 程远方，李友志，时贤，等. 页岩气体积压裂缝网模型分析及应用. 天然气工业，2013，33(9)：53-59.

[6] Abdassah D, Ershaghi I. Triple-porosity system for representing naturally fractured reservoirs[J]. SPE Formation Evaluation, 1986, 4: 113-127, SPE-13409-PA.

[7] El-Banbi A H. Analysis of tight gas well performance [D]. Texas A&M University, 1998.

[8] Zhao Y L, Zhang L H, Zhao J Z, *et al*. Triple porosity modeling of transient well test and rate decline analysis for multi-fractured horizontal well in shale gas reservoirs[J]. Journal of Petroleum Science and Engineering, 2013, 110: 253-262.

[9] Shi J T, Zhang L, Yan S L, *et al*. Diffusion and flow mechanisms of shale gas through matrix pores and gas production forecasting[J]. Paper SPE 167226 presented at the SPE Unconventional Resource Conference held in Calgary, Albert, Canada, 5-7 November 2014.

[10] Wang H T. Performance of multiple fractured horizontal wells in shale gas reservoirs with consideration of multiple mechanisms[J]. Journal of Hydrology, 2014, 5(10): 299-312.

[11] Javadpour F, Fisher D, Unswort M. Nanoscale gas flow in shale gas sediments[J]. Journal of Canadian Petroleum Technology, 2007, 46(10): 16-21.

[12] Javadpour F. Nanopores and apparent permeability of gas flow in mudrocks (shale and siltstone) [J]. Journal of Canadian Petroleum Technology, 2009, 48(8): 16-21.

[13] 谈慕华,黄蕴元. 表面物理化学[M]. 北京:中国建筑工业出版社,1985.

页岩气分段压裂水平井渗流机理及试井分析

刘晓旭[1]　杨学锋[1]　陈远林[2]　吴建发[1]　冯　曦[1]

(1. 中国石油西南油气田公司勘探开发研究院;2. 中国石油西南油气田公司开发部)

摘要:近年来分段压裂水平井在国内的应用越来越广泛,但对其渗流机理及渗流特征的认识还不够明确,又鉴于页岩气储层的特殊性及页岩气井特殊的生产方式,国内外对页岩气分段压裂水平井试井分析的研究基本处于空白状态。本文首先研究了水平井分段压裂产生的裂缝形态,系统分析了分段压裂水平井在开发过程中的主要渗流特征及其在双对数曲线上的特征表现,考虑页岩储层的特殊性,给出了页岩气井生命期内通常表现出的渗流形态。结合中国第一口页岩气分段压裂水平井——W201-H1 井的压力恢复测试数据,分析该井在压力恢复双对数图上表现出的渗流特征,明确了该井在测试期表现出的径向流应属早期径向流,并验证了该井在未来 1 年内的生产均处于复合线性流阶段。通过前后三次压力恢复试井曲线的对比分析,指出了要展现页岩气分段压裂水平井完整的渗流流态及其演化过程,获取准确的储层与裂缝参数,需要较长的关井时间,确保压力计尽量靠近水平段位置,并选择合适的关井点来最大程度降低井储和井筒积液的影响。

关键词:页岩气藏;水平井;分段压裂;渗流特征;试井分析

美国页岩气开发已经取得了巨大的成功,水平井分段压裂技术的突破是推动页岩气革命的关键因素之一[1,2]。分段压裂水平井渗流机理既不同于压裂直井,也不同于传统的单裂缝压裂水平井,又鉴于页岩气储层的特殊性,页岩气分段压裂水平井渗流机理极其复杂[3,4]。目前国内外关于页岩气分段压裂水平井渗流机理的研究基本上是建立在理论模型的基础上开展的,未能结合页岩气井的生产实际情况[4-9]。国内页岩气井生产期较短[10],压力恢复试井是研究页岩气井渗流特征最主要的手段之一。中国石油西南油气田公司于 2012 年开展了中国第一口页岩气分段压裂水平井——W201-H1 井的压力恢复测试。本文从理论上分析了分段压裂水平井的渗流特征及流态演化过程,结合 W201-H1 井的试井数据,探讨页岩气分段压裂水平井的渗流机理及渗流特征。

1　裂缝形态

通常水平井分段压裂裂缝存在两种极限情况,即纵向和横向(图 1)。当水平井井筒方位与最大水平主应力方位垂直时,可产生横向裂缝,当两者方位一致时,可产生轴向裂缝(纵向裂缝),当两者夹角处于其他情况时,裂缝形态会非常复杂。由于水平井轴向裂缝有效泄流面积相对横向裂缝来说小很多,一般希望水平井压裂产生横向裂缝作为裂缝网络的主裂缝。页岩气藏最佳的压裂效果是产生垂直于水平井段的横向主裂缝的同时,还产生垂直于横向裂缝的二级裂缝,以提高裂缝与地层的接触面积,有利于页岩气的解吸和气体在储层中的流动。据统计,水平井分段压裂的裂缝以横向裂缝为主[3-5]。

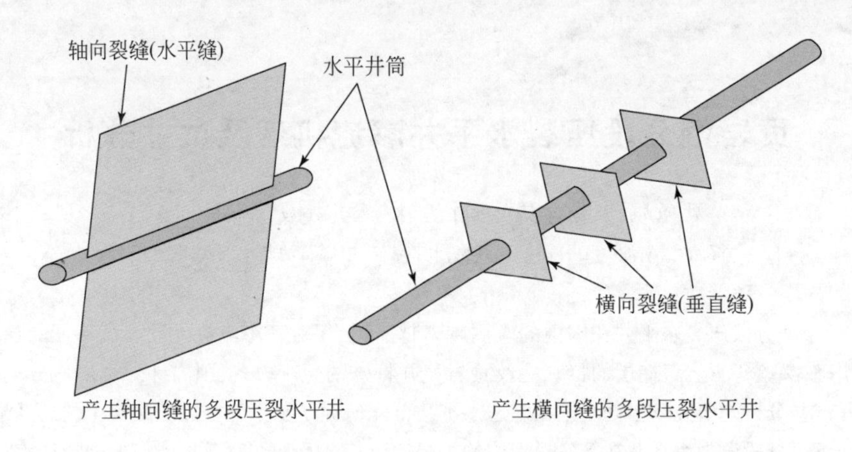

图1　水平井纵向裂缝和横向裂缝示意图

2　分段压裂水平井流态特征

由于页岩储层水平井分段压裂通常以横向缝(垂直缝)为主,下面以产生横向缝的分段压裂水平井生产为例,从理论上分析页岩储层分段压裂水平井的渗流特征及流态演化过程。

2.1　井储效应

井储效应出现在气井开井或关井的短时间内,其影响持续时间较短。对分段压裂水平井而言,井储效应又分为井筒存储效应和裂缝存储效应,根据井筒体积和裂缝体积的大小不同,两者在井储效应中发挥的作用不同。

2.2　早期线性流

分段压裂水平井早期第一个明显的流态是垂直于裂缝的线性流动,即早期线性流(图2)。如果裂缝导流能力与储层渗流能力相当,通常还会在早期线性流之前出现双线性流,即气体由裂缝流向井筒的线性流和由储层流向裂缝的线性流;如果裂缝导流能力远大于储层的渗流能力,则通常只会出现气体由储层流向裂缝的线性流[11]。早期线性流的持续时间取决于裂缝规模。

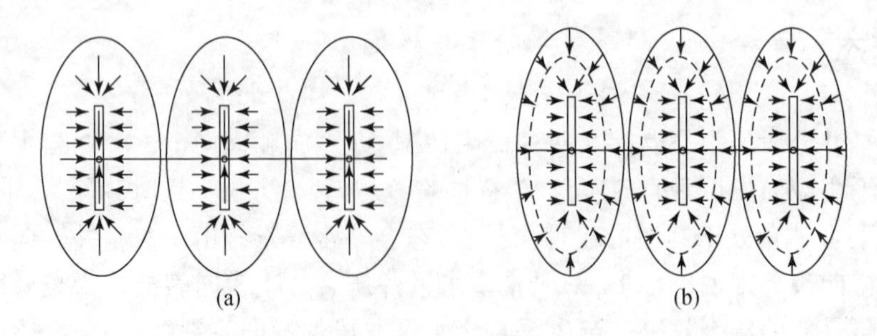

图2　早期线性流(a)与早期径向流(b)示意图

2.3 早期径向流

如果裂缝间距够大,早期线性流之后就会出现早期径向流。分段压裂水平井的早期径向流发生在单个裂缝内部,类似于未压裂水平井的系统径向流,而垂直缝就类似于未压裂的水平段。早期径向流的持续时间取决于裂缝间距和渗透率的大小,如若裂缝间距较小,渗流率较高,则该流态特征不明显(被其他流态掩盖)或者不会出现此种流态[12-14]。

2.4 裂缝边界流

随着气井生产的进行,压力波进一步向储层内部传递。在某一时间点,相邻裂缝间的压力扰动前缘汇合,这样就会产生裂缝干扰效应,此时对应的流动为裂缝边界流。

2.5 复合线性流

裂缝边界流之后,压力扰动将逐渐覆盖储层有效压裂体积(SRV)的整个范围,并进入复合线性流,其特征如同气体流入一条大的裂缝。该流态不是一个单纯的线性流,而是垂直于水平井筒的线性流为主导,水平井段两端的流线则呈椭圆形,两端的椭圆流的效应小于垂直水平井筒的线性流。

2.6 系统径向流和外边界效应

如果页岩气井生产时间够长,随着泄流面积的增大,整个水平井段和裂缝系统就如同一口影响范围扩大了的直井,在距离水平井段和裂缝系统较远的储层内就会出现系统径向流。对于页岩气井而言,由于渗透率较低,系统径向流出现的时间很晚,在页岩气井的整个生命期内一般不会出现系统径向流[15]。

在复合线性流后,压力扰动会进一步向周围储层传播。基于储层的几何形状和边界条件,可能会出现以下 3 种情况:封闭边界(拟稳态)、定压边界或无限大边界(图 3)。

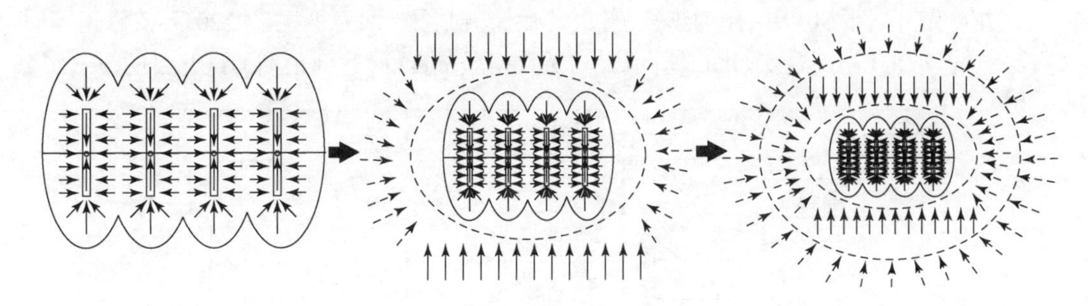

图 3 裂缝边界流→复合线性流→系统径向流示意图

对一特定的页岩储层来说,分段压裂水平井只能产出有效压裂体积(SRV)内小范围的气体,且由于储层渗透率极低[3,8,16],压力扰动传播速度非常慢,在水平井的生产过程中,除了井储效应外,一般会出现图 4(a)中的流态:早期线性流→早期径向流→裂缝边界流→复合线性流,而系统径向流在页岩气井的生命期内通常不会出现,而理论计算得出的外边界效应要在上百年之后才能看到。按照流态出现的顺序,双对数曲线上的压力导数曲线示意图如图 4(b)所示。

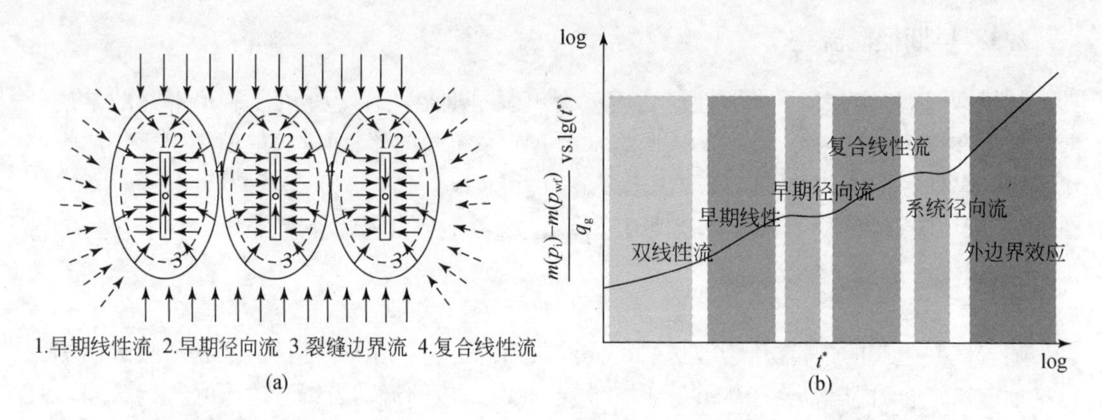

图4　分段压裂水平井示流态示意图(有限导流裂缝)(a)与其对应的压力导数曲线示意图(b)

3　压力恢复试井分析

国内页岩气井生产时间较短,不能从生产动态数据来完整地揭示页岩气藏分段压裂水平井的渗流特征,而压力恢复试井则能用相对较短的关井时间来展现其完整的流态演化,下面就从中国第一口页岩气水平井——W201-H1井的压力恢复试井开展分析。

3.1　气井基本情况及测试数据质量评价

W201-H1井于2011年7月完成水平段11级压裂施工,创造了当时国内页岩气水平井压裂段数最多、单井用液量最大、施工排量最大、连续施工时间最长等记录。

W201-H1井压裂液返排程度较低,存在较为严重的井筒积液。该井投产时控制井口压力生产,初始日产气$1.34\times10^4 m^3/d$,关井前气井日均产气$1.13\times10^4 m^3$。

该井首先开展了30天的压力恢复测试,此后又开展了“两开两关”共20天的修正等时试井,测试获取的压力、温度如图5所示。测试数据质量良好,达到预期的试井目的。

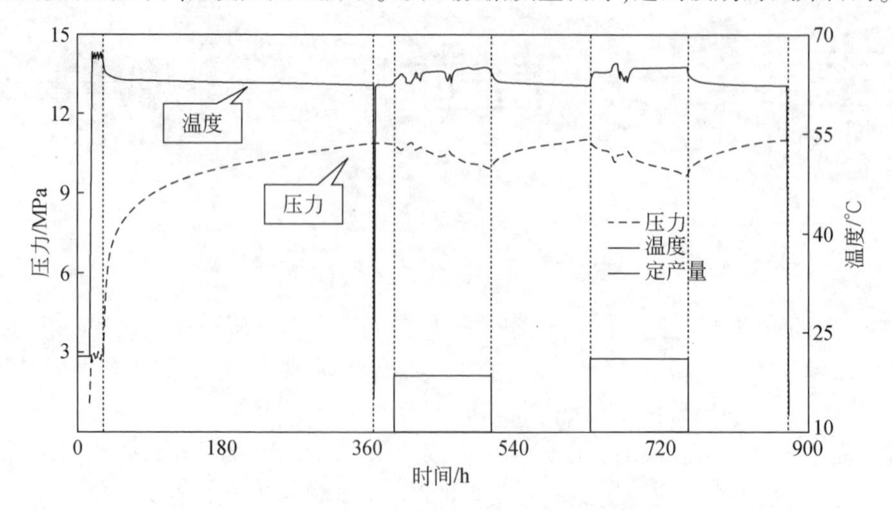

图5　W201-H1井压力-温度实测数据图

3.2 压力恢复试井曲线诊断与特征分析

第一次关井(关井时间30天)压力恢复双对数图如图6所示。

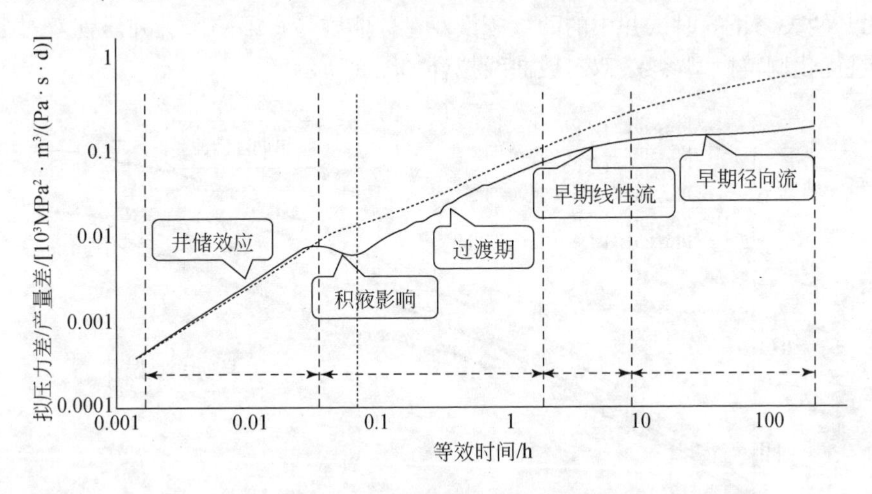

图6 第一次关井压力恢复双对数曲线图

W201-H1井水平井段较长,压裂规模较大,在双对数曲线图上表现为井储效应明显。结合测试数据(图5)可知,第一次关井起始点压力较低,跨度范围较大,并且井底积液严重,图6中的压力导数曲线出现了负斜率,这种现象应归因于在关井初期可能出现了井筒积液回流至裂缝和储层。此后出现了径向流阶段(压力导数曲线接近水平)。对于出现的线性流和径向流,笔者认为应该是早期线性流和早期径向流。

为了验证分析中的线性流和径向流分别是早期线性流和早期径向流,应用最新试井分析软件中的分段压裂水平井模块,通过试井解释参数开展一个1年预测期的试井设计(设计气井日产气$1.0\times10^4m^3$),以全面展现气井渗流特征,如图7所示。

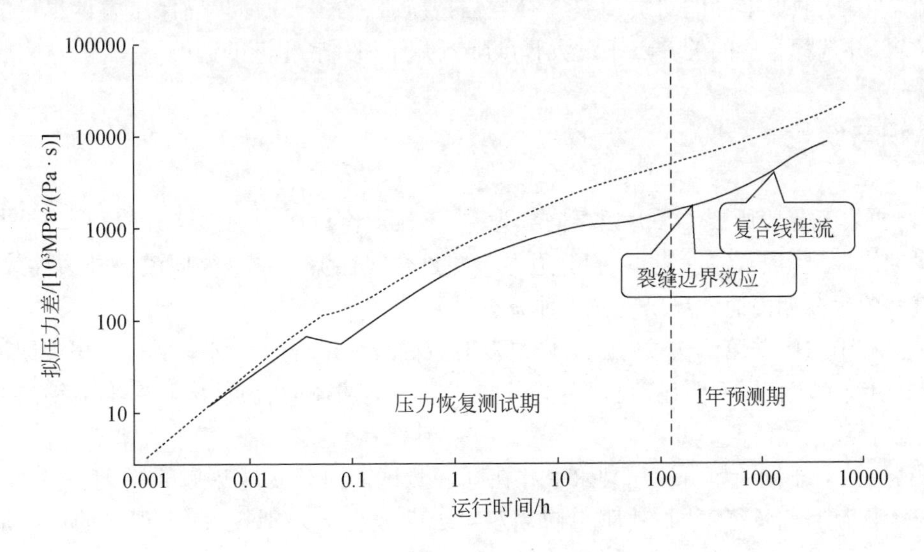

图7 1年期试井设计双对数图

从图 7 中的 1 年预测期可以看出,压力恢复测试期过后表现为裂缝边界流和复合线性流特征,气井生产 1 年仍然没有出现系统径向流和外边界效应,同时也验证了测试期出现的线性流和径向流分别为早期线性流和早期径向流。

将两开两关修正等时试井中的两次压恢双对数曲线与将第一次关井压恢双对数曲线相叠加,通过优化调整模型参数,双对数曲线拟合如图 8 所示。

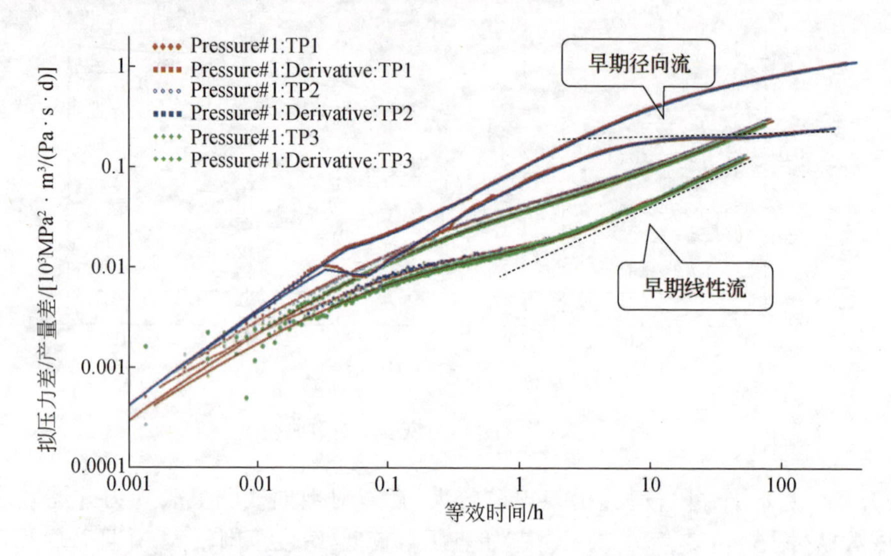

图 8　三次关井压力恢复试井双对数曲线数拟合叠加分析图

从图 8 中可以看出,后两次关井压力恢复双对数图明显不同于第一次关井压力恢复双对数图,这归因于后两次关井压力起始点较高,压力跨度范围较小,井储效应和井底积液的影响大大减小,压力导数曲线没有出现负斜率。后两次关井由于关井时间较短,仅有 5 天,主要表现出早期线性流特征,没有出现径向流。因此,要展现页岩气分段压裂水平井比较完整的流态特征,获取准确的储层和裂缝参数,需要较长的关井时间,确保压力计尽量靠近水平段位置,并选择合适的关井点来最大程度降低井储和井底积液的影响。

4　认识与结论

(1)页岩气分段压裂水平井以横向缝为主,渗流形态依次表现为:(井筒与裂缝)井储效应→早期线性流(或双线性流)→裂缝边界流→复合线性流→系统径向流→外边界效应,在页岩气井的生命期内通常只出现前 4 种流态。

(2)W201-H1 井第一次关井压力恢复双对数曲线中出现的负斜率是井底积液回流所造成的,早期线性流特征不明显,表现出的径向流应为早期径向流,且该井未来 1 年内的生产均处于复合线性流阶段。

(3)即使对于同一口分段压裂水平井在同一测试期内开展试井测试,由于关井点不同,关井压力恢复时间长短不同,其压力恢复双对数曲线也会差别较大,表现出的渗流形态也不同。

(4)要展现页岩气分段压裂水平井比较完整的流态特征,获取准确的储层和裂缝参数,需要较长的关井时间,确保压力计尽量靠近水平段位置,并选择合适的关井点来最大程度降低井储和井底积液的影响。

参考文献

[1] Curtis J B. Fractured shale-gas systems[J]. AAPG,2002,86(11):1921-1938.

[2] Gale J F W,Reed R M,Holder J. Natural fractures in the Barnett Shale and their importance for hydraulic fracture treatment [J]. American Association of Petroleum Geologists Bulletin,2007,91(4):603-622.

[3] 张小涛,吴建发,冯曦,等. 页岩气藏水平井分段压裂渗流特征数值模拟[J]. 天然气工业,2013,33(3):47-52.

[4] 李道伦,徐春元,卢德唐,等. 多段压裂水平井的网格划分方法及其页岩气流动特征研究[J]. 油气井测试,2013,22(1):13-16.

[5] 任俊杰,郭平,王德龙,等. 页岩气藏压裂水平井产能模型及影响因素[J]. 东北石油大学学报,2012,36(6):76-81.

[6] 张士诚,牟松茹,崔勇. 页岩气压裂数值模型分析[J]. 天然气工业,2011,31(12):81-84.

[7] 段永刚,李建秋. 页岩气无限导流压裂井压力动态分析[J]. 天然气工业,2010,30(10):26-29.

[8] 李建秋,曹建红,段永刚,等. 页岩气井渗流机理及产能递减分析[J]. 天然气勘探与开发,2011,2(34):34-37.

[9] 程远方,董丙响,时贤,等. 页岩气藏三孔双渗模型的渗流机理[J]. 天然气工业,2012,9(32):44-47.

[10] 李欣,段胜楷,侯大力,等. 多级压裂页岩气水平井的不稳定生产数据分析[J]. 天然气工业,2012,32(12):44-48.

[11] Cipolla C L, Lolon E P, Erdle J C. Modeling well performance in shale-gas reservoirs[C]. Paper SPE 125532 presented at the SPE/EAGE Reservoir Characterization and Simulation Conference, Abu Dhabi, UAE, 19-21 October 2009.

[12] Medeiros F, OzkanE. Productivity and drainage area of fractured horizontal wells in tight gas reservoirs [C]. Paper SPE 108110 presented at the Rocky Mountain Oil & Gas Technology Symposium, Denver, Colorado, USA, 16-18 April 2008.

[13] Baihly J, Altman R, Malpani R, *et al*. Shale gas production decline trend comparison over time and basins [C]. Paper SPE 135555 presented at the SPE Annual Technical Conference and Exhibition, Florence, Italy, 19-22 September 2010.

[14] Ahmadi H A, Almarzooq A, Wattenbarger R. Application of linear flow analysis to shale gas wells—field cases[C]. Paper SPE 130370 presented at the SPE Unconventional Gas Conference, Pittsbergh, Pennsylvania, USA, 23-25 February 2010.

[15] Bell R O, Wattenbarger R A. Rate transient analysis in naturally fractured shale gas reservoirs[C]. Paper SPE 114591 presented at the CIPC/SPE Gas Technology Symposium 2008 Joint Conference, Calgary, Alberta, Canada, 16-19 June 2008.

[16] 李治平,李智峰. 页岩气纳米级孔隙渗流动态特征[J]. 天然气工业,2012,32(4):50-53.

A Semi-analytical Solution for Multiple-trilinear-flow Model with Asymmetry Configuration in Multifractured Horizontal Well

Wang Junlei Jia Ailin Wei Yunsheng

(Research Institute of Petroleum Exploration and Development, PetroChina)

Abstract: Trilinear-flow model presented by Brown in 2009 are commonly used to simulate the pressure-transient and production behaviors of multiple fractured horizontal wells (MFHW) in unconventional reservoirs. Brown's model was established on the symmetry hypothesis of MFHW, i. e., single-fracture-based configuration. The principle focus of this work is on proposing a novel semi-analytical solution to fill the gap between trilinear-flow model on the symmetry hypothesis of MFHW and actual asymmetry multi-fractured configuration. The new solution is generated by incorporating the principle of pressure superposition, which could take the effect of fracture-production interaction and property heterogeneity of fractures into account. The results obtained from new solution are also validated by comparing with numerical simulation in two specific cases: identical properties of fractures in the unequally-spaced configuration and different properties of fractures in the equally-spaced configuration. Meanwhile, this paper provides approximated relationships to associate existing single-fracture-based solution with our multi-fractures-based solution. A synthetic sample is also presented to illustrate the improvement of new solution towards pressure transient analysis. The proposed model could provide a practical alternative to the solutions for MFHW model based on Green's function method, and also extend existing multilinear-flow (e. g., enhanced five-region model) and multi-porosity model (e. g., dual/triple porosity and adsorption effect) to multiple-fractured configuration.

Key words: trilinear-flow model; multiple fractures; pressure superposition; fracture interaction; shunt line; approximated solution

1 Introduction

Over the last two decades, rate transient analysis (RTA) or pressure transient analysis (PTA) of unconventional fractured reservoirs has become invaluable tool in determining the productivity and predicting the future production rate[1,2]. Compared with Green's function method and source/sink method[3-9], it is advantage for linear-flow method to account for the finite conductivity of hydraulic fracture without discretizing hydraulic fracture into numerous segments. Therefore, linear-flow model is an important alternative for simulating production behavior in fractured reservoirs. Based on the significant contributions of Cinco-Ley and Samaniego[10], Lee and Brockenbrough[11], Wattenbarger et al.[12] firstly studied the pressure/rate characteristic of linear fracture system to analyze long-term production performance of tight gas wells. This formed the basis of the subsequent works, namely linear dual-porosity model presented by Bello et al. [13] and trilinear flow model presented by Brown et al. [14]. In their work, the whole drainage area is decomposed into three contiguous flow regions: outer reservoir (neglected in the work of Bello et al.), inner reservoir and hydraulic fractures. Their models are simple but versatile enough to study the pressure response characteristics of fractured horizontal well within stimulated reservoir volume (SRV).

Recently, various modified multi-linear flow models were presented to describe stimulated region of limit extent. Al-Ahmadi et al. [15] presented a triple porosity model to account for extra linear flow depletion from matrix to macrofracture network in inner complex reservoirs. To relax the assumption of sequential-depletion from matrix to MF network and from MF network to hydraulic fractures, Obinna and Hassan[16] further proposed a quasilinear flow model to allow simultaneous matrix-MF network and matrix-HF depletion. To simulate the flow mechanism in shale matrix, Tian et al. [17] established the mechanism model of dual diffusion incorporating Langmuir adsorption Equation and Fick's law to account for the desorption and diffusion in matrix. As for the effect of hydraulic-fracture configuration, Ambrose et al. [18] developed a hybrid model for heterogeneous completion (unequal hydraulic fracture length) without relying on the assumption of a homogeneous completion, where the hydraulic fractures are evenly spaced and are all of the same length.

In addition, to simulate the characteristics of more-complex fracture network within SRV, Stalgorova and Mattar[19] developed enhanced-fracture-region model with a higher permeability region near each hydraulic fracture and unstimulated region in the bulk of the space between fractures. Considering the region beyond the fracture tips, Stalgorova and Mattar[20] further represented a more advanced five-region model that encompasses both the existing trilinear-flow model and previous enhanced-fracture-region model. An induced permeability field within SRV was presented by Fuentes-Cruz et al. [21] to improve the continuous assumption of linear-composite permeability model. This model was subsequently extended to the dual porosity idealization considering transient fluid transfer from matrix to natural fractures[22].

It is noted that most of multi-linear flow models are established on the assumption of multiple-fracture with symmetry hypothesis. Anderson et al. [23] concluded that multiple fracture system could be equivalent to a series of single-fracture models in closed system with different geometries as seen in Figure 1, which was generated by production-interaction effect between adjacent fractures. Generally, it is convenient to account for fracture-production interaction by incorporating the superposition principle[24,25]. To be my best known, Meyer et al. [26] was one of the few researchers to provide first-order approximations for explaining the relationship between multi-fracture solution and single-fracture solution by using the trilinear model and reservoir/fracture domain resistivity model of Meyer and Jacot[27]. Besides, Deng et al. [28] presented an enhanced five-region model considering different respective properties (fracture conductivity, permeability, etc.). However, the drainage area controlled by individual fracture is regarded as a constant which is the half-value area between adjacent fractures.

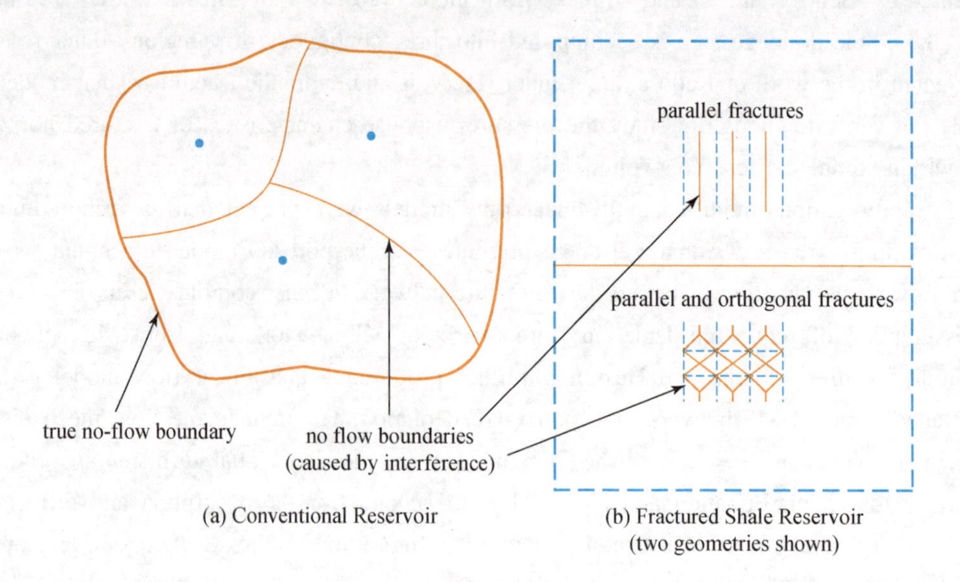

Figure 1　No-flow effect between adjacent wells/fractures in conventional reservoir and fractured reservoir[2]

As mentioned above, few studies are focused on the combination of linear-flow method based on symmetry hypothesis with the exact superposition principle to establish a rigorous multiple-fracture model. In this paper, we proposed a semi-analytical solution for multiple-trilinear-flow model, which was derived in Laplace-transform domain and numerically inverted to the real time domain using the algorithm proposed by Stehfest[29]. The effect of unequally-spaced configuration and the heterogeneity of fracture property on fracture-production interaction is discussed in detail. In addition, more rigorous transformed approximations from multiple-fracture solution into single-fracture solution are provided under different operating conditions.

2　Mathematical Model

2.1　Model description

Based on the research of Brown et al. [14] in Figure 2, a systematic model consisting of multipletrilinear-flow elements is built. In this paper, the reservoir volume contributing hydrocarbons to production is divided into three regions as shown in Figure 3: hydraulic fracture volume (HFV), stimulated reservoir volume (SRV), and external reservoir volume (XRV).

Generally, HFV consists of a set of discrete finite-conductivity hydraulic fractures along horizontal wellbore, SRV with higher permeability is created in the stimulated vicinity of HFV, and XRV indicates the unstimulated region beyond the tips of HFV. The flow process is successional and consequent: fluid flows from XRV to SRV, from SRV to HFV, and finally along HFV to wellbore.

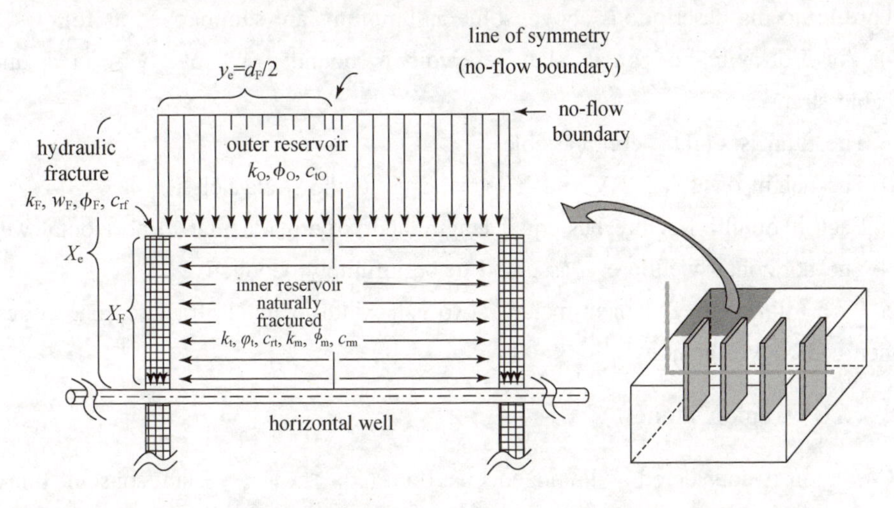

Figure 2 Trilinear-flow model representing three contiguous regions for MFHW[14]

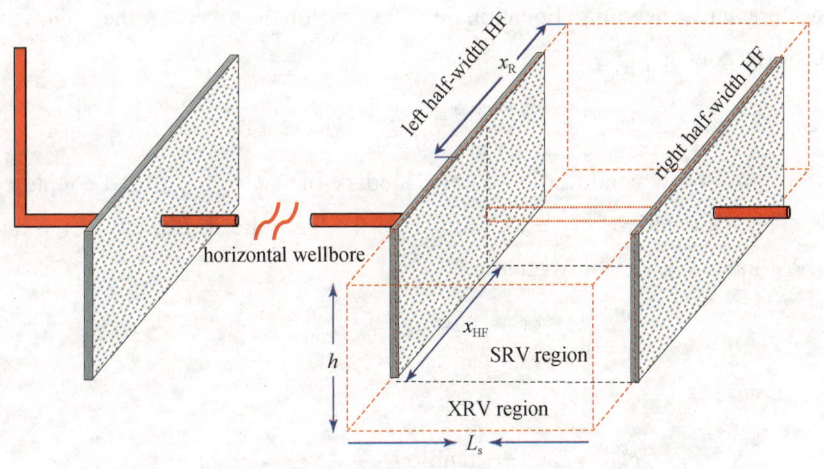

(a) Horizontal well with n_f hydraulic fractures

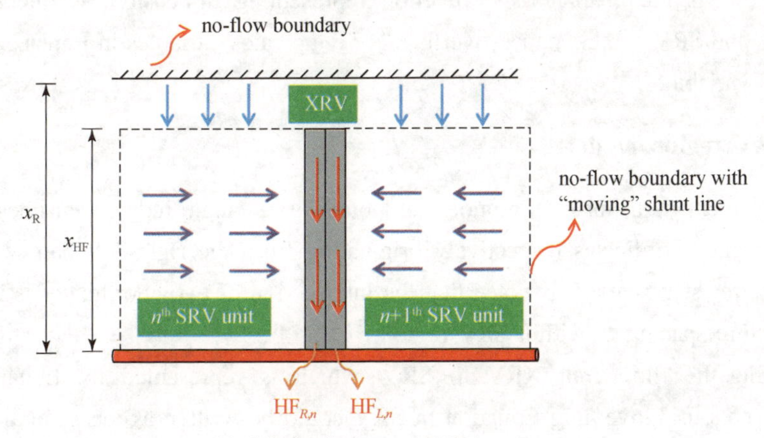

(b) n^{th} hydraulic fracture (HF) element

Figure 3 Schematic of multiple HF-units along a horizontal well

According to the descriptions above, some assumptions are summarized as follows:

(1) An isotropic, horizontal, slap reservoir is bounded by overlying and underlying impermeable strata.

(2) The fluid is slightly compressible.

(3) The height of HFV, SRV and XRV is same as the strata height.

(4) Each hydraulic fracture has equal length, and is perpendicular to horizontal wellbore.

(5) The horizontal wellbore is assumed to be infinitely conductive.

(6) Each trilinear-flow element includes two half-width hydraulic fractures, one corresponding SRV region and XRV region.

2.2 XRV region model

XRV region is considered as homogeneous, finite($x_{HF} \leqslant x \leqslant x_R$), slab, constant thickness h. Flow in XRV can be described in the 1D orthogonal coordinate originating from the tip of hydraulic fracture (seen in Figure 3). After using dimensionless definitions in Appendix A, the dimensionless pressure governing Equation in XRV region is given as the following forms in Laplace-transform domain,

$$\frac{\partial^2 \tilde{p}_{XRVD}}{\partial x_D^2} = \frac{s}{\eta_{XRVD}} = \tilde{p}_{XRVD} \tag{1}$$

According to no-flow condition on outer boundary of XRV($x=x_R$) and couple condition on interface between XRV and SRV($x=x_{HF}$) similar to the literature of Brown[14], the dimensionless flux from XRV into SRV can be written as

$$\left(\frac{\partial \tilde{p}_{XRVD}}{\partial x_D}\right)\bigg|_{x_D=x_{HFD}} = -\tilde{F}_{SRV}^{XRV}(\tilde{p}_{XRVD})\,|_{x_D=x_{HFD}} \tag{2}$$

Where

$$F_{SRV}^{XRV} = \sqrt{\frac{s}{\eta_{XRVD}}}\tanh\left[\sqrt{\frac{s}{\eta_{XRVD}}}(x_{RD}-x_{HFD})\right] \tag{3}$$

Equation (3) is the characteristic function representing dimensionless interchanging flux between XRV and SRV. In this paper, overline " ~ " indicates variables in Laplace domain and s is the Laplace variable.

2.3 SRV region model

For the elementary trilinear-flow model as noted by rectangle red broken lines in Figure 3, two 1D orthogonal coordinates respectively originating from the right/left half-width hydraulic fracture plane are established. It is noted that the intersection of two coordinates is located on the middle of fracture spacing L_s within SRV region.

Considering the flux from XRV to SRV, which is represented by Equation (1), the dimensionless pressure governing Equation in SRV could be written as the general form,

$$\frac{\partial^2 \tilde{p}_{SRVD}}{\partial y_D^2} = \left(\frac{s}{\eta_{SRVD}} + \lambda_{SRV}^{XRV} F_{SRV}^{XRV}\right)\tilde{p}_{SRVD} \tag{4}$$

Noted that the flux and pressure are continuous on the interface between right/left systems, which are satisfied as

$$(\tilde{p}^{\mathrm{L}}_{\mathrm{SRVD}})\big|_{y_{\mathrm{D}}=L_{\mathrm{sD}}/2}=(\tilde{p}^{\mathrm{R}}_{\mathrm{SRVD}})\big|_{y_{\mathrm{D}}=L_{\mathrm{sD}}/2} \tag{5a}$$

$$\left(\frac{\partial\tilde{p}^{\mathrm{L}}_{\mathrm{SRVD}}}{\partial y_{\mathrm{D}}}\right)\Bigg|_{y_{\mathrm{D}}=L_{\mathrm{sD}}/2}+\left(\frac{\partial\tilde{p}^{\mathrm{R}}_{\mathrm{SRVD}}}{\partial y_{\mathrm{D}}}\right)\Bigg|_{y_{\mathrm{D}}=L_{\mathrm{sD}}/2}=0 \tag{5b}$$

Where, variables mentioned in Equation (5) are distinguished by superscript of R and L to represent different coordinates. L_{s} is the distance between two half-width hydraulic fractures.

Combining Equation (4) and Equation (5) with couple condition between SRV and hydraulic fracture in Brown et al. [14], dimensionless flux rate from SRV into right/left half-width hydraulic fracture is obtained, which is respectively given as

$$\left(\frac{\partial\tilde{p}^{\mathrm{L}}_{\mathrm{SRVD}}}{\partial y_{\mathrm{D}}}\right)\Bigg|_{y_{\mathrm{D}}=w^{\mathrm{L}}_{\mathrm{HFD}}/2}=F^{\mathrm{SRV}}_{\mathrm{HF,R_1}}\tilde{p}^{\mathrm{R}}_{\mathrm{HFD}}-F^{\mathrm{SRV}}_{\mathrm{HF,L_1}}\tilde{p}^{\mathrm{L}}_{\mathrm{HFD}} \tag{6a}$$

$$\left(\frac{\partial\tilde{p}^{\mathrm{R}}_{\mathrm{SRVD}}}{\partial y_{\mathrm{D}}}\right)\Bigg|_{y_{\mathrm{D}}=w^{\mathrm{R}}_{\mathrm{HFD}}/2}=F^{\mathrm{SRV}}_{\mathrm{HF,L_2}}\tilde{p}^{\mathrm{R}}_{\mathrm{HFD}}-F^{\mathrm{SRV}}_{\mathrm{HF,R_2}}\tilde{p}^{\mathrm{L}}_{\mathrm{HFD}} \tag{6b}$$

Where, dimensionless flux is the function with regard to pressure in hydraulic fracture. In addition, Equation (7) represents the characteristics of flow depletion from XRV and SRV:

$$F^{\mathrm{SRV}}_{\mathrm{HF,L_1}}=F^{\mathrm{SRV}}_{\mathrm{HF,R_2}}=\Omega\left\{1+\exp\left[-\sqrt{\frac{s}{\eta_{\mathrm{SRVD}}}+\lambda^{\mathrm{XRV}}_{\mathrm{SRV}}F^{\mathrm{XRV}}_{\mathrm{SRV}}}\,(2L_{\mathrm{sD}}-w^{\mathrm{R}}_{\mathrm{HFD}}-w^{\mathrm{L}}_{\mathrm{HFD}})\right]\right\} \tag{7a}$$

$$F^{\mathrm{SRV}}_{\mathrm{HF,L_2}}=F^{\mathrm{SRV}}_{\mathrm{HF,R_1}}=2\Omega\exp\left[-\sqrt{\frac{s}{\eta_{\mathrm{SRVD}}}+\lambda^{\mathrm{XRV}}_{\mathrm{SRV}}F^{\mathrm{XRV}}_{\mathrm{SRV}}}\left(L_{\mathrm{sD}}-\frac{w^{\mathrm{R}}_{\mathrm{HFD}}+w^{\mathrm{L}}_{\mathrm{HFD}}}{2}\right)\right] \tag{7b}$$

$$\Omega=\frac{\sqrt{s/\eta_{\mathrm{SRVD}}+\lambda^{\mathrm{XRV}}_{\mathrm{SRV}}F^{\mathrm{XRV}}_{\mathrm{SRV}}}}{1-\exp\{-\sqrt{s/\eta_{\mathrm{SRVD}}+\lambda^{\mathrm{XRV}}_{\mathrm{SRV}}F^{\mathrm{XRV}}_{\mathrm{SRV}}}\,[2L_{\mathrm{sD}}-(w^{\mathrm{L}}_{\mathrm{HFD}}+w^{\mathrm{R}}_{\mathrm{HFD}})]\}} \tag{7c}$$

In our work, the fracture-production interaction effect can be decomposed into two parts: interaction between adjacent fractures within individual trilinear-flow element and interaction among trilinear-flow elements. Thus, Equation (7a) and (7b) only takes the effect of interaction between adjacent fractures into account. By contrast, the other effect would be presented in section 3 by incorporating the superposition principle. The method of two-partly-consideration mentioned above is significantly different from the fully-consideration of interaction effect presented in numerous literatures [3-7].

2.4 Hydraulic fracture model

The vertical hydraulic fracture is considered with finite half-length x_{HF} and width w_{HF}. The flow in hydraulic fracture could be described in the 1D orthogonal coordinate originating from the wellbore. Therefore, the basic dimensionless pressure governing Equation in right/left half-width hydraulic fracture is given as

$$\frac{\partial^2\tilde{p}^{\zeta}_{\mathrm{HFD}}}{\partial x^2_{\mathrm{D}}}+2\lambda^{\mathrm{SRV}}_{\mathrm{HF},\zeta}\left(\frac{\partial\tilde{p}_{\mathrm{SRVD}}}{\partial y_{\mathrm{D}}}\right)_{y_{\mathrm{D}}=\frac{w^{\zeta}_{\mathrm{HFD}}}{2}}=s\,\frac{\tilde{p}^{\zeta}_{\mathrm{HFD}}}{\eta^{\zeta}_{\mathrm{HFD}}} \tag{8}$$

The dimensionless flux rate from half-width hydraulic fracture into wellbore can be expressed as

$$\left(\frac{\partial \tilde{p}_{\mathrm{HFD}}^{\zeta}}{\partial x_{\mathrm{D}}}\right)\bigg|_{x_{\mathrm{D}}=0}=-\frac{2\pi}{C_{\mathrm{HFD}}^{\zeta}}\tilde{q}_{\mathrm{HFD}}^{\zeta} \tag{9a}$$

The no-flow assumption on the fracture tip gives the boundary condition as

$$\left(\frac{\partial \tilde{p}_{\mathrm{HFD}}^{\zeta}}{\partial x_{\mathrm{D}}}\right)\bigg|_{x_{\mathrm{D}}=x_{\mathrm{HFD}}}=0 \tag{9b}$$

Where ζ = L or R. After using Fourier Cosine Transformation seen in Appendix B, the dimensionless pressure distribution along hydraulic fracture is respectively given as follows,

$$\tilde{p}_{\mathrm{HFD}}^{\mathrm{L}}(x_{\mathrm{D}})=\mathrm{AX}_{\mathrm{SRV}}^{\mathrm{L}}(x_{\mathrm{D}})\tilde{q}_{\mathrm{HFD}}^{\mathrm{L}}+\mathrm{BX}_{\mathrm{SRV}}^{\mathrm{L}}(x_{\mathrm{D}})\tilde{q}_{\mathrm{HFD}}^{\mathrm{R}} \tag{10a}$$

$$\tilde{p}_{\mathrm{HFD}}^{\mathrm{R}}(x_{\mathrm{D}})=\mathrm{AX}_{\mathrm{SRV}}^{\mathrm{R}}(x_{\mathrm{D}})\tilde{q}_{\mathrm{HFD}}^{\mathrm{R}}+\mathrm{BX}_{\mathrm{SRV}}^{\mathrm{R}}(x_{\mathrm{D}})\tilde{q}_{\mathrm{HFD}}^{\mathrm{L}} \tag{10b}$$

It is noted that the solutions above are presented based on vertical fracture, which is different from transverse fracture intersecting horizontal wellbore. Within transverse fracture, radial flow pattern would result in additional pressure drop due to the convergence of fluid into horizontal wellbore. The effect of flow convergence is regarded as a skin factor; this is, the convergence flow skin (chocking skin) presented by Mukherjee and Economides[30]. Appendix C provides Equations to modify Equation (10) to account for the additional pressure drop in detail.

3 Solution for multiple-trilinear-flow model

The multiple-fractures system is divided into several independent elements as mentioned in Section 2. To integrate the solutions of single-trilinear-flow element together, the number of hydraulic fracture and SRV element is denoted as seen in Figure 4.

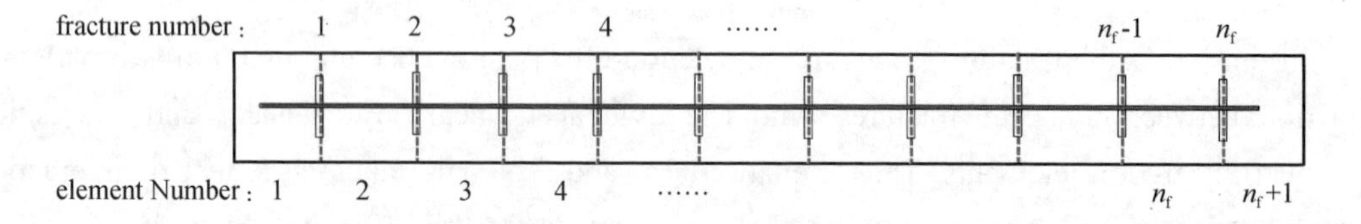

Figure 4 Multiple fractured horizontal well is separated into several elements and every element is a trilinear-flow model

According to the superposition principle, pressure distribution caused by n_{f} fractures can be written in Laplace domain as follows,

$$\tilde{p}_{\mathrm{D}j,i}=\sum_{i=1}^{n_{\mathrm{f}}}\tilde{q}_{\mathrm{HFD},i}^{\mathrm{L}}\Delta\tilde{p}_{\mathrm{D}j,i}^{\mathrm{L}}+\sum_{i=1}^{n_{\mathrm{f}}}\tilde{q}_{\mathrm{HFD},i}^{\mathrm{R}}\Delta\tilde{p}_{\mathrm{D}j,i}^{\mathrm{R}} \tag{11}$$

Where subscript "j,i" indicates dimensionless pressure response of the j-th fracture caused by the production of the i-th fracture. Based on the assumption of infinite conductivity wellbore, the dimensionless pressure on the interface between each hydraulic fracture and wellbore is equal to wellbore pressure p_{wD}.

$$\tilde{p}_{\mathrm{wD},1}=\tilde{p}_{\mathrm{wD},2}=\cdots=\tilde{p}_{\mathrm{wD},n_{\mathrm{f}}}=\tilde{p}_{\mathrm{wD}} \tag{12}$$

In the condition of constant flow rate for multiple-fracture system, the additional condition yields the following expression in Laplace domain,

$$\sum_{i=1}^{n_f} \tilde{q}_{\mathrm{HFD},i}^{\mathrm{L}} + \sum_{i=1}^{n_f} \tilde{q}_{\mathrm{HFD},i}^{\mathrm{R}} = \frac{1}{s} \tag{13}$$

Thus, the following matrix by combining Equation (10) ~ Equation (13) is obtained, which is

$$\begin{bmatrix} \boldsymbol{A} & -\boldsymbol{B} \\ \boldsymbol{B}^{\mathrm{1}} & \boldsymbol{0} \end{bmatrix} \cdot \begin{bmatrix} \boldsymbol{X} \\ s\tilde{p}_{\mathrm{wD}} \end{bmatrix} = \begin{bmatrix} \boldsymbol{0} \\ 1 \end{bmatrix} \tag{14}$$

Where bold types in Equation (14) are used for vectors and matrixes. The matrix $\boldsymbol{A}$ is a coefficient matrix of dimension $2n_f \times 2n_f$, and i-th element in $\boldsymbol{A}$ represents the i-th elementary trilinear-flow model (Figure 4), which can be further expressed as,

$$A_i = \begin{cases} AX_{\mathrm{SRV}}^{\mathrm{R},1}, i=1 \\ \begin{bmatrix} AX_{\mathrm{SRV}}^{\mathrm{L},i} & BX_{\mathrm{SRV}}^{\mathrm{L},i} \\ BX_{\mathrm{SRV}}^{\mathrm{R},i} & AX_{\mathrm{SRV}}^{\mathrm{R},i} \end{bmatrix}, 2 \leqslant i \leqslant n_f \\ AX_{\mathrm{SRV}}^{\mathrm{L},n_f+1}, i=n_f+1 \end{cases} \tag{15a}$$

Where element $\boldsymbol{B}$ represents the constraint of Equation (13), which is given as

$$B_i = \begin{cases} 1, i=1 \text{ and } n_f+1 \\ [1 \quad 1]^{\mathrm{T}}, 2 \leqslant i \leqslant n_f \end{cases} \tag{15b}$$

In addition, rate vector $\boldsymbol{X}$ of dimensionless individual half-width fracture is expressed as

$$X_i = \begin{cases} s\tilde{p}_{\mathrm{HFD}}^{\mathrm{R},1}(s), i=1 \\ [\, s\tilde{p}_{\mathrm{HFD}}^{\mathrm{L},i-1}(s) \quad s\tilde{p}_{\mathrm{HFD}}^{\mathrm{R},i}(s)\,]^{\mathrm{T}}, 2 \leqslant i \leqslant n_f \\ s\tilde{p}_{\mathrm{HFD}}^{\mathrm{L},n_f}(s), i=n_f+1 \end{cases} \tag{15c}$$

Noted that the two outermost elements should be established.

By solving Equation (14), the unknown wellbore pressure and the flow rate distribution of hydraulic fractures in real time domain could be obtained by applying Stehfest numerical-inversion algorithm[29].

4 Validation

To validate semi-analytical solutions in this paper, synthetic data generated from Ecrin-Saphir numerical simulator are selected. The basic parameters used in this section are list in Table 1.

Table 1 Parameters in validated model

Model parameters	value	Model parameters	value
Boundary width/m	440	Initialization pressure/Pa	3.44738×10^{7}
Boundary length/m	1200	Wellbore radius/m	0.1
Reservoir thickness/m	10	Fracture half-length/m	200
Reservoir porosity/%	10	Fracture number	3
Reservoir permeability/mD	100	Fracture conductivity/mD · m	5×10^{4}
Total compressibility/Pa^{-1}	4.35×10^{-10}	Production rate/(m^3/s)	1.15741×10^{-4}

In the configuration of equally-spaced fractures with identical properties (width and conductivity), the rate of each individual fracture keeps constant during the whole flow period[4,5]. Besides this special configuration, the rate of individual fracture would vary with production time increasing.

In this section, two cases are introduced to validate our model: unequally-spaced fractures and multiple fractures with different properties (mainly referring to conductivity). We resort to the numerical model of multiple-well with finite-conductivity fracture in Ecrin-Saphir.

4.1 Validation of unequally-spaced fractures

In this subsection, the parameters of fracture spacing are list in Table 2. As mentioned above, the rate of individual fracture is different from each other during the whole production period. Thus, Duhamel's principle is used to deal with the effect of varied rate on pressure response, which is satisfied as

$$p_i - p_w = \sum_{i=1}^{N_f} \int_0^t q_{u,i}(\tau) p'_{u,i}(t - \tau) d\tau \tag{16}$$

Table 2　Fracture spacing in multiple-fractures system

Fracture spacing	F_{s1}/m	F_{s2}/m	F_{s3}/m	F_{s4}/m
value	400	200	200	400

Firstly, solving Equation (14) provides the rate history of each individual fracture, which provides data source of production rate for individual well in the numerical model. And then, numerical model in Ecrin-Saphir could yield synthetic data of wellbore pressure, based on Duhamel's principle of Equation (16).

As seen from Figure 5, the model shows excellent agreement with numerical solutions in terms of pressure and pressure derivatives. It is necessary to further emphasize that the rate history of individual fracture is provided by solving Equation (14). In addition, it is noted that the pressure response for each individual well in numerical model is identical with each other. This phenomenon is also consistent with the assumption of infinite-conductivity wellbore.

4.2 Validation of multiple fractures with heterogeneity conductivity

In this subsection, the parameters of fracture conductivities are list in Table 3. Different from the case of identical fracture properties in Section 4.1, the rate of individual fracture is proportional to fracture conductivity before fracture-production interaction occurs. The pressure and derivatives obtained in this paper with the numerical solutions is compared, where the ratio of maximum conductivity to minimum F_{cmax}/F_{cmin} equals to 10. It is shown in Figure 6 that the semi-analytical solutions match exactly the numerical solutions during the whole time domain.

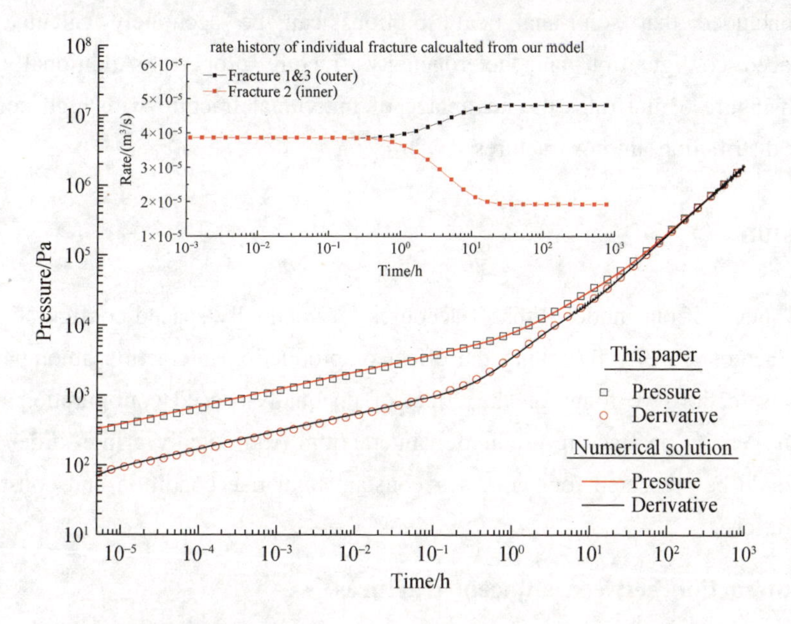

Figure 5 Comparison of pressure and derivate with numerical results in unequally-spaced configuration

Table 3 Distribution of fracture conductivity in multiple-fractures system

Fracture Conductivity	$F_{c1}/(\mathrm{mD \cdot m})$	$F_{c2}/(\mathrm{mD \cdot m})$	$F_{c3}/(\mathrm{mD \cdot m})$
value	50000	500000	50000

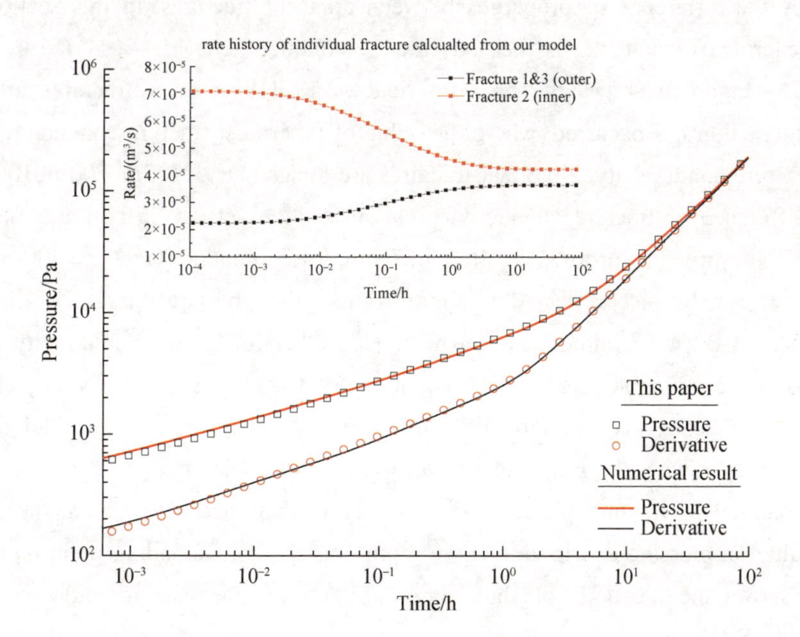

Figure 6 Comparison of pressure and derivate with numerical results for the case of heterogeneity of fracture property

It is concluded that semi-analytical solutions can be accurately calculated for both unequally-spaced configuration and heterogeneity fracture property. Additionally, comparing Figure 5 with Figure 6 illustrates the advantage of individual fracture with high conductivity in terms of rate distribution among fractures.

5 Discussion

The advantage of our model is that it accounts for unequally-spaced configuration and heterogeneity of fracture property by taking the effect of production interaction among fractures into account. In this section, we focus on the effect of unequally-spaced configuration and fracture-property heterogeneity on fracture-production interaction. Additionally a high-order approximate expression would be presented for two cases: constant total rate condition and constant wellbore pressure condition.

5.1 Interaction between adjacent fractures

Generally, multiple-fractures system in closed drainage area could be divided into several single-fracture systems occupying different sub drainage area with closed boundary. These virtual no-flow boundaries generated by fracture-production interaction between adjacent fractures are designed as "shunt line" by Wang et al. [31].

Shunt line is physically a special streamline during the fracture-production interaction period, which represents no flowing fluid along shunt line. The location of shunt line is determined by the difference of properties between adjacent fractures. In this section, the most representative property parameter is focused: fracture conductivity.

Figure 7 ~ Figure 9 shows the pressure field of MFHW with 3 fractures after fracture-production interaction has occurred, where the value of outermost fracture conductivity is smaller than inner fracture conductivity. Note that fractures are located at y=200m, 600m, 1000m. As expected, the difference of fracture conductivity has an obvious effect on pressure field. As seen from Figure 7, the pressure drop along fracture with lower conductivity (y = 200m, 1000m) is more serious. It can be also reflected in Figure 9 that the pressure curves on three different locations (A-A′, B-B′, C-C′) almost overlap with each other for higher-conductivity fracture (y = 600m). Thus the production rate of high-conductivity fracture is relatively higher. Note that black line A-A′ indicates the pressure distribution in vertical intersection parallel to horizontal wellbore. Put another way, the shape of pressure contour in SRV region more approximates to straight line parallel to fracture plane as seen in Figure 8. It needs to be emphasized that the maximum value on pressure distribution curve represents the corresponding point on shunt line in Figure 9. It shows the locations of shunt lines are not on the spacing midpoint of adjacent fractures within SRV region.

Figure 10 shows the evolution of shunt line as production time varies. It is noted that the shunt line on the intersection between fracture and wellbore (x = 0) is fixed on the middle of fracture spacing between adjacent fractures($y=200$m). With time increasing, shunt line gradually approaches towards low-conductivity fracture. It indicates that high-conductivity fracture would occupy larger drainage area. However, the moving rate of shunt line gradually slows down. After reaching pseudo steady state, shunt line would be fixed. In addition, the more close to fracture tip, the more obvious the phenomenon is.

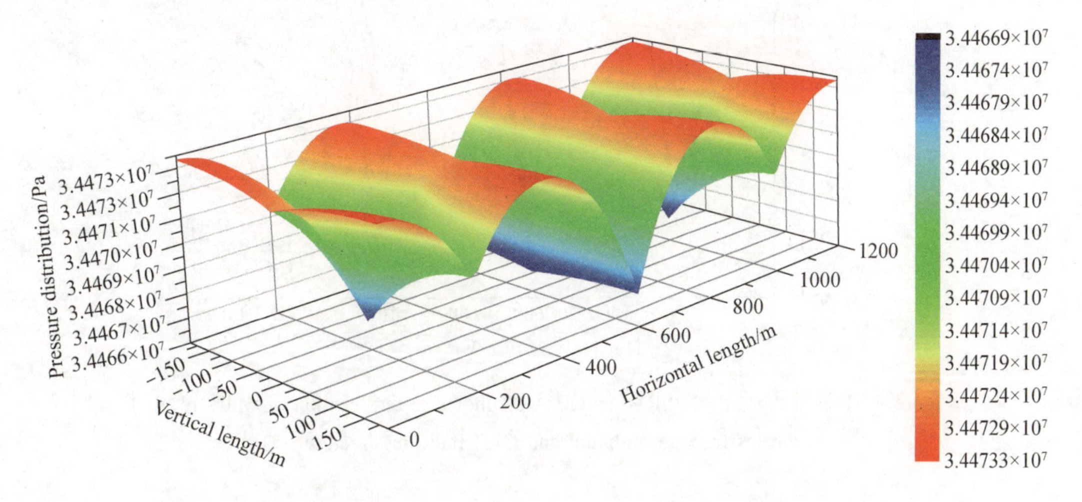

Figure 7 3D pressure field of multiple fractures with different conductivities ($t=0.272$h, $F_{c1}=F_{c3}=50000$mD · m, $F_{c2}=500000$mD · m)

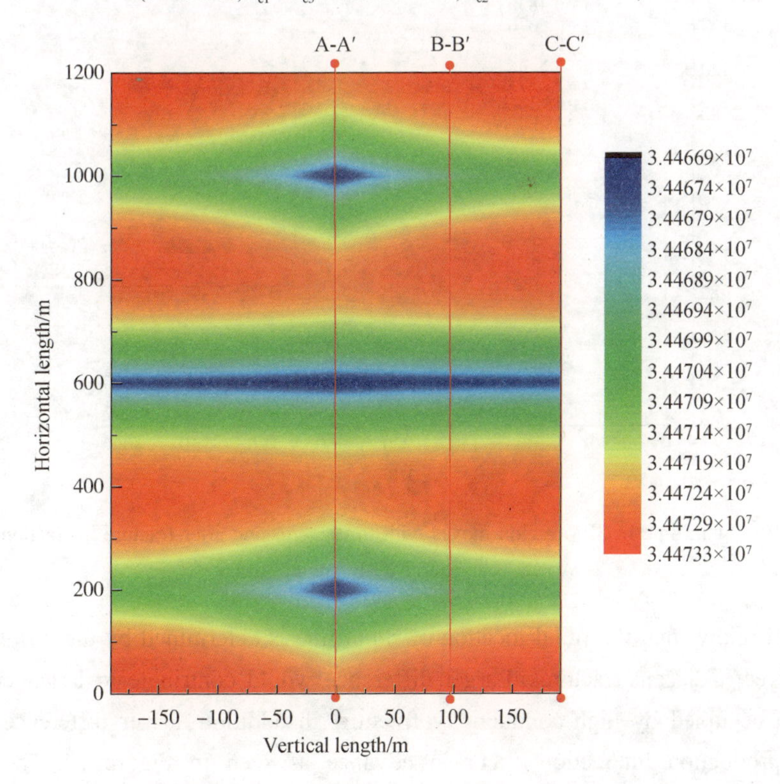

Figure 8 Contour of pressure in the system of multiple fractures with different conductivities

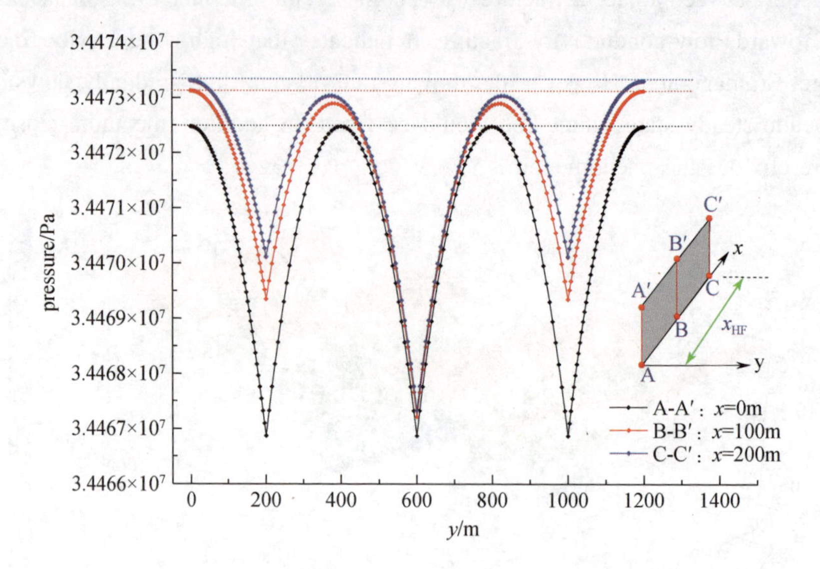

Figure 9　Pressure profile in SRV region (A-A′ indicates the location of wellbore, B-B′ indicates fracture midpoint and C-C′ indicates fracture tip)

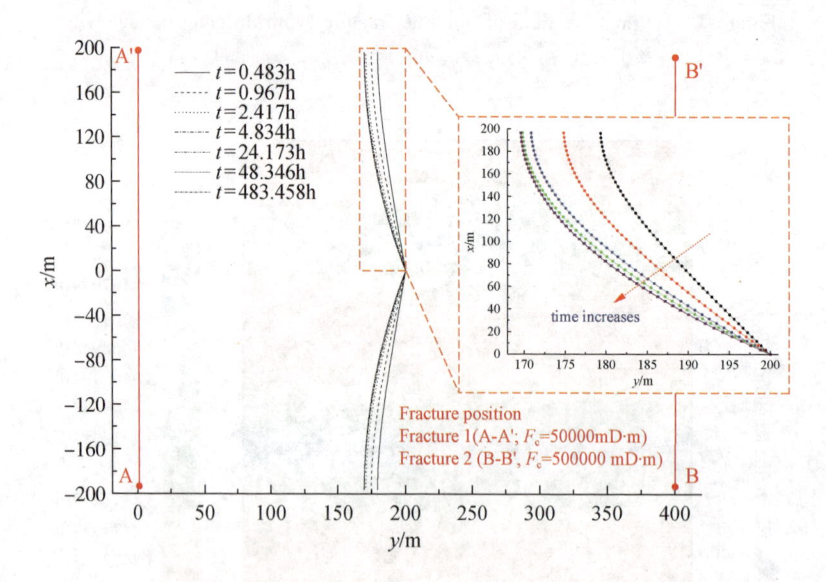

Figure 10　Shunt line in SRV region moves with time increasing after fracture interaction occurs (A-A′indicates fracture 1, B-B′indicates fracture 2)

Figure 11 shows that the finial location of shunt line is determined by the difference of conductivity between adjacent fractures. Larger difference would contribute to larger corresponding drainage area occupied by high-conductivity fracture. In addition, larger difference could make the fracture-production interaction occur in advance as seen in Figure 12. The duration of fracture-production interaction would elongate as the difference of heterogeneous property increases. Petroleum engineers would generally prefer to delay interaction as long as possible to obtain higher production rate during early-time period, so the case of heterogeneous properties among fractures should be avoided in practice.

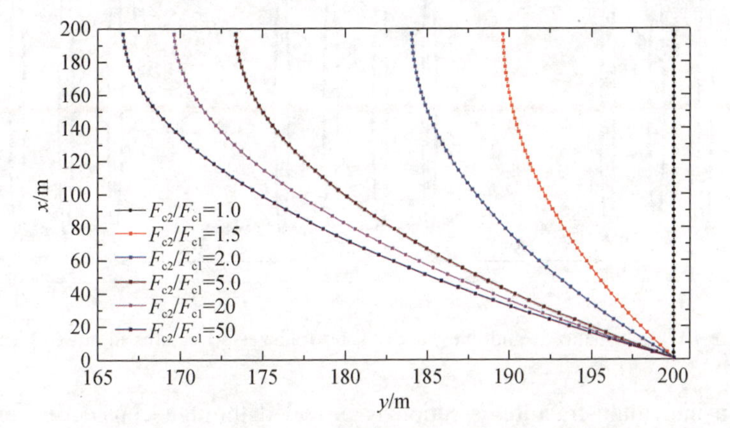

Figure 11 Effect of fracture conductivity on final shunt line within spacing between adjacent fractures

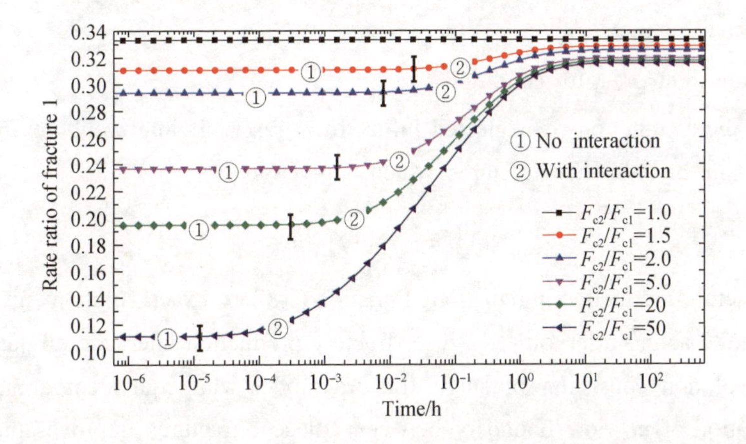

Figure 12 Effect of property heterogeneity on the occasion of fracture interaction

(vertical bar indicates the occurring time of fracture-production interaction)

5.2 Transformation single-fracture into multiple-fracture solution

As seen in Figure 13 from the research of Meyer et al. [26], the system includes multiple stage/clusters with n_s equally spaced stages and n_c equally spaced cluster per stage in closed rectangular reservoir, which satisfies the following Equation,

$$y_e = \Delta y_c(n_c - 1)n_s + \Delta y_i(n_s - 1) + \Delta y_e \tag{17}$$

Meyer et al. [26] provided first-order approximations, not exact solutions for the general configuration of multiple-fractures system. It is built on the assumption that shunt lines are fixedly located on the middle of spacing between adjacent fractures.

Based on the conclusions in Section 5.1, approximated expression transforming single-

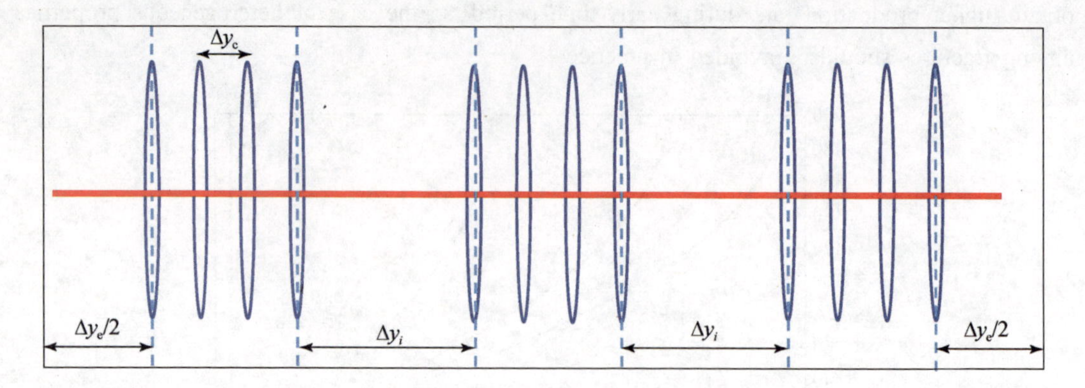

Figure 13　Schematic of multiple stage/cluster transverse fractures in closed reservoir

fracture solution into multi-fractures solution is provided through considering exact shunt lines generated by fracture-production interaction. Here, approximations are confirmed by comparing actual results from Equation (14) under two cases: rate response in the condition of constant pressure and the pressure response in the condition of constant total rate for the configuration of MFHW with three fractures.

5.2.1　Constant pressure condition

Based on the assumption of neglected influx from reservoir into wellbore, the total rate of MFHW is the sum of individual fractures, which is given as

$$q = \sum_{i=1}^{n_f} q(t;y_i) \tag{18}$$

Before fracture-production interaction, Equation (18) is exact solution equivalent to the solution of infinite-acting reservoir[8,25]. After fracture-production interaction, Equation (18) is an approximated solution consisting of single-fracture model with equivalent drainage area. The exact consideration of no-flow boundary between adjacent fractures improves the order of approximating to exact solution, which can be obtained explicitly.

Figure 14 shows the comparison results of individual fracture rate generated from approximated and exact results during the whole production period. Excellent agreement between exact solution of Equation (14) and approximated solution of Equation (18) indicates the accuracy and reliability of approximated solution.

5.2.2　Constant rate condition

For the special case of multiple fractures with identical properties in equally-spaced configuration, the pressure response of multi-fracture system will be equivalent to that of single-fracture multiplied with $1/n_f$[32]. For the case of heterogeneous properties in unequally-spaced configuration, the general form can be expressed as follows

$$\frac{1}{p_w} = \sum_{i=1}^{n_f} \frac{1}{p_w(t;y_i)} \tag{19}$$

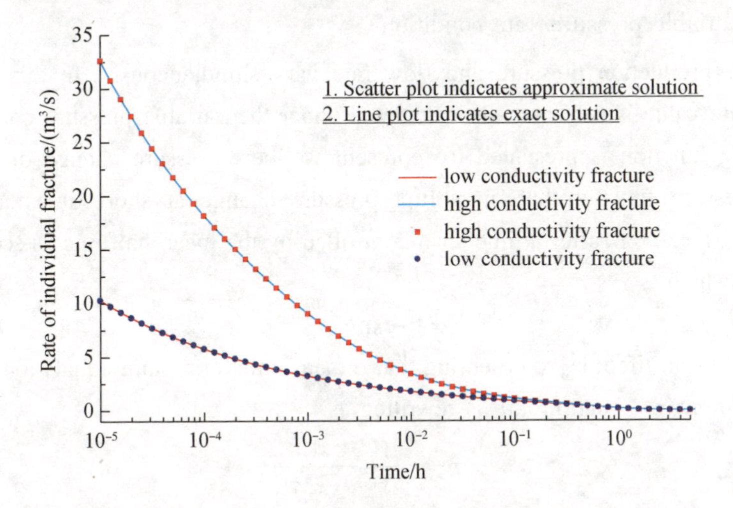

Figure 14 Comparison result of exact and approximated solutions under constant pressure condition ($F_{cmax}/F_{cmin}=10$)

Figure 15 shows the comparison results of exact and approximated solutions in the case of different differences of heterogeneity property. It indicates that Equation (19) is suiTable to different properties of fractures in unequally-spaced configuration during the whole production period.

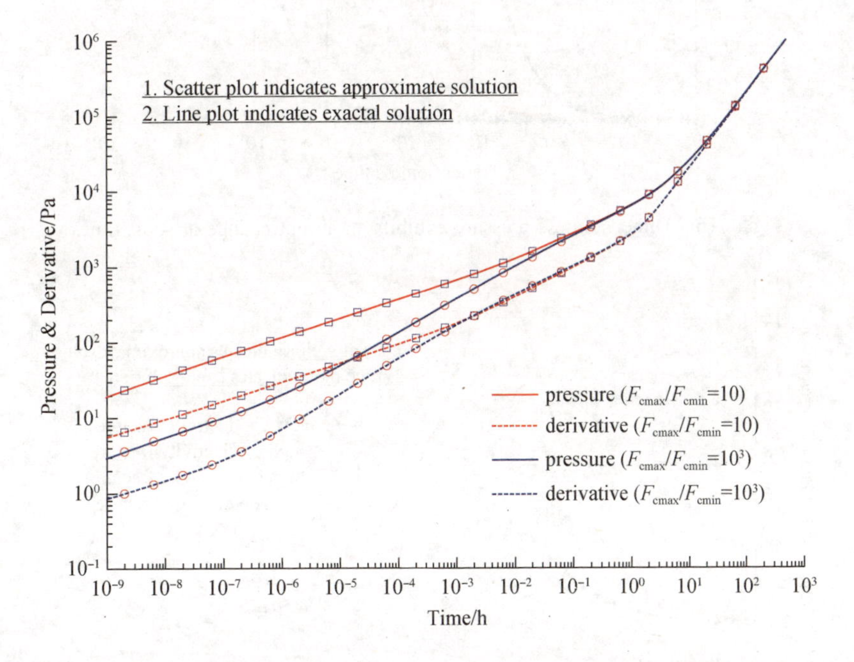

Figure 15 Comparison result of exact solutions and approximated solutions under constant rate condition

5.2.3　Variable pressure/rate condition

In practice, production pressure and flow rate vary simultaneously. In this subsection, the transient pressure value is subsequently calculated under the variable pressure condition.

A decaying function is presented to represent wellbore-pressure change during production period. The pressure function has an abrupt pressure change at short time, and reaches the stabilized value(i. e. , =1) after a time span controlled by the mean halftime as seen in Figure 16 and Figure 17, which is

$$p_{wD}=1-\exp(-t_D/\tau_D) \tag{20}$$

Where τ_D is the mean lifetime, representing that dimensionless pressure equals to 0.63 when $t_D=\tau_D$. In Laplace domain, Equation(20) is rewritten as

$$\tilde{p}_{wD}=\frac{1}{s}-\frac{1}{s+1/\tau_D} \tag{21}$$

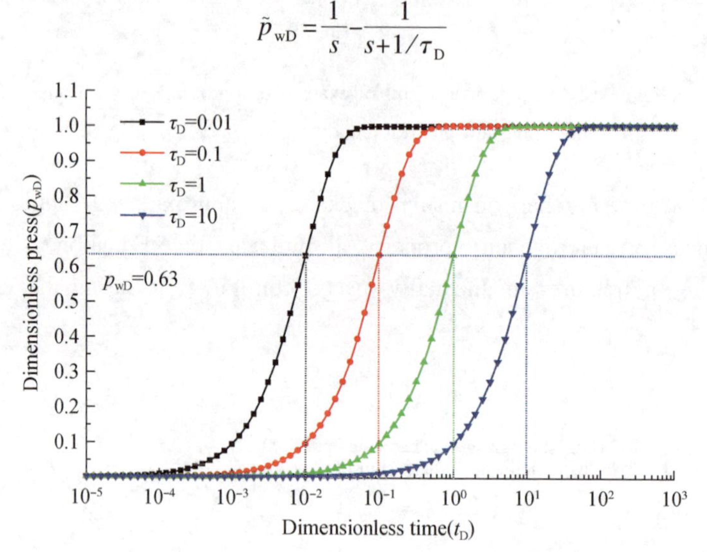

Figure 16　Dimensionless pressure exhibits an abrupt change at short times

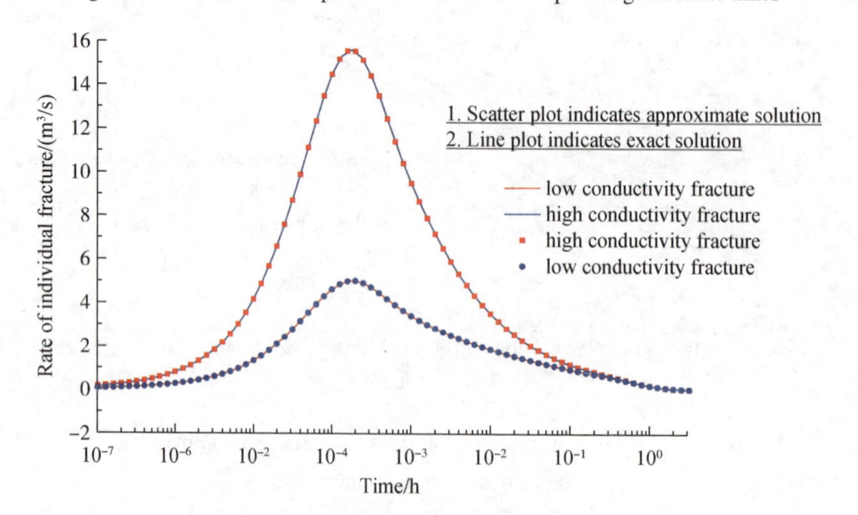

Figure 17　Rate comparison of exact and approximated solutions under variable pressure condition($F_{cmax}/F_{cmin}=10$)

Similar to Section 5. 2. 1, the total rate of MFHW is the summation of respective fractures. Figure 16 and Figure 17 shows the comparison results of exact and approximated solutions, where production rate increases from zero time and then declines after reaching a peak value. It shows the approximated results obtained in this subsection are consistent with exact semi-analytical results. So this approximation is also suiTable to variable pressure/rate condition.

Note that in Equation (18) and Equation (19), variable y_i represents the drainage area occupied by i-th fracture, which is determined by the fracture spacing and the property difference of adjacent fractures. Originating from the results of Figure 11, the explicit solution of drainage area is presented, which is presented in the form of mean value of integral with regard to shunt line path, given in Equation(22):

$$y_i = \frac{1}{x_{\mathrm{HF}}}\int_0^{x_{\mathrm{HF}}} y(x; C_{\mathrm{HFD},i}^{\mathrm{L}}, C_{\mathrm{HFD},i}^{\mathrm{R}})\,\mathrm{d}x \tag{22}$$

5.3 Improvement towards PTA & RTA

The theory behind rate transient analysis (RTA) is exactly analogous to pressure transient analysis (PTA), and the techniques used for RTA are very similar to PTA[33,34]. Hence this subsection takes PTA for example to illustrate the improvement of the exact solution in this paper towards PTA. For the problem of MFHW based on multilinear flow model, the semi-analytical results are generally obtained by using the traditional method[27,28,35]. For traditional method, PTA is identical to pressure transient behavior of single variable-rate fracture with fixed closed drainage area; this is, $x_{n,1,2}$ = constant and $x_{n,0,2}$ = constant, as seen in Figure 18.

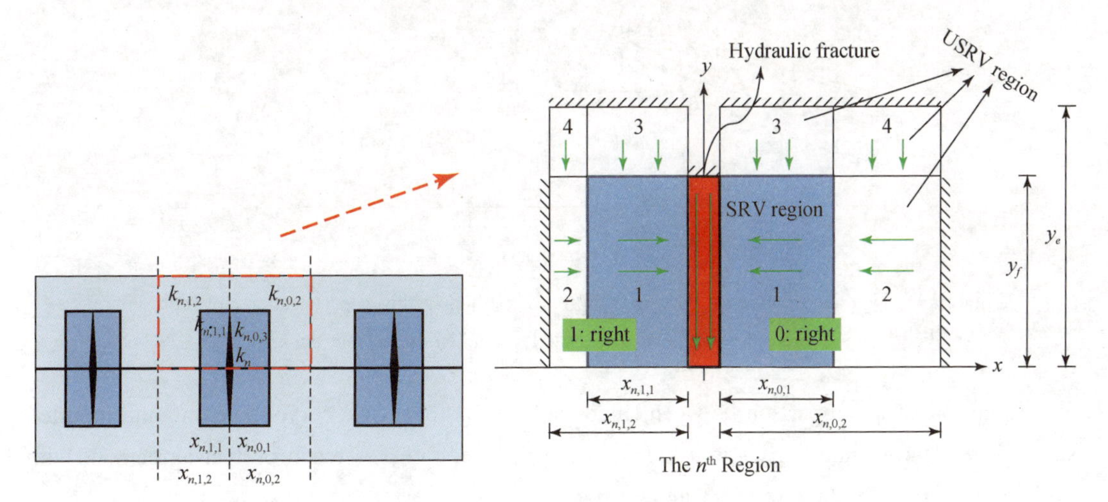

Figure 18 Schematic and dimensions for MHFW and nth individual region the enhanced five-region model[28], where MFHW is divided several single fractures with fixed closed boundary

A synthetic case has been generated using numerical simulator to confirm the improvement of exact solution in this paper. The basic data in Table 1 are introduced to simulate the test with 3 fractures in equally spaced configuration (L_s = 400m, x_R = 220m). The conductivity of outermost fracture is minimum to be set as 20000mD · m and the conductivity of middle fracture is maximum ($C_{HFD,max} = 10C_{HFD,min}$).

5.3.1 New type curve matching

Figure 19 shows the matching result of synthetic data and type curve of new method in this paper, where the best matching of the synthetic data is obtained when $C_{HFD,min} = 1$ for two methods. By type-curve matching calculation, we could obtain reservoirs permeability and fracture half-length,

$$k_{SRV} = \frac{\mu Bq_{sc}[p_{wD}]_{MP}}{2\pi h[\Delta p_w]_{MP}}, x_{HF} = \sqrt{\frac{k_{SRV}}{\varphi_{SRV}\mu c_t}\frac{[t]_{MP}}{[t_D]_{MP}}} \tag{23}$$

and subsequent fracture conductivity, formation half-width and fracture half-length,

$$F_{cmin} = C_{HFDmin}k_{SRV}x_{HF}, x_R = x_{RD}x_{HF}, L_s = L_{sD}x_{HF} \tag{24}$$

It is noted that matching points are selected as: $[p_{wD}]_{MP} = 0.01$, $[\Delta p_w]_{MP} = 186\text{Pa}$, $[t]_{MP} = 2.85\times10^{-4}$ hour, $[t_D]_{MP} = 6\times10^{-5}$.

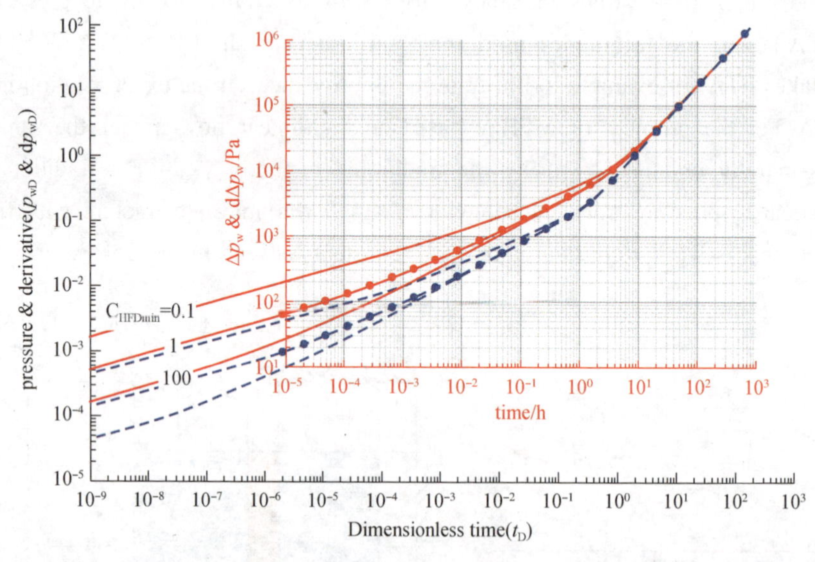

Figure 19　Type curve analysis by using improved method in this paper(scatter plot represents synthetic data), where the synthetic data exactly matches the correct curve of $C_{HFDmin} = 1$

Table 4 shows comparison between calculated results from type curve matching and inputted parameters in the simulation. It indicates that the new type curves reach perfect agreement with synthetic data during the whole production period.

Table 4　Comparison between new type-curve results and inputted parameters

Compared parameters	New type curve	Numerical simulation inputted
Fracture spacing/m	396.4	400
Reservoir permeability/mD	99.1	100
Fracture half-length/m	198.2	200

续表

Compared parameters	New type curve	Numerical simulation inputted
Minimum conductivity/mD · m	1.964×10^4	2×10^4
Maximum conductivity/mD · m	1.964×10^5	2×10^5

5.3.2 Already available type curve matching

Figure 20 shows the matching results of synthetic data and type curves of already available type curve. As seen in Figure 20, there are three periods. ①Before fracture-interaction period: the available type curves exactly match with numerical simulation. ② During the fracture-interaction period: the available type curves cannot match very well. ③During pseudo steady state (PSS): the available type curves can match by adjusting fracture spacing ($L_{sD}=2.45$).

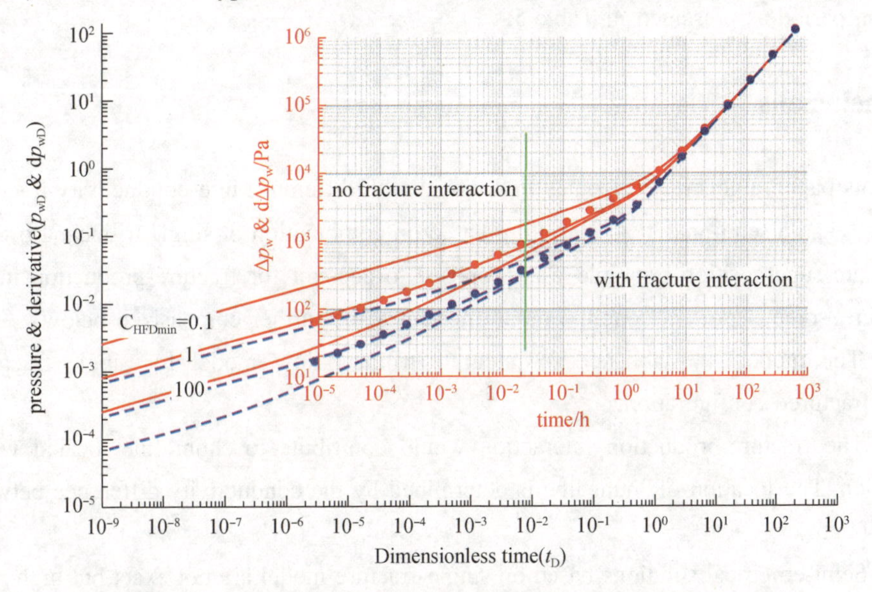

Figure 20 Type curve analysis by using traditional method, where synthetic data fails to match the correct curve of $C_{HFDmin}=1$ very well

Table 5 shows comparison between calculated results from type curve matching and inputted parameters in the simulation. It is noted that the calculated results are consistent with inputted parameters when using matching point in no-interaction period. But there is significant error when using matching point in fracture-interaction period and PSS period, where matching points are selected as $[p_{wD}]_{MP}=0.23$, $[\Delta p_w]_{MP}=3000\text{Pa}$, $[t]_{MP}=0.15\text{h}$, $[t_D]_{MP}=0.1$.

Table 5 Comparison between available type-curve results and inputted parameters

Compared parameters	Matching points before interaction	Matching points after interaction	Numerical simulation inputted
Fracture spacing/m	N/A	324.38	400
Reservoir permeability/mD	102.4	141.2	100

续表

Compared parameters	Matching points before interaction	Matching points after interaction	Numerical simulation inputted
Fracture half-length/m	194.7	132.4	200
Minimum conductivity/mD · m	1.993×10^4	1.870×10^4	2×10^4
Maximum conductivity/mD · m	1.993×10^5	1.870×10^5	2×10^5

In terms of rate distribution of individual fracture, the fracture distribution keeps constant at early time period, then would redistribute during fracture interaction and remain constant during the PSS period. In already available method, the fracture distribution would keep constant during the whole flow period, which contributes to error when using matching point in fracture-interaction period as presented in Table 5.

6 Conclusions

In this paper, a general horizontal well model with different finite-conductivity fractures was established, and a semi-analytical algorithm based on combination of single trilinear-flow solution and pressure-superposition principle was proposed to account for fracture-production interaction effect. Here, some important conclusions are needed to be further emphasized below:

(1) The model is established to integrate trilinear-flow model for single fracture with multiple-fractured configuration.

(2) The fracture-production interaction would contribute to shunt line located within the SRV region. The location of shunt line is determined by the conductivity difference between two adjacent fractures.

(3) Semi-empirical solutions based on single-fracture model are not exact but high-order approximations. For the case of equally-spaced fractures with identical properties, the solution of multi-fractures is that of single-fracture multiplied by the number of hydraulic fracture.

(4) This model could get rid of the constraint on equally-spaced configuration and homogeneity properties among fractures, which makes it more convenient to apply this model to conduct RTA or PTA from multi-fractured horizontal well.

(5) Some modified linear-flow model[15-20] could be extended to multi-fractured configuration according to the method presented in our paper.

Nomenclature

Dimensionless variables

C_{HFD}　　Dimensionless fracture conductivity

t_D	Dimensionless time
p_D	Dimensionless pressure

Field variables

C	compressibility, Pa^{-1}
F_c	Fracture conductivity, $\times 10^{-3} \mu m^2 \cdot m$
F_s	Fracture spacing, m
K	permeability, m^2
P	pressure, Pa
Q	rate, m^3/s
H	formation thickness, m
T	time, s
w_{HF}	fracture width, m
S	Laplace variable, dimensionless
L_s	fracture spacing, m
x_e	formation width, m
y_e	formation length, m
x_{HF}	Fracture half-length, m
x_R	Half-width formation, m
ϕ	porosity, fraction
M	fluid viscosity, $Pa \cdot s$
Λ	flow-capacity ratio, dimensionless
Z	symbol indicating right or left part within SRV unit
H	diffusion capacity, dimensionless

Special subscripts/superscripts

HF	Fracture property
SRV	Stimulated reservoir volume property
XRV	External reservoir volume property

W Wellbore property

References

[1] Nobakht M, Clarkson C R, Kaviani D. New and improved methods for performing rate-transient analysis of shale gas reservoirs[C]. Paper SPE 147869 presented at the SPE Asia Pacific Oil and Gas Conference and Exhibition, Jakarta, 20-22 September 2011.

[2] Anderson D M, Nobakht M, Mattar L. Analysis of production data from fractured shale gas wells[C]. Paper SPE 131787 presented at the SPE Unconventional Gas Conference, Pennsylvania, USA, 23-25, February 2010.

[3] Larsen L, Hegre T M. Pressure-transient behavior of horizontal wells with finite-conductivity vertical fractures[C]. Paper SPE 22076 presented at the International Arctic Technology Conference, Anchorage, Alaska, 29-31 May 1991.

[4] Guo G L, Evans R D. Pressure transient behavior and inflow performance of horizontal wells intersecting discrete fractures [c]. Paper SPE 26446 presented at the 68th Annual Technology Conference and Exhibition, Houston, USA, 3-6 October 1993.

[5] Hegre T M, Larsen L. Productivity of multifractured horizontal wells[C]. Paper SPE 28845 presented at the European Petroleum Conference, London, UK, 25-27 October 1994.

[6] Kuchuk F J, Habusky T M. Pressure behavior of horizontal wells with multiple fractures[C]. Paper SPE 27971 presented at the University of Tulsa Centennial Petroleum Engineering Symposium, Tulsa, USA, 29-31 August 1994.

[7] Chen C C, Raghavan R. A multiply-fractured horizontal well in rectangular drainage region[J]. SPE Journal,12: 455-465 1997.

[8] Zerzar A, Bettam Y. Interpretation of multiple hydraulically fractured horizontal wells in closed systems [C]. Paper SPE 84888 presented at the SPE International Improved Oil Recovery Conference in Asia Pacific, Kuala Lumpur, Malaysia, 20-21 October 2003.

[9] Luo W J, Tang C F, Wang X D. Pressure transient analysis of a horizontal well intercepted by multiple non-planar vertical fractures[J]. Journal of Petroleum Science and Enigneering, 2014,124: 232-242.

[10] Cinco-Ley H, Samaniego-V F. Transient pressure analysis for fractured wells[J]. Journal of Petroleum Technology, 1989, 9: 1748-1765.

[11] Lee S, Brockenbrough J R. A new approximate analytic solution for finite-conductivity vertical fractures [J]. SPE Formation Evaluation, 1986, 2: 75-94.

[12] Wattenbarger R A, El-Banbi H, Villegas M E. Production analysis of linear flow into fractured tight gas wells[C]. Paper SPE 39931 presented at the 1998 SPE Rock Mountain Regional/Low Permeability Reservoirs Symposium and Exhibition, Colorado, USA, 5-8 April 1998.

[13] Bello O, Wattenbarger R A. Rate transient analysis in naturally fractured shale gas reservoirs[C]. Paper SPE114591 presented at the CIPC/SPE Gas Technology Symposium 2008 Joint Conference, Alberta, Canada, 16-19 June 2008.

[14] Brown M, Ozkan E, Raghavan R, *et al*. Practical solutions for pressure-transient responses of fractured horizontal wells in unconventional shale reservoirs[C]. Paper SPE 12503 presented at the SPE Annual Technical Conference and Exhibition, New Orleans, 4-7 October 2009.

[15] Al-Ahmadi A H, Wattenbarger R A. Triple-porosity models: One further step towards capturing fractured

reservoirs heterogeneity [C]. Paper SPE 149054 presented at the SPE/DGS Saudi Arabia Section Technical Symposium and Exhibition, Al-Khobar, Saudi Arabia, 15-18 May 2001.

[16] Obinna E D, Hassan D. A model for simultaneous matrix depletion into natural and hydraulic fracture networks[J]. Journal of Natural Gas Science and Engineering, 2014, 16: 57-69.

[17] Tian L, Xiao C, Liu M J, *et al.* Well testing model for multi-fractured horizontal well for shale gas reservoirs with consideration of dual diffusion in matrix [J]. Journal of Natural Gas Science and Engineering, 2014, 21: 283-295.

[18] Ambrose R J, Clarkson C R, Youngblood J, *et al.* Life-cycle decline curve estimation for tight/shale gas reservoirs[C]. Paper SPE 140519 presented at the SPE Hydraulic Fracturing Technology Conference and Exhibition, Texas, USA, 24-26 January 2011.

[19] Stalgorova E, Mattar L. Practical analytical model to simulate production of horizontal wells with branch fractures[C]. Paper SPE 162515 presented at the SPE Canadian Unconventional Resources Conference, Alberta, Canada, 30 October-1 Noverber 2012.

[20] Stalgorova E, Mattar L. Analytical model for unconventional multifractured composite systems[C]. Paper SPE 162516 presented at the SPE Canadian Unconventional Resources Conference, Alberta, Canada, 30 October-1 Noverber 2012.

[21] Fuentes-Cruz G, Gildin E, Valko P. Analyzing production data from hydraulically fractured wells: the concept of induced permeability field[J]. SPE Reservoir Evaluation & Engineering, 2014, 5: 220-232.

[22] Fuentes-Cruz G, Gildin E, Valko P. On the analysis of production data: Practical approaches for hydraulically fractured wells in unconventional reservoirs [J]. Journal of Petroleum Science and Engineering, 2014, 119: 54-68.

[23] Anderson D M, Thompson J M. How reliable is production data analysis [C]. Paper SPE 169013 presented at the Unconventional Resources Conference, Texas, USA, 1-3 April 2014.

[24] Valko P P, Amini S. The method of distributed volumetric sources for calculating the transient and pseudosteady state productivity of complex well-fracture configurations[C]. Paper SPE 106279 presented at the 2007 SPE Hydraulic Fracturing Technology Conference, Texas, USA, 29-31 January 2007.

[25] Rbeawi S A, Tiab D. Predicting productivity index of hydraulically fractured formations[J]. Journal of Petroleum Science and Engineering, 2013, 112: 185-197.

[26] Meyer B R, Bazan L W, Jacot R H, *et al.* Optimization of multiple transverse hydraulic fractures in horizontal wellbores[C]. Paper SPE131732 presented at the SPE Unconventional Gas Conference, Pennsylvania, USA, 23-25 February 2010.

[27] Meyer B R, Jacot R H. Pseudosteady-state analysis of finite-conductivity vertical fractures[C]. Paper 95941 presented at the 2005 SPE Annual Technical Conference and Exhibition, Texas, USA, 9-12 October 2005.

[28] Deng Q, Nie R S, Jia Y L, *et al.* A new analytical model for non-uniformly distributed multi- fractured system in shale gas reservoirs[J]. Journal of Natural Gas Science and Engineering, 2015, 9: 1-19.

[29] Stehfest H. Numerical inversion of Laplace transform. Comm[J]. ACM, 1970, 13 (1): 47-49.

[30] Mukherjee H, Economides M J. A parametric comparison of horizontal and vertical well performace[J]. SPE Formation and Evaluation, 1991, 7: 209-216.

[31] Wang J L, Jia A L. A general productivity model for optimization of multiple fractures with heterogeneous properties[J]. Journal of Natural Gas Science and Engineering, 2014, 21: 608-624.

[32] Raghavan R S, Chen C C, Agarwal B. An analysis of horizontal wells intercepted by multiple fractures [C]. Paper SPE 27652 presented at the 1994 SPE Permian Basin Oil and Gas Recovery Conference,

Texas, USA, 16-18 March 1994.

[33] Clarkson C R. Production data analysis of unconventional gas wells: Review of theory and best practices [J]. International Journal of Coal Geology, 2013,109: 101-146.

[34] Clarkson C R. Production data analysis of unconventional gas wells: Workflow[J]. International Journal of Coal Geology,2013, 109:147-157.

[35] Su Y L, Wang W D, Zhou S Y, *et al.* Trilinear flow model and fracture arrangement of volume-fractured horizontal well[J]. Oil & Gas Geology, 2014, 35 (3):435-440.

[36] Wang X D, He S L, Wang F. Productivity analysis of horizontal wells intercepted by multiple finite-conductivity fractures[J]. Petroluem Science, 2010,7: 367-371.

[37] Wang X D, Luo W J, Hou X C, *et al.* Pressure transient analysis of multi-stage fractured horizontal wells in boxed reservoirs[J]. Petroleum Exploration and Development, 2014,41(1):82-87.

Appendix A: Dimensionless definitions

For simplicity, the physical variables in Equations are converted into dimensionless variables. The dimensionless pressure in hydraulic fracture volume, stimulated reservoir volume (SRV) and external reservoir volume (XRV) is given as

$$p_{\xi D}=\frac{2\pi k_{ref}h(p_i-p_\xi)}{\mu Bq_{ref}},\xi=HF,SRV,XRV \tag{A1}$$

The dimensionless time

$$t_D=\frac{k_{ref}t}{\phi_{SRV}\mu c_t L_{ref}^2} \tag{A2}$$

In the hydraulic fracture, the dimensionless diffusivity, conductivity, width and permeability are given as follows:

$$\eta_{HFD}=\frac{k_{HF}\phi_{ref}c_{ref}}{k_{ref}\phi_{HF}c_{HF}},C_{HFD}=\frac{k_{HF}w_{HF}}{k_{ref}L_{ref}},w_{HFD}=\frac{w_{HF}}{L_{ref}},k_{HFD}=\frac{k_{HF}}{k_{ref}} \tag{A3}$$

The corresponding dimensionless variables in SRV are given as

$$\eta_{SRVD}=\frac{k_{SRV}\phi_{ref}c_{ref}}{k_{ref}\phi_{SRV}c_{SRV}},\lambda_{HF}^{SRV}=\frac{k_{SRV}L_{ref}}{k_{HF}w_{HF}},x_{HFD}=\frac{x_{HF}}{L_{ref}} \tag{A4}$$

The dimensionless variables in XRV are given as

$$\eta_{XRVD}=\frac{k_{XRV}\phi_{ref}c_{ref}}{k_{ref}\phi_{XRV}c_{XRV}},\lambda_{SRV}^{XRV}=\frac{k_{XRV}L_{ref}}{k_{SRV}x_{HF}},x_{RD}=\frac{x_R}{L_{ref}} \tag{A5}$$

Appendix B: Solution in SRV region with interaction effect

According to the work presented by Wang and Jia[31], Fourier cosine transformations are given as

$$F_x[f(x_D)]=\int_0^{x_{HFD}}f(x_D)\cos(\beta_m x_D)\,dx_D=\bar{f}(\beta_m) \tag{B1}$$

where

$$\beta_m = \frac{m\pi}{x_{\mathrm{HFD}}}, m = 0,1,2,3\cdots \tag{B2}$$

Inversion Fourier cosine transformation are given as

$$\tilde{m}_{\mathrm{HFD}}^{\zeta}(x_{\mathrm{D}}) = \sum_{m=0}^{\infty} \frac{\cos(\beta_m x_{\mathrm{D}})}{N(m)} \hat{\tilde{m}}_{\mathrm{HFD}}^{\zeta} \tag{B3}$$

Where

$$N(m) = \int_0^{x_{\mathrm{HFD}}} \cos^2(\beta_m x_{\mathrm{D}})\,\mathrm{d}x_{\mathrm{D}} = \begin{cases} x_{\mathrm{HFD}}/2, m > 0 \\ x_{\mathrm{HFD}}, m = 0 \end{cases} \tag{B4}$$

Taking Fourier transformation through Equation (8) to Equation (9b), the Fourier solution for single half-width fracture within individual element is respectively given in Laplace transformation domain as

$$\begin{bmatrix} \left(\beta_m^2 + 2\lambda_{\mathrm{HF,L}}^{\mathrm{SRV}} F_{\mathrm{HF,L_1}}^{\mathrm{SRV}} + \frac{s}{\eta_{\mathrm{HFD}}^{\mathrm{L}}}\right) & (-2\lambda_{\mathrm{HF,L}}^{\mathrm{SRV}} F_{\mathrm{HF,R_1}}^{\mathrm{SRV}}) \\ (-2\lambda_{\mathrm{HF,R}}^{\mathrm{SRV}} F_{\mathrm{HF,L_2}}^{\mathrm{SRV}}) & \left(\beta_m^2 + 2\lambda_{\mathrm{HF,R}}^{\mathrm{SRV}} F_{\mathrm{HF,R_2}}^{\mathrm{SRV}} + \frac{s}{\eta_{\mathrm{HFD}}^{\mathrm{R}}}\right) \end{bmatrix} \cdot \begin{bmatrix} \tilde{p}_{\mathrm{HFD}}^{\mathrm{L}} \\ \tilde{p}_{\mathrm{HFD}}^{\mathrm{R}} \end{bmatrix} = \begin{bmatrix} \frac{2\pi \tilde{q}_{\mathrm{HFD}}^{\mathrm{L}}}{C_{\mathrm{HFD}}^{\mathrm{L}}} \\ \frac{2\pi \tilde{q}_{\mathrm{HFD}}^{\mathrm{R}}}{C_{\mathrm{HFD}}^{\mathrm{R}}} \end{bmatrix} \tag{B5}$$

Fortunately, Equation (B5) can be solved analytically. Solving Equation (B5) by using (inversion) Fourier transformation of Equation (B1) ~ Equation (B4) could contribute to the dimensionless pressure distribution along hydraulic fracture, which is written as follows:

$$\tilde{p}_{\mathrm{HFD}}^{\mathrm{L}}(x_{\mathrm{D}}) = AX_{\mathrm{SRV}}^{\mathrm{L}}(x_{\mathrm{D}})\,\tilde{q}_{\mathrm{HFD}}^{\mathrm{L}} + BX_{\mathrm{SRV}}^{\mathrm{L}}(x_{\mathrm{D}})\,\tilde{q}_{\mathrm{HFD}}^{\mathrm{R}} \tag{B6}$$

$$\tilde{p}_{\mathrm{HFD}}^{\mathrm{R}}(x_{\mathrm{D}}) = AX_{\mathrm{SRV}}^{\mathrm{R}}(x_{\mathrm{D}})\,\tilde{q}_{\mathrm{HFD}}^{\mathrm{R}} + BX_{\mathrm{SRV}}^{\mathrm{R}}(x_{\mathrm{D}})\,\tilde{q}_{\mathrm{HFD}}^{\mathrm{L}} \tag{B7}$$

where

$$AX_{\mathrm{SRV}}^{\mathrm{L}} = \sum_{m=0}^{\infty} 2\pi \frac{\cos(\beta_m x_{\mathrm{D}})}{N(m) C_{\mathrm{HFD}}^{\mathrm{L}} \Lambda_m} \left(\beta_m^2 + 2\lambda_{\mathrm{HF,R}}^{\mathrm{SRV}} F_{\mathrm{HF,R_2}}^{\mathrm{SRV}} + \frac{s}{\eta_{\mathrm{HFD}}^{\mathrm{R}}}\right) \tag{B8}$$

$$BX_{\mathrm{SRV}}^{\mathrm{L}} = \sum_{m=0}^{\infty} 4\pi \frac{\cos(\beta_m x_{\mathrm{D}})}{N(m) C_{\mathrm{HFD}}^{\mathrm{R}} \Lambda_m} \lambda_{\mathrm{HF,L}}^{\mathrm{SRV}} F_{\mathrm{HF,R_1}}^{\mathrm{SRV}} \tag{B9}$$

$$AX_{\mathrm{SRV}}^{\mathrm{R}} = \sum_{m=0}^{\infty} 2\pi \frac{\cos(\beta_m x_{\mathrm{D}})}{N(m) C_{\mathrm{HFD}}^{\mathrm{R}} \Lambda_m} \left(\beta_m^2 + 2\lambda_{\mathrm{HF,L}}^{\mathrm{SRV}} F_{\mathrm{HF,L_1}}^{SRV} + \frac{s}{\eta_{\mathrm{HFD}}^{\mathrm{L}}}\right) \tag{B10}$$

$$BX_{\mathrm{SRV}}^{\mathrm{R}} = \sum_{m=0}^{\infty} 4\pi \frac{\cos(\beta_m x_{\mathrm{D}})}{N(m) C_{\mathrm{HFD}}^{\mathrm{L}} \Lambda_m} \lambda_{\mathrm{HF,R}}^{\mathrm{SRV}} F_{\mathrm{HF,L_2}}^{\mathrm{SRV}} \tag{B11}$$

with

$$\Lambda_m = \left(\beta_m^2 + 2\lambda_{\mathrm{HF,L}}^{\mathrm{SRV}} F_{\mathrm{HF,L_1}}^{\mathrm{SRV}} + \frac{s}{\eta_{\mathrm{HFD}}^{\mathrm{L}}}\right)\left(\beta_m^2 + 2\lambda_{\mathrm{HF,R}}^{\mathrm{SRV}} F_{\mathrm{HF,R_2}}^{\mathrm{SRV}} + \frac{s}{\eta_{\mathrm{HFD}}^{\mathrm{R}}}\right) - 4\lambda_{\mathrm{HF,L}}^{\mathrm{SRV}} \lambda_{\mathrm{HF,R}}^{\mathrm{SRV}} F_{\mathrm{HF,L_2}}^{\mathrm{SRV}} F_{\mathrm{HF,R_1}}^{\mathrm{SRV}} \tag{B12}$$

Appendix C: Consideration of convergence flow effect within transverse hydraulic fracture

According to the description of Mukherjee and Economides[30], chocking skin due to the convergence of fluids is expressed as the following dimensionless form

$$s_{\mathrm{c}} = \frac{k_{\mathrm{SRVD}} h_{\mathrm{D}}}{k_{\mathrm{HFD}} w_{\mathrm{HFD}}} \left[\ln \frac{h_{\mathrm{D}}}{2 r_{\mathrm{wD}}} - \frac{\pi}{2}\right] \tag{C1}$$

In essence, skin indicates the dimensionless pressure drop per production rate[36], which is defined as

$$s_c = \frac{2\pi kh}{q\mu B}(\Delta p_R - \Delta p_L) \tag{C2}$$

Noted that variable s_c is dimensionless, Δp_R indicates the steady-state pressure drop due to the radial flow pattern, and Δp_L indicates the steady-state pressure drop due to the linear flow pattern. Therefore, the wellbore pressure drop is equal to the sum of normal pressure drop and the additional pressure drop caused by skin.

Here, it needs to be emphasized that skin only has effect on the pressure drop of corresponding flow system, not on other flow systems[36,37]. Thus, Equation (10) could be modified with chocking skin to account for the convergence characteristic within transverse fracture intersecting horizontal wellbore, which is

$$\tilde{p}_{\mathrm{wHFD}}^{\mathrm{L}} = [AX_{\mathrm{SRV}}^{\mathrm{L}}(0) + s_c^{\mathrm{L}}]\tilde{q}_{\mathrm{HFD}}^{\mathrm{L}} + BX_{\mathrm{SRV}}^{\mathrm{L}}(0)\tilde{q}_{\mathrm{HFD}}^{\mathrm{R}} \tag{C3}$$

$$\tilde{p}_{\mathrm{wHFD}}^{\mathrm{R}} = [AX_{\mathrm{SRV}}^{\mathrm{R}}(0) + s_c^{\mathrm{R}}]\tilde{q}_{\mathrm{HFD}}^{\mathrm{R}} + BX_{\mathrm{SRV}}^{\mathrm{R}}(0)\tilde{q}_{\mathrm{HFD}}^{\mathrm{L}} \tag{C4}$$

A Coupled Model for Fractured Shale Reservoirs with Characteristics of Continuum Media and Fractal Geometry

Wei Yunsheng　He Dongbo　Wang Junlei　Qi Yadong

(Research Institute of Petroleum Exploration and Development, PetroChina)

Abstract: The principle focus of this work is on proposing a new model to evaluate the production performance of complexly-fractured well in shale reservoirs for conducting rate transient analysis (RTA) and pressure transient analysis (PTA). The model is established on the trilinear-flow idealization presented by Brown et al. for fractured horizontal well, and associated with continuum geometry, fractal characterization and anomalous diffusion. The coupled model could take into account the non-uniform of fracture network within stimulated reservoir volume (SRV) and property heterogeneity among hydraulic fractures in the configuration of multiple fractured horizontal well (MFHW). The second focus is put on developing a novel semi-analytical solution to fill the gap between trilinear-flow model for single-fracture hypothesis and actual MFHW configuration. The new solution could account for the effect of fracture-production interaction and property heterogeneity among fractures. Calculative results show that the type, sequence and duration of flow regimes are determined by fractal characteristics, associated anomalous diffusion, fracture number/spacing, fracture conductivity, etc. Hence, it is possible to observe the feature of long-period linear flow trends of fractured shale well and explain the presentation of new flow regimes which are generally not identified by conventional trilinear-flow model. In addition, approximate solutions for corresponding flow regimes have also been proposed to interpret the production performance during different periods. Furthermore, a field example from Sichuan Shale in China is presented to demonstrate the practical application of the new model for interpreting production data analysis of MFHW. The model provides new knowledge and insight into understanding production performance and allows us to correctly predict well performance in fractured shale reservoirs.

Key words: fractured horizontal well; continuum media; fractal geometry; anomalous diffusion; coupled model

1　Introduction

The investigation of production behavior in fractured shale reservoirs is an entirely challenging problem because of the inherent complexity of coupled natural-induced fracture network within the region of stimulated reservoir volume (SRV). Therefore, direct application of conventional gas-well performance method would unfortunately result in erroneous evaluation and production predictions[1].

To simulate hydrocarbon production from fractured reservoirs, numerical approaches and analytical approaches were simultaneously utilized. Common numerical models include direct numerical reservoir-simulation methods[2,3] and dual continuum model with discrete fracture model[4,5]. Although numerical model provide an accurate insight into the production performance of complex fracture network[6,7], but the special numerical-simulation knowledge requirement and the associated huge time-consuming computation make this method less attractive and practicable. Comparably, analytical model is a major simplification of actual physical system, which was presented in the form of multi-continuum model[8,9] and multi-linear-flow model[1,10,11]. This approach has no capacity of providing accurate solutions in the presence of SRV, but is advantageous in terms of computational convenience. It is versatile enough to incorporate the fundamental production characteristics of fractured reservoirs.

As analyzed from the literature mentioned above, an efficient and convenient production-simulation method is still lacking, which contributes to bad-quality and inefficient history matches of production for fractured reservoirs. Hence, a new semi-analytical model is presented to attempt to fill the gap in this work. The solution is an alternative to more rigorous but computationally intensive and time-consuming numerical solutions, and also improves the application effect of analytical solution.

2 Model description

2.1 Background

Many cases in Eagle Ford, Woodford and Marcellus have confirmed that unconventional gas well generally behaves transient linear flow regime for a long duration[12-14]. It indicates that it is possible and reasonable to establish an extended model based on fundamental multi-linear model.

Multi-linear/porosity model is an important alternative for simulating production behavior of fractured horizontal well in addition to instantaneous source model[9,15-17]. Common multi-linear models are list in Figure 1. Linear flow model in fractured tight gas well was specially studied by Wattenbarger et al.[18] and EI-Banbi et al.[19] Linear dual porosity model was presented to approximate the production performance of shale gas well by Bello et al.[10]. The trilinear-flow model was developed by Brown et al.[1] based on three contiguous flow regions to represent sufficient details of fluid and reservoir characteristics. Subsequently, Stalgorova and Mattar[20,21] further subdivided the fractured reservoirs into five regions instead of three to simulate a fracture that is surrounded by a stimulated region of limited extent (fracturing branching). These works

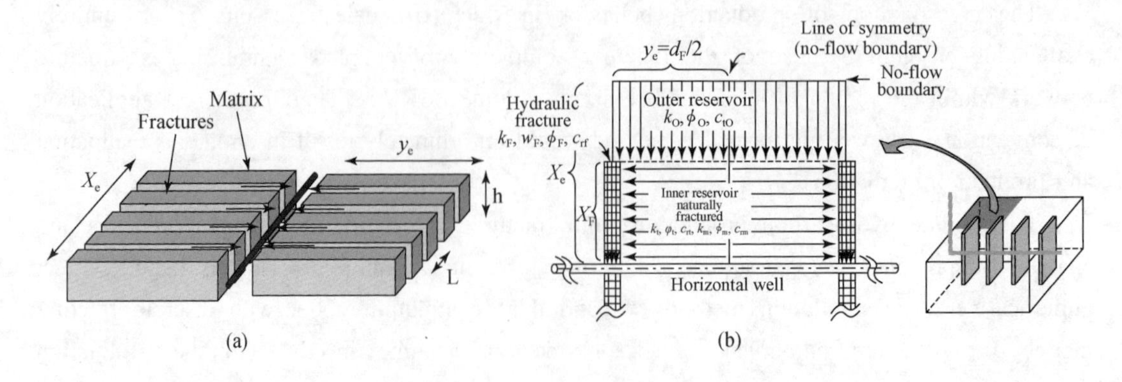

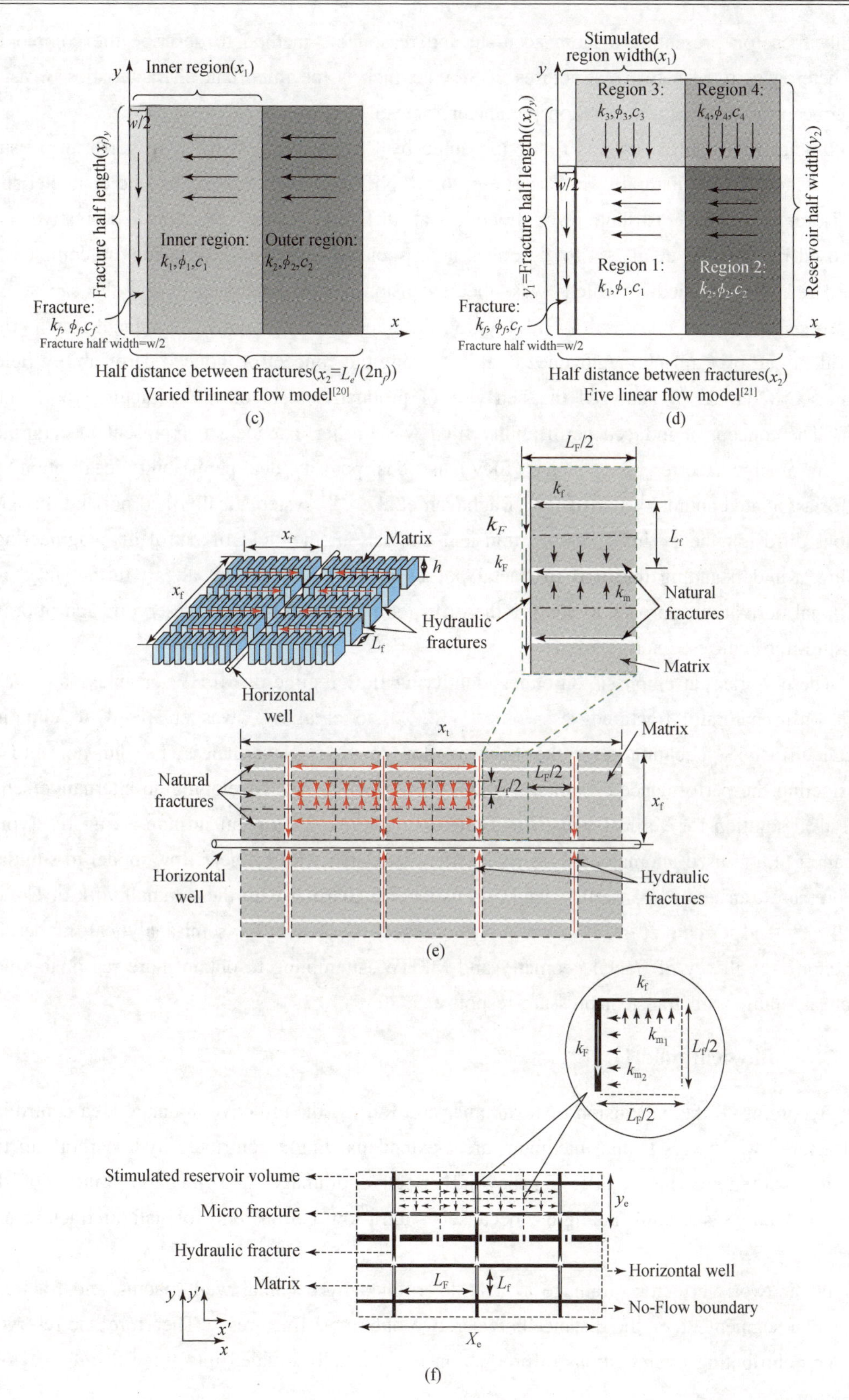

Figure 1 Common multi-linear/porosity flow model of fractured horizontal well with SRV characteristics

mainly focus on presenting a more accurate and reasonable method to describe the complexity and heterogeneity of network structures in SRV, which is the important characteristics of using the concept of SRV as a correlation parameter for well performance.

Fractal diffusion theory was firstly presented by Chang and Yortsos[22] to perform pressure transient analysis for naturally fractured reservoirs (NFRs), which represented the characteristics of disorderly spatial distribution of fractures at different scales. Fractional derivative was proposed by Metzler et al.[23] to describe the associated anomalous diffusion in complex or disordered fractal media, which was the extension of O'Shaugnessy and Procaccia[24]. Subsequently, the rate response of NFRs during the transient-and boundary-dominant flow period was identified by Camacho-Velάzquez et al.[25] Using the concept of induced permeability field, Fuentes-Cruz et al.[26] analyzed the behavior of production data of multi-fractured horizontal well. The concept of induced permeability filed was further extended to represent the complex structure of stimulated reservoir volume (SRV) in a dual porosity/dual permeability idealization[27]. As for associated anomalous diffusion, Raghavan et al.[28-31] systematically incorporated the anomolous diffusion theory into pressure transient analysis in the field of petroleum engineering, aiming at understanding transport in fractal porous media. Razminia et al.[32] further used the fractional derivative approach to analyze the diffusion process in fractured reservoirs and improve pressure-transient-test interpretation.

These works put emphasis on either identifying flow regime of MFHW or analyzing fractal diffusion in naturally fractured reservoirs[22,25]. Cossio et al.[33] was creatively to combine fractal diffusivity equation with the trilinear flow model. Nevertheless, he did not aim at considering the performance of wells in fractal-like structure, but developing an alternative semi-analytical solution for vertical well with finite conductivity fracture in infinite-acting reservoir. Ozcan et al.[34] used anomalous diffusion model associated with trilinear flow model to simulate the pressure transient response in fractal porous media. Starting with the seminal work of Cossio et al.[33] and Ozcan et al.[34], we are motivated to develop a semi-analytical model by combining the theory of fractal geometry and MFHW, attempting to obtain more reasonable and practical solutions of transient-pressure response.

2.2 Physical model

According to the conclusion of Jayakumar and Rai[14], the effective drainage area controlled by fractured well refers to the maximum areal extent that might contribute hydrocarbons to the well in its lifetime. In the area far beyond effective drainage area, the contribution of the reservoir volume is usually negligible because of ultra-low permeability of natural fracture and matrix.

In our work, effective drainage area is determined by eventual well spacing and fractured horizontal segment (i. e. the distance between two outermost fractures). Therefore, the reservoir volume contributing hydrocarbons to production is potentially divided into three distinct regions in Figure 2.

(1) Hydraulic Fractured Volume (HFV). HFV consists of a set of discrete finite-conductivity hydraulic fractures intersected by horizontal wellbore. The proppant is concentrated in these dominant hydraulic fractures that are connected to an unpropped complex fracture network in reservoirs. HFV contributes much higher conductivity and better connection with reservoir.

(2) Stimulated Reservoir Volume (SRV). SRV is created in the vicinity of HFV (i. e., the effective drainage area impacted by fracturing treatment). SRV is related to the non-uniformly-distributed fracture network with fractal characteristic on a wide range of scales. Generally, its areal extension is determined by perforated lateral length and dominant HF length.

(3) External Reservoir Volume (XRV). XRV refers to the part of effective drainage area in addition to HFV and SRV. XRV is slightly disturbed by fracturing treatment but could contribute hydrocarbons in production lifetime. It locates in the vicinity beyond HF tips, external to SRV region.

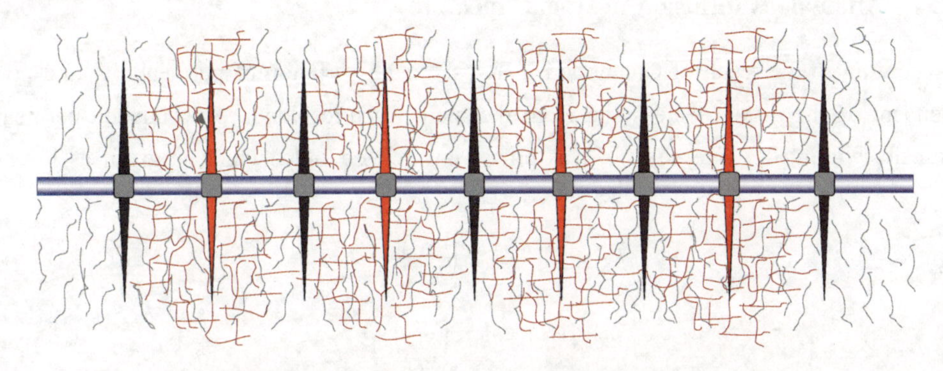

(a) Complexly-distributed fracture network within SRV region

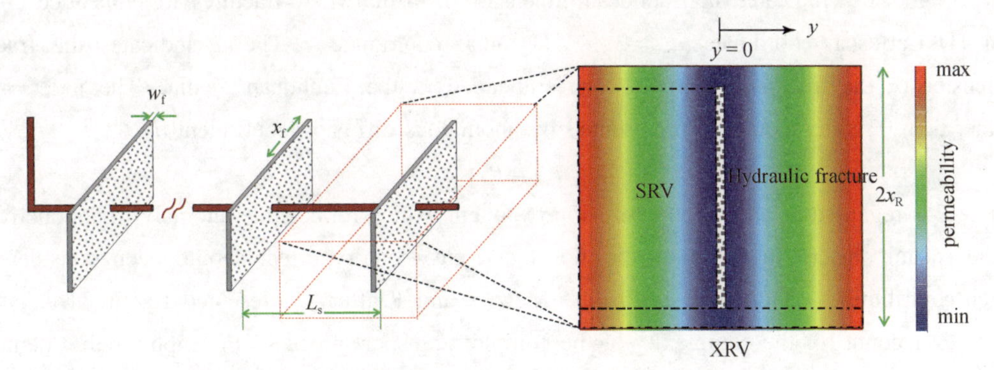

(b)Schematic of multi-fractured horizontal well in effective drainage area

Figure 2 Multi-fracture model in fractured reservoir

It is noted that the model in this work is established on the single-porosity concept, and the consideration of dual-medium (double-porosity) is presented in Appendix E.

2.3 Mathematical Assumption

According to the descriptions above, some idealizations and simplifying assumptions are summarized as follows:

(1) An isotropic, horizontal, slap gas reservoir is bounded by overlying and underlying impermeable strata.

(2) The height of HFV, SRV and XRV is assumed to be same as the strata height.

(3) The petrophysical properties in XRV are homogeneous and identical.

(4) SRV is described as weak-interconnected fracture network with fractal characteristic to account for permeability and porosity alteration around HFV.

(5) HFV connects to horizontal wellbore, and HF wings are symmetrically spaced with regard to horizontal wellbore. Correspondingly, Production from the surface of horizontal wellbore might be negligible.

(6) The production process is isothermal, and matrix and fracture network are considered incompressible compared to gas compression.

2.4 Anomalous diffusion in fractal medium

According to the result of Chang and Yortsos[22], the variation of permeability and porosity is dependent on the distance measured from the main fracture plane. The induced permeability and porosity within SRV are described as power-law-related field, which is given as in another form:

$$k_{SRV}(y) = k_{SRV}^{ref}\left(\frac{y}{L_{ref}}\right)^{d_f-d_e-\theta} \tag{1}$$

$$\phi_{SRV}(y) = \phi_{ref}\left(\frac{y}{L_{ref}}\right)^{d_f-d_e} \tag{2}$$

The index d_e indicates the Euclidean dimension of media where fractures are embedded ($d_e = 1$ in 1D Cartesian coordinate, $d_e = 2$ in 1D radial coordinate). The d_f indicates the fractal dimension of the fractal-fracture network embedded in the Euclidean media. The index θ is greater than 0, which could be represented by anomalous diffusion coefficient d_w[30]:

$$d_w = \theta + 2 \tag{3}$$

Where θ is regarded as anomalous diffusion exponent, which reflects the anomaly in conductivity in the fractal object. In fractal media, non-local and memory effect should been considered, which contributes to anomalous diffusion. Metzler and Klafter[23] presented fractional calculus (FC) to account for the complexity of the complex transport process. FC captures the memory of a dynamical phenomenon, and the modified Darcy rate is given by the gradient law in a convolved form:

$$\vec{v}_{SRV} = -\frac{k_{SRV}}{\mu}\frac{\partial}{\partial t}\int_0^t \frac{\nabla p}{(t-t')^{1-\gamma}}dt' \tag{4}$$

Where the constant γ plays a role analogous to d_w or θ, and the relations is given by

$$\gamma = \frac{2}{d_w} \quad \text{or} \quad \gamma = \frac{2}{2+\theta} \tag{5}$$

The relationship between gradient law and fractional derivative was presented by

$$\frac{\partial^{\gamma}}{\partial t^{\gamma}}f(t)=\frac{1}{\Gamma(1-\gamma)}\int_{0}^{t}\frac{1}{(t-t')^{\gamma}}\frac{\partial f(t')}{\partial t'}\mathrm{d}t' \tag{6}$$

Taking the Caputo fractional operator into Equation(4), Darcy law is given in the form of fractional derivative as

$$\vec{v}_{\mathrm{SRV}}=-\frac{k_{\mathrm{SRV}}}{\mu}\frac{\partial^{1-\gamma}(\nabla p)}{\partial t^{1-\gamma}} \tag{7}$$

3 Mathematical model

The mathematical model is described as seen in Figure 3. Our fundamental assumption is analogous to the model presented by Fuentes-Cruz et al. [26,27] The new formulation of MFHW is established based on multiple trilinear-flow models with fractal geometry. Each individual trilinear-flow model corresponds to an element/unit mentioned in the following Section *Stimulated reservoir volume with fractal fracture network*.

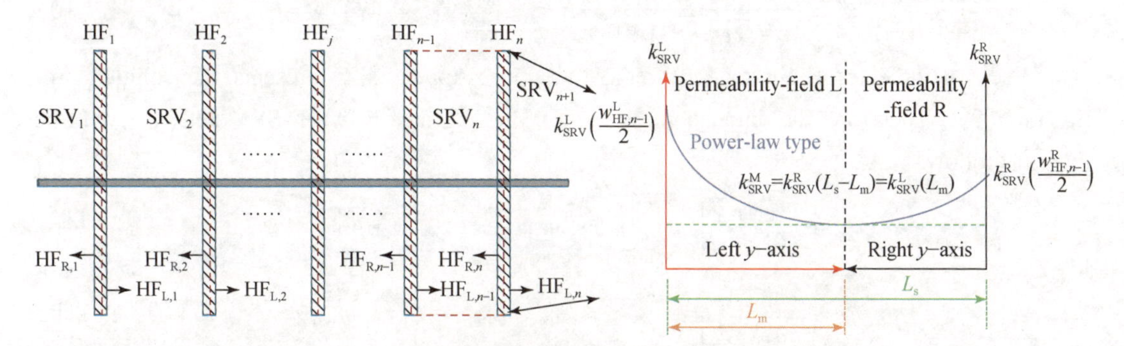

Figure 3 Permeability-distributed field in elementary unit within SRV

3.1 External Reservoir Volume with continuum

Within XRV, the volume is less slightly fractured compared with SRV due to the decrement of fracturing treatment. Hence, fracture network within XRV is modeled as continuum media, where fractures interconnect with each other.

After incorporating dimensionless definitions in Appendix A, the dimensionless pressure governing equation in Laplace-transform domain is satisfied as follows:

$$\frac{\partial^{2}\tilde{m}_{\mathrm{XRVD}}}{\partial x_{\mathrm{D}}^{2}}=\frac{s}{\eta_{\mathrm{XRVD}}}\tilde{m}_{\mathrm{XRVD}} \tag{8}$$

Appendix B provides the details of derivation for Equation (7). Thus we could obtain the dimensionless flux on interface between SRV and XRV, which is given as

$$\left.\left(\frac{\partial\tilde{m}_{\mathrm{XRVD}}}{\partial x_{\mathrm{D}}}\right)\right|_{x_{\mathrm{D}}=x_{\mathrm{HFD}}}=-\tilde{F}_{\mathrm{SRV}}^{\mathrm{XRV}}(\tilde{m}_{\mathrm{SRVD}})\big|_{x_{\mathrm{D}}=x_{\mathrm{HFD}}} \tag{9}$$

3.2 Stimulated reservoir volume with fractal fracture network

SRV accounts for the presence of fractures with a non-uniform spatial distribution at different scales. We assume that the highest permeability position is located at the hydraulic fracture plane. As the distance to the fracture plane increases, the permeability enhancement monotonically reduces to a minimal value, which is the result of energy dissipation of fracturing treatment operation.

In this subsection, a representative unit of SRV is selected as the element of MFHW. The unit consists of two half-width hydraulic fractures and one SRV region with fractal geometry, as noted by the red rectangle in Figure 3. In the elementary unit, variables related to HF are distinguished by superscript denoted as *R* and *L*.

After incorporating Laplace transformation, the flow equation in SRV is given in the following form of dimensionless pressure, this is

$$\frac{\partial^2 \tilde{m}_{\mathrm{SRVD}}^{\xi}}{\partial y_{\mathrm{D}}^2}+\frac{d_{\mathrm{f}}^{\xi}-d_{\mathrm{e}}^{\xi}-\theta^{\xi}}{y_{\mathrm{D}}}\frac{\partial \tilde{m}_{\mathrm{SRVD}}^{\xi}}{\partial y_{\mathrm{D}}}=y_{\mathrm{D}}^{\theta}\left(\frac{s^{\gamma_\xi}}{\eta_{\mathrm{SRVD}}^{\xi}}+\frac{\lambda_{\mathrm{MAF},\xi}^{\mathrm{MIF}}F_{\mathrm{SRV},\xi}^{\mathrm{XRV}}}{s^{1-\gamma_\xi}}\right)\tilde{m}_{\mathrm{SRVD}}^{\xi} \tag{10}$$

Where, ξ indicates Right or Left region in unit. Appendix C presents the solutions for Equation(9). Therefore, the dimensionless rate on the interface between SRV and HF could be given as follows:

$$\left(\frac{\partial \tilde{m}_{\mathrm{SRVD}}^{\mathrm{L}}}{\partial y_{\mathrm{D}}}\right)\Bigg|_{y_{\mathrm{D}}=w_{\mathrm{HFD}}/2}=\tilde{F}_{\mathrm{HF,R1}}^{\mathrm{SRV}}\tilde{m}_{\mathrm{HFD}}^{\mathrm{R}}-\tilde{F}_{\mathrm{HF,L1}}^{\mathrm{SRV}}\tilde{m}_{\mathrm{HFD}}^{\mathrm{L}} \tag{11a}$$

$$\left(\frac{\partial \tilde{m}_{\mathrm{SRVD}}^{\mathrm{R}}}{\partial y_{\mathrm{D}}}\right)\Bigg|_{y_{\mathrm{D}}=w_{\mathrm{HFD}}/2}=\tilde{F}_{\mathrm{HF,R2}}^{\mathrm{SRV}}\tilde{m}_{\mathrm{HFD}}^{\mathrm{R}}-\tilde{F}_{\mathrm{HF,L2}}^{\mathrm{SRV}}\tilde{m}_{\mathrm{HFD}}^{\mathrm{L}} \tag{11b}$$

It is noted that fracture-production interaction in this work is decomposed into two parts: interaction between adjacent fractures within single unit and interaction among adjacent units. This Section only accounts for the effect of interaction between adjacent fractures. The other interaction effect would be presented in Section 3.4.

3.3 Hydraulic fractured volume with discrete fractures

HFs are discretely distributed along horizontal wellbore. Most studies on multi-linear/porosity model are established on the hypothesis of single-fracture hypothesis[1,10], which is inherently assumed identical properties among HFs and SRV.

Utilizing modified Darcy law with anomalous diffusion, the pressure governing equation in HF is given in the following dimensionless form.

$$\frac{\partial^2 \tilde{m}_{\mathrm{HFD}}^{\xi}}{\partial x_{\mathrm{D}}^2}+2\frac{\lambda_{\mathrm{HF},\xi}^{\mathrm{SRV}}}{s^{\gamma_\xi-1}}\left(\frac{\partial \tilde{m}_{\mathrm{SRVD}}^{\xi}}{\partial y_{\mathrm{D}}}\right)\Bigg|_{y_{\mathrm{D}}=w_{\mathrm{fD}}/2}=\frac{s}{\eta_{\mathrm{HFD}}^{\xi}}\tilde{m}_{\mathrm{HFD}}^{\xi} \tag{12}$$

Appendix D provides the derivation of Section 3.3 in detail. The dimensionless pressure distribution along hydraulic fracture is given by

$$\tilde{m}^{\mathrm{L}}_{\mathrm{HFD}}(x_{\mathrm{D}}) = \tilde{q}^{\mathrm{L}}_{\mathrm{HFD}} A X^{\mathrm{L}}_{\mathrm{SRV}}(x_{\mathrm{D}}) + \tilde{q}^{\mathrm{R}}_{\mathrm{HFD}} B X^{\mathrm{L}}_{\mathrm{SRV}}(x_{\mathrm{D}}) \tag{13a}$$

$$\tilde{m}^{\mathrm{R}}_{\mathrm{HFD}}(x_{\mathrm{D}}) = \tilde{q}^{\mathrm{L}}_{\mathrm{HFD}} B X^{\mathrm{R}}_{\mathrm{SRV}}(x_{\mathrm{D}}) + \tilde{q}^{\mathrm{R}}_{\mathrm{HFD}} A X^{\mathrm{R}}_{\mathrm{SRV}}(x_{\mathrm{D}}) \tag{13b}$$

It is noted that the solutions above are presented based on vertical fracture, which is different from transverse fracture intersecting horizontal wellbore. Within transverse fracture, radial flow pattern would result in additional pressure drop due to the convergence of fluid into horizontal wellbore. The effect of flow convergence is regarded as a skin factor, namely the convergence flow skin (chocking skin) presented by Mukherjee and Economides[34], which is given as

$$s_{\mathrm{c}} = \frac{k_{\mathrm{SRVD}} h_{\mathrm{D}}}{k_{\mathrm{HFD}} w_{\mathrm{HFD}}} \left[\ln \frac{h_{\mathrm{D}}}{2 r_{\mathrm{wD}}} - \frac{\pi}{2} \right] \tag{14}$$

Where $k_{\mathrm{SRVD}} = k_{\mathrm{SRV}}/k_{\mathrm{ref}}$, $k_{\mathrm{HFD}} = k_{\mathrm{HF}}/k_{\mathrm{ref}}$, $h_{\mathrm{D}} = h/L_{\mathrm{ref}}$, $r_{\mathrm{wD}} = r_{\mathrm{w}}/L_{\mathrm{ref}}$.

Thus, Equation (13) is modified with chocking skin to account for the convergence characteristic within transverse fracture intersecting horizontal wellbore, which is

$$\tilde{m}^{\mathrm{L}}_{\mathrm{wD}} = \tilde{q}^{\mathrm{L}}_{\mathrm{HFD}} [A X^{\mathrm{L}}_{\mathrm{SRV}}(0) + s^{\mathrm{L}}_{\mathrm{c}}] + \tilde{q}^{\mathrm{R}}_{\mathrm{HFD}} B X^{\mathrm{L}}_{\mathrm{SRV}}(0) \tag{15a}$$

$$\tilde{m}^{\mathrm{R}}_{\mathrm{wD}} = \tilde{q}^{\mathrm{R}}_{\mathrm{HFD}} [A X^{\mathrm{R}}_{\mathrm{SRV}}(0) + s^{\mathrm{R}}_{\mathrm{c}}] + \tilde{q}^{\mathrm{L}}_{\mathrm{HFD}} B X^{\mathrm{R}}_{\mathrm{SRV}}(0) \tag{15b}$$

3.4 Solution for multiple-trilinear-flow model

To integrate the solutions of single-trilinear-flow element together, we incorporate the principle of pressure superposition. It indicates that pressure distribution caused by n_{f} fractures can be written in Laplace domain as follows,

$$\tilde{m}_{\mathrm{D}j,i} = \sum_{i=1}^{n_{\mathrm{f}}} \tilde{q}^{\mathrm{L}}_{\mathrm{HFD},i} \Delta \tilde{m}^{\mathrm{L}}_{\mathrm{D}j,i} + \sum_{i=1}^{n_{\mathrm{f}}} \tilde{q}^{\mathrm{R}}_{\mathrm{HFD},i} \Delta \tilde{m}^{\mathrm{R}}_{\mathrm{D}j,i} \tag{16}$$

Where subscript "j,i" indicates dimensionless pressure response of the j-th fracture caused by the production of the i-th fracture. Based on the assumption of infinite conductivity wellbore, the dimensionless pressure on the interface between each hydraulic fracture and wellbore is equal to wellbore pressure m_{wD}.

$$\tilde{m}_{\mathrm{wD},1} = \tilde{m}_{\mathrm{wD},2} = \cdots = \tilde{m}_{\mathrm{wD},n_{\mathrm{f}}} = \tilde{m}_{\mathrm{wD}} \tag{17}$$

In the condition of constant flow rate for multiple-fracture system, the additional condition yields the following expression in Laplace domain

$$\sum_{i=1}^{n_{\mathrm{f}}} \tilde{q}^{\mathrm{L}}_{\mathrm{HFD},i} + \sum_{i=1}^{n_{\mathrm{f}}} \tilde{q}^{\mathrm{R}}_{\mathrm{HFD},i} = \frac{1}{s} \tag{18}$$

Thus, we can obtain the following matrix by combining Equation (14) ~ Equation (18), which is

$$\begin{bmatrix} \boldsymbol{A} & -\boldsymbol{B} \\ \boldsymbol{B}^{\mathrm{T}} & \boldsymbol{0} \end{bmatrix} \cdot \begin{bmatrix} \boldsymbol{X} \\ s\tilde{p}_{\mathrm{wD}} \end{bmatrix} = \begin{bmatrix} \boldsymbol{0} \\ 1 \end{bmatrix} \tag{19}$$

Where bold types in Equation (14) are used for vectors and matrixes. The matrix $\boldsymbol{A}$ is a coefficient matrix of dimension $2n_{\mathrm{f}} \times 2n_{\mathrm{f}}$, and i-th element in $\boldsymbol{A}$ represents the i-th elementary trilinear-flow model, which can be further expressed as

$$A_i = \begin{cases} AX_{\mathrm{SRV}}^{\mathrm{R},1}, i = 1 \\ \begin{bmatrix} AX_{\mathrm{SRV}}^{\mathrm{L},i} & BX_{\mathrm{SRV}}^{\mathrm{L},i} \\ BX_{\mathrm{SRV}}^{\mathrm{R},i} & AX_{\mathrm{SRV}}^{\mathrm{R},i} \end{bmatrix}, 2 \leqslant i \leqslant n_{\mathrm{f}} \\ AX_{\mathrm{SRV}}^{\mathrm{L},n_f+1}, i = n_{\mathrm{f}} + 1 \end{cases} \tag{20a}$$

Where element $\boldsymbol{B}$ represents the constraint of Equation(18), which is given as

$$B_i = \begin{cases} 1, i = 1 \text{ and } n_{\mathrm{f}} + 1 \\ [1 \quad 1]^{\mathrm{T}}, 2 \leqslant i \leqslant n_{\mathrm{f}} \end{cases} \tag{20b}$$

In addition, rate vector $\boldsymbol{X}$ of dimensionless individual half-width fracture is expressed as

$$X_i = \begin{cases} s\tilde{q}_{\mathrm{HFD}}^{\mathrm{R},1}(s), i = 1 \\ [s\tilde{q}_{\mathrm{HFD}}^{\mathrm{L},i-1}(s) \quad s\tilde{q}_{\mathrm{HFD}}^{\mathrm{R},i}(s)]^{\mathrm{T}}, 2 \leqslant i \leqslant n_{\mathrm{f}} \\ s\tilde{q}_{\mathrm{HFD}}^{\mathrm{L},n_f}(s), i = n_{\mathrm{f}} + 1 \end{cases} \tag{20c}$$

By solving Equation(19), we can obtain the unknown wellbore pressure and the flow rate distribution of hydraulic fractures in real time domain by applying Stehfest numerical-inversion algorithm[36].

4 Model validation

In this section, two alternative simulations are selected to validate our model, where the parameters used for the validation of this model are list in Table 1.

Table 1 Basic parameters used for model validation in the form of oil phase

Basic model parameter	Value	Basic model parameter	Value
Formation width, x_{R}/m	440	Formation porosity, φ/%	10
Formation thickness, h/m	10	Production rate, q_{sc}/(m^3/s)	3.4722×10^{-4}
Formation permeability, k_{m}/m^2	10^{-13}	Initial pressure, p_{i}/Pa	3.4471×10^{7}
Fracture spacing, L_{s}/m	400	Viscosity of fluid, μ/(Pa · s)	0.001
Fracture half-length, x_{HF}/m	200	Compressibility, c_{t}/Pa^{-1}	4.3511×10^{-10}

In this work, the homogeneity case generally refers to Euclidean dimension and classical diffusion in the symmetry configuration with equally-spaced fractures, which is given as

$$d_{\mathrm{e}} = 1, d_{\mathrm{f}} = 1, \gamma = 1, \theta = 0, L_{\mathrm{mD}} = 0.5L_{\mathrm{sD}} \tag{21}$$

When the parameters in Left region are identical to the Right within a SRV unit in a homogeneity case, the multi-fracture solution in this work completely converges to conventional MFHW solution[15,16]. It needs to be emphasized that multi-linear model has no capacity of simulating pseudo radial flow regime in infinite-acting reservoirs, so the numerical results must be generated in a closed reservoir. The constraint condition is generally set to be $0.7 \leqslant (x_{\mathrm{HF}}/x_{\mathrm{R}})$

$\leqslant 1$. It indicates boundary-dominant flow would emerge in advance before pseudo radial flow regime occurs. Song et al. [37] had affirmed that constraint condition is reasonable and reliable in the practice of fracturing treatment.

For homogeneity case, we use Ecrin-Saphir numerical simulator to generate synthetic data of multi-fractured horizontal well on the constraint condition of $0.7 \leqslant (x_{HF}/x_R) \leqslant 1$. Figure 4 indicates the case of equally-spaced configuration with 5 fractures. Excellent correlations in Figure 4 indicate that the new algorithm presented in Section 3.4 is reliable to simulate the production performance of MFHW.

For the heterogeneity case, we resort to single-well model with infinite conductivity fracture in Saphir simulator. The heterogeneity characteristic of SRV is described with permeability-porosity alternation fields, which are created in accordance with power-law regulation presented by Equation (1) and Equation (2). Then corresponding transient pressure data are generated by using numerical simulator. As shown in Figure 5, there is a good agreement between our solutions and the results from numerical simulator. It is verified that the semi-analytical with fractal geometry is calculated accurately.

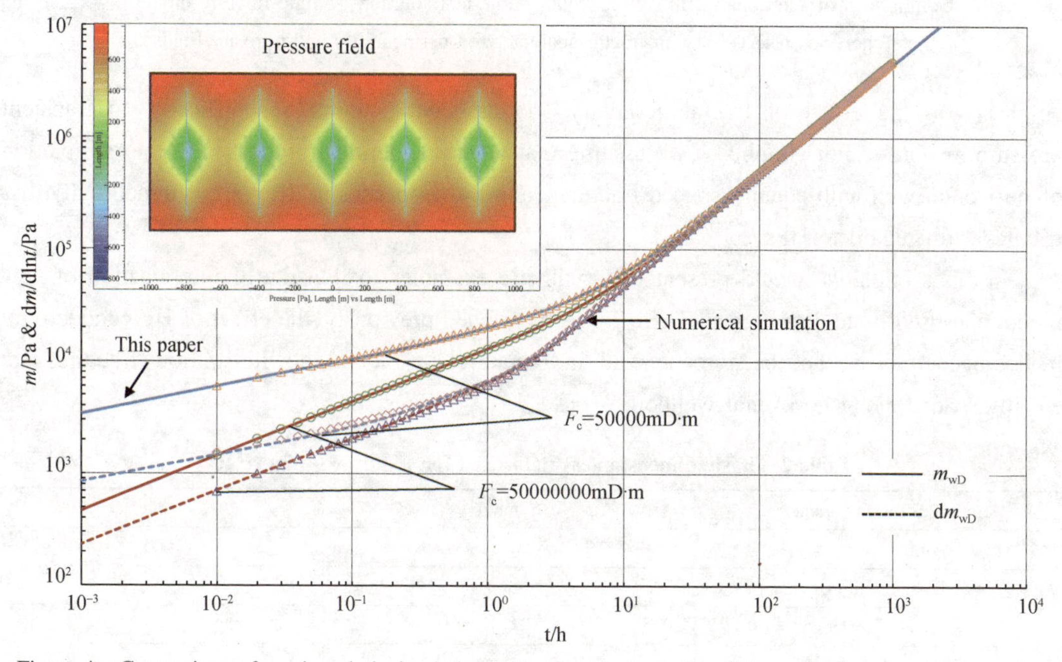

Figure 4 Comparison of semi-analytical solution with numerical solution (The multi-fracture model consists of 5 fractures with identical properties in homogeneous reservoirs)

5 Results and discussions

The advantage of coupled model is to account for the heterogeneity and arbitrary of fracture system of SRV and the varied properties of HF (width, permeability, HF number and HF spacing, etc.). There are many factors affecting transient-pressure characteristics. Here we put

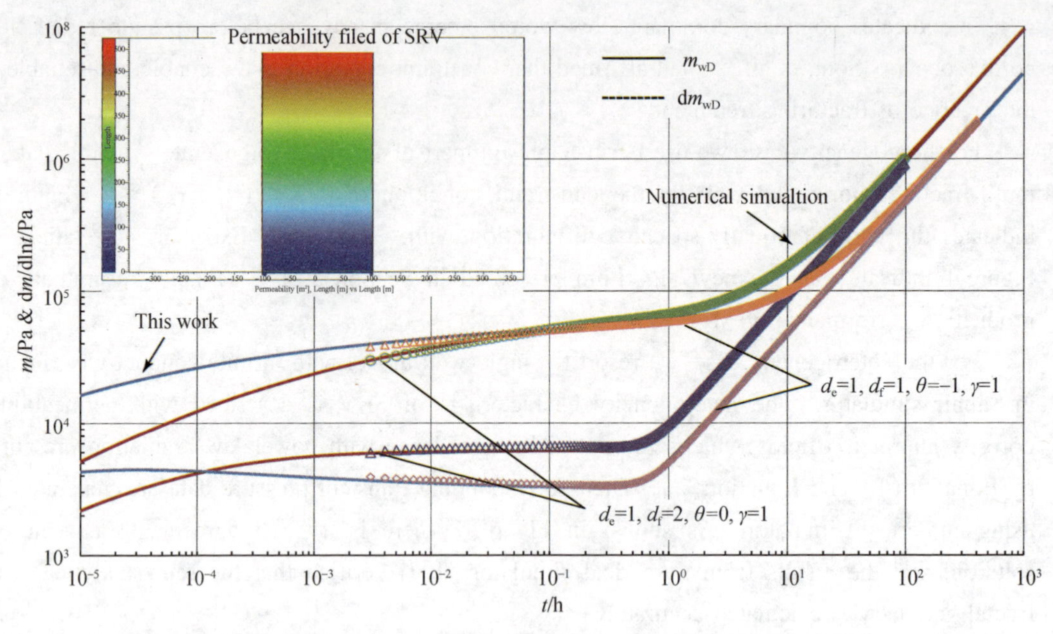

Figure 5　Comparison of semi-analytical solution with numerical solution (Single-fracture model is located in heterogeneous reservoirs with fractal-characterized permeability and porosity fields)

emphasis on the effect of fractal geometry and associated anomalous diffusion on transient pressure response. For simplicity, we assume that HFs has identical properties in the completion of horizontal well with equally-spaced fractures, and the properties in SRV are symmetrically distributed with regard to HFs.

In this section, we would present two synthetic examples to illustrate the application of new model and used basic data used in Table 2. As mentioned previously, the effect of HF conductivity in homogeneity case has been recognized in literature[9,15,16]. In this illustration, fractures are equally spaced along horizontal wellbore.

Table 2　Basic dimensionless data used for Figure 6 ~ Figure 10

Basic dimension parameter	Symbol	Value
Formation width	x_{RD}	1000
Fracture number	n_f	5
Fracture spacing	L_s	2
Fracture half-length	x_{HFD}	1
Fracture width	w_{HFD}	0.001
Fracture conductivity	C_{HFD}	$10^{-2}, 10^{-1}, \cdots, 10^{4}$
Hydraulic fracture diffusivity	η_{HFD}	$10^{1}, 10^{0}, \cdots, 10^{6}$
SRV diffusivity	η_{SRVD}	1
XRV diffusivity	η_{XRD}	0.05
Flow capacity ratio of SRV-HF	λ_{HF}^{SRV}	$10^{-4}, 10^{-3}, \cdots, 10^{2}$
Flow capacity ratio of XRV-SRV	λ_{SRV}^{XRV}	0.5

续表

Basic dimension parameter	Symbol	Value
Production rate of MFHW	q_{wD}	1
Euclidean dimension	d_e	1
Fractal dimension	d_f	0.5,0.6,…,1
Scaling variable	θ	0
Fractional derivative order	γ	0.5,0.6,…,1

5.1 Effect of fracture conductivity

Figure 6 and Figure 7 respectively show the effect of HF conductivity on transient-pressure behavior in heterogeneity case, including fractal case and anomalous-diffusion case. Here, dashed lines represent the pressure derivatives for homogeneity case (Euclidian dimension and classical diffusion) in different conditions of HF conductivity. It is found that all curves can be divided into three parts: early-time flow regime, intermediate-time flow regime and late-time flow regime. At early-time period, the transient response in HF is dominant without SRV participation, while the transient response in XRV would be dominant at late-time period after the responses in HF and SRV both reach the boundary-dominant state. Noted that early and late-time flow regimes are not affected by HF conductivity, and HF conductivity mainly affects the type, sequence and duration of intermediate-time flow regimes.

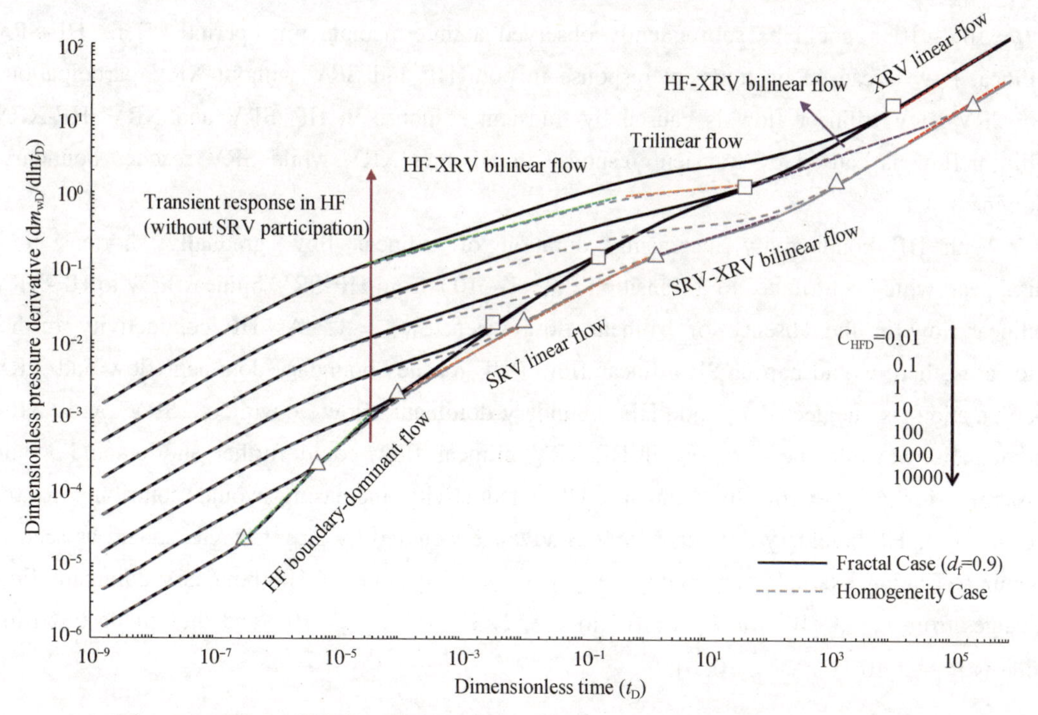

Figure 6 Effect of HF conductivity on pressure derivative in fractal case (d_f = 0.9)

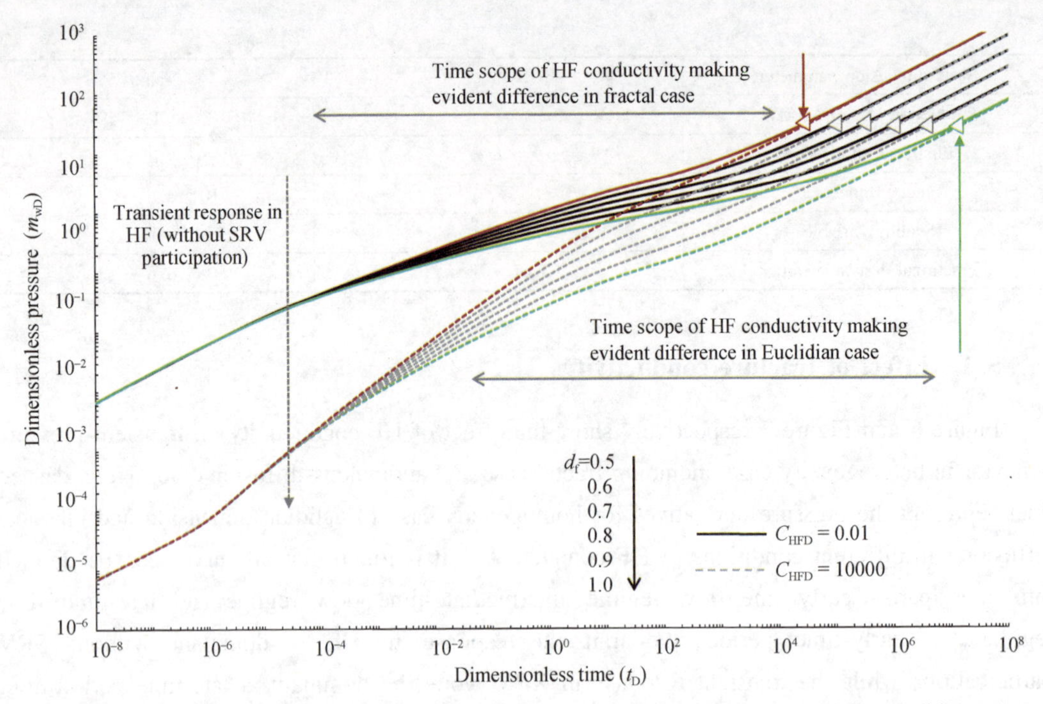

Figure 7　Effect of fractal dimension df on pressure in large/small HF conductivity condition

For smaller HF conductivity (e. g. C_{HFD} = 0. 01 in homogeneity case), HF-SRV bilinear flow ($t_D = 10^{-4} \sim 0.5$), HF-SRV-XRV trilinear flow ($t_D = 1 \sim 10$), and HF-XRV bilinear linear ($t_D = 10^2 \sim 10^4$) could be subsequently observed at intermediate time period. Here, HF-SRV bilinear flow is caused by transient response in both HF and SRV without XRV participation; HF-SRV-XRV trilinear flow is caused by transient response in HF, SRV and XRV; HF-XRV bilinear flow is caused by transient response in HF and XRV while SRV reaches boundary-dominant state.

With HF conductivity increasing, duration of trilinear flow gradually shortens until disappear, which contributes to a transition ($t_D = 1 \sim 10$) from HF-SRV bilinear flow to HF-XRV bilinear flow in the absence of trilinear flow when C_{HFD} = 1. As HF conductivity further increases, there would appear SRV linear flow (HF reaches boundary-dominant flow and XRV participation is neglected) and HF boundary-dominant flow (without SRV and XRV participation), while the duration of HF-XRV bilinear flow could further shorten and would disappear for $C_{HFD} \geqslant 10$. In summary, HF conductivity increasing would contribute to the occurring of HF boundary-dominant flow in advance, denoted by gray triangle/square as seen in Figure 6; Taking example for homogeneity case, starting time of HF boundary-dominant flow changes from $t_D \approx 3\times10^5$ for $C_{HFD} = 10^4$ to $t_D \approx 2\times10^3$ for $C_{HFD} = 10^3$, and the smallest starting time is $t_D \approx 3\times10^{-7}$ for $C_{HFD} = 0.01$.

5.2　Effect of fractal Case

Figure 6 shows the effect of HF conductivity on pressure derivative in the condition of fractal case ($d_f<1$) incorporating the weak-interconnection of SRV, denoted by unbroken lines. During early-time period, the transient response in SRV is not taken into account, so the pressure derivatives for fractal case overlap with homogeneity case as mentioned above. As soon as the fluid in SRV participates in the flowing process, the derivatives would deviate from homogeneity case. During the deviation period, the fractal case contributes to greater pressure depletion, which is identified by the characterized straight with larger slope compared with homogeneity case. During the late-time period, pressure derivative for different conditions of HF conductivity overlaps again with each other. In addition, the starting time of HF boundary-dominant flow for fractal case (denoted by gray squares) accelerates to occur, and the time gap between homogeneity case and fractal case would be more obvious as HF conductivity decreases.

In order to further discuss the effect of fractal characteristic, Figure 7 shows an illustration including six values of fractal dimension index d_f. We find that the smaller fractal dimension index is, the greater pressure depletion is, which is identified by a bigger-slope straight. Besides, the gray triangles represent the starting time of late-time SRV linear flow for smaller HF conductivity. It indicates that changing HF conductivity makes evident difference on the pressure depletion within early time scope in the case of stronger fractal structure (smaller value of d_f). Put another way, the HF conductivity plays a more important role at early time period in fractal-like structured reservoirs.

5.3 Effect of anomalous diffusion

Figure 8 shows the effect of HF conductivity on pressure for the case of $\gamma<1$ incorporating the influence of memory or anomalous diffusion. Compared with Section 5.2, we find that there are intersects between anomalous diffusion ($\gamma<1$) and classical diffusion ($\gamma=1$), which is consistent with the conclusion of Raghavan and Chen[31] that "the characteristic of intersect is typical of solutions that are governed by fractional diffusion". At early-time period when HF linear flow period finishes, the pressure depletion of anomalous diffusion is smaller than classical diffusion, but would subsequently become greater at the time scope where starting time is the intersect time. Therefore, anomalous diffusion is advantage in the terms of reducing pressure depletion during early-time period.

Figure 9 further displays the influence of anomalous diffusion. The most important results for anomalous diffusion are two fold: ① the smaller the anomalous diffusion index γ is, the early the pressure response deviates from classical diffusion case. For example, the ending time of HF linear flow in the condition of larger HF conductivity is at $t_D \approx 4\times10^{-7}$ for $\gamma=1$, but SRV has fully participated in the flowing process at the same time for $\gamma=0.5$. ② the duration of intermediate-time period decreases as anomalous diffusion index γ decreases. For instance, intermediate-time flow regimes ($C_{HFD}=10^4$, $\gamma=0.5$), consisting of HF boundary-dominant flow, SRV linear flow and SRV-XRV bilinear flow, are not identified obviously. Thus it appears possible to explain the feature of long linear trends in unconventional gas reservoirs; for example, the duration of linear flow lasts from $t_D \approx 2\times10^{-5}$ to $t_D \approx 10$ for $\gamma=0.5$ in the condition

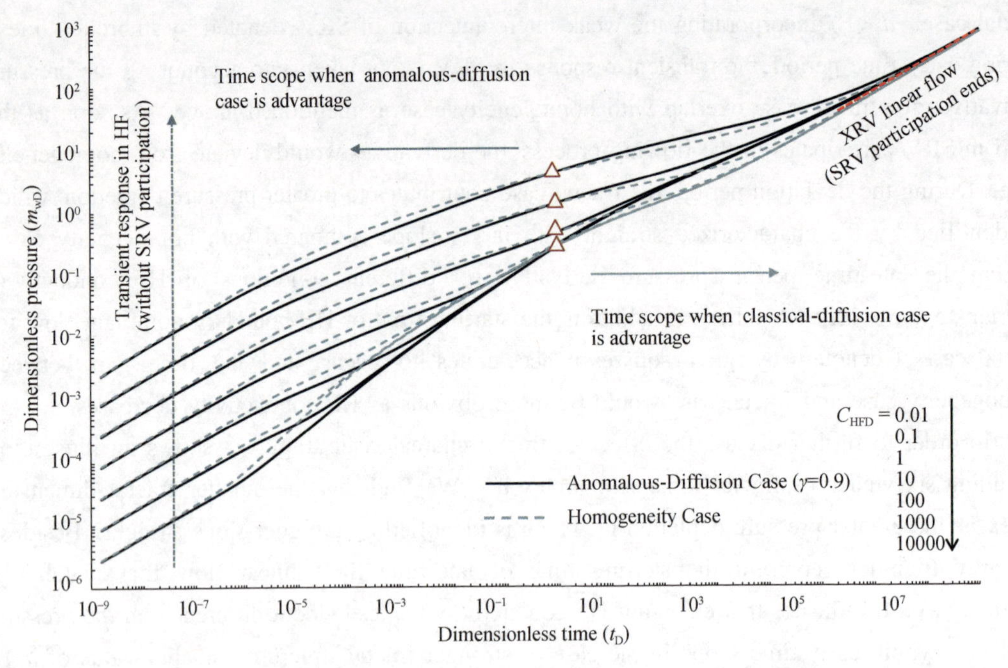

Figure 8　Effect of HF conductivity on pressure in anomalous-diffusion case

of larger HF conductivity.

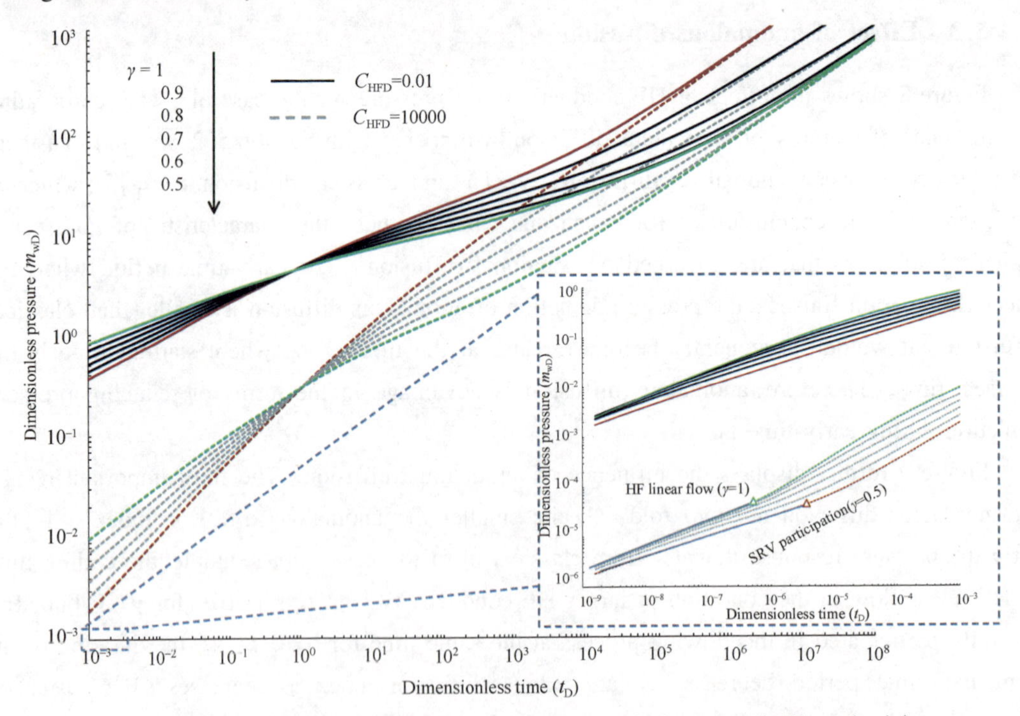

Figure 9　Effect of anomalous-diffusion constant on pressure in large/small HF conductivity case

5.4 Asymptotic approximations

As mentioned by Clarkson[38], there are two main production data analysis methods that have been commonly applied including: type-curve method as shown in Figure 6 ~ Figure 8, and straight-line (or flow regime) analysis method based on asymptotic approximations. In the following, we detail asymptotic approximations for corresponding flow regimes.

For the case of equally-spaced configuration, Equation(19) could be simplified in the form of single-fracture solution multiplied by fracture number n_{f}, as follows:

$$\tilde{m}_{\mathrm{wD}} = \frac{1}{n_{\mathrm{f}}} \frac{2\pi}{sC_{\mathrm{HFD}}} \left(\frac{s}{\eta_{\mathrm{HFD}}} + \frac{2\lambda_{\mathrm{HF}}^{\mathrm{SRV}} \tilde{F}_{\mathrm{HF}}^{\mathrm{SRV}}}{s^{\gamma-1}} \right)^{-0.5} \coth\left[\left(\frac{s}{\eta_{\mathrm{HFD}}} + \frac{2\lambda_{\mathrm{HF}}^{\mathrm{SRV}} \tilde{F}_{\mathrm{HF}}^{\mathrm{SRV}}}{s^{\gamma-1}} \right)^{-0.5} x_{\mathrm{HFD}} \right] \tag{22}$$

5.4.1 Short-time approximations

1) Linear flow in HF

At very short time, the transient response only occurs in hydraulic fracture. The variables in Laplace domain is considered to be large enough ($s \to \infty$), which indicates that $\tilde{F}_{\mathrm{HF}}^{\mathrm{SRV}} = 0$. Therefore Equation(22) in Laplace domain can be simplified as

$$\tilde{m}_{\mathrm{wD}} = \frac{1}{n_{\mathrm{f}}} \frac{4\pi}{C_{\mathrm{HFD}} s \sqrt{s/\eta_{\mathrm{HFD}}}} \tag{23}$$

Using the inversion theorem for the Laplace transformation, we obtain the transient linear flow in HF

$$m_{\mathrm{wD}} = \frac{1}{n_{\mathrm{f}}} \frac{4\pi \eta_{\mathrm{HFD}}^{0.5}}{C_{\mathrm{HFD}}} \frac{t_{\mathrm{D}}^{0.5}}{\Gamma(1.5)} \tag{24}$$

2) Bilinear flow in HF & SRV

If the transient response in SRV is also taken into account, Equation (22) can be approximately given by

$$m_{\mathrm{wD}} = \frac{1}{n_{\mathrm{f}}} \frac{4\pi}{C_{\mathrm{HFD}} s} \left(\frac{s}{\eta_{\mathrm{HFD}}} + \frac{2\lambda_{\mathrm{HF}}^{\mathrm{SRV}} \tilde{F}_{\mathrm{HF}}^{\mathrm{SRV}}}{s^{\gamma-1}} \right)^{-0.5} \tag{25a}$$

where $\tilde{F}_{\mathrm{HF}}^{\mathrm{SRV}}$ represents the transient response in SRV, which is given by

$$\tilde{F}_{\mathrm{HF}}^{\mathrm{SRV}} = bc \left(\frac{w_{\mathrm{HFD}}}{2} \right)^{c-1} \frac{K_{n-1}[c(0.5w_{\mathrm{HFD}})^c]}{K_n[c(0.5w_{\mathrm{HFD}})^c]} \tag{25b}$$

and b represents the diffusion capacity in SRV, which is given by $b = \frac{2}{\theta+2} \left(\frac{s^{\gamma}}{\eta_{\mathrm{SRVD}}} \right)^{0.5}$.

Integrating Equation (25a) and Equation (25b) and approximated Bessel function, the duple transient flow in HF and SRV in real time domain is given as follows:

$$m_{\mathrm{wD}} = \frac{1}{n_{\mathrm{f}}} \frac{\dfrac{2\pi t_{\mathrm{D}}^{[1+\gamma(n-1)]/2}}{C_{\mathrm{HFD}} \Gamma\{0.5[3+\gamma(n-1)]\}}}{\left\{ \dfrac{c\lambda_{\mathrm{HF}}^{\mathrm{SRV}} (0.5w_{\mathrm{HFD}})^{2nc-1}}{(\theta+2)^{2n} \eta_{\mathrm{SRVD}}^{n}} \dfrac{\Gamma(1-n)}{\Gamma(n)} \right\}^{0.5}} \tag{26}$$

Equation(26) is considered as the counterpart of the bilinear-flow regime approximation in the case of $\gamma = 1$ and $d_f = 1$, identified by a 1/4 slope straight line on log-log of pressure and derivative responses.

3) Trilinear flow in HF & SRV & XRV

If the transient response in XRV is simultaneously taken into account, it indicates that transient response in SRV is affected by XRV, which is reflected by diffusion capacity function b,

$$b = \frac{2}{\theta + 2}\left[\frac{s^{\gamma}}{\eta_{\mathrm{SRVD}}} + \frac{\lambda_{\mathrm{SRV}}^{\mathrm{XRV}}}{s^{1-\gamma}}\left(\frac{s}{\eta_{\mathrm{XRVD}}}\right)^{0.5}\right]^{0.5} \tag{27}$$

Therefore, we obtain the triple transient flow in HF, SRV and XRV as follows:

$$m_{\mathrm{wD}} = \frac{1}{n_{\mathrm{f}}} \frac{\dfrac{2\pi t_{\mathrm{D}}^{0.5[(1-n)(1-\gamma)+0.5n]}}{C_{\mathrm{HFD}}\Gamma\{0.5[(1-n)(1-\gamma)+0.5n+2]\}}}{\left[\dfrac{c\lambda_{\mathrm{HF}}^{\mathrm{SRV}}(\lambda_{\mathrm{SRV}}^{\mathrm{XRV}})^{n}(0.5w_{\mathrm{HFD}})^{2nc-1}}{(\theta+2)^{2n}\eta_{\mathrm{XRVD}}^{0.5n}} \dfrac{\Gamma(1-n)}{\Gamma(n)}\right]^{0.5}} \tag{28}$$

For $\gamma = 1$ and $d_f = 1$, we would recover Equation (28) into the trilinear flow regime identified by a 1/8 slope straight line on log-log plots of pressure and derivative responses.

5.4.2 Long-time approximations

1) Boundary-dominant (BD) flow in HF

For long enough time, the response in hydraulic fracture reaches the state of boundary dominant flow. We obtain the approximated equation in Laplace domain as:

$$\tilde{m}_{\mathrm{wD}} \approx \frac{4\pi}{C_{\mathrm{HFD}}s}\frac{1}{Ex_{\mathrm{HFD}}^{2}} + \frac{1}{3}\frac{4\pi}{C_{\mathrm{HFD}}s} \tag{29a}$$

Where E represents the effect of SRV and XRV on pressure response, given by

$$E = \frac{s}{\eta_{\mathrm{HFD}}} + \frac{2\lambda_{\mathrm{HF}}^{\mathrm{SRV}}\tilde{F}_{\mathrm{HF}}^{\mathrm{SRV}}}{s^{\gamma-1}} \tag{29b}$$

When the response in HF reaches BD flow without the effect of transient response in SRV and XRV, then $\tilde{F}_{\mathrm{HF}}^{\mathrm{SRV}} = 0$. We could obtain the BD flow in HF by using inversion Laplace transformation with regard to Equation (29a), as follows:

$$m_{\mathrm{wD}} = \frac{1}{n_{\mathrm{f}}}\left\{\frac{4\pi\eta_{\mathrm{HFD}}}{C_{\mathrm{HFD}}x_{\mathrm{HFD}}s}t_{D} + \frac{4\pi x_{\mathrm{HFD}}}{3C_{\mathrm{HFD}}}\right\} \tag{30}$$

2) Linear flow in SRV

If the transient response in SRV is also taken into account, substituting Equation (25b) ~ Equation (29b) into Equation (29a) could yield duple transient flow (i. e. BD flow in HF, transient flow in SRV, neglected XRV):

$$m_{\mathrm{wD}} = \frac{1}{n_{\mathrm{f}}}\left\{\frac{\pi\eta_{\mathrm{SRVD}}^{n}(\lambda_{\mathrm{HF}}^{\mathrm{SRV}})^{-1}(\theta+2)^{2n}}{c(0.5w_{\mathrm{HFD}})^{2nc-1}C_{\mathrm{HFD}}x_{\mathrm{HFD}}}\frac{\Gamma(n)t_{\mathrm{D}}^{\gamma(n-1)+1}}{\Gamma(1-n)\Gamma[2+\gamma(n-1)]} + \frac{4\pi x_{\mathrm{HFD}}}{3C_{\mathrm{HFD}}}\right\} \tag{31}$$

3) Bilinear flow in SRV&XRV

If transient response in XRV is simultaneously taken into account, substituting Equation (25b) and Equation (27) into Equation (29a) yield bilinear flow in SRV and XRV (Figure 10), given by

$$m_{\mathrm{wD}}=\frac{1}{n_{\mathrm{f}}}\left\{\frac{\pi\,(\theta+2)^{2n}\eta_{\mathrm{XRVD}}^{0.5n}\Gamma(n)\Gamma^{-1}[(n-1)(\gamma-1)+0.5n+1]}{c\lambda_{\mathrm{HF}}^{\mathrm{SRV}}C_{\mathrm{HFD}}\,(\lambda_{\mathrm{SRV}}^{\mathrm{XRV}})^{n}\,(0.5w_{\mathrm{HFD}})^{2nc-1}x_{\mathrm{HFD}}\Gamma(1-n)}t^{(n-1)(\gamma-1)+0.5n}+\frac{4\pi x_{\mathrm{HFD}}}{3C_{\mathrm{HFD}}}\right\} \tag{32}$$

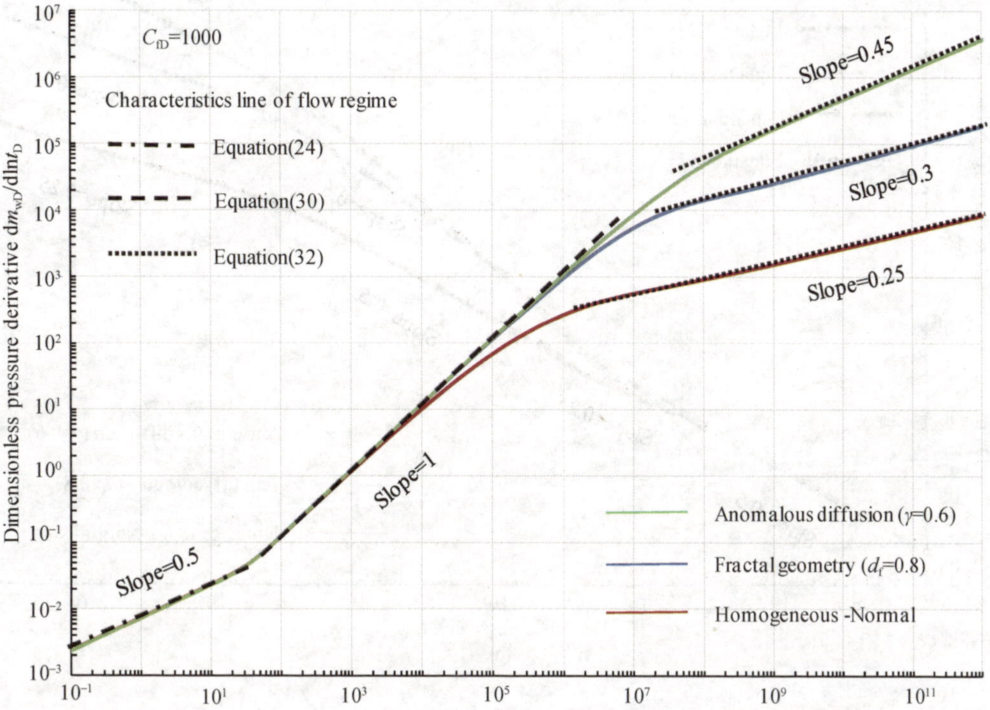

(a)

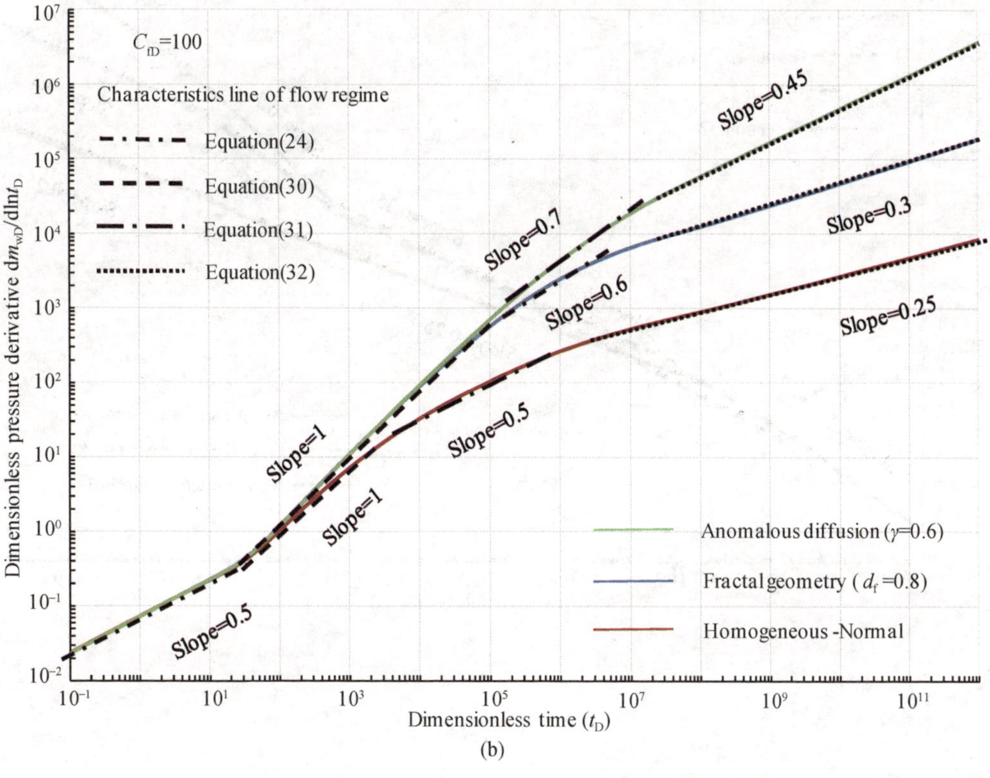

(b)

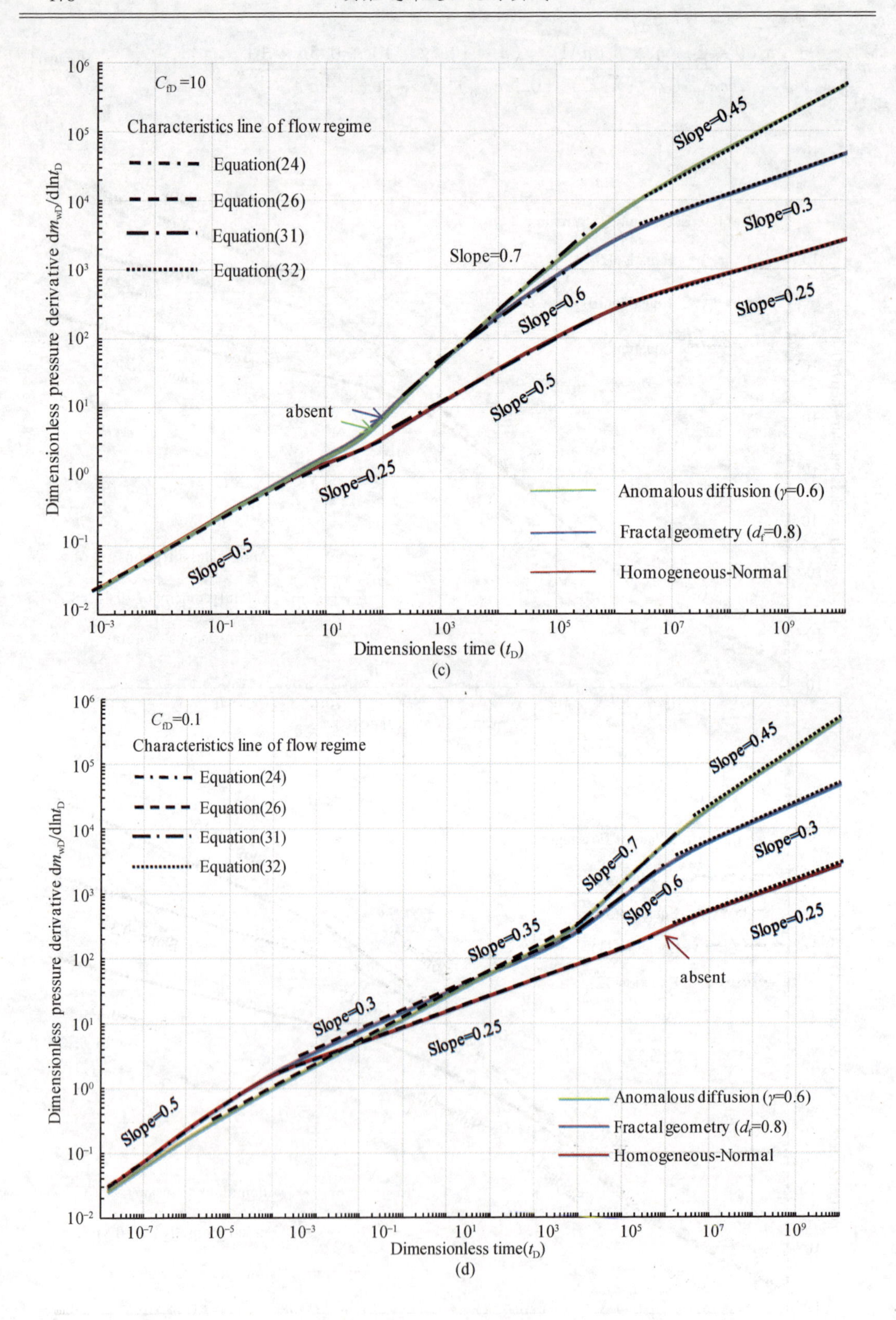

C_{fD}=10
Characteristics line of flow regime
Equation(24)
Equation(26)
Equation(31)
Equation(32)
Slope=0.45
Slope=0.3
Slope=0.7
Slope=0.6
Slope=0.25
Slope=0.5
absent
Slope=0.25
Slope=0.5
Anomalous diffusion (γ=0.6)
Fractal geometry (d_f=0.8)
Homogeneous-Normal
Dimensionless pressure derivative $dm_{wD}/d\ln t_D$
Dimensionless time (t_D)
(c)
C_{fD}=0.1
Characteristics line of flow regime
Equation(24)
Equation(26)
Equation(31)
Equation(32)
Slope=0.45
Slope=0.3
Slope=0.7
Slope=0.6
Slope=0.25
Slope=0.35
absent
Slope=0.3
Slope=0.25
Slope=0.5
Anomalous diffusion (γ=0.6)
Fractal geometry (d_f=0.8)
Homogeneous-Normal
Dimensionless pressure derivative $dm_{wD}/d\ln t_D$
Dimensionless time(t_D)
(d)

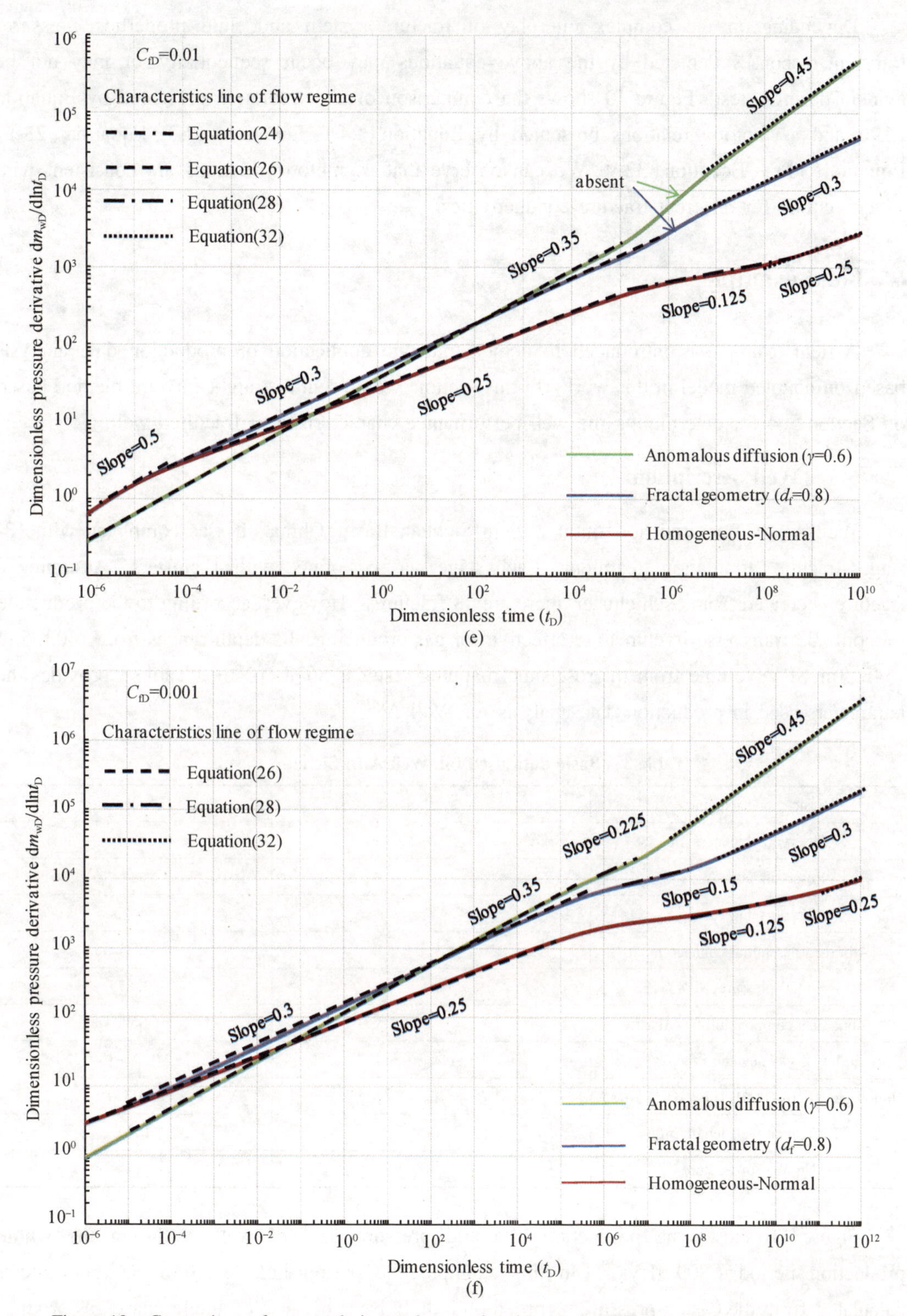

Figure 10 Comparison of exact solution and approximations under different values of conductivity

Depending on the complex interplay of fracture system and fluid properties, pressure-transient responses depicted by the above equations may occur sequentially, or may not be evident during a test. Figure 10 shows the comparison of exact solution presented by Equation (19) and asymptotic relations presented by Equation (24), Equation (26), Equation (28), Equation(30) ~ Equation(32). We can observe that asymptotic relations are consistent with exact solution for different fracture-conductivities.

6 Field example

A field example is introduced to demonstrate the application of production data analysis based on coupled model in this work. In this section, we present a straight forward method based on Section 5.4 for determining the well performance characteristic of fractured well.

6.1 Well description

Well A is a fractured horizontal well in Sichuan Basin, China. It was completed with 12-stages fracture stimulation treatment. Each stage was perforated with 4 clusters. Assuming a fracture is created from each cluster, there are 48 fractures. However, according to the production log, only 33 transverse fractures are effective for gas producing. Its depth ranges from 1503.6 to 1543.3m; SRV volume from microseismic mapping result is $8678\times10^4 m^3$. Table 3 provides the parameters used in production data analysis for Well A.

Table 3 Basic data used for Well A in Sichuan Basin

Basic model parameter	symbol	value
Reservoir thickness, m	h	39.7
Reservoir porosity, %	φ	8
Reservoir temperature, K	T	364.5
Adsorbed gas initial connect, m^3/t	V_E	2.31
Initial pressure, MPa	p_i	13.875
Initial gas compressibility, MPa^{-1}	c_{gi}	0.06904
Initial gas deviation factor	Z_{gi}	0.7915
Initial gas viscosity, cP($1cP=10^{-3}Pa\cdot s$)	μ_{gi}	0.01731
Horizontal well length, m	L_h	1079
Number of fracture	n_f	33

Figure 11 shows the production rate and pressure data of well A during the whole production period (809 days). In this example, the bottom-hole pressure can be directly calculated from the casing pressure recorded on a daily basis because of single-phase gas production.

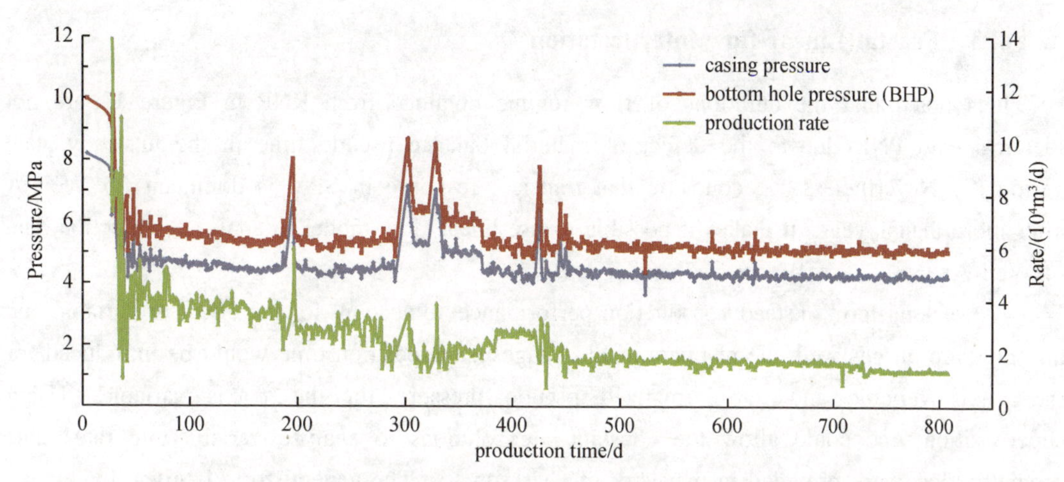

Figure 11 Production history profile: gas rate, casing pressure and calculative BHP

6.2 Flow regime identification

The pseudo functions are used to account for the pressure-dependent gas properties. The effect of adsorbed gas and apparent permeability in shale gas well could be considered according to the modified pseudo functions (pressure, time and material-balance-time) presented by Nobakht et al. [39].

Figure 12 shows two log-log plots, including rate-normalized pseudo pressure (RNP) versus material balance pseudo time (t_{mba}), pseudo pressure-normalized time average rate (PNR) versus t_{mba}. Noted that time average rate is the integral of $q_{sc}/\Delta m_w$ with regard to t_a, which contributes to a smooth rate curve. Based on Figure 12, three subsequent flow regimes are identified: ① transient linear flow in HF. ② bilinear flow in both HF and SRV. ③ transient linear flow in SRV. In summary, Well A is still determined by the combination of linear flow regimes.

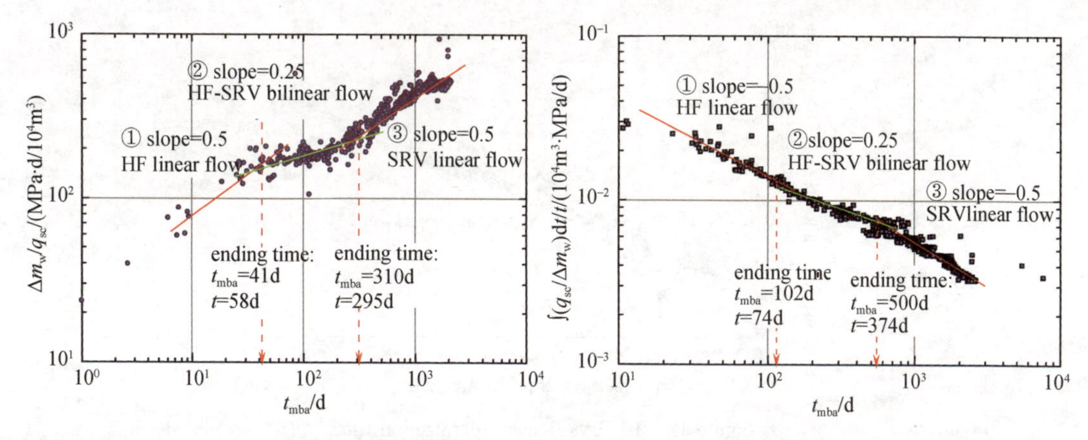

Figure 12 RNP vs t_{mba} and PNR vs t_{mba} for Well A

6.3 Fractal Linear-flow interpretation

It is noted that the durations of flow regime obtained from RNP in Figure 12 are not identical with PNR due to the defect of material balance pseudo time in the unsteady state period[40]. Nevertheless, we conclude that transient response in SRV is dominant for Well A after more than 1 year. It makes it possible to use linear flow model to analyze production data of Well A.

As for long-term unsteady production performance, to account for the effect of variable rate and pressure on gas well due to operational changes, superposition time would be introduced for the x-axis variable and rate-normalized pseudo pressure for the y-axis variable. Using superposition time could allow the constant-rate solutions to analyze variable-rate rate. The specialty plots were provided in the work of Clarkson[38]. The general form is given by

$$\frac{m_i - m_w(t_n)}{q_{sc,n}} = \sum_{j=1}^{n} (q_{sc,j} - q_{sc,j-1}) \frac{m_u(t_n - t_{j-1})}{q_{sc,n}} \tag{33}$$

Using linear superposition time to deal with production data, rate transient analysis plot gives a straight line to the corresponding superposition time function, as shown in Figure 13. The complexity of fracture network in SRV results in a bad-quality linear correlation of straight line. Based on the observation in Figure 12, there is an apparent skin due to the effect of finite conductivity and flow convergence within HF. To enhance the linear correlation in straight forward method, we incorporate fractal equation of Equation (31) to simulate the transient response in SRV as shown in Figure 14. The exponent of time variable in Equation (31) is identified as 0.35, which is the hybrid of 0.25 for bilinear flow and 0.5 for linear flow. The correlation of linear of RNP vs superposition time is improved. The matches for RNP using linear flow model and fractal linear flow model are shown in Figure 15. Contrast two match results in Figure 15, it can be found that new model reaches perfect agreement with field data, which is an important alternative tool of rate transient analysis for fractured shale gas well.

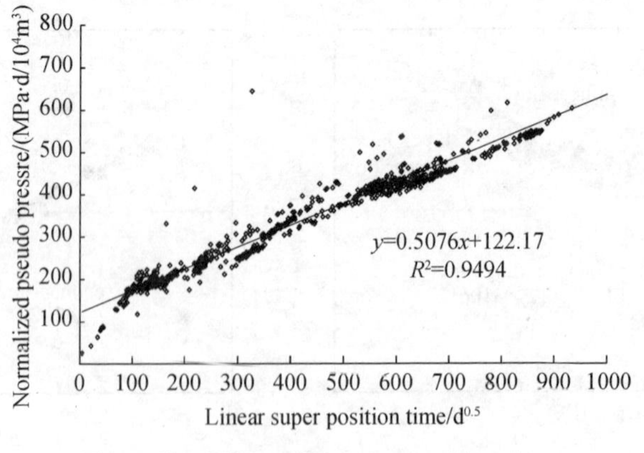

Figure 13　RNP vs linear superposition time

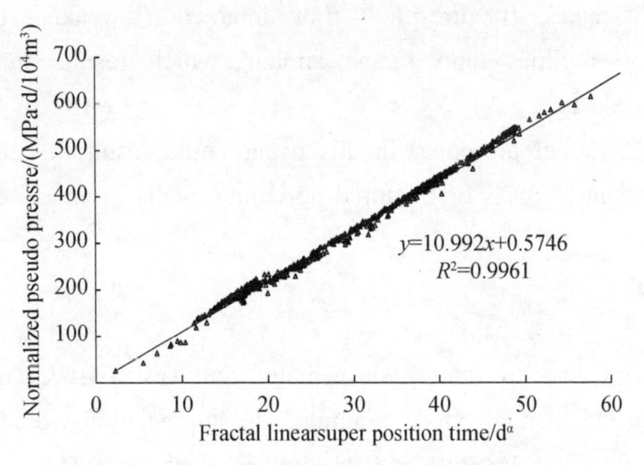

Figure 14 RNP vs fractal linear superposition time

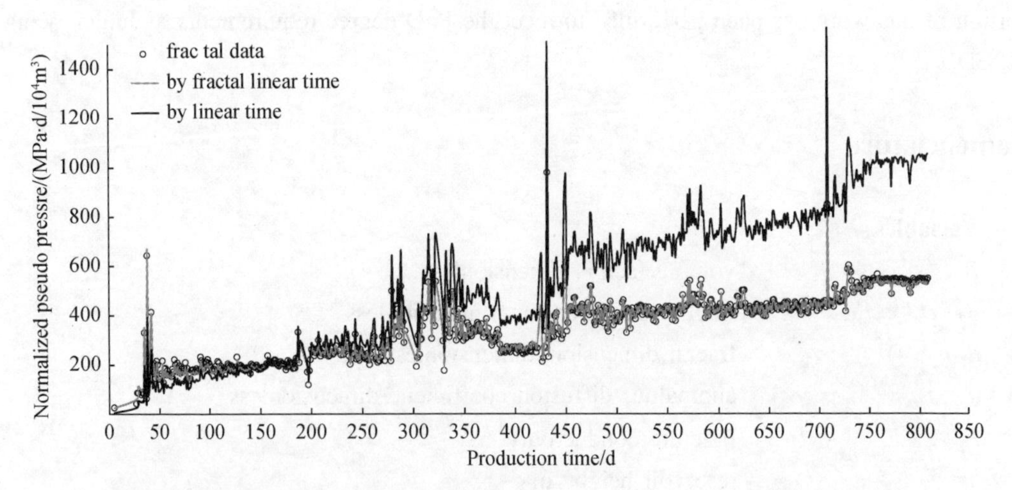

Figure 15 Comparison of RNP matches using linear model and fractal linear model

7 Conclusion

In our work, a coupled model for multi-fractured horizontal well with fractal geometry is established, and an accurate algorithm is proposed to interpret the transient pressure response. In summary, several principal contributions are further summarized below:

(1) The semi-analytical solution accounts for the effect of permeability alteration in SRV and arbitrary properties of HFV. The newly-presented algorithm is complete, rigorous, computationally stable and accurate.

(2) Fractal geometry contributes to greater pressure depletion during the whole flow period, while anomalous diffusion contribute to smaller pressure depletion at the early-flow period but becomes larger at the late-flow period.

(3) In the fractured reservoir with fractal geometry and anomalous diffusion, the effect of HF conductivity on pressure behavior become weaker. The existence of fractal characteristic and

anomalous diffusion makes fracture-production interaction weaker; the characteristic of intermediate-time flow regimes tend to be similar, which results in long linear-trending production period in field test.

(4) The coupled model presented in this paper could result in better history match for pressure and rate-transient analysis of fractured horizontal wells in unconventional reservoirs.

Acknowledgement

The authors would like to thank PetroChina Southwest Oil & Gasfield Company for providing field data in the Sichuan Shale example. The appreciation would be also expressed to Professor Jia Ailin in Research Institute of Petroleum Exploration & Development (RIPED) and Zhou Yingfang in University of Birmingham for their contribution and support to our research. Portion of this work has been performed to meet the PhD degree requirements of Junlei Wang at RIPED.

Nomenclature

Variables

B	volume factor, dimensionless
d_e	Euclidean dimension, dimensionless
d_f	fractal dimension, dimensionless
d_w	anomalous diffusion coefficient, dimensionless
F_c	Fracture conductivity, m^3
h	reservoir height, m
$I_v(x)$	modified Bessel function of the first kind with order v
$K_v(x)$	modified Bessel function of the second kind with order v
k_{SRV}	anomalous-diffusion permeability, $m^2 \cdot s^{1-\gamma}$
k_{SRV}^{ref}	anomalous-diffusion reference permeability, $m^2 \cdot s^{1-\gamma}$
k	permeability, m^2
L_m	distance measured from left-hydraulic-fracture plane, m
L_s	fracture spacing, m
n_f	number of hydraulic fracture
q	production rate, m^3/s
s	dimensionless variable in Laplace-transform domain, dimensionless
T	temperature, K
t	time variable, s
x	spatial variable perpendicular to horizontal well, m
x_{HF}	half-hydraulic-fracture length, m
x_R	drainage area width, m

y	spatial variable parallel to horizontal well, m
Z	gas deviation factor, dimensionless
$\Gamma(x)$	gamma function
γ	fractional derivative order, dimensionless
η	diffusion capacity, m^2
θ	scaling variable, dimensionless
μ	viscosity, Pa · s
λ	flow-capacity ratio, dimensionless
ζ	symbol indicating right or left part within SRV unit
ρ	density, g/m^3

φ = porosity, dimensionless

Subscript

D	dimensionless
HF	hydraulic fracture
i	Initial state
ref	reference parameter
SRV	stimulated reservoir volume
XRV	external reservoir volume
w	Horizontal wellbore

Superscript

L	left half-hydraulic fracture in elementary SRV unit
R	right half-hydraulic fracture in elementary SRV unit
M	middle position on the interface between right and left unit
~	Laplace transform
^	Fourier cosine transform

References

[1] Brown M, Ozkan E, Raghavan R, *et al.* Practical solutions for pressure-transient responses of fractured horizontal wells in unconventional shale reservoirs [C]. Paper SPE 12503 presented at the SPE Annual Technical Conference and Exhibition, New Orleans, 4-7 October 2009.

[2] Cippla C L, Fitzpatrick T, Williams M J, *et al.* Seismic-to-simulation for unconventional reservoir development [C]. Paper SPE 146876 presented at the SPE Reservoir Characterisation and Simulation Conference and Exhibition, Abu Dhabi, UAR, 9-11 October 2011.

[3] Weng X, Kresse O, Cohen C, *et al.* Modeling of hydraulic-fracture-network propagation in a naturally fractured formation[J]. SPE Production & Operation, 2011, 26(4): 368-380.

[4] Wu Y S, Ehlig-Economides C, Qin G, *et al.* A triple-continuum pressure-transient model for a naturally fractured vuggy reservoir[C]. Paper SPE 110044 presented at the 2007 SPE Annual Technical Conference and Exhibition, Anaheim, USA, 11-14 November 2007.

[5] Monifar A M, Varavei A, Johns R T, *et al.* Development of a coupled dual continuum and discrete fracture model for the simulation of unconventional reservoirs[C]. Paper SPE 163647 presented at the SPE Reservoir

Simulation Symposium, Woodlands, USA, 18-20 February 2013.

[6] Cipolla C L. Modeling production and evaluating fracture performance in unconventional gas reservoirs[J]. Journal of Petroleum Technology, 2009, 9: 84-90.

[7] Xie J, Yang C D, Gupata N, *et al*. Integration of shale-gas-production data and microseismic for fracture and reservoir properties with the fast marching method[J]. SPE Journal, 2014: 1-13.

[8] Tivayanonda V, Apiwathanasorn S, Ehlig-Economides C, *et al*. Alternative interpretation of shale gas/oil rate behavior using a triple porosity model[C]. Paper SPE 59703 presented at the SPE Annual Technical Conference and Exhibition, Texas, USA, 8-10 October 2012.

[9] Zhao Y L, Zhang L H, Luo J X, *et al*. Performance of fractured horizontal well with stimulated reservoir volume in unconventional gas reservoir[J]. Journal of Hydrology, 2014, 512: 447-456.

[10] Bello R O, Wattenbarger R A. Modeling and analysis of shale gas production with a skin effect[J]. Journal of Canadian Petroleum Technology, 2010, 49(12): 37-48.

[11] Obinna E D, Hassan D. A model for simultaneous matrix depletion into natural and hydraulic fracture networks[J]. Journal of Natural Gas Science and Engineering, 2014, 16: 57-69.

[12] Samandarli O, Al-Ahmadi H, Wattenbarger R A. A new method for history matching and forecasting shale gas reservoir production performance with a dual porosity model[C]. Paper SPE 144335 presented at the SPE North American Unconventional Gas Conference and Exhibition, Texas, USA, 12-16 June 2011.

[13] Xu B X, Haghighi M, Cooke D, *et al*. Production data analysis in Eagle Ford shale gas reservoir[C]. Paper SPE 153072 presented at the SPE European Unconventional Resources Conference and Exhibition, Vienna, Austria, 20-22 March 2012.

[14] Jayakumar R, Rai R. Impact of uncertainty in estimation of shale-gas-reservoir and completion properties on EUR forecast and optimal development planning: A Marcellus case study[J]. SPE Reservoir Evaluation & Engineering, 2014, 17(1): 60-73.

[15] Chen C C, Raghavan R. A multiply-fractured horizontal well in rectangular drainage region[J]. SPE Journal, 1997, 2(11): 455-465.

[16] Zerzar A, Bettam Y. Interpretation of multiple hydraulically fractured horizontal wells in closed systems [C]. Paper SPE 84888 presented at the SPE International Improved Oil Recovery Conference, Kuala Lumpur, Malaysia, 2003.

[17] Luo W J, Tang C F. Pressure-transient analysis of multiwing fractures connected to a vertical wellbore[J]. SPE Journal, 2014, 8: 1-8.

[18] Wattenbarger R A, EI-Banbi A H, Villegas M E, *et al*. Production analysis of linear flow into fractured tight gas wells[C]. Paper SPE 39931 presented the 1998 Rocky Mountain Regional Symposium and Exhibition, Denver, Colorado, 5-8 April 1998.

[19] EI-Banbi A H, Wattenbarger R A. Analysis of linear flow in gas well production[C]. Paper SPE 39972 presented at the 1998 SPE Gas Technology Symposium, Calgary, Canada, 15-18 March 1998.

[20] Stalgorova E, Mattar L. Practical analytical model to simulate production of horizontal wells with branch fractures[C]. Paper SPE 162515 presented at the SPE Canadian Unconventional Resources Conference, Calgary, Canada, 30 October-1 November 2012.

[21] Stalgorova E, Mattar L. Analytical model for unconventional multfractured composite systems[J]. SPE Reservoir Evaluation & Engineering, 2013, 8: 246-256.

[22] Chang J C, Yortsos Y C. Pressure-transient analysis of fractal reservoirs[J]. SPE Formation Evaluation, 1990, 3: 31-38.

[23] Metzler R, Klafter J. The random walk's guide to anomalous diffusion: a fractional dynamics approach

[J]. Physics Reports,2000,339:1-77.

[24] O' Shaughnessy B,Procaccia I. Analytical solutions for diffusion on fractal objects[J]. Physical Review Letters,1985,54(5):455-458.

[25] Camacho-Velázquez R,Fuentes-Cruz G,Vásquez-Cruz M. Decline-curve analysis of fractured reservoirs with fractal geometry[J]. SPE Reservoir Evaluation & Engineering,2008,6: 606-619.

[26] Fuentes-Cruz G,Gildin E,Valkó P P. Analyzing production data from hydraulically fractured wells: the concept of induced permeability field[J]. SPE Reservoir Evaluation & Engineering2014,5: 220-232.

[27] Fuentes-Cruz G, Gildin E, Valkó P P. On the analysis of production data: Practical approaches for hydraulically fractured wells in unconventional reservoirs [J]. Journal of Petroleum Science and Engineering,2014,119:54-68.

[28] Raghavan R. Fractional derivatives: application to transient flow[J]. Journal of Petroleum Science and Engineering,2012,80:7-13.

[29] Raghavan R. Fractional derivatives: performance of fractured well[J]. Journal of Petroleum Science and Engineering2012,92-93:167-173.

[30] Raghavan R,Chen C. Fractional diffusion in rocks produced by horizontal wells with multiple,transverse fractures of finite conductivity[J]. Journal of Petroleum Science and Engineering,2013,109:133-143.

[31] Raghavan R,Chen C. Fractured-well performance under anomalous diffusion[J]. SPE Reservoir Evaluation & Engineering,2013,8:237-245.

[32] Razminia K,Razminia A,Machado J A. Analysis of diffusion process in fractured reservoirs using fractional derivative approach[J]. Communications in Nonlinear Science & Numerical Simulation,2014,19:3161-3170.

[33] Cossio M,Moridis G J,Blasingame T A. A semianalytic solution for flow in finite-conductivity vertical fractures by use of fractal theory[J]. SPE Journal,2013,2:83-96.

[34] Ozcan O,Sarak H,Ozkan E,*et al.* A trilinear flow model for a fractured horizontal well in a fractal unconventional reservoir[C]. Paper SPE-170971-MS presented at the SPE Annual Technical Conference and Exhibition,Amsterdam,Netherlands,27-29 October 2014.

[35] Mukherjee H,Economides M J. A Parametric Comparison of Horizontal and Vertical Well Performance [J]. SPE Formation Evaluation,1911,7:209-216.

[36] Stehfest H. Numerical inversion of Laplace transforms algorithm[J]. Communications of the ACM,1970, 13(1): 47-49.

[37] Song B,Economides M J,Ehlig-Economides C. Design of multiple transverse fracture horizontal wells in shale gas reservoirs [C]. Paper SPE-140555 presented at the SPE Hydraulic Fracturing Technology Conference and Exhibition,Texas,USA,24-26 January 2011.

[38] Clarkson C R. Production data analysis of unconventional gas wells: review of theory and best practices [J]. International Journal of Coal Geology,2013,109: 101-146.

[39] Nobakht M,Clarkson C,Kaviani D. New and improved method for performing rate-transient analysis of shale gas reservoirs[C]. Paper 147869 presented at the PSE Asia Pacific Oil and Gas Conference and Exhibition,Jakarta,20-22 September 2011.

[40] Nobakht M,Clarkson C. A new analytical method for analyzing linear flow in tight/shale gas reservoirs: constant-flow-pressure boundary condition [C]. Paper SPE 143989 presented at the Americans Unconventional Gas Conference,Texas,12-16 June 2011.

[41] Bowman F. Introduction to Bessel functions[M]. New York:Dover Publications Edition,1958.

[42] Ertekin T,Sung W. Pressure transient analysis of coal seams in the presence of multi-mechanistic flow and sorption phenomena[C]. Paper SPE 19102 presented at the SPE Gas Technology Symposium,Dallas,

Texas, 7-9 June 1989.
[43] Anbarci K, Ertekin T. A comprehensive study of pressure transient analysis with sorption phenomena for single-phase gas flow in coal seams [C]. Paper SPE 20568 presented by the 65th Annual Technical Conference and Exhibition, New Orleans, 23-26 September 1990.
[44] Ozkan E, Raghavan R, Apaydin O. Modeling of fluid transfer from shale matrix to fracture network [C]. Paper 134830 presented at the SPE Annual Technical Conference and Exhibition, Florence, Italy, 19-22 October 2010.

Appendix A: Dimensionless definitions

The dimensionless pressure, time respectively are given as,

$$m_{\xi D} = \frac{2\pi k_{ref} h [m_i - m_\xi]}{q_{ref} \mu_o c_o}, \xi = HF, SRV, XRV \tag{A1}$$

$$t_D = \frac{k_{ref}}{\varphi_{ref} c_i L_{ref}^2} t \tag{A2}$$

In the hydraulic fracture model, the dimensionless fracture conductivity, C_{HFD}, is

$$C_{HFD} = \frac{k_{HF} w_{HF}}{r_{ref} L_{ref}} \tag{A3}$$

Note that the dimensionless fracture conductivity is defined with the reference length and reference permeability.

The dimensionless hydraulic fracture diffusivity is given as

$$\eta_{HFD} = \frac{k_{HF}/\phi_{HF}}{k_{ref}/\phi_{ref}} \tag{A4}$$

The dimensionless diffusivity of SRV is given as

$$\eta_{SRVD} = \frac{k_{SRV}^{ref}}{k_{ref}} \left(\frac{\varphi_{ref} \mu_{gi} c_{gi} L_{ref}^2}{k_{ref}} \right)^{\gamma - 1} \tag{A5}$$

The dimensionless diffusivity of XRV is given as

$$\eta_{XRVD} = \frac{k_{XRV}/\phi_{XRV}}{k_{ref}/\phi_{ref}} \tag{A6}$$

The flow capacity ratio between SRV and HF is given,

$$\lambda_{HF}^{SRV} = \frac{k_{SRV}^{ref} L_{ref}}{k_{HF} w_{HF}} \left(\frac{w_{HF}}{2L_{ref}} \right)^{d_f - d_e - \theta} \left(\frac{\phi_{ref} \mu_i c_i L_{ref}^2}{k_{ref}} \right)^{\gamma - 1} \tag{A7}$$

The flow capacity ratio between XRV and SRV is given,

$$\lambda_{SRV}^{XRV} = \frac{k_{XRV} L_{ref} \phi_{ref}}{k_{SRV}^{ref} x_{HF}} \left(\frac{k_{ref}}{\phi_{ref} \mu_i c_i L_{ref}^2} \right)^{\gamma - 1} \tag{A8}$$

The dimensionless production rate of hydraulic fracture is defined as,

$$q_{HFD} = \frac{q_{HF}}{q_{ref}} \tag{A9}$$

The dimensionless length in 1D coordinates is

$$\zeta_D = \frac{\zeta}{L_{ref}}, \zeta = x, y, x_R, x_f, L_s, w_{HF} \tag{A10}$$

Appendix B: Flow model and solution for XRV

The fundamental partial differential equation that governs fluid flow into XRV is given in the 1D coordinate system by

$$-\frac{\partial}{\partial x}(\rho\vec{v}_{\mathrm{XRV}})=\varphi_{\mathrm{XRV}}\frac{\partial\rho}{\partial t} \tag{B1}$$

In dimensionless terms defined in Appendix A, Equation (B1) can be rewritten in Laplace domain

$$\frac{\partial^2\tilde{m}_{\mathrm{XRVD}}}{\partial x_{\mathrm{D}}^2}=\frac{s}{\eta_{\mathrm{XRVD}}}\tilde{m}_{\mathrm{XRVD}} \tag{B2}$$

According to no-flow condition on outer boundary of XRV ($x=x_{\mathrm{R}}$) and coupling condition on interface between XRV and SRV ($x=x_{\mathrm{HF}}$), similar to Equation(20) and Equation(21) in the work of Brown et al. [1], we could obtain the dimensionless flux on the interface between SRV and XRV region, as follows:

$$\left(\frac{\partial\tilde{m}_{\mathrm{XRVD}}}{\partial x_{\mathrm{D}}}\right)\bigg|_{x_{\mathrm{D}}=x_{\mathrm{HFD}}}=-\tilde{F}_{\mathrm{SRV}}^{\mathrm{XRV}}\left(\tilde{m}_{\mathrm{SRVD}}\big|_{x_{\mathrm{D}}=x_{\mathrm{HFD}}}\right) \tag{B3}$$

Where $\tilde{F}_{\mathrm{SRV}}^{\mathrm{XRV}}$ is defined as the characteristic function representing the dimensionless flux from XRV to SRV, which is given as

$$\tilde{F}_{\mathrm{SRV}}^{\mathrm{XRV}}=\sqrt{\frac{s}{\eta_{\mathrm{XRVD}}}}\tanh\left[\sqrt{\frac{s}{\eta_{\mathrm{XRVD}}}}(x_{\mathrm{RD}}-x_{\mathrm{HFD}})\right] \tag{B4}$$

Appendix C: Flow model and solution for SRV

Two coordinate systems are established with respect to hydraulic fracture plane. In the elementary unit of SRV, the flow equation within Right/Left half-width-HF region is given as,

$$\frac{\partial(\rho\vec{v}_{\mathrm{SRV}}^{\xi})}{\partial y}+\frac{\partial(\rho\varphi_{\mathrm{SRV}}^{\xi})}{\partial t}=0 \tag{C1}$$

Using fractal-geometry equation presented by Equation(1) and Equation(2) and modified Darcy equation presented by Equation(3), we could obtain pressure governing equation as,

$$\frac{\partial^{1-\gamma_\xi}}{\partial t^{1-\gamma_\xi}}\left[\frac{\partial k_{\mathrm{SRV}}^{\xi}}{\partial y}\frac{\partial m_{\mathrm{SRV}}^{\xi}}{\partial y}+k_{\mathrm{SRV}}^{\xi}\frac{\partial^2 m_{\mathrm{SRV}}^{\xi}}{\partial y^2}\right]=-\varphi_{\mathrm{SRV}}\left[\frac{k_{\mathrm{XRV}}}{x_{\mathrm{HF}}}\left(\frac{\partial m_{\mathrm{XRV}}}{\partial x}\right)\bigg|_{x=x_{\mathrm{HF}}}-\mu c\frac{\partial m_{\mathrm{SRV}}^{\xi}}{\partial t}\right] \tag{C2}$$

The dimensionless pressure governing equation in Laplace-transform domain is rewritten as follows:

$$\frac{\partial^2\tilde{m}_{\mathrm{SRVD}}^{\xi}}{\partial y_{\mathrm{D}}^2}+\frac{d_{\mathrm{f}}^{\xi}-d_{\mathrm{e}}-\theta_\xi}{y_{\mathrm{D}}}\frac{\partial\tilde{m}_{\mathrm{SRVD}}^{\xi}}{\partial y_{\mathrm{D}}}=y_{\mathrm{D}}^{\theta_\xi}\left(\frac{s^{\gamma_\xi}}{\eta_{\mathrm{SRVD}}^{\xi}}+\frac{\lambda_{\mathrm{SRV},\xi}^{\mathrm{XRV}}\tilde{F}_{\mathrm{SRV},\xi}^{\mathrm{XRV}}}{s^{1-\gamma_\xi}}\right)\tilde{m}_{\mathrm{SRVD}}^{\xi} \tag{C3}$$

Where, ξ represents the Right/Left SRV regions or Middle point between right and left region in elementary unit.

Boundary Condition 1. The dimensionless pressure is continuous in the interface between

SRV and HFV, which is respectively satisfied as,

For left region: $\tilde{m}_{\text{SRVD}}^{\text{L}}(w_{\text{HFD}}/2)=\tilde{m}_{\text{HFD}}^{\text{L}}$; $\tilde{m}_{\text{SRVD}}^{\text{L}}(L_{\text{mD}})=\tilde{m}_{\text{HFD}}^{\text{M}}$ (C4)

For right region: $\tilde{m}_{\text{SRVD}}^{\text{R}}(w_{\text{HFD}}/2)=\tilde{m}_{\text{HFD}}^{\text{R}}$; $\tilde{m}_{\text{SRVD}}^{\text{R}}(L_{\text{sD}}-L_{\text{mD}})=\tilde{m}_{\text{HFD}}^{\text{M}}$ (C5)

Boundary Condition 2. the flux continuity on the interface between Right and Left SRV region is given as

$$\left(\frac{\partial \tilde{m}_{\text{SRVD}}^{\text{L}}}{\partial y_{\text{D}}}\right)\Bigg|_{y_{\text{D}}=L_{\text{mD}}}+\left(\frac{\partial \tilde{m}_{\text{SRVD}}^{\text{R}}}{\partial y_{\text{D}}}\right)\Bigg|_{y_{\text{D}}=L_{\text{sD}}-L_{\text{mD}}}=0 \tag{C6}$$

Equation(C3) is a general form of the modified Bessel differential equation. According to the general solution presented by Bowman[41], the dimensionless flux on interface between SRV and HF is respectively given by incorporating Equation(C4) ~ Equation(C6), as follows:

$$\left(\frac{\partial \tilde{m}_{\text{SRVD}}^{\text{L}}}{\partial y_{\text{D}}}\right)\Bigg|_{y_{\text{D}}=w_{\text{HFD}}/2}=\tilde{F}_{\text{HF,R1}}^{\text{SRV}}\tilde{m}_{\text{HFD}}^{\text{R}}-\tilde{F}_{\text{HF,L1}}^{\text{SRV}}\tilde{m}_{\text{HFD}}^{\text{L}} \tag{C7}$$

$$\left(\frac{\partial \tilde{m}_{\text{SRVD}}^{\text{R}}}{\partial y_{\text{D}}}\right)\Bigg|_{y_{\text{D}}=w_{\text{HFD}}/2}=\tilde{F}_{\text{HF,L2}}^{\text{SRV}}\tilde{m}_{\text{HFD}}^{\text{L}}-\tilde{F}_{\text{HF,R2}}^{\text{SRV}}\tilde{m}_{\text{HFD}}^{\text{R}} \tag{C8}$$

Where

$$\tilde{F}_{\text{HF,R1}}^{\text{SRV}}=-\left(\frac{w_{\text{HFD}}}{2}\right)^{a_{\text{L}}+c_{\text{L}}-1-a_{\text{R}}}\frac{(L_{\text{sD}}-L_{\text{mD}})^{a_{\text{R}}+c_{\text{R}}-1}}{L_{\text{mD}}^{a_{\text{L}}}}\frac{\Psi_{\text{R1}}}{\Theta} \tag{C9a}$$

$$\tilde{F}_{\text{HF,L2}}^{\text{SRV}}=-\left(\frac{w_{\text{HFD}}}{2}\right)^{a_{\text{R}}+c_{\text{R}}-1-a_{\text{L}}}\frac{L_{\text{mD}}^{a_{\text{L}}+c_{\text{L}}-1}}{(L_{\text{sD}}-L_{\text{mD}})^{a_{\text{R}}}}\frac{\Psi_{\text{L2}}}{\Theta} \tag{C9b}$$

$$\tilde{F}_{\text{HF,L1}}^{\text{SRV}}=-\left(\frac{w_{\text{HFD}}}{2}\right)^{c_{\text{L}}-1}\left(E_{\text{L1}}+L_{\text{mD}}^{c_{\text{L}}-1}\frac{\Psi_{\text{L1}}}{\Theta}\right) \tag{C9c}$$

$$\tilde{F}_{\text{HF,R2}}^{\text{SRV}}=-\left(\frac{w_{\text{HFD}}}{2}\right)^{c_{\text{R}}-1}\left(E_{\text{R2}}+(L_{\text{sD}}-L_{\text{mD}})^{c_{\text{R}}-1}\frac{\Psi_{\text{R2}}}{\Theta}\right) \tag{C9d}$$

When $i=1$, ζ denotes Left and $L_{\zeta\text{D}}=L\times10^{-3}\,\mu\text{m}^2$; when $i=2$, ζ denotes Right and $L_{\zeta\text{D}}=L_{\text{sD}}-L\times10^{-3}\,\mu\text{m}^2$. Relative relationships above are list as follows:

$$\Theta=\sum_{i=1}^{2}b_{\xi}c_{\xi}L_{\text{D}\xi}^{c_{\xi}-1}\frac{K_{n_{\xi}}(\delta_{\xi1})I_{n_{\xi-1}}(\delta_{\xi2})+K_{n_{\xi-1}}(\delta_{\xi2})I_{n_{\xi}}(\delta_{\xi1})}{K_{n_{\xi}}(\delta_{\xi2})I_{n_{\xi}}(\delta_{\xi1})-K_{n_{\xi}}(\delta_{\xi1})I_{n_{\xi}}(\delta_{\xi2})} \tag{C10}$$

$$\psi_{\text{R1}}=\prod_{i=1}^{2}b_{\xi}c_{\xi}\frac{K_{n_{\xi}}(\delta_{\xi i})I_{n_{\xi-1}}(\delta_{\xi i})+K_{n_{\xi-1}}(\delta_{\xi i})I_{n_{\xi}}(\delta_{\xi i})}{K_{n_{\xi}}(\delta_{\xi2})I_{n_{\xi}}(\delta_{\xi1})-K_{n_{\xi}}(\delta_{\xi1})I_{n_{\xi}}(\delta_{\xi2})} \tag{C11a}$$

$$\psi_{\text{L2}}=\prod_{i=1}^{2}b_{\xi}c_{\xi}\frac{K_{n_{\xi}}(\delta_{\xi k})I_{n_{\xi-1}}(\delta_{\xi k})+K_{n_{\xi-1}}(\delta_{\xi k})I_{n_{\xi}}(\delta_{\xi k})}{K_{n_{\xi}}(\delta_{\xi2})I_{n_{\xi}}(\delta_{\xi1})-K_{n_{\xi}}(\delta_{\xi1})I_{n_{\xi}}(\delta_{\xi2})},k=\begin{cases}2,\text{if } i=1\\1,\text{if } i=2\end{cases} \tag{C11b}$$

$$\psi_{\text{L1}}=\prod_{i=1}^{2}b_{\text{L}}c_{\text{L}}\frac{K_{n_{\text{L}}}(\delta_{\text{L}i})I_{n_{\text{L}-1}}(\delta_{\text{L}i})+K_{n_{\text{L}-1}}(\delta_{\text{L}i})I_{n_{\text{L}}}(\delta_{\text{L}i})}{K_{n_{\text{L}}}(\delta_{\text{L2}})I_{n_{\text{L}}}(\delta_{\text{L1}})-K_{n_{\text{L}}}(\delta_{\text{L1}})I_{n_{\text{L}}}(\delta_{\text{L2}})} \tag{C11c}$$

$$\psi_{\text{R2}}=\prod_{i=1}^{2}b_{\text{R}}c_{\text{R}}\frac{K_{n_{\text{R}}}(\delta_{\text{R}i})I_{n_{\text{R}-1}}(\delta_{\text{R}i})+K_{n_{\text{R}-1}}(\delta_{\text{R}i})I_{n_{\text{L}}}(\delta_{\text{R}i})}{K_{n_{\text{R}}}(\delta_{\text{R2}})I_{n_{\text{R}}}(\delta_{\text{R1}})-K_{n_{\text{R}}}(\delta_{\text{R1}})I_{n_{\text{R}}}(\delta_{\text{R2}})} \tag{C11d}$$

where

$$E_{\text{L1}}=b_{\text{L}}c_{\text{L}}\frac{K_{n_{\text{L}}}(\delta_{\text{L2}})I_{n_{\text{L}-1}}(\delta_{\text{L1}})+K_{n_{\text{L}-1}}(\delta_{\text{L1}})I_{n_{\text{L}}}(\delta_{\text{L2}})}{K_{n_{\text{L}}}(\delta_{\text{L2}})I_{n_{\text{L}}}(\delta_{\text{L1}})-K_{n_{\text{L}}}(\delta_{\text{L1}})I_{n_{\text{L}}}(\delta_{\text{L2}})} \tag{C12a}$$

$$E_{R2} = b_R c_R \frac{K_{n_R}(\delta_{R2}) I_{n_{R-1}}(\delta_{R1}) + K_{n_{R-1}}(\delta_{R1}) I_{n_R}(\delta_{R2})}{K_{n_R}(\delta_{R2}) I_{n_R}(\delta_{R1}) - K_{n_R}(\delta_{R1}) I_{n_R}(\delta_{R2})} \tag{C12b}$$

and

$$\delta_{L1} = \frac{b_L c_L w_{HFD}}{2}, \delta_{L2} = b_L c_L L_{mD}, \delta_{R1} = \frac{b_R c_R w_{HFD}}{2}, \delta_{R2} = b_R c_R (L_{sD} - L_{mD}) \tag{C13a}$$

$$a_\xi = \frac{\theta_\xi + d_e - d_f^\xi + 1}{2}, c_\xi = \frac{\theta_\xi + 2}{2}, n_\xi = \frac{a_\xi}{c_\xi} \tag{C13b}$$

$$b_\xi = \left(\frac{2}{\theta_\xi + 2}\right)\sqrt{s^{\gamma_\xi - 1}\left(\frac{s}{\eta_{SRVD}^\xi} + \lambda_{SRV,\xi}^{XRV} \tilde{F}_{SRV,\xi}^{XRV}\right)} \tag{C13c}$$

Appendix D: Flow model and solution for HFV

In single unit of SRV, the flow equations in the right or left half-HF is given by,

$$\frac{\partial(\rho \vec{v}_{HF}^\xi)}{\partial x} + \frac{2}{w_{HF}} (\rho \vec{v}_{SRV}^\xi)\big|_{y = w_{HF}/2} + \frac{\partial(\rho \varphi_{HF}^\xi)}{\partial t} = 0 \tag{D1}$$

Incorporating fractal-geometry equation presented by Equation(1) and Equation(2) and modified Darcy equation presented by Equation (3), we use dimensionless definition as Appendix A to deal with Equation(D1). Hence, the dimensionless pressure is given in Laplace-transform domain as,

$$\frac{\partial^2 \tilde{m}_{HFD}^\xi}{\partial x_D^2} + 2\frac{\lambda_{HF,\xi}^{SRV}}{s^{\gamma_\xi - 1}}\left(\frac{\partial \tilde{m}_{SRVD}^\xi}{\partial y_D}\right)\Bigg|_{y_D = w_{HFD}/2} = \frac{s\tilde{m}_{HFD}}{\eta_{HFD}} \tag{D2}$$

Boundary Condition 1. The flux rate from the region beyond hydraulic fracture tip into HFV is neglected ($x_D = x_{HFD}$),

$$\left.\frac{\partial \tilde{m}_{HFD}^\xi}{\partial x_D}\right|_{x_D = x_{HFD}} = 0 \tag{D3}$$

Boundary Condition 2. The flux rate along half-width fracture into horizontal wellbore is given by

$$\left(\frac{\partial \tilde{m}_{HFD}^\xi}{\partial x_D}\right)_{x_D = 0} = -\frac{2\pi}{C_{HFD}^\xi}\tilde{q}_{HFD}^\xi \tag{D4}$$

Where ζ = Left or Right.

Taking Fourier-cosine transformation and inversion transformation, we can obtain the solution for Equation(D2), as follows:

$$m_{HFD}^L(x_D) = AX_{SRV}^L(x_D)\tilde{q}_{HFD}^L + BX_{SRV}^L(x_D)\tilde{q}_{HFD}^R \tag{D5a}$$

$$m_{HFD}^R(x_D) = AX_{SRV}^R(x_D)\tilde{q}_{HFD}^R + BX_{SRV}^R(x_D)\tilde{q}_{HFD}^L \tag{D5b}$$

Where

$$AX_{SRV}^L = 2\pi \sum_{m=0}^{\infty} \frac{\cos(\beta_m x_D)}{N_m C_{HFD}^L \Lambda_L}\left(\beta_m^2 + 2\frac{\lambda_{HF,R}^{SRV} \tilde{F}_{HF,R2}^{SRV}}{s^{\gamma_L - 1}} + \frac{s}{\eta_{HFD}^R}\right) \tag{D6a}$$

$$BX_{\mathrm{SRV}}^{\mathrm{L}} = 4\pi\sum_{m=0}^{\infty}\frac{\cos(\beta_m x_{\mathrm{D}})}{N_m C_{\mathrm{HFD}}^{\mathrm{L}}\Lambda_{\mathrm{L}}}\frac{\lambda_{\mathrm{HF,L}}^{\mathrm{SRV}}\tilde{F}_{\mathrm{HF,R1}}^{\mathrm{SRV}}}{s^{\gamma_{\mathrm{L}}-1}} \tag{D6b}$$

$$AX_{\mathrm{SRV}}^{\mathrm{R}} = 2\pi\sum_{m=0}^{\infty}\frac{\cos(\beta_m x_{\mathrm{D}})}{N_m C_{\mathrm{HFD}}^{\mathrm{R}}\Lambda_{\mathrm{R}}}\left(\beta_m^2 + 2\frac{\lambda_{\mathrm{HF,L}}^{\mathrm{SRV}}\tilde{F}_{\mathrm{HF,L1}}^{\mathrm{SRV}}}{s^{\gamma_{\mathrm{R}}-1}} + \frac{s}{\eta_{\mathrm{HFD}}^{\mathrm{L}}}\right) \tag{D6c}$$

$$BX_{\mathrm{SRV}}^{\mathrm{R}} = 4\pi\sum_{m=0}^{\infty}\frac{\cos(\beta_m x_{\mathrm{D}})}{N_m C_{\mathrm{HFD}}^{\mathrm{R}}\Lambda_{\mathrm{R}}}\frac{\lambda_{\mathrm{HF,R}}^{\mathrm{SRV}}\tilde{F}_{\mathrm{HF,L2}}^{\mathrm{SRV}}}{s^{\gamma_{\mathrm{R}}-1}} \tag{D6d}$$

and

$$\Lambda_{\mathrm{L}} = \left(\beta_m^2 + 2\frac{\lambda_{\mathrm{HF,L}}^{\mathrm{SRV}}\tilde{F}_{\mathrm{HF,L1}}^{\mathrm{SRV}}}{s^{\gamma_{\mathrm{L}}-1}} + \frac{s}{\eta_{\mathrm{HFD}}^{\mathrm{L}}}\right)\left(\beta_m^2 + 2\frac{\lambda_{\mathrm{HF,R}}^{\mathrm{SRV}}\tilde{F}_{\mathrm{HF,R2}}^{\mathrm{SRV}}}{s^{\gamma_{\mathrm{L}}-1}} + \frac{s}{\eta_{\mathrm{HFD}}^{\mathrm{R}}}\right) - 4\frac{\lambda_{\mathrm{HF,L}}^{\mathrm{SRV}}\lambda_{\mathrm{HF,R}}^{\mathrm{SRV}}\tilde{F}_{\mathrm{HF,R1}}^{\mathrm{SRV}}\tilde{F}_{\mathrm{HF,L2}}^{\mathrm{SRV}}}{s^{2\gamma_{\mathrm{L}}-2}} \tag{D7a}$$

$$\Lambda_{\mathrm{R}} = \left(\beta_m^2 + 2\frac{\lambda_{\mathrm{HF,L}}^{\mathrm{SRV}}\tilde{F}_{\mathrm{HF,L1}}^{\mathrm{SRV}}}{s^{\gamma_{\mathrm{R}}-1}} + \frac{s}{\eta_{\mathrm{HFD}}^{\mathrm{L}}}\right)\left(\beta_m^2 + 2\frac{\lambda_{\mathrm{HF,R}}^{\mathrm{SRV}}\tilde{F}_{\mathrm{HF,R2}}^{\mathrm{SRV}}}{s^{\gamma_{\mathrm{R}}-1}} + \frac{s}{\eta_{\mathrm{HFD}}^{\mathrm{R}}}\right) - 4\frac{\lambda_{\mathrm{HF,L}}^{\mathrm{SRV}}\lambda_{\mathrm{HF,R}}^{\mathrm{SRV}}\tilde{F}_{\mathrm{HF,R1}}^{\mathrm{SRV}}\tilde{F}_{\mathrm{HF,L2}}^{\mathrm{SRV}}}{s^{2\gamma_{\mathrm{R}}-2}} \tag{D7b}$$

and

$$\beta_m = \frac{m\pi}{x_{\mathrm{HFD}}}, m = 0,1,2,3\cdots; N_m = \int_0^{x_{\mathrm{HFD}}}\cos^2(\beta_m x_{\mathrm{D}})\mathrm{d}x_{\mathrm{D}} = \begin{cases}0.5x_{\mathrm{HFD}}, m > 0\\ x_{\mathrm{HFD}}, m = 0\end{cases} \tag{D8}$$

Appendix E: Consideration of dual continuum

In this section, the consideration of dual continuum is established on Fick's law in matrix and (modified) Darcy flow in fracture network of XRV and SRV.

The relative formations could be referred to the works presented by Ertekin and Sung[42] and Anbarci and Ertekin[43]: Langmuir adsorption isotherm is used to describe the equilibrium sorption, while Fick's law of diffusion is used to model the non-equilibrium diffusion within matrix block.

In XRV region, $\tilde{F}_{\mathrm{SRV}}^{\mathrm{XRV}}$, corresponding to Equation (B4), is modified as follows

$$\tilde{F}_{\mathrm{SRV}}^{\mathrm{XRV}} = \sqrt{\frac{s}{\eta_{\mathrm{XRVD}}} + 2\lambda_{\mathrm{XRV}}^{\mathrm{m}}\tilde{F}_{\mathrm{XRV}}^{\mathrm{m}}}\tanh\left[\sqrt{\frac{s}{\eta_{\mathrm{XRVD}}} + 2\lambda_{\mathrm{XRV}}^{\mathrm{m}}\tilde{F}_{\mathrm{XRV}}^{\mathrm{m}}}(x_{\mathrm{RD}} - x_{\mathrm{HFD}})\right] \tag{E1}$$

Where $\tilde{F}_{\mathrm{XRV}}^{\mathrm{m}}$ reflects the dimensionless flux from matrix into fracture network in XRV, which is satisfied as

$$\tilde{F}_{\mathrm{XRV}}^{\mathrm{m}} = \Xi[\sqrt{s\tau}\coth(\sqrt{s\tau}) - 1] \tag{E2}$$

$$\lambda_{\mathrm{XRV}}^{\mathrm{m}} = \frac{3\pi k_{\mathrm{ref}}DhL_{\mathrm{ref}}^2}{k_{\mathrm{XRV}}q_{\mathrm{ref}}r_{\mathrm{m}}^2}, \tau = \frac{k_{\mathrm{ref}}}{L_{\mathrm{ref}}^2\varphi_{\mathrm{ref}}\mu_{\mathrm{gi}}c_{\mathrm{gi}}}\frac{r_{\mathrm{m}}^2}{D}, \Xi = \frac{q_{\mathrm{ref}}\mu_{\mathrm{gi}}B_{\mathrm{gi}}}{2\pi k_{\mathrm{ref}}h}\frac{m_{\mathrm{L}}V_{\mathrm{L}}}{(m_{\mathrm{L}} + m_{\mathrm{i}})(m_{\mathrm{L}} + m)} \tag{E3}$$

In SRV region, b, corresponding to Equation (C13c), is modified as follows

$$b = \left(\frac{2}{\theta + 2}\right)\sqrt{s^{\gamma-1}\left(\frac{s}{\eta_{\mathrm{SRVD}}} + \lambda_{\mathrm{SRV}}^{\mathrm{XRV}}\tilde{F}_{\mathrm{SRV}}^{\mathrm{XRV}} + 2\lambda_{\mathrm{SRV}}^{\mathrm{m}}\tilde{F}_{\mathrm{SRV}}^{\mathrm{m}}\right)} \tag{E4}$$

$$\tilde{F}_{\mathrm{SRV}}^{\mathrm{m}} = \Xi[\sqrt{s\tau}\coth(\sqrt{s\tau}) - 1] \tag{E5}$$

Where

$$\lambda_{\mathrm{SRV}}^{\mathrm{m}} = \frac{k_{\mathrm{m}} L_{\mathrm{ref}}^2 \varphi_{\mathrm{ref}}}{k_{\mathrm{SRV}}^{\mathrm{ref}} h_{\mathrm{SRV}} r_{\mathrm{m}}} \left(\frac{k_{\mathrm{ref}}}{\varphi_{\mathrm{ref}} \mu_{\mathrm{gi}} c_{\mathrm{gi}} L_{\mathrm{ref}}^2} \right)^{\gamma - 1} \tag{E6}$$

Where r_{m} is the radius of matrix block, m; D is diffusion coefficient of gas, m^2/s; h_{SRV} is the uniform average thickness of fracture volume enveloping the matrix block, m[44]; V_{L} is the total sorption capacity, $\mathrm{m}^3/\mathrm{m}^3$; m_{L} is the Langmuir pseudo pressure, Pa; τ is dimensionless desorption time; Ξ is the dimensionless desorption coefficient.

Pressure Transient Analysis of Multi-stage Fracturing Horizontal Wells with Finite Fracture Conductivity in Shale Gas Reservoirs

Dai Yu[1] Ma Xinhua[1,2] Jia Ailin[1] He Dongbo[1] Wei Yunsheng[1] Xiao Cong[3]

(1. Research Institute of Petroleum Exploration and Development, PetroChina; 2. Southwest Oil and Gasfield Company, PetroChina; 3. China University of Petroleum)

Abstract: A new analytical solution of pressure transient analysis is proposed for multi-stage fracturing horizontal well (MFHW) with finite-conductivity transverse hydraulic fractures in shale gas reservoirs. The effects of absorption, diffusion, viscous flowing, stress sensitivity, flow-convergence, skin damage and wellbore storage are simultaneously considered as well in this paper. Laplace transformation, source sink function, perturbation method and superposition principle are respectively employed to solve related mathematical models of reservoir system and hydraulic fracture system. And then boundary element method (BEM) is applied to couple reservoir system and hydraulic fracture system. The transient pressure is inverted from Laplace space into real time space with Stehfest numerical inversion algorithm. Based on this new solution, the distribution of transient pseudo-pressure for multi-stage fracturing horizontal well with multiple finite-conductivity transverse hydraulic fractures are obtained. Different flowing regimes are identified, and the effects of relevant parameters are analyzed as well. The essence of this paper is considering the effects of transient gas flowing occurrence in finite-conductivity hydraulic fractures with boundary element method. Compared with some existing models and numerical simulation model of shale gas reservoirs, this proposed new model can provide a relative more accurate analysis of the relevant parameters, especially for the fracture conductivity. In conclusion, this new model provides the relative accurate and comprehensive evaluation results for multi-stage fracturing horizontal technology.

Key words: shale gas; multi-fractured horizontal well; finite-conductivity; flow-convergence; stress sensitivity; type curves

1 Introduction

Compared with conventional reservoirs, the shale gas reservoirs have its unique features, such as ultra-low permeability, ultra-low porosity, multi-scale pores and so on. These characters cause some special flowing patterns in the shale gas reservoirs. Based on the research of some scholars [1-3], the type of flowing occurrence in nano-scale matrix pores does meet diffusion principle instead of Darcy's law. At present, some scholars have conducted a large number of researches about transient pressure analysis for shale gas wells, and some analytical and semi-analytical solutions are developed. The shale gas reservoir is the classical naturally fracturing reservoir (NFR) which contains complex natural fractures and ultra-low permeability. In terms of those kinds of reservoirs, Barenblatt[4], Warren and Root[5] originally proposed the dual-porosity models which were assumed pseudo-steady state fluid transferring between the matrix and fractures, and then Kazemi[6], de Swaan[7] and Ozkan et al. [8] developed some other dual-porosity models for shale gas reservoirs to enrich the former productivity models, which were assumed unsteady-state (transient) flow condition between matrix and fractures.

Multi-stage fracturing horizontal well (MFHW) technology currently has been proved to be the most effective way to produce shale gas, and this method can not only create several high-conductivity hydraulic fractures, but also activate and connect existing natural fractures so as to form large spacious network system[9]. Similarly, some analytical and semi-analytical models came out with the researches of flow in the shale gas reservoirs which contain high conductivity hydraulic fractures. These models can be divided into two categories according to the conductivity of hydraulic fractures: ① Infinite-conductivity hydraulic fractures. These models assume that there is no pressure drop along the hydraulic fractures and the pressure is equal to bottom-hole pressure. In another words, the hydraulic fractures can be regarded as the parts of the wellbore. Zhao et al[10-12], Wang[13] and so on have derived its semi-analytical solutions in Laplace domain using Laplace transformation, besides, absorption and diffusion are considered in these model. The source function or line source function, coupled with superposition principle, is applied to solve the governing equations so that the interference among hydraulic fractures can be analyzed. ② Finite-conductivity hydraulic fractures. Similarly, these models assume that there is continuous pressure drop along the hydraulic fractures from the tip of hydraulic fractures to the wellbore, and the pressure distribution is distinct throughout the whole hydraulic fracture. El-Banbi[14], Al-Ahmadi and Wattenbarger[15], Buhidma et al.[16] and Xu et al.[17] simplified the shale gas reservoirs as linear fractured reservoirs, they assumed that the reservoir fluid continually deplete from one media to another (from matrix to natural fractures, and then from natural fractures to hydraulic fractures). Therefore, the governing equation of every media can be derived; however, the interference among hydraulic fractures is ignored. Some scholars, including Nashawi[18], Larsen and Hegre[19,20], Guo and Evans[21], Horne and Temeng[22], Chen and Raghavan[23], presented some transient analytical solutions in Laplace domain for the fracturing horizontal well with multiple finite-conductivity by using sink-source integral. Some other scholars, including Al-Kobaisi and Ozkan[24], Valk and Amini[25], Brown et al.[26], coped with this issue using numerical difference method. However, the process of solution is extremely complex and is short of practical application. Riley[27] and Wang et al.[28] considered the influence of conductivity with introduction of conductivity influence function, this method simultaneously can extremely reduce the complexity of calculation and obtain accurate results. However, this model is suitable for conventional homogeneous reservoirs, desorption and diffusion are not considered as well.

Klkanl and Horne[29] analyzed the transient pressure response in an arbitrary shaped heterogeneous oil reservoir with multiple sources/sinks using BEM. It pointed out that the divergence and grid orientation problems were no longer exist because the basic solution of BEM had the characteristics of analytical solution. Zhang and Zeng[30] studied the unstable pressure dynamic of a vertical well in arbitrary shaped dual media reservoirs in Laplace space. Using BEM, Wang and Zhang[31] researched the transient pressure response of a vertical well in an arbitrary shaped reservoir with partial impermeable region and corresponding type curves were drawn and analyzed. Based on the theory and advantage of BEM, we initially found that the finite hydraulic fractures system and reservoir system may can be coupled using BEM so that the shortage of current pressure transient model, which just can consider the hydraulic fractures as infinite, can be overcome.

In view of this, the pressure transient analysis model of multi-stage fracturing horizontal well for shale gas reservoirs with finite-conductivity hydraulic fractures has not been established comprehensively. Therefore, it is necessary to establish a relevant model associated with the special properties, such as dual-porosity media, absorption, diffusion and so on. Besides, in terms of NFR, the stress sensitivity of natural fractures has certain influence on well performance[32-34], especially for these kinds of ultra-low permeability, therefore, the stress sensitivity is also considered in this paper. Finally, the flow-convergence effects and formation damage by fracturing fluid are also considered with the conception of flow-convergence skin[35,36] and hydraulic fracture skin.

In conclusion, this paper firstly establish a new model with comprehensive consideration of multiple mechanisms, such as adsorption, diffusion, viscous flow, the hydraulic fracture conductivity, flow-convergence, stress sensitivity, skin factor and wellbore storage. Laplace transformation, perturbation technology, dispersion method and boundary element method (BEM) are employed to solve this new model. The pressure transient response is inverted into real time space with Stehfest numerical inversion algorithm. Type curves are plotted, and different flow regimes in shale gas reservoirs are identified. The effects of relevant parameters are analyzed as well, especially for the conductivity of hydraulic fractures.

2 Physical model for multi-fractured horizontal wells in shale gas reservoir

The schematic illustration in Figure 1 depicts a multi-stage fracturing horizontal wells located in a boundless shale reservoir.

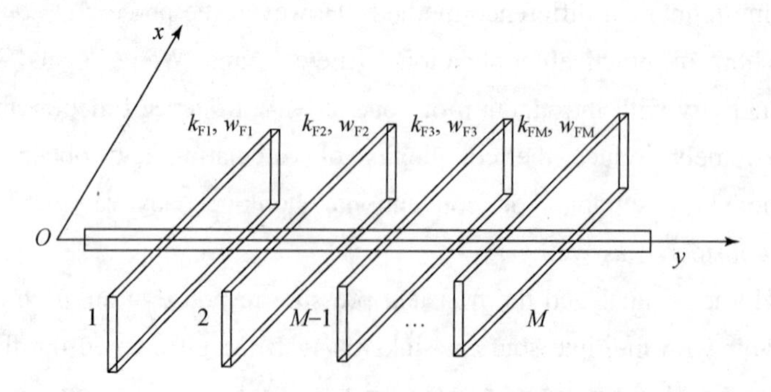

Figure 1 The illustration of multi-fractured horizontal well with finite conductivity fractures

The horizontal well is intercepted by some finite-conductivity hydraulic fractures. The number of hydraulic fractures is M, the length and width of the i-th $(i=1,2,3,\cdots,M)$ fracture respectively are equal to X_{Fi} and W_{Fi}. The other assumptions are as follows:

(1) The initial pressure distributed in the reservoir is uniform and equals to P_i.

(2) The shale gas reservoir is considered to be dual-porosity system, containing natural fractures and matrix.

(3) The gas seepage in the natural fracture meets Darcy's law. The shape of matrix is simplified as sphere and gas flow in matrix is assumed to be diffusive flow driven by concentration difference.

(4) The hydraulic fractures are perpendicular to the horizontal well and assumed to be penetrated fully, and the permeability of natural fractures is stress-dependent.

(5) Only reservoir fluid transferring from hydraulic fractures into horizontal wellbore is considered; the horizontal well produces at a constant rate, and the rate of each hydraulic fracture is different and the function of time.

(6) Compressibility coefficient of the slightly compressible shale gas is constant.

(7) The impacts of gravity and capillary are neglected.

(8) The gas absorption and diffusion respectively meets the Langmuir isotherm equation and the first law of diffusion.

(9) The wellbore storage and skin factor are considered.

The pseudo-function was used to account for the pressure dependent gas properties, the formulas of pseudo-pressure and pseudo-time are as follows[1]:

$$\text{Pseudo-pressure: } \psi = 2\int_0^p \frac{p}{\mu Z}\mathrm{d}p \text{ ; Pseudo-time: } t_a = \int_0^t \frac{(\mu C_t)_i}{\mu(\bar{p})C_t(\bar{p})}\mathrm{d}t \tag{1}$$

3 Mathematical model

In this section, the whole establishment process of the mathematical model can be divided into the following steps: ①seepage model of one single infinite-conductivity hydraulic fracture located in the boundless shale gas reservoir; ②effects of flow-convergence and fracture damage; ③seepage model in finite-conductivity hydraulic fractures.

3.1 Seepage model for infinite-conductivity fractures

In terms of seepage model for infinite-conductivity hydraulic fractures, at the present time, the source function is the most classical method to deal with this problem. The flow in the natural fractures is assumed to be radial flow, thus, based on this assumption, the line sink solution is applied to derive the seepage model in the natural fracture system. Combined with the material conservation equation and relative motion equation, the governing equations of fractures and matrix are as follows.

Natural fracture:

$$\frac{1}{r}\frac{\partial}{r}\left(r\rho\frac{k_f}{\mu}\frac{\partial P_f}{\partial r}\right) = \frac{\partial(\rho\varphi_f)}{3.6\partial t} + \rho_{sc}\frac{\partial V}{3.6\partial t} \tag{2}$$

According to the Fick's first law, pseudo-steady diffusion in shale matrix can be presented as

Matrix:

$$\frac{\partial V}{3.6\partial t}=\frac{6D\pi^2}{R_m^2}(V_E-V) \tag{3}$$

The pseudo-function was used to account for the pressure dependent gas properties, the governing equation with the formulas of pseudo-pressure and pseudo-time are as follows.

Natural fracture:

$$\frac{1}{r}\frac{\partial}{r}\left(rk_f\frac{\partial \psi_f}{\partial r}\right)=(\varphi\mu_i C_{ti})_f\frac{\partial \psi_f}{3.6\partial t_a}+2\frac{P_{sc}T}{T_{sc}}\frac{\partial V}{3.6\partial t_a} \tag{4}$$

Matrix:

$$\frac{\partial V}{3.6\partial t_a}=\frac{6D\pi^2}{R_m^2}(V_E-V) \tag{5}$$

Where, k_f is the permeability of natural fractures, D; ψ_f is the pseudo-pressure in fracture system, $MPa^2/(mPa\cdot s)$; μ_i is the initial viscosity of shale gas, $mPa\cdot s$; φ_f is porosity of fracture system, fraction; C_{ti} is the total initial compressibility coefficient, MPa^{-1}; V is the shale gas concentration in matrix, sm^3/m^3; t_a is the pseudo-time, h; r is distance, m; P_{sc} is the pressure at standard condition, MPa; T_{sc} is the temperature at standard condition, K; T is the formation temperature, K; D is gas diffusion coefficient, m^2/s; R_m is the radius of sphere matrix, m; V_E is the initial concentration in matrix, sm^3/m^3.

The feature of permeability of stress-dependent is described via introducing a permeability modular ζ, the relationship between permeability and pseudo pressure is as follows (Pedrosa, 1986; Wang, 2013):

$$k_f=k_{fi}e^{-\zeta(\Psi_i-\Psi_f)} \tag{6}$$

Where, k_{fi} is initial permeability of natural fracture under initial pressure condition, D; ζ is permeability modular, $(mPa\cdot s)/MPa^2$; Ψ_i is the initial pseudo-pressure, $MPa^2/(mPa\cdot s)$.

Submitting Equation (6) into Equation (4)、Equation(5), the final formation of governing formula can be transformed to the following formula.

Natural fracture:

$$\frac{\partial^2\psi_f}{\partial r^2}+\frac{1}{r}\frac{\partial \psi_f}{\partial r}+\zeta\left(\frac{\partial \psi_f}{\partial r}\right)^2=e^{\zeta(\psi_i-\psi_f)}\left[(\varphi\mu_i C_{ti})_f\frac{\partial \psi_f}{3.6\partial t_a}+2\frac{P_{sc}T}{T_{sc}}\frac{\partial V}{3.6\partial t_a}\right] \tag{7}$$

Matrix:

$$\frac{\partial V}{3.6\partial t_a}=\frac{6D\pi^2}{R_m^2}(V_E-V) \tag{8}$$

The initial condition:

$$\psi_f=\psi_i,V=V_i,t_a=0 \tag{9}$$

Based on the theory of line sink, the inner condition is presented as follow:

$$e^{-\zeta(\psi_i-\psi_f)}r\frac{\partial \psi_f}{\partial r}\bigg|_{r\to 0}=\frac{q(t_a)P_{sc}T}{271.44k_{fi}T_{sc}} \tag{10}$$

Where, $q(t_a)$ is the production rate of line sink, sm^3/d; h is the thickness of the reservoir, m; k_{fi}

is the initial permeability of natural fractures, D.

The reservoir is assumed to be infinite, and the outer boundary is as follow:

$$\psi_f(r \to \infty, t_a) = \psi_i \tag{11}$$

For the convenience of solution, some dimensionless variables are defined (Table 1).

Table 1 The definition of dimensionless variables

Dimensionless pseudo-pressure	$\psi_{jD} = \dfrac{271.44k_{fi}hT_{sc}(\psi_i - \psi_j)}{q_{sc}TP_{sc}}, (j = f)$
Dimensionless total rate	$Q_{Dsc} = \dfrac{q_{sc}TP_{sc}}{271.44k_{fi}hT_{sc}(\psi_i - \psi_{wf})}$
Dimensionless time	$t_D = \dfrac{3.6k_{fi}t_a}{\Lambda\mu L_{ref}}$
Dimensionless storativity ratio	$\omega = \dfrac{(\varphi C\,ti)f}{\Lambda}, \Lambda = (\varphi C_{ti})_f + \dfrac{2\pi k_{fi}h}{q_{sc}\mu_i}$
Dimensionless stress sensitivity coefficient	$\zeta_D = \dfrac{P_{sc}Tq_{sc}}{271.44k_{fi}hT_{sc}}\zeta$
Dimensionless distance	$r_D = \dfrac{r}{L_{ref}} \quad \lambda = \dfrac{6\pi^2 D\mu_i\Lambda L_{ref}^2}{k_{fi}R_m}$
Dimensionless gas concentration	$V_D = V_i - V$
Dimensionless rate of line sink	$q_D = \dfrac{q(t_a)}{q_{sc}}$

With the definition of these dimensionless variables, the governing Equation (7) ~ Equation (11) using the dimensionless formation can be presented as follows:

$$\frac{\partial^2\psi_{fD}}{\partial r_D^2} + \frac{1}{r_D}\frac{\partial\psi_{fD}}{\partial r_D} - \zeta_D\left(\frac{\partial\psi_{fD}}{\partial r_D}\right)^2 = e^{\zeta_D\psi_{fD}}\left[\omega\frac{\partial\psi_{fD}}{\partial t_{aD}} + (1-\omega)\frac{\partial V_D}{\partial t_{aD}}\right] \tag{12}$$

$$\frac{\partial V_D}{\partial t_{aD}} = \lambda(V_{ED} - V_D) \tag{13}$$

Dimensionless initial and boundary conditions:

$$\psi_{fD} = 0, V_D = 0, t_{aD} = 0 \qquad \psi_{fD}(r_D \to \infty, t_{aD}) = 0$$

$$e^{-\zeta_D\psi_{fD}} r_D \frac{\partial\psi_{fD}}{\partial r_D}\bigg|_{r_D\to 0} = -\frac{q(t_a)}{q_{sc}} = -q_D(t_{aD}) \tag{14}$$

Where, q_{sc} is the total production rate of horizontal well, sm^3/d.

It's apparently found that Equation (2) contains strong nonlinearity and it's mostly hardly possible to obtain the analytical solution with conventional methods, thus, the perturbation technology and the Presoda transformation are applied to linearize the equations:

$$\psi_{fD}(r_D, t_{aD}) = -\frac{1}{\zeta_D}\ln[1 - \zeta_D\eta(r_D, t_{aD})] \tag{15}$$

According to the theory conducted by Wang[1] and Presoda[33], performing a parameter perturbation in ζ_D by defining the following series:

$$\eta = \eta_0 + \zeta_D\eta_1 + \zeta_D^2\eta_2 + \zeta_D^3\eta_3 + \cdots \tag{16a}$$

$$-\frac{1}{\zeta_D}\ln[1-\zeta_D\eta(r_D,t_{aD})]=\eta(r_D,t_{aD})+\frac{1}{2}\zeta_D\eta^2(r_D,t_{aD})+\frac{1}{6}\zeta_D\eta^3(r_D,t_{aD})+\cdots \tag{16b}$$

$$\frac{1}{1-\zeta_D\eta(r_D,t_{aD})}=1+\zeta_D\eta(r_D,t_{aD})+\zeta_D^2\eta(r_D,t_{aD})+\zeta_D^3\eta(r_D,t_{aD})+\cdots \tag{16c}$$

and considering that the value of ζ_D, dimensionless stress sensitivity coefficient, is always small, the zero-order perturbation solution can greatly meets the requirements, therefore, the final formation of Equation(12) and Equation(13) are as follow:

$$\frac{\partial^2\eta}{\partial r_D^2}+\frac{1}{r_D}\frac{\partial\eta}{\partial r_D}=\omega\frac{\partial\eta}{\partial t_{aD}}+(1-\omega)\frac{\partial V_D}{\partial t_{aD}} \tag{17}$$

$$\frac{\partial V_D}{\partial t_{aD}}=\lambda(V_{ED}-V_D) \tag{18}$$

Dimensionless initial and boundary conditions:

$$\eta(r_D,t_{aD}=0)=0\qquad \eta(r_D\to\infty,t_{aD})=0 \tag{19}$$

$$r_D\frac{\partial\eta}{\partial r_D}\Big|_{r_D\to 0}=-q_D(t_{aD}) \tag{20}$$

The Laplace transformation is used to deal with Equation (17) ~ Equation(20), and we can obtain the formation of governing equation of fracture system in the Laplace domain:

$$\frac{d^2\bar{\eta}}{dr_D{}^2}+\frac{1}{r_D}\frac{d\bar{\eta}}{dr_D}=\omega u\bar{\eta}+(1-\omega)u\overline{V_D} \tag{21}$$

$$u\overline{V_D}=\lambda(\overline{V_{ED}}-\overline{V_D}) \tag{22}$$

Dimensionless boundary conditions:

$$\bar{\eta}(r_D\to\infty,u)=0 \tag{23}$$

$$r_D\frac{d\bar{\eta}}{dr_D}\Big|_{r_D\to 0}=-\overline{q_D} \tag{24}$$

Where, V_{ED} represents the dimensionless gas concentration on the surface the matrix, which can be determined by the fracture pressure. Based on the Langmuir isotherm function, the absorption behavior of shale gas can be described as follows:

$$V_{ED}=V_L\frac{\psi_i}{\psi_i+\psi_L}-V_L\frac{\psi_f}{\psi_f+\psi_L}=\sigma\psi_{fD} \tag{25}$$

Where, $\sigma=\dfrac{P_{sc}q_{sc}T}{\pi k_{fi}hT_{sc}}\dfrac{\psi_L V_L}{(\psi_L+\psi_i)^2}$.

Finally, substituting Equation (22) into Equation (21), one can get the following system of governing equation and boundary conditions:

$$\frac{d^2\bar{\eta}}{dr_D^2}+\frac{1}{r_D}\frac{d\bar{\eta}}{dr_D}=f(u)\bar{\eta} \tag{26}$$

$$\bar{\eta}(r_D\to\infty,u)=0 \tag{27}$$

$$r_D\frac{d\bar{\eta}}{dr_D}\Big|_{r_D\to 0}=-\overline{q_D} \tag{28}$$

Where, $f(u)=\omega u+\dfrac{(1-\omega)\mu\sigma\lambda}{\lambda+u}$.

The typical solution of Equation (26) combined with Equation (27) and Equation (28) can be written as

$$\bar{\eta} = \overline{q_{\mathrm{D}}} K_0(\sqrt{f(u)}\, r_{\mathrm{D}}) \tag{29}$$

Under the more general condition that the line sink is located in arbitrary position $(x_{\mathrm{w}}, y_{\mathrm{w}})$ instead of the center of shale gas reservoirs, the pressure response at $(x_{\mathrm{D}}, y_{\mathrm{D}})$ caused by this line sink can be presented as the following formation:

$$\bar{\eta}(x_{\mathrm{D}}, y_{\mathrm{D}}, u) = \overline{q_{\mathrm{D}}} K_0\left(\sqrt{f(u)}\, \sqrt{(x_{\mathrm{D}} - x_{\mathrm{WD}})^2 + (y_{\mathrm{D}} - y_{\mathrm{WD}})^2}\right) \tag{30}$$

Now, we are just considering one single finite-conductivity hydraulic fracture located in the laterally infinite shale gas reservoir. The flow rate in hydraulic fractures is the function of time and position. Generally speaking, all of the hydraulic fractures approximately have the same length. Therefore, the distribution of the dimensionless pseudo-pressure caused by this whole hydraulic fracture can be obtained with the integration of Equation (30):

$$\overline{\eta_1}(x_{\mathrm{D}}, y_{\mathrm{D}}, u) = \int_{-\frac{x_{\mathrm{F}}}{L_{\mathrm{ref}}}}^{\frac{x_{\mathrm{F}}}{L_{\mathrm{ref}}}} \overline{q_{\mathrm{Dr}}}(\varepsilon, u) K_0\left(\sqrt{f(u)}\, \sqrt{(x_{\mathrm{D}} - \varepsilon)^2 + (y_{\mathrm{D}} - y_{\mathrm{WD}})^2}\right) \mathrm{d}\varepsilon \tag{31}$$

Where, $\overline{q_{\mathrm{Dr}}}(x_{\mathrm{D}}, u)$ represents the dimensionless flux density.

The region that contains the hydraulic fractures and its vicinity is supported by the proppants. Therefore, the stress sensitivity can be ignored in this region. The dimensionless pseudo-pressure distributed throughout this region in Laplace domain is as follow:

$$\overline{\varphi_{\mathrm{FD}}}(x_{\mathrm{DF}}, y_{\mathrm{DF}}, u) = \overline{\eta_1}(x_{\mathrm{DF}}, y_{\mathrm{DF}}, u) \tag{32}$$

Where, $(x_{\mathrm{DF}}, y_{\mathrm{DF}})$ represents the position located in the hydraulic fractures.

3.2 The effect of radial flow and skin factor in the hydraulic fractures

On the one hand, due to the high conductivity of hydraulic fractures, the flow is assumed to be pure linear flow throughout the whole hydraulic fractures [seen from Figure 2(a), the linear flow is from the tip of hydraulic fractures to the wellbore]. The above assumption may be reasonable and can obtain an agreeable result when the site where the linear flow occurs is far away from the wellbore. However, as a matter of fact, the phenomenon of flow-convergence occurs in the vicinity of horizontal wellbore [Figure 2(b)], and it seems like the radial flow. Because of the variation of streamline, an additional pressure drop can occur. It is apparent that solving the problem of transverse hydraulic fractures of multi-stage fracturing horizontal well needs to append a flow-convergence skin factor S_{c}, and the definition of those kinds of skin factor is presented as follow[35]:

$$S_{\mathrm{c}} = \frac{k_{\mathrm{fi}} h}{k_{\mathrm{F}} w_{\mathrm{F}}}\left(\ln \frac{h}{2r_{\mathrm{w}}} - \frac{\pi}{2}\right) \tag{33}$$

Where, r_{w} is the radius of horizontal wellbore, m.

On the other hand, the fracturing fluid can leak into formation during the process of hydraulic fracturing stimulation treatment. This action of leakage can lead to another type of pressure drop. For the convenience of calculation, a skin factor of hydraulic fracture denoted as S_{F} is introduced in this paper.

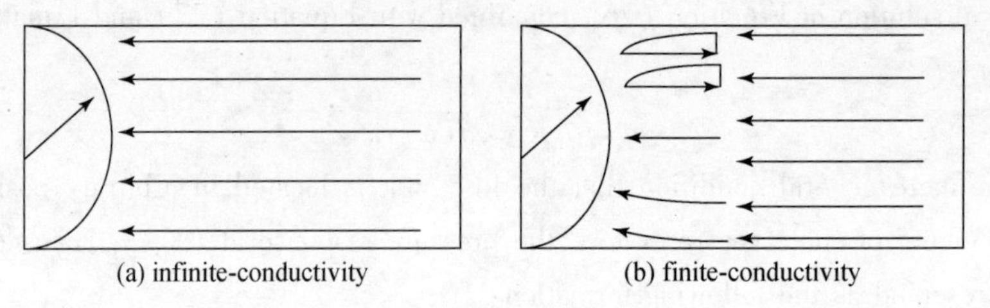

Figure 2 The streamline under different conductivity of hydraulic fracture

Above all, according to the relationships between the pressure and skin factor, the total pressure drop induced by this two kinds of skin factors (flow-convergence skin and hydraulic fracture skin) can be expressed as follows:

$$\varphi_{SD} = q_D \cdot (S_F + S_c) \tag{34}$$

Similarly, the region that contains the hydraulic fractures and its vicinity is supported by the proppants. Therefore, the stress sensitivity can be ignored in this region. The final formation of dimensionless pseudo-pressure in Laplace domain is as follows:

$$\overline{\eta_{SD}} = \overline{q_D} \cdot (S_F + S_c) \tag{35}$$

3.3 The effect of finite-conductivity hydraulic fractures

The issue of the effect of finite-conductivity fractures has been studied by some scholars. In the one word, many of them deal with the problem with the method of analytical integration, however, the process is a little complex, in the another word, many of other scholars utilize finite different methods to solve the equations of finite-conductivity fractures, however, these methods cannot obtain a more general and practical solution. In this paper, the boundary element theory is applied to disperse the hydraulic fractures, and then the superposition principle is utilized to consider the impact of interference of different hydraulic fractures.

To begin with, due to the existence of skin factor, the relationships based on Equation(32) ~ Equation(35) between the natural fracture pressure and hydraulic fracture pressure for one single hydraulic fracture can be presented as

$$\overline{\varphi_{RFD}(x_{FD}, y_{FD}, u)} = \overline{\eta_1}(x_{FD}, y_{FD}, u) + \overline{q_D} \cdot (S_F + S_c) \tag{36}$$

Meanwhile, the interference among hydraulic fractures is considered. In terms of multi-stage fracturing horizontal well, we assume that the numbers of hydraulic fractures equal to M. In terms of every single hydraulic fracture, the dimensionless pseudo-pressure drop can be expressed using Equation(36). However, it is worth to emphasizing that q_D equals to the dimensionless production rate of each hydraulic fracture and they can be different from each other. When the M hydraulic fractures produce simultaneously, the dimensionless pseudo-pressure drop at any location (x_{FD}, y_{FD}) in the shale gas reservoir is equal to the sum of dimensionless pressure drop of all single hydraulic fractures according to the potential of superposition principle. Thus, the dimensionless pseudo-pressure of the j-th hydraulic fracture can be obtained with the superposition principle. Similarly, the region that contains the hydraulic fractures and its vicinity is supported by the proppant. Therefore, the stress sensitivity can be ignored in this region. The final formation of dimensionless pseudo-pressure of the j-th hydraulic fracture in Laplace domain is as follow:

$$\overline{\eta_{\mathrm{RFD}j}}(x_{\mathrm{FD}}, y_{\mathrm{FD}j}, u) = \sum_{i=1}^{M} \bar{E}(\kappa_i, \kappa_j, u) + \overline{q_{\mathrm{D}j}}(x_{\mathrm{D}}, y_{\mathrm{FD}j}, u)(S_{\mathrm{F}} + S_{\mathrm{c}}) \tag{37}$$

Where

$$\kappa_i = (x_{\mathrm{FD}}, y_{\mathrm{FD}i}), \kappa_j = (x_{\mathrm{FD}}, y_{\mathrm{FD}j})$$

$$\bar{E}(\kappa_i, \kappa_j, u) = \int_{-\frac{x_{\mathrm{F}}}{L_{\mathrm{ref}}}}^{\frac{x_{\mathrm{F}}}{L_{\mathrm{ref}}}} \overline{q_{\mathrm{D}ri}}(u) K_0\left(\sqrt{f(u)}\sqrt{(x_{\mathrm{FD}} - \varepsilon)^2 + (y_{\mathrm{FD}j} - y_{\mathrm{FD}i})^2}\right) \mathrm{d}\varepsilon \tag{38}$$

The one dimensional linear flow is assumed in the hydraulic fractures, the compressibility of fractures and gas is ignored, and the flow state in hydraulic fractures meets continuous pseudo-state theory. The governing equation with the dimensionless formation in the j-th hydraulic fracture is as follow (seen from Appendix A):

$$\frac{\partial^2 \overline{\eta_{\mathrm{RFD}j}}}{\partial x_{\mathrm{D}}^2} + \frac{2}{C_{\mathrm{FD}j}} \frac{\partial \overline{\eta_{\mathrm{RFD}j}}}{\partial y_{\mathrm{D}}}\bigg|_{y_{\mathrm{D}} = y_{\mathrm{FD}j}} = 0, (0 < x_{\mathrm{D}} < 1) \tag{39}$$

The boundary conditions are as follows:

$$\overline{q_{\mathrm{D}j}}(x_{\mathrm{FD}}, y_{\mathrm{FD}j}, u) = -\frac{2}{\pi} \frac{\partial \overline{\eta_{\mathrm{RFD}j}}}{\partial y_{\mathrm{D}}}\bigg|_{y = y_{\mathrm{FD}j}} \tag{40}$$

$$\frac{\partial^2 \overline{\eta_{\mathrm{RFD}j}}}{\partial x_{\mathrm{D}}}\bigg|_{x_{\mathrm{FD}} = 0} = -\frac{\pi \overline{q_{\mathrm{FD}j}}}{C_{\mathrm{FD}j}} \tag{41}$$

Where, $\overline{q_{\mathrm{D}j}}(x_{\mathrm{FD}}, y_{\mathrm{FD}j}, u)$ represents the dimensionless rate at the position $(x_{\mathrm{FD}}, y_{\mathrm{FD}j})$ in the j-th hydraulic fracture. $\overline{q_{\mathrm{FD}j}}$ represents the total dimensionless rate of the j-th hydraulic fracture.

4 Model solution

Based on the principle of superstition, the total dimensionless pseudo-pressure of the j-th hydraulic fracture caused by above various factors, such as finite-conductivity fracture, flow convergence and fracture damage skin factor, can be obtained by combining Equation (35) ~ Equation (38). Through integrating the Equation (39) from certain position of hydraulic fracture to the wellbore, we can get the final integration formation coupled with the Equation (40) and Equation (41):

$$\overline{\eta_{\mathrm{WD}j}} - \overline{\eta_{\mathrm{RFD}j}}(x_{\mathrm{FD}}, y_{\mathrm{FD}j}, u) = \frac{\pi \overline{q_{\mathrm{FD}j}}}{C_{\mathrm{FD}j}} x_{\mathrm{D}} - \frac{\pi}{C_{\mathrm{FD}j}} \int_0^{x_{\mathrm{D}}} \int_0^{x'} \overline{q_{\mathrm{D}j}}(x'', y_{\mathrm{FD}j}, u) \mathrm{d}x'' \tag{42}$$

Submitting Equation (31) into Equation (36), one can obtain a general formula:

$$\begin{aligned} &\overline{\eta_{\mathrm{WD}j}} - \sum_{i=1}^{M} \frac{1}{2} \int_{-1}^{1} \overline{q_{\mathrm{D}i}}(\varepsilon, u) K_0\left(\sqrt{f(u)}\sqrt{(x_{\mathrm{FD}} - \varepsilon)^2 + (y_{\mathrm{FD}j} - y_{\mathrm{FD}i})^2}\right) \mathrm{d}\varepsilon = \\ &\overline{q_{\mathrm{D}j}}(x_{\mathrm{D}}, y_{\mathrm{FD}j}, u) \cdot (S_{\mathrm{F}} + S_{\mathrm{c}}) + \frac{\pi \overline{q_{\mathrm{FD}j}}}{C_{\mathrm{FD}j}} x_{\mathrm{D}} - \frac{\pi}{C_{\mathrm{FD}j}} \int_0^{x_{\mathrm{D}}} \int_0^{x'} \overline{q_{\mathrm{D}j}}(x'', y_{\mathrm{FD}j}, u) \mathrm{d}x'' \end{aligned} \tag{43}$$

Where, $\overline{q_{\mathrm{FD}j}} = \sum_{v=1}^{M} q_{\mathrm{D}(j,v)}$ represents the dimensionless conductivity of the j-th hydraulic fracture.

Considering the wellbore infinite-conductivity, the flow pressure of each hydraulic fracture is approximately the same and equal to the bottom-hole flowing pressure, then one can get the following equation:

$$\overline{\eta_{\mathrm{WD}}} - \sum_{i=1}^{M} \frac{1}{2}\int_{-1}^{1} \overline{q_{\mathrm{D}i}}(\varepsilon,u)K_0\left(\sqrt{f(u)}\sqrt{(x_{\mathrm{FD}}-\varepsilon)^2+(y_{\mathrm{FD}j}-y_{\mathrm{FD}i})^2}\right)\mathrm{d}\varepsilon = \overline{q_{\mathrm{D}j}}(x_{\mathrm{D}},y_{\mathrm{FD}j},u)\cdot(S_{\mathrm{F}}+S_{\mathrm{c}}) + \frac{\pi\overline{q_{\mathrm{FD}j}}}{C_{\mathrm{FD}j}}x_{\mathrm{D}} - \frac{\pi}{C_{\mathrm{FD}j}}\int_0^{x_{\mathrm{D}}}\int_0^{x'}\overline{q_{\mathrm{D}j}}(x'',y_{\mathrm{FD}j},u)\mathrm{d}x'' \tag{44}$$

To solve the Equation (44), the dispersion method is applied for every hydraulic fracture. Each fracture is artificially divided into N equivalent elements (Figure 3). Thus, Equation (38) can be presented as

$$\overline{\eta_{\mathrm{WD}}} - \frac{1}{2}\sum_{i=1}^{M}\sum_{k=1}^{N}\int_{x_{\mathrm{FD}(i,k)}-\Delta x_{\mathrm{D}}}^{x_{\mathrm{FD}(i,k)}+\Delta x_{\mathrm{D}}}\overline{q_{\mathrm{D}(i,k)}}K_0\left(\sqrt{f(u)}\sqrt{(x_{\mathrm{FD}(j,v)}-\varepsilon)^2+(y_{\mathrm{FD}(j,v)}-y_{\mathrm{FD}(i,k)})^2}\right)\mathrm{d}\varepsilon - \overline{q_{\mathrm{D}(j,v)}}\cdot(S_{\mathrm{F}}+S_{\mathrm{c}}) + \frac{\pi}{C_{\mathrm{FD}j}}\left\{\sum_{l=1}^{v-1}\overline{q_{\mathrm{D}(j,l)}}\left[\frac{\Delta x_{\mathrm{D}}^2}{2}+\Delta x_{\mathrm{D}}(x_{\mathrm{FD}(j,v)}-l\Delta x_{\mathrm{D}})\right]+\frac{\Delta x_{\mathrm{D}}^2}{8}\overline{q_{\mathrm{D}(j,v)}}\right\} = \frac{\pi\overline{q_{\mathrm{FD}j}}}{C_{\mathrm{FD}j}}x_{\mathrm{FD}(j,v)} \tag{45}$$

where, $q_{\mathrm{FD}j}$ represents the total dimensionless rate of the j-th hydraulic fracture, it is able to be denoted as: $C_{\mathrm{FD}j} = \frac{k_{\mathrm{F}j}w_{\mathrm{F}j}}{k_{\mathrm{f}}x_{\mathrm{F}j}}$. $q_{\mathrm{D}(i,k)}$ represents the dimensionless rate of the i-th fracture's k-th element; $x_{\mathrm{FD}(j,v)}$, $y_{\mathrm{FD}(j,v)}$ respectively represent the dimensionless x and y coordinate of the j-th fracture's v-th element; x_{D} represents the dimensionless length of every element.

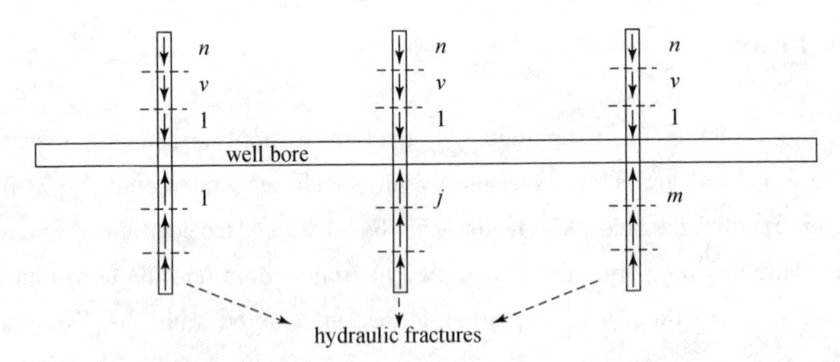

Figure 3 The illustration of hydraulic fractures discretion

By rewriting Equation (45) for all elements, we can get $N*M$ equations with $(N*M+1)$ unknowns parameters which are $q_{\mathrm{D}(j,v)}$ and η_{wD}. One more equation is needed to solve the equations. We assume that the well produces at the constant production rate condition, thus

$$\Delta x_{\mathrm{D}}\sum_{j=1}^{M}\sum_{v=1}^{N}\overline{q_{\mathrm{D}(j,v)}} = \frac{1}{u} \tag{46}$$

Equation (45) and Equation (46) compose a $(N*M+1)$ order system linear algebraic equations which can be presented as the following matrix form:

$$\begin{bmatrix} A_{1,1} & \cdots & A_{1,k} & \cdots & A_{1,N*M} & -1 \\ \cdots & \cdots & \cdots & \cdots & \cdots & -1 \\ A_{k,1} & \cdots & A_{k,k} & \cdots & A_{k,N} & -1 \\ \vdots & & \vdots & \vdots & \vdots & -1 \\ A_{N*M,1} & \cdots & A_{N*M,k} & \cdots & A_{N*M,N*M} & -1 \\ 1 & \cdots & 1 & \cdots & 1 & 0 \end{bmatrix} \begin{bmatrix} \overline{q_{D1}} \\ \vdots \\ \overline{q_{DN*M}} \\ \overline{\eta_{WD}} \end{bmatrix} = \begin{bmatrix} 0 \\ \vdots \\ 0 \\ 1/\Delta x_D u \end{bmatrix} \tag{47}$$

$$A_{i,j} = \begin{cases} \displaystyle\int_{x_{FDi}-\Delta x_D}^{x_{FDi}+\Delta x_D} \overline{q_{Di}} K_0\left(\sqrt{f(u)}\sqrt{(x_{FDj}-\varepsilon)^2+(y_{FDj}-y_{FDi})^2}\right)d\varepsilon \\ \displaystyle -\frac{\pi}{C_{FD}}\left[\frac{\Delta x_D^2}{2}+\Delta x_D(x_{FDj}-i\Delta x_D)\right]+\frac{\pi}{C_{FD}}x_{FDi}, & (i<j) \\ \displaystyle\int_{x_{FDi}-\Delta x_D}^{x_{FDi}+\Delta x_D} \overline{q_{Di}} K_0\left(\sqrt{f(u)}\sqrt{(x_{FDj}-\varepsilon)^2+(y_{FDj}-y_{FDi})^2}\right)d\varepsilon+\frac{\pi}{C_{FD}}x_{FDi}, & (i>j) \\ \displaystyle\int_{x_{FDi}-\Delta x_D}^{x_{FDi}+\Delta x_D} \overline{q_{Di}} K_0\left(\sqrt{f(u)}\sqrt{(x_{FDj}-\varepsilon)^2+(y_{FDj}-y_{FDi})^2}\right)d\varepsilon \\ \displaystyle +(S_F+S_c)-\frac{\pi}{C_{FD}}\frac{\Delta x_D^2}{8}+\frac{\pi}{C_{FD}}x_{FDj}, & (i=j) \end{cases}$$

Due to the solution given in Laplace domain, the wellbore storage can be easily added into the above solution with Duhamel's theory. Therefore, the final pseudo-pressure solution with the consideration of wellbore storage is as follow:

$$\bar{\eta}_{wfin} = \frac{\bar{\eta}_{wD}}{1+u^2 C_D \bar{\eta}_{wD}} \tag{48}$$

Where, C_D is the dimensionless wellbore storage which can be defined as

$$C_D = \frac{C}{2\pi L_{ref}^2 h\Lambda} \tag{49}$$

Using the Stehfest numerical invention algorithm[37] to obtain the pseudo-pressure solution in real space, namely, converting $\overline{\eta_{Wfin}} \rightarrow \eta_{Wfin}$. According to Equation (15), one can get the bottom hole pressure for multi-stage fracturing horizontal well in the shale gas reservoir which takes the stress sensitivity of natural fractures into consideration.

$$\varphi_{WD} = -\frac{1}{\zeta_D}\ln(1-\zeta_D\eta_{Wfin}) \tag{50}$$

5 Discussion

In this section, to discuss this proposed new model, some relevant basic parameters are listed in the Table 2.

Table 2 The input value of relevant basic parameters in shale gas reservoir

parameters	value
Initial pressure, P_i, MPa	20

续表

parameters	value
Gas rate, Q_{sc}, m^3/d	50000
Formation temperature, T, K	330
Horizontal length, L, m	1000
Formation thickness, h, m	50
Diffusion coefficient, D, m^2/s	3.0×10^{-7}
Total compressibility of fracture, C_{tfi}, MPa^{-1}	2.0×10^{-4}
Porosity of natural fracture φ_f, fraction	0.005
The number of hydraulic fracture M	8
Permeability of natural fracture k_f, D	10^{-5}
Langmuir Pressure P_L, MPa	10
Langmuir volume V_L, sm^3/m^3	5
Matrix radiu R_m, m	0.01
Hydraulic fracture width W_F, m	0.005
Hydraulic fracture width k_F, D	5
Hydraulic fracture length L_F, m	150
Initial gas viscosity μ_i, mPa · s	0.25
Average gas compressibility Z	0.998

5.1 Flow regimes

The essence of this paper is the comprehensive consideration of seepage in the hydraulic fractures system and in the shale gas formation system. The type curves are plotted using Stehfest numerical inversion algorithm. The division of flow regimes and corresponding analysis of related parameters will be discussed in the following sections. The bold lines represent the dimensionless pressure curves; the broken lines represent the dimensionless derivative pressure curves (Figure 4).

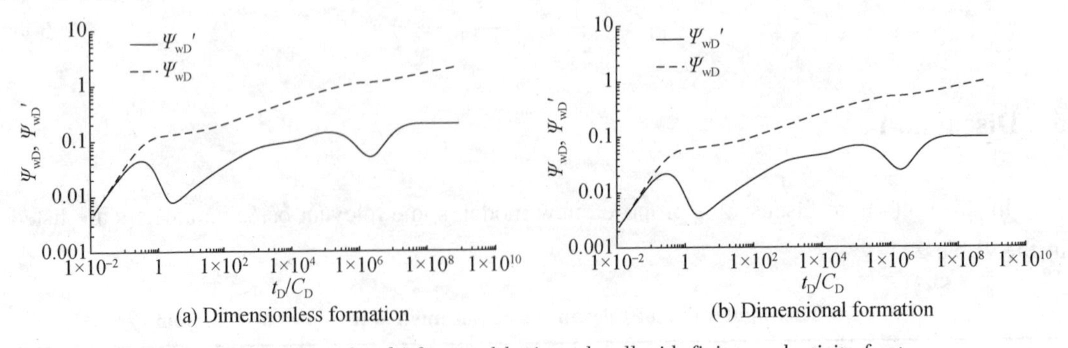

(a) Dimensionless formation (b) Dimensional formation

Figure 4 Type curve of multi-fractured horizontal well with finite-conductivity fractures

Figure 4(a) represents the type curves under certain value of relevant parameters listed in Table 2, the whole flow process can be divided into the following eight regimes.

(1) Regime Ⅰ. The pure wellbore storage period regime. Dimensionless derivative pseudo-pressure curve and pseudo-pressure curve align, and the values of curves slope are equal to 1. This stage is mainly controlled by wellbore storage effect.

(2) Regime Ⅱ. The transition flow and linear flow in finite-conductivity hydraulic fractures. The early stage of this regime derives from the straight line which has unit slope, and then, the linear flow occurs. The value of the slope of those two curves is approximately equal to 0.5.

(3) Regime Ⅲ. The bi-linear flow regime. The seepage flow simultaneously occurs in the hydraulic fractures and shale gas formation (mainly natural fractures), the value of the slope of the derivative dimensionless pseudo-pressure is equal to 0.25. Because the permeability of hydraulic fractures is much bigger than natural fractures, this regime is ambiguous.

(4) Regime Ⅳ. The early pseudo-radial flow regime. When the pressure wave reaches the tip of every hydraulic fracture, this stage can be observed from the type curves. This regime is mainly dominated by the hydraulic fracture length and spacing. It seems to be a horizontal line, and the value of the line is equal to $1/2M$ (where, M is 8, therefore, the value of this horizontal line is equal to 0.0625).

(5) Regime Ⅴ. The intermediate-time linear flow regime. When the pressure wave reaches beyond all of the hydraulic fractures. The interference among different hydraulic fractures occurs. The value of this type curve is equal to 0.5, similarly, because of the ultra-permeability of natural fractures, this stage is not so obvious.

(6) Regime Ⅵ. The intermediate-time transition flow regime. This stage happens between the linear flow regime and pseudo-diffusion regime . In this paper, the stress sensitivity is considered, so this phenomenon occurs firstly during this stage.

(7) Regime Ⅶ. The pseudo-steady diffusion regime between the matrix and natural fractures. When the pressure reaches to the absorption pressure, the shale gas which originally adsorbs in the surface of matrix will desorb from matrix. The concentration difference between the matrix and natural fractures causes the diffusion flow. When we consider the pseudo-diffusion between matrix and fractures, the main feature is marked by a "dip" in the derivation dimensionless pseudo-pressure curves.

(8) Regime Ⅷ. The late-time pseudo-radial flow regime. We assume that the shale gas reservoir is boundless, so the hydraulic fractures can be seen as a point compared with the whole reservoir. The shape of type curves is derivation from dimensionless pseudo-pressure curve is like the horizontal line. The value of this horizontal well equals to 0.5.

Universally, the well testing type curve is mainly used to match with real parameters. Therefore, it is necessary to plot and analyze the dimensional formation of type curves. We have defined the definition of pseudo-time, which has certain relationship with total compressibility and gas viscosity. Therefore, it is important for us to convert dimensionless time to real time. When we have obtained the relationship between dimensionless pseudo-time and dimensionless pseudo-pressure, we can firstly convert the dimensionless pseudo-pressure into dimensional pseudo-pressure based on the definition so that we can obtain the relationship between dimensionless pseudo-time and dimensional pseudo-pressure. When we have known the pseudo-pressure, the related total compressibility and gas viscosity also can be calculated. Finally, based on the definition of pseudo-time, the real time can be calculated using iterative method. Seen from the Figure 4(b), when we convert the dimensionless type curve into dimensional formation, the mainly eight flow regime still can be observed.

5.2 Model validation

This proposed new model comprehensively considers the multiple mechanisms, such as adsorption, diffusion, viscous flow, the hydraulic fracture conductivity, flowing convergence, stress sensitivity, skin factor and wellbore storage. However, it is not difficult to find that the new model can be simplified to be some existing simple model when some certain parameters are equal to special value. For example, when we ignore the flow convergence skin and fracture conductivity, it can be consistent with the model proposed by Wang et al[34]. Therefore, to validate my new model, it will be compared with the model proposed by Wang et al[34]. The basic parameters are listed in Table 2. Seen from Figure 5, when we ignore the flow convergence skin and fracture conductivity, this new model can perfectly match with model proposed by Wang et al[34]. On the contrary, when the flow convergence skin and fracture conductivity are considered proposed by this new model, the regime of transition flow and linear flow in finite-conductivity hydraulic fractures is different from the model proposed by Wang et al[34]. From the analysis of basic theory, fracture conductivity and flow convergence skin can cause additional pressure loss. Therefore, this proposed new model is reasonable.

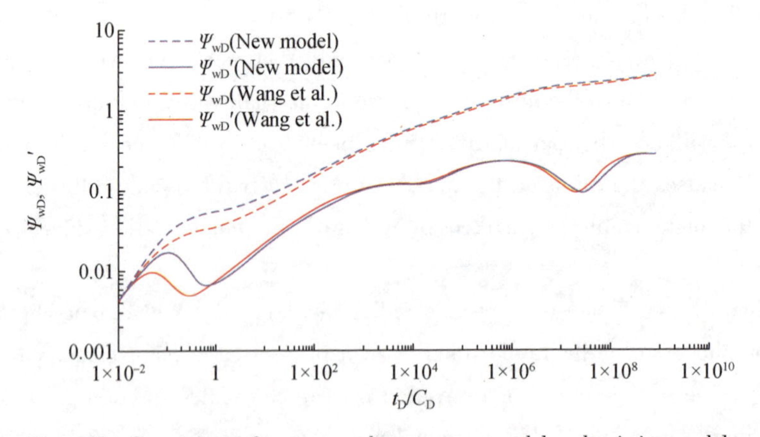

Figure 5 Comparison of type curve between new model and existing model

Besides, a numerical simulation is also conducted to compare with the models (the input data are listed in Table 2, and the water saturation is equal to the connate water saturation 0.25 so that the influence of water can be ignored). The commercial simulation software- CMG 2010 is selected to verify the derived analytical solutions. The model considers the flow from a horizontal well in a dual-porosity reservoirs and GEM model is selected to simulate the proposed flow process. Seen from Figure 6, one representative segment is modeled which represents one part of the reservoir volume around a hydraulic fracture. This segment contains twenty four hydraulic fractures orthogonal to the horizontal well at 250m fracture spacing. The model is a 2-D model with 55 grid cells in the x-direction, 55 grid cells in y-direction and only one grid cell in the z-direction. The multi-porosity method is applied to subdivided the matrix so that the transient flow in matrix can be simulated and desorption model is instant desorption. The local grid refinement (LGR) method is employed to form the hydraulic fractures so that the multi-stage hydraulic fracturing can be simulated. ROCKTAB keyword is used to indicate the stress sensitivity. It is known that the grid can be refined to characterize the hydraulic fracture so that numerical simulation can consider the fracture conductivity, which can provide a good comparison and verification for our new model. Figure 7 demonstrates the comparison between new proposed model, existing model and numerical simulation. We can find that the new proposed model has more agreements with the numerical simulation than that of existing model. There are still some deviations between new proposed model and numerical simulation because CMG cannot consider the flow convergence effects. Therefore, this proposed new model can provide a relative more accurate analysis of the relevant parameters, especially for the fracture conductivity.

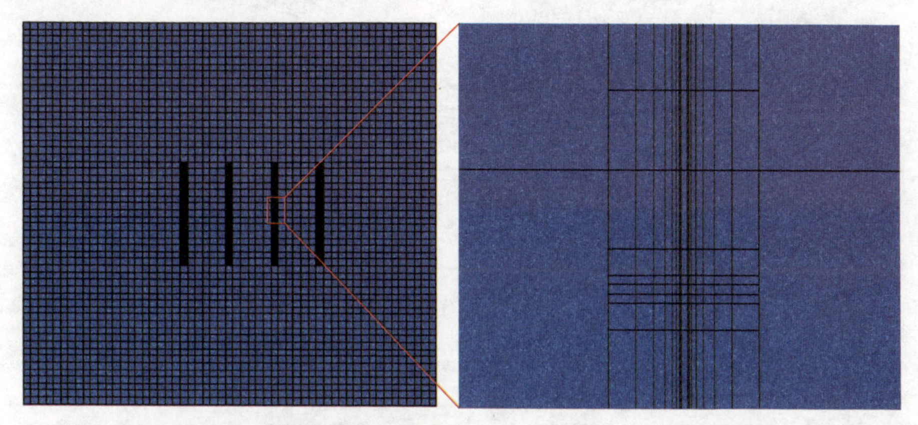

Figure 6 Establishment of shale gas reservoir model using CMG

5.3 Sensitivity analysis

Seen from Figure 8, showing the effect of dimensionless stress sensitivity coefficient, ζ_D, on pressure transient responses of a multiple fracturing horizontal well with finite-conductivity hydraulic fractures in shale gas reservoirs, the type curves are plotted with the different values of ζ_D, $\zeta_D = 0.05, 0.10, 0.15$, and other characteristic parameters are as follows: $\lambda = 0.002$, $\omega = 0.0032$, $\sigma = 1.5$, $S_c = 0.5$, $S_F = 0.5$, $C_{FD} = 18.7$, $M = 8$. According to the Figure 6, the dimensionless stress sensitivity coefficient mainly affects the late-time flow regime, including intermediate-time radial flow, transition flow, pseudo-diffusion and late-time radial flow. The larger the value of ζ_D is, the larger the value of dimensionless pseudo-pressure is. The type curve will ascend

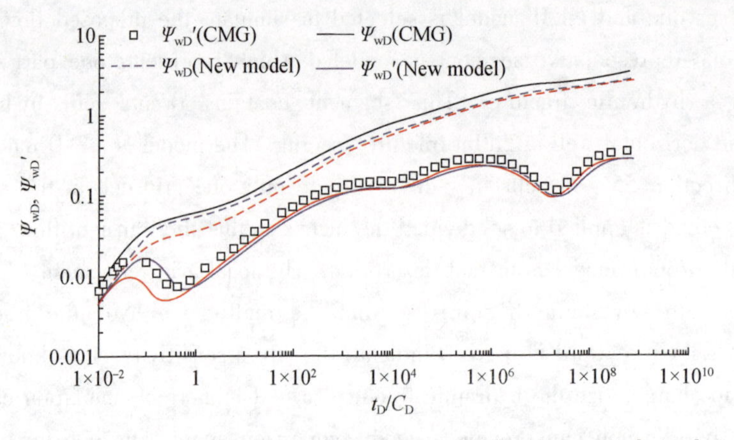

Figure 7　Comparison of type curve between new model and numerical simulation

with the increment of ζ_D. Based on its micro-mechanism, the natural fractures can be closed with the deduction of pressure, as a result, the fracture permeability simultaneously decreases, and the flow resistance will increase, therefore, the larger pressure drop is required to keep the constant rate.

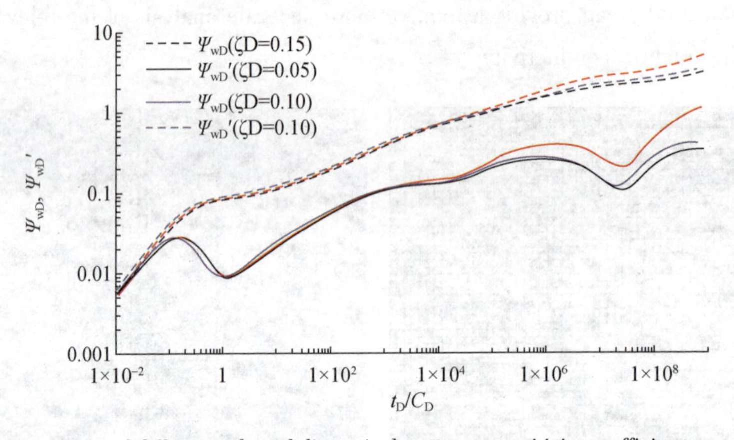

Figure 8　The effects of different value of dimensionless stress sensitivity coefficient on type curves

Being illustration from Figure 9, it shows the effects of dimensionless fracture conductivity C_{FD} on pressure transient responses of a multiple fracturing horizontal well with finite-conductivity hydraulic fractures in the shale gas reservoirs, the type curves are plotted with different dimensionless fracture conductivity, $C_{FD}=10, 15, 20$, other parameters are as follows: $\lambda=0.002$, $\omega=0.0032$, $\sigma=1.5$, $S_c=0.5$, $S_F=0.5$, $\zeta_D=0.05$, $M=8$. This parameter mainly affects the wellbore storage and linear flow regime in hydraulic fractures, the type curves during those two regimes will descend with the increment of dimensionless conductivity. The larger the value of dimensionless conductivity, the earlier the pure wellbore storage ends, and the earlier the pure linear flow in hydraulic fractures occurs. Conductivity means the flow ability of hydraulic fractures, and the larger this parameter is, the stronger the flow ability is.

According to the Figure 10, it can find the effects of fracture number M on pressure transient responses of a multiple fracturing horizontal well with finite-conductivity hydraulic fractures in

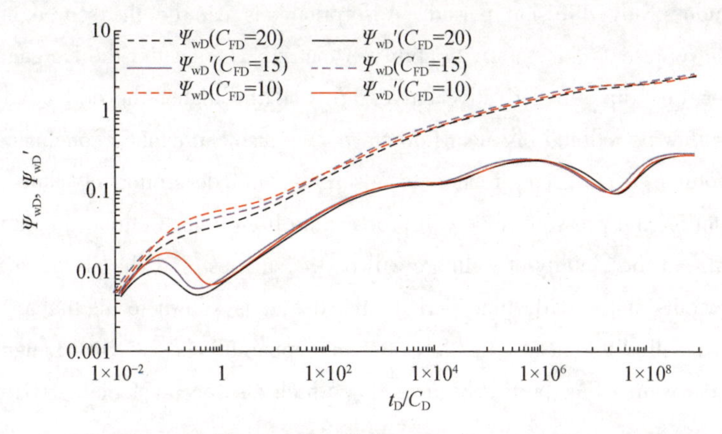

Figure 9 The effects of different value of dimensionless conductivity on type curves

shale gas reservoirs, the type curves are plotted with different dimensionless fracture space, $M=6$, 7, 8, and other parameters are as follows: $\lambda=0.002$, $\omega=0.0032$, $\sigma=1.5$, $S_c=0.5$, $S_F=0.5$, $\zeta_D=0.05$, M, $C_{FD}=18.7$, $\zeta_D=0.05$. Actually, the fracture number reflects the fracturing spacing. The more the fracture is, the shorter is fracturing spacing. We can observe that the fracture number mainly has effect on the existence and duration time of intermediate-time pseudo-radial flow period. When the fracture length keeps constant, larger fracture spacing corresponds to longer intermediate-time pseudo-radial flow period. It should be pointed out that when the fractures are close to each other enough, the intermediate-time pseudo-radial flow period may not be observed.

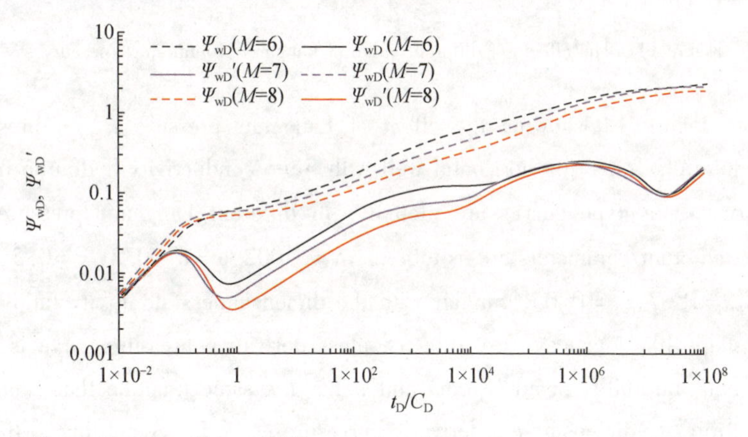

Figure 10 The effects of different value of fracture number on type curves

Figure 11 shows the effect of Langmuir volume V_L on pressure transient responses of a multiple fracturing horizontal well with finite-conductivity hydraulic fractures in the shale gas reservoirs, the type curves are plotted with different Langmuir volume $V_L=4\text{sm}^3/\text{m}^3$, $6\text{sm}^3/\text{m}^3$, $8\text{sm}^3/\text{m}^3$. In addition to the Langmuir volume V_L, the values of other relevant parameters are listed as follows: $\lambda=0.002$, $\omega=0.0032$, $S_c=0.5$, $S_F=0.5$, $\zeta_D=0.05$, $M=8$, $C_{FD}=18.7$, $\zeta_D=0.05$. The Langmuir volume has main effects on the early-time linear flow period, bi-linear flow, and intermediate-time pseudo-radial flow around each hydraulic fracture, intermediate-time linear flow period and pseudo-steady diffusion period. Adsorption gas can be the source of gas; therefore, Langmuir volume represents the storativity ability of matrix. The smaller the Langmuir volume is, the deeper and wider the "dip" in derivative curves are, and the higher the derivative curve is during early-time linear flow period and bi-linear flow stage. We also can explain conclusion from Figure 11 that Langmuir volume, a parameter related to adsorption and desorption of shale gas. During the process of production, the pressure wave will propagation from inner wellbore to outer boundary, as a result, the pressure at the vicinity of wellbore will be the smallest and sharply decrease to desorption pressure which occurs at the early-time period. Besides, it is known to all that pressure derivative can magnify the small difference. Therefore, we can apparent observe that Langmuir volume has main effects on early-time flow period, bilinear flow which also occur at the early-time period.

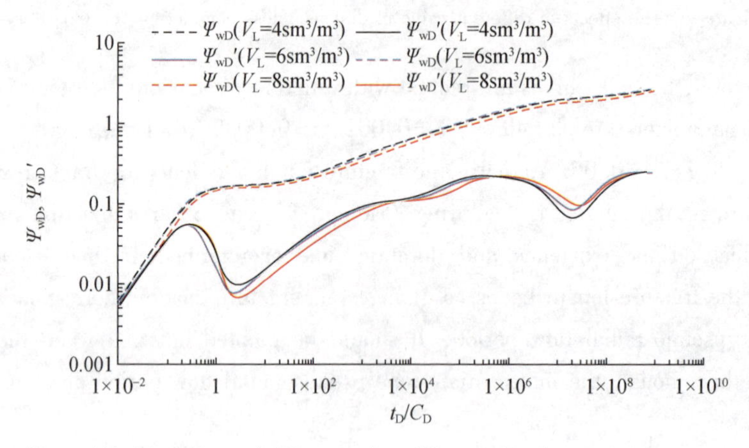

Figure 11 The effects of different value of Langmuir volume on type curves

Seen from Figure 12, showing the effect of Langmuir pressure P_L on pressure transient responses of a multiple fracturing horizontal well with finite-conductivity hydraulic fractures in the shale gas reservoirs, the type curves are plotted with different Langmuir pressure $P_L = 6$MPa, 8MPa, 10MPa, and other parameters are as follows: $\lambda=0.002$, $\omega=0.0032$, $S_c=0.5$, $S_F=0.5$, $\zeta_D=0.05$, $M=8$, $C_{FD}=18.7$, $\zeta_D=0.05$. Similar with the dimensionless storativity ratio which controls how deep the shape like "concave" of dimensionless derivative pressure curve is, the Langmuir pressure represents the ability of adsorption, and it has the same function that controls how deep the shape like "dip" of dimensionless derivative pressure curve is. According to the definition of dimensionless storativity coefficient, when the pressure decreases in matrix, the larger the value of the P_L is, the earlier adsorption happens and the larger the value of dimensionless storativity coefficient. As a result, with the decreasing of Langmuir pressure, the shape like "dip" in the derivative curve becomes deeper and wider. Due to the adsorption of shale gas in matrix, the gas concentration in matrix gradually increases so that the concentration gap between the matrix and fractures increases, as a result, the larger the concentration gap is, the earlier the diffusion occurs (the type curves wholly move left).

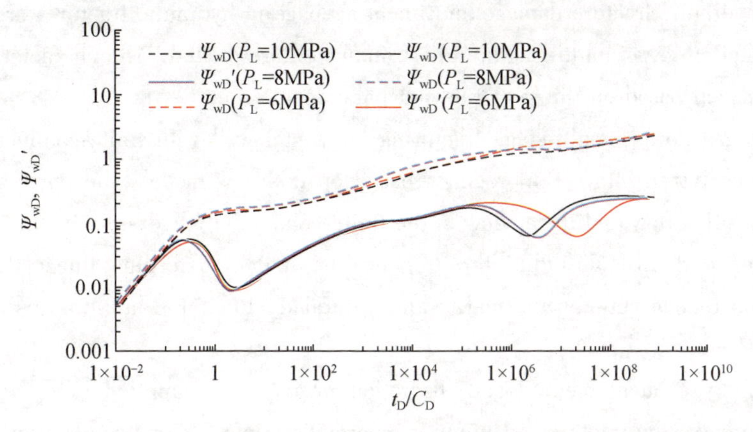

Figure 12 The effects of different value of Langmuir pressure on type curves

Based on the depiction from Figure 13, it is easily to observe the effect of flow-convergence skin, S_c, on pressure transient responses of a multiple fracturing horizontal well with finite-conductivity hydraulic fractures in the shale gas reservoirs, the type curves are plotted with different flow-convergence skin, $S_c = 0.3, 0.5, 0.8$, and other parameters are as follows: $\lambda = 0.002$, $\omega = 0.0032$, $S_c = 0.5$, $S_F = 0.5$, $\zeta_D = 0.05$, $M = 8$, $C_{FD} = 18.7$, $\zeta_D = 0.05$. According to the definition of flow-convergence skin, this parameter has an inverse relationship with the fracture conductivity. That is to say, the larger the value of fracture permeability or fracture width is, the smaller the value of flow-convergence skin is. Therefore, the influence of flow-convergence skin on pressure response is similar with the dimensionless conductivity. When we are analyzing and estimating the value of the conductivity of hydraulic fractures, this flow parameter may have certain interference.

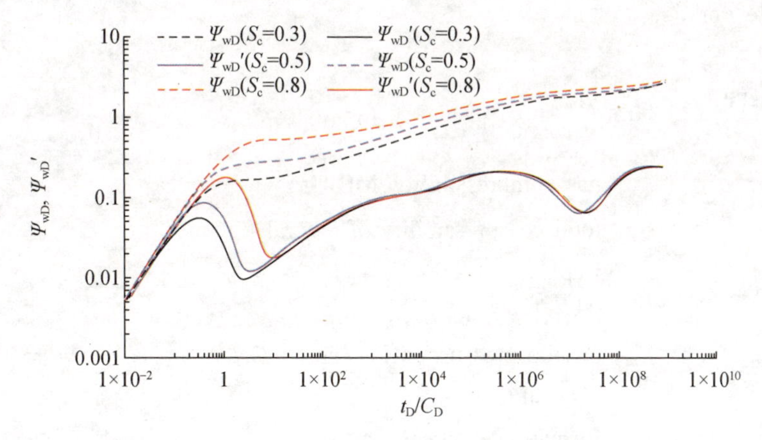

Figure 13 The effects of different value of flow convergence skin on type curves

6 Conclusion

We establish a new model for multi-stage fracturing horizontal well with finite-conductivity hydraulic fractures in the shale gas reservoirs. Adsorption, diffusion, stress sensitivity, flow-convergence, hydraulic fracture damage and linear seepage in hydraulic fractures are considered as well. The typical curves of multi-fractured horizontal well are plotted. The characteristics of typical curves are analyzed based on different relevant parameters.

The shale gas flow of multi-stage fracturing horizontal well with finite-conductivity hydraulic fractures can be divided into 8 flow regimes based on this new model: the pure wellbore storage regime; the transition flow and linear flow in the finite-conductivity hydraulic fractures; the bi-linear flow regime; the early pseudo-radial flow regime; the intermediate-time linear flow regime; the pseudo-diffusion regime between the matrix and natural fractures; the late-time pseudo-radial flow regime.

The boundary element method and discretion principle are applied to solve the governing equation in the hydraulic fractures. The most important two parameters, conductivity and flow-convergence skin, are analyzed via plotting the relevant type curves. It is found that those two parameters have the similar significant influence on the early-linear flow regime in the hydraulic fractures. Therefore, one of the parameter has certain interference with another parameter when we need to estimate one of the parameters. For example, if the flow-convergence skin is ignored, the estimation result of hydraulic fractures has certain error.

Acknowledgment

The authors acknowledge a fund from the National Natural Science Foundation (NNSF) of China (No. 2011ZX05015). And this paper is supported by the fund NNSF of China (No. 2011ZX05015). I would also like to thanks for the editor and reviewers because of their laborious and responsible work.

Nomenclature

C_g	gas compressibility, MPa^{-1}
C_{tf}	total compressibility of natural fractures
u	Laplace variable
t	time, h
ψ_j	pseudo pressure, MPa^2/cp, $j=F, f$
μ	viscosity, cp
h	formation thickness, m
L_{ref}	reference length, m
t_D	dimensionless time
q_D	dimensionless rate
φ_f	porosity of natural fractures, fraction
x_F	hydraulic fracture half length, m
x_D, y_D	dimensionless space

q_{DF}	total dimensionless rate of one fracture
S_c	flow-convergence skin
S_F	fracture damage skin
k_F, k_f	permeability of hydraulic fractures and natural fractures, D
k_{fi}	initial permeability of natural fractures, D
D	diffusion coefficient, m^2/s
q_{sc}	well production rate, m^3/d
T	reservoir temperature, K
T_{SC}	standard temperature, K
P_{SC}	standard pressure, MPa
P_L	Langmuir pressure, MPa
V_L	Langmuir volume, sm^3/m^3
M	the numbers of hydraulic fracture
x, y	coordination, m
r	radial distance, m
r_D	dimensionless radial distance
ε	integration variable
x'	integration variable
x''	integration variable
C_{FD}	dimensionless conductivity
C	wellbore storage coefficient
q_{Dr}	dimensionless flux density
subscript	
D	dimensionless
superscript	
—	Laplace transform
′	derivative

References

[1] Javadpour F. Nanopores and apparent permeability of gas flow in mudrocks (shale and siltstone) [J]. Journal of Canadian Petroleum Technology, 2009, 48(8): 16-21.

[2] Javadpour F, Fisher D, Unswort M. Nanoscale gas flow in shale gas sediments [J]. Journal of Canadian Petroleum Technology, 2007, 46(10): 55-61.

[3] Xiao J, Wei J. Diffusion mechanism of hydrocarbons in zeolites-I. Theory [J]. Chemical Engineering Science, 1992, 47 (5): 1123-1141.

[4] Barenblatt G I, Zheltov I P, Kochina I N. Basic concepts in the theory of seepage of homogeneous liquids in fissured rocks [strata] [J]. Journal of Applied Mathematics and Mechanics, 1960, 24(5): 1286-1303.

[5] Warren J E, Root P J. The behavior of naturally fractured reservoirs [C]. SPE 426, 1963.

[6] Kazemi H. Pressure transient analysis of naturally fractured reservoirs with uniform fracture distribution [C]. SPE 2156, 1969.

[7] De Swaan O A. Analytic solutions for determining naturally fractured reservoir properties by well testing [C]. Paper SPE 5346 presented at the SPE-AIME 45th Annual California Regional Meeting, Ventura, California, USA, 2-4 April 1975.

[8] Ozkan E, Brown M L, Raghavan R, *et al.* Comparison of fractured-horizontal-well performance in tight sand and shale reservoirs[C]. SPE 121290, 2011.

[9] Clarkson C R . Production data analysis of unconventional gas wells: Review of theory and best practices [J]. International Journal of Coal Geology, 2013, 109: 101-146.

[10] Zhao Y L, Zhang L H, Wu F. Pressure transient analysis for multi-fractured horizontal well in shale gas reservoirs[J]. Journal of Petroleum Science and Engineering, 2012, 91(3): 31-38.

[11] Zhao Y L, Zhang L H, Wu F, *et al.* Analysis of horizontal well pressure behavior in fractured low permeability reservoirs with consideration of the threshold pressure gradient[J]. Journal of Geophysics & Engineering, 2013, 10(3): 5-14.

[12] Zhao Y L, Zhang L H, Zhao J, *et al.* Triple porosity modeling of transient well test and rate decline analysis for multi-fractured horizontal well in shale gas reservoirs [J]. Journal of Petroleum Science and Engineering, 2013, 110: 253-262.

[13] Wang H T. Performance of multiple fractured horizontal wells in shale gas reservoirs with consideration of multiple mechanisms[J]. Journal of Hydrology, 2014, 510: 299-312.

[14] El-Banbi A H . Analysis of Tight Gas Well performance[D]. Texas A&M University, 1998.

[15] Al-Ahmadi A H, Wattenbarger R A . Triple-porosity models: One further step towards capturing fractured reservoirs heterogeneity[C]. Paper SPE 149054 presented at the SPE/DGS Saudi Arabia Section Technical Symposium and Exhibition, Al-Khobar, Saudi Arabia, 15-18 May 2011.

[16] Buhidma I M, Habbtar A, Rahim Z. Practical considerations for pressure transient analysis of multi-stage fractured horizontal wells in tight sands[C]. Paper 168096 presented at the SPE Saudi Arabia Section Technical Symposium and Exhibition, Al-Khobar, Saudi Arabia, 19-22 May 2013.

[17] Xu B X, Haghighi M, Li X F, *et al.* Development of new type curves for production analysis in naturally fractured shale gas/tight gas reservoirs[J]. Journal of Petroleum Science and Engineering, 2013, 105(1): 107-115.

[18] Nashawi I S . Constant-pressure well test analysis of finite conductivity hydraulically fractured gas wells influenced by non-Darcy flow effects [J]. Journal of Petroleum Science and Engineering, 2006, 53 (3): 225-238.

[19] Larsen L, Hegre T M. Pressure-transient behavior of horizontal wells with finite-conductivity vertical fractures[C]. Paper SPE 22076 presented at the International Arctic Technology Conference, Anchorage, Alaska, USA, 29-31 May 1991.

[20] Larsen L, Hegre T M. Pressure transient analysis of multifractured horizontal wells[C]. Paper SPE 28389 presented at the SPE Annual Technical Conference and Exhibition, New Orleans, Louisiana, USA, 25-28 September 1994.

[21] Guo G, Evans R D. Pressure-transient behavior and inflow performance of horizontal wells intersecting discrete fractures[C]. Paper SPE 26446 presented at the SPE 68th Annual Technology Conference and Exhibition, Houston, USA, 3-6 October 1993.

[22] Horne R N, Temeng K O. Relative productivities and pressure transient modeling of horizontal wells with multiple fractures[C]. Paper SPE 29891 presented at the SPE Middle East Oil Show, Bahrain, 11-14 March 1995.

[23] Chen C C, Rajagopal R. A multiply-fractured horizontal well in a rectangular drainage region[J]. SPE

Journal, 1997, 2(11): 455-465.

[24] Al-Kobaisi M, Ozkan E, Kazemi H. A hybrid numerical analytical model of finite-conductivity vertical fractures intercepted by a horizontal well[R]. SPE92040, 2006.

[25] Valko P P, Amini S. The method of distributed volumetric sources for calculating the transient and pseudosteady state productivity of complex well-fracture configurations[C]. Paper SPE 106279 presented at the 2007 SPE Hydraulic Fracturing Technology Conference, Texas USA, 29-30 January 2007.

[26] Brown M, Ozkan E, Raghavan R, *et al.* Practical solutions for pressure-transient responses of fractured horizontal wells in unconventional shale reservoirs[C]. Paper SPE 125043 presented at the SPE Annual Technical Conference and Exhibition, New Orleans, USA, 4-7 October 2009.

[27] Riley M F, Brigham W E, Horne R N. Analytic solutions for elliptical finite-conductivity fractures[C]. Paper SPE 22656 presented at the SPE 66th Annual Technical Conference and Exhibition, Dallas, Texas USA, 6-9 October 1991.

[28] Wang X D, Luo W J, Hou X C, *et al.* Pressure transient analysis of multi-stage fractured horizontal wells in boxed reservoirs[J]. Petroleum Exploration & Development, 2014, 41(1): 82-87.

[29] Kikani J, Horne R N. Pressure-transient analysis of arbitrarily shaped reservoirs with the boundary-element method[J]. SPE Formation Evaluation, 1992, 3: 53-60.

[30] Zhang W, Zeng P. A boundary element method applied to pressure transient analysis of irregularly shaped double-porosity reservoir[C]. SPE 25284, 1992.

[31] Wang H, Zhang L. A boundary element method applied to pressure transient analysis of geometrically complex gas reservoirs[C]. SPE 122055, 2009.

[32] Samaniego V F, Cinco L H. Production rate decline in pressure sensitive reservoirs [J]. Journal of Canadian Petroleum Technology, 1980, 19 (3): 335-387.

[33] Pedrosa O A. Pressure transient response in stress-sensitive formations[C]. SPE 15115, 1986.

[34] Archer R A. Impact of stress sensitive permeability on production data analysis[R]. SPE 114166, 2008.

[35] Mukherjee H, Economies M J. A parametric comparison of horizontal and vertical well performance[J]. SPE Formation Evaluation, 1991, 6: 209-215.

[36] Stalgorova K, Mattar L. Analytical model for unconventional multifractured composite systems[J]. SPE Reservoir Evaluation & Engineering, 2013, 16 (3): 246-256.

[37] Stehfest H. Algorithm 368: Numerical inversion of Laplace transforms[J]. Communications of the ACM, 1970, 13(1): 47-49.

Appendix A

The one dimensional linear flow is assumed in the hydraulic fractures, the compressibility of fractures and gas is ignored, and the flow state in hydraulic fractures meets the continuous pseudo-state theory. Thus, the governing equation is presented as

$$-\frac{\partial(\rho v)}{\partial x}+\rho\frac{2k_{\mathrm{fi}}}{\mu w_{\mathrm{F}}}\frac{\partial p_{\mathrm{f}}}{\partial y}\bigg|_{y=y_{\mathrm{F}}}=0,(0<x<x_{\mathrm{F}}) \tag{A1}$$

Based on the gas state equation, the gas density can be presented as follow:

$$\rho=\frac{pM}{ZRT} \tag{A2}$$

Meanwhile, the gas pseudo-pressure is defined as the following formation:

$$\varphi = \int_{p0}^{p} \frac{p}{Z\mu} \mathrm{d}p \tag{A3}$$

Submitting the Equation(A2) and Equation(A3) into Equation(A1), one can obtain:

$$\frac{\partial^2 \varphi_F}{\partial x^2} + \frac{2k_{fi}}{k_F w_F} \frac{\partial \varphi}{\partial y}\Big|_{y=y_F} = 0, (0 < x < x_F) \tag{A4}$$

In this paper, we assume that the flow rate along the hydraulic fractures is variable. At arbitrary position (x_F, y_F), the flow rate from formation into hydraulic fractures can be obtained applied the Darcy's law:

$$\frac{p_{sc} T}{271.44 k_{fi} h T_{sc}} q(x_F, y_F) = \Delta x \frac{\partial \varphi_f}{\partial y}\Big|_{y=y_F} \tag{A5}$$

Where, Δx represents the micro-elements, we make an assumption that the flux density is constant in this micro-element.

In terms of the inner boundary condition, we assume that the well keeps constant production rate condition. Similarly, the inner boundary condition can be obtained with Darcy's law:

$$w_F \frac{\partial \varphi_F}{\partial x}\Big|_{x_F=0} = q_F \frac{p_{sc} T}{271.44 k_{Fi} h T_{sc}} \tag{A6}$$

Where, q_F is the total production rate of the hydraulic fractures, sm^3/d.

With the definition of these dimensionless variables in Table 1, the above equations from Equation(A4) ~ Equation(A6) with the formula of dimensionless can be presented as follows:

$$\frac{\partial^2 \varphi_{FD}}{\partial x_D^2} + \frac{2}{C_{FD}} \frac{\partial \varphi_D}{\partial y_D}\Big|_{y_D=y_{FD}} = 0, (0 < x_D < 1) \tag{A7}$$

And the boundary conditions [Equation(A5) and Equation(A6)] also can be converted into the dimensionless formation:

$$q_D(x_{FD}, y_{FD}) = -\frac{2}{\pi} \frac{\partial \varphi_D}{\partial y_D}\Big|_{y=y_{FD}} \tag{A8}$$

$$\frac{\partial \varphi_{FD}}{\partial x_D}\Big|_{x_{FD}=0} = -\frac{\pi q_{FD}}{C_{FD}} \tag{A9}$$

The region that contains the hydraulic fractures and its vicinity is supported by proppants. Therefore, the stress sensitivity can be ignored in this region. The dimensionless pseudo-pressure distributed throughout this region in Laplace domain is as follow:

$$\frac{\partial^2 \overline{\eta_{FD}}}{\partial x_D^2} + \frac{2}{C_{FD}} \frac{\partial \overline{\eta_D}}{\partial y_D}\Big|_{y_D=y_{FD}} = 0, (0 < x_D < 1) \tag{A10}$$

The boundary conditions [Equation (A8) and Equation(A9)] also can be converted into the dimensionless formation:

$$\overline{q_D}(x_{FD}, y_{FD}) = -\frac{2}{\pi} \frac{\partial \overline{\eta_D}}{\partial y_D}\Big|_{y=y_{FD}} \tag{A11}$$

$$\frac{\partial \overline{\eta_{FD}}}{\partial x_D}\Big|_{x_{FD}=0} = -\frac{\pi \overline{q_{FD}}}{C_{FD}} \tag{A12}$$

Rate Decline Analysis of Multiple Fractured Horizontal Well in Shale Reservoir with Triple Continuum

Wang Junlei[1] Yan Cunzhang[2] Jia Ailin[1] He Dongbo[1] Wei Yunsheng[1] Qi Yadong[1]

(1. Research Institue of Petroleum Exploration & Development, PetroChina;

2. Foreign Cooperative Company, PetroChina)

Abstract: Multiple fractured horizontal well (MFHW) is widely applied in the development of shale gas. To investigate the gas flow characteristics in shale, based on a new dual mechanism triple continuum model, an analytical solution for MFHW surrounded by stimulated reservoir volume (SRV) was presented. Pressure and pressure derivative curves were used to identify the characteristics of flow regimes in shale. Blasingame type curves were established to evaluate the effects of sensitive parameters on rate decline curves, which indicate that the whole flow regimes could be divided into transient flow, feeding flow, pseudo steady state flow. In feeding flow regime, the production of gas well is gradually fed by adsorbed gases in sub matrix, and free gases in matrix. The proportion of different gas sources to well production is determined by such parameters as storability ratios of triple continuum, transmissibility coefficients controlled by dual flow mechanism and fracture conductivity.

Key words: triple continuum; desorption and diffusion; darcy flow; multiple finite conductivity fractures; pressure; blasingame type curves

1 Introduction

Shale is referred to as extraordinary fine-grain sediments, which does not possess the value of commercial exploration and development without stimulating. With technological advancements in horizontal drilling, multistage hydraulic fracturing, and stimulating reservoir volume (SRV), shale gas has been becoming commercial hydrocarbon production targets. Meanwhile, these "difficult to produce" reservoirs will play an increasingly important role in China because they are showing the potential to offset declining in conventional gas production[1,2].

Gas production from shale gas reservoirs is entirely challenging, because the pore network and mechanisms of fluid flow in shale are significantly different from those in conventional reservoirs[3,4]. Some researches show that the pores in shale rocks are dominantly nanometer and micrometer in scale [5,6]. In pores, gas flows through a network of pores with different diameters ranging from nanometers to micrometers, which is controlled by different flow mechanisms. Gas flow through micropores is described by Darcy equation, while that through nanopores is simulated by using either the continuum or the molecular approach.

In the development of shale gas reservoirs, hydraulic fracturing is an effective technique for productivity enhancement because of the nature of tight shale. Recently, pressure/rate transient analysis and productivity evaluation on MFHW are attracting more and more attention with the rapid development of unconventional gas reservoirs [7,8]. Published papers were repeatedly establishing mathematical models based on the theory of single porosity media [9,10] or dual porosity media[11-13]. However, some limitations have been shown in reliably forecasting shale gas production, because natural or induced micro fractures usually serve as hidden pathways for the communication between matrix and fracture network. Consequently, triple porosity models need to be presented to couple micro fractures into dual porosity models [14,15]. Unfortunately, few papers succeed in applying multiple flow mechanisms to multiple porosity model in MFHW analysis [16,17].

Most of the above literatures studied on pure flow mechanism in shale[18,19], or on pressure transient analysis of horizontal well with infinite conductivity fractures in a dual porosity model without taking SRV into account[11,12]. According to the viewpoint introduced by Dehghanpour and Shirdel[14], based on dual laws of Fick diffusion and Darcy flow, this paper presents an analytical model. The model covers triple-continuum-medium including sub matrix, matrix, macro fracture network. The flow regimes of MFHW surrounded by SRV are presented. The effects of dual mechanism triple continuum parameters on rate decline curves are analyzed in detail.

2 Conceptual model

Hydraulic fracturing can induce an altered stress field in the surrounding formation, and thus generate second fractures, including fracture networks or branched networks. SRV is just the reservoir volume affected by this stimulation. Hydraulic fractures and their interaction with induced natural fractures create a complex macro fracture network in SRV. For fractured horizontal wells in shale gas reservoirs with matrix permeability in the range of microdaray or below, the contribution of the reservoir beyond the stimulated reservoir volume (SRV) is usually negligible[20]. The ultimate drainage area is also constrained to the created SRV in super-tight shale gas reservoirs. Within SRV the fracture network is described as a discrete set of high-permeability fractures in two orthogonal directions. Therefore, the complex SRV can be abstracted as a region with closed boundaries composing of macro fracture networks and matrix blocks [Figure 1(a)].

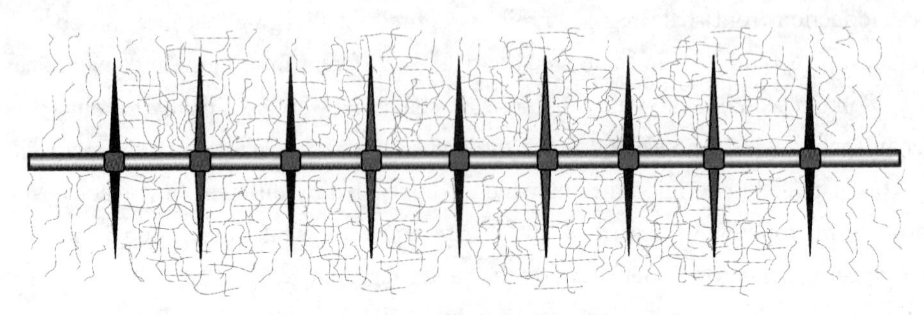

(a) Actual fracture network and matrix within SRV

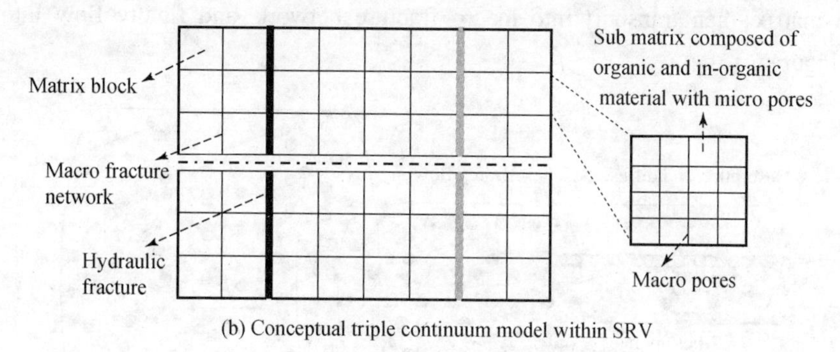

(b) Conceptual triple continuum model within SRV

Figure 1 Dual mechanism and triple continuum model in shale gas reservoirs

Given induced micro fractures within matrix serve as hidden pathways, it is necessary to further subdivide matrix blocks. Based on description above, the spaces of gas flow in SRV are categorized as micro pores in sub matrix, macro pores or micro fractures in matrix, macro fracture network and hydraulic fracture in MFHW. In summary, shale formation can be considered to be triple media, i. e. , matrix block, sub matrix block, organic matter, and three flow pores, i. e. , macro fracture network with milli Darcy permeability, micro fracture/macro pore with micro Darcy permeability, micro pore with nano Darcy permeability Figure 1(b). Through the above analysis, the physical model can be described as:

(1) An isotropic, horizontal, slap shale reservoir is bounded by overlying and underlying impermeable strata, and is around by impermeable rectangular boundaries of SRV, $x_e \times y_e$: the length of SRV x_e is that of fractured horizontal well, and the width of SRV y_e is the fracture length plus a 1/2 length of fracture spacing D_f [21].

(2) The shale gas reservoir possesses uniform thickness h.

(3) The shale gas production is assumed to be an isothermal process. Flow in micro pores is controlled by desorption and diffusion. Darcy flow is valid in macro pores and macro fracture network. Linear flow occurs in the hydraulic fracture with a finite conductivity.

(4) Shale gas is produced through a horizontal well with transverse hydraulic fractures, where all fractures are fully penetrating formation.

(5) MFHW is comprised of n_f hydraulic fractures with symmetric wings, and possesses an infinite conductivity wellbore. All fractures possess uniform construct parameters including half length x_{hf}, width w_{hf}, permeability k_{hf}, and porosity φ_{hf}.

(6) Fracture spaces in MFHW are uniform and fixed, and fractures are central-symmetric within SRV.

(7) MFHW produces in a constant rate condition.

3 Mathematical model

3.1 Pressure governing equation in shale

During the process of the development, adsorbed gases desorb and diffuse into micro fractures in matrix, then transport into macro fracture network, and finally flow into hydraulic fractures (Figure 2).

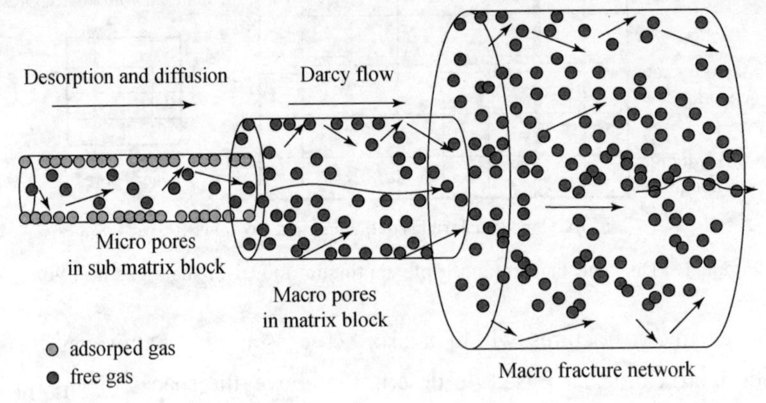

Figure 2　Different flow mechanisms and connection of adjacent continuums

Macro fracture network is mainly fed by gases from matrix. Therefore, according to Darcy law, the partial differential equation governing gas flow in macro fracture network is given in the rectangular coordinate system by

$$k_{\mathrm{f}} \frac{\partial}{\partial x}\left(\frac{p_{\mathrm{f}}}{\mu_{\mathrm{g}} Z_{\mathrm{g}}} \frac{\partial p_{\mathrm{f}}}{\partial x}\right)+k_{\mathrm{f}} \frac{\partial}{\partial y}\left(\frac{p_{\mathrm{f}}}{\mu_{\mathrm{g}} Z_{\mathrm{g}}} \frac{\partial p_{\mathrm{f}}}{\partial y}\right)+q_{\text{Darcy}}=\frac{\varphi_{\mathrm{f}} c_{\mathrm{g}} p_{\mathrm{f}}}{Z_{\mathrm{g}}} \frac{\partial p_{\mathrm{f}}}{\partial t} \tag{1}$$

Where p_{f} is the pressure in macro fracture network, Pa; t is time, s; k_{f} is macro fracture permeability, m^2; μ_{g} is Gas viscosity, Pa · s; Z_{g} is Gas compressibility factor, dimensionless; c_{g} is gas compressibility, Pa^{-1}; φ_{f} is macro fracture porosity, fraction; q_{Darcy} is gas flux from matrix to macro fracture, m^3/s.

Because the permeability of macro fracture k_{f} is orders of magnitude higher than that of the matrix k_{m}, so Darcy flow effect in matrix is negligible. Directly fed by sub matrix, the continuity equation in matrix is given by

$$-q_{\text{Darcy}}=\frac{\varphi_{\mathrm{m}} c_{\mathrm{g}} p_{\mathrm{m}}}{Z_{\mathrm{g}}} \frac{\partial p_{m}}{\partial t}+\frac{p_{\mathrm{sc}} T}{T_{\mathrm{sc}}} q_{\text{Fick}} \tag{2}$$

Where p_{m} is the pressure in matrix, Pa; p_{sc} is standard state pressure, Pa; φ_{m} is matrix porosity, fraction; T is reservoir temperature, K; T_{sc} is standard state temperature, K; q_{Fick} is the flux from sub matrix to matrix, m^3/s.

In sub matrix, using Fick's second law and spherical model, the partial differential equations describing transient state diffusive flow is determined by

$$\frac{1}{r_{\mathrm{m}}^{2}} \frac{\partial}{\partial r_{\mathrm{m}}}\left(D r_{\mathrm{m}}^{2} \frac{\partial V_{\mathrm{m}}}{\partial r_{\mathrm{m}}}\right)=\frac{\partial V_{\mathrm{m}}}{\partial t} \tag{3}$$

Where r_{m} is radial coordinate in sub matrix, m; D is diffusion coefficient, m^2/s; V_{m} is gas concentration, $\mathrm{m}^3/\mathrm{m}^3$.

In the next section, we rewrite Equation (1) and Equation (2) in terms of gas pseudo pressure defined as

$$m(p_j)=\frac{\mu_{gi}Z_{gi}}{p_i}\int_{p0}^{p_f}\frac{p_j}{\mu_g Z_g}\mathrm{d}p \quad (j=\mathrm{m},\mathrm{f}) \tag{4}$$

For simplicity, pseudo steady state interflows are employed as follows:

$$q_{\mathrm{Fick}}=\frac{\partial V}{\partial t}=\frac{6D\pi^2}{R^2}(V_E-V) \tag{5}$$

$$q_{\mathrm{Darcy}}=\alpha\frac{p_i}{\mu_{gi}Z_{gi}}[m(p_m)-m(p_f)] \tag{6}$$

$$\alpha=\frac{k_m A T_{sc} Z_{sc}\rho_{sc}}{l\cdot p_{sc}T} \tag{7}$$

Where p_i is initial reservoir pressure, Pa; μ_{gi} is initial gas viscosity, Pa · s; Z_{gi} is gas compressibility factor under the initial condition, dimensionless; R is external radius of organic material block, m; V_E is equilibrium gas concentration, m^3/m^3.

For the convenience of calculation, a set of dimensionless variables are defined in Table 1. Where B_{gi} is volume factor under initial condition, dimensionless; q_{ref} is reference production, m^3/s; L_{ref} is reference length, m.

Table 1 Definitions of the dimensionless variable

Parameters	Formula
Reference variable	$\Lambda=\varphi_m\mu_g c_g+\frac{p_{sc}T}{T_{sc}}\frac{Z_i}{p_i}\frac{2\pi k_m h}{\mu_{gi}q_{ref}B_{gi}}$
Desorption time	$\tau=\frac{R^2}{\pi^2 D}$
Storability ratio of matrix block	$\omega_2=\frac{\varphi_m\mu_g c_g}{\Lambda}$
Interporosity transmissibility coefficient between matrix and sub matrix	$\lambda_2=\frac{k_f}{\Lambda L_{ref}^2}\frac{\tau}{6}$
Storability ratio of macro fracture network	$\omega_1=\frac{\varphi_f\mu_g c_g}{\Lambda}$
Interporosity transmissibility coefficient between matrix to macro fracture network	$\lambda_1=\frac{\alpha L_{ref}^2 k_m}{k_f}$
Dimensionless gas concentration	$V_D=V_i-V \quad V_{ED}=V_i-V_E$
Dimensionless pseudo pressure	$m_D(p_{jD})=\frac{2\pi k_m h[m(p_i)-m(p_j)]}{q_{ref}B_{gi}\mu_{gi}}$
Dimensionless time	$t_D=\frac{k_f t}{\Lambda L_{ref}^2}$
Dimensionless distance	$x_D=\frac{x}{L_{ref}}$

Respectively substituting Equation(5) and Equation (6) into Equation(1) and Equation (2), and taking the Laplace transform in time domain,

$$\tilde{h}(s) = L[h(t)] = \int_0^{\infty}[e^{-st}h(t)]dt, \quad 0 \leqslant t \leqslant \infty$$

The dimensionless governing equations for triple continuum are rewritten as

$$\nabla_D^2 \tilde{m}_{fD}(p_{fD}) + \lambda_1[\tilde{m}_{mD}(p_{mD}) - \tilde{m}_{fD}(p_{fD})] = \omega_1 s\tilde{m}_{fD}(p_{fD}) \tag{8}$$

$$-\lambda_1[\tilde{m}_{mD}(p_{mD}) - \tilde{m}_{fD}(p_{fD})] = \omega_2 s\tilde{m}_{mD}(p_{mD}) - (1-\omega_2)s\tilde{V}_D \tag{9}$$

$$\lambda_2 s\tilde{V}_D = \tilde{V}_{ED} - \tilde{V}_D \tag{10}$$

Substituting Langmuir's equilibrium sorption equation into Equation(10), we have

$$\tilde{V}_{ED} = L\left[\frac{V_L m_m(p_m)}{m_L(p_m) + m_m(p_m)} - \frac{V_L m(p_i)}{m_L(p_i) + m(p_i)}\right] = L[\sigma \cdot m_{mD}(p_{mD})] \tag{11}$$

Assuming that the adsorption index, σ is a constant, which is written as

$$\sigma = \frac{q_{ref}\mu_{gi}}{2\pi k_m h}\frac{V_L m_L(p_m)}{[m_L(p_m) + m_m(p_m)][m_L(p_i) + m(p_i)]} \tag{12}$$

Substituting Equation(11) into Equation(9) contributes to the dimensionless pseudo pressure relationship between matrix block and macro fracture.

$$\tilde{m}_{mD}(p_{mD}) = \frac{1}{\left[\frac{\omega_2 s}{\lambda_1} + \frac{(1-\omega_2)s\sigma}{(1+\lambda_2 s)\lambda_1} + 1\right]}\tilde{m}_{fD}(p_{fD}) \tag{13}$$

After substituting Equation(13) into Equation(8), the dimensionless governing equation in macro fracture network is given as

$$\nabla_D^2 \tilde{m}_{fD}(p_{fD}) = \tilde{f}(s)s\tilde{m}_{fD}(p_{fD}) \tag{14}$$

Where

$$\tilde{f}(s) = \omega_1 + \frac{\lambda_1}{s}\frac{\omega_2 + \frac{(1-\omega_2)\sigma}{(1+\lambda_2 s)}}{\omega_2 + \frac{(1-\omega_2)\sigma}{(1+\lambda_2 s)} + \frac{\lambda_1}{s}} \tag{15}$$

3.2 Solution for single hydraulic fracture

When the following conditions be satisfied, $x_{wD} - x_{fD} \leqslant x_D \leqslant x_{wD} + x_{fD}, y_D = y_{wD}$, dimensionless flux distribution along uniform flux fracture is defined below.

$$q_{fxD}(x_D, y_D, t_D) = q_{fD}(t_D)/(2x_{fD}) \tag{16}$$

After taking dimensionless flux of fracture into account, the governing equation of single fracture is given as,

$$\frac{\partial^2 \tilde{m}_{fD}(p_{fD})}{\partial x_D^2} + \frac{\partial^2 \tilde{m}_{fD}(p_{fD})}{\partial y_D^2} + \tilde{q}_{fxD}\delta(y_D - y_{wD}) = \tilde{f}(s)\tilde{m}_{fD}(p_{fD}) \tag{17}$$

Where x_{wD} and y_{wD} are dimensionless midpoint coordinates of hydraulic fracture, m; q_{fD} is dimensionless production of fracture. The outer boundary conditions of SRV are described as

$$\frac{\partial \tilde{m}_{\mathrm{fD}}(0,y_{\mathrm{D}})}{\partial x_{\mathrm{D}}}=\frac{\partial \tilde{m}_{\mathrm{fD}}(x_{\mathrm{eD}},y_{\mathrm{D}})}{\partial x_{\mathrm{D}}}=0 \tag{18}$$

$$\frac{\partial \tilde{m}_{\mathrm{fD}}(x_{\mathrm{D}},0)}{\partial y_{\mathrm{D}}}=\frac{\partial \tilde{m}_{\mathrm{fD}}(x_{\mathrm{D}},y_{\mathrm{eD}})}{\partial y_{\mathrm{D}}}=0 \tag{19}$$

Taking superposition principle[22] to deal with Equation (17) ~ Equation (19), the final pressure solution for single fracture in Laplace domain is given as the following form:

$$\frac{\tilde{m}_{\mathrm{D}}(x_{\mathrm{D}},y_{\mathrm{D}},s,x_{\mathrm{wD}},y_{\mathrm{wD}})}{\tilde{q}_{\mathrm{fD}}(s)}=\Delta\tilde{m}_{\mathrm{D}}=\Delta\tilde{m}_{\mathrm{Dinf}}+\Delta\tilde{m}_{\mathrm{Db1}}+\Delta\tilde{m}_{\mathrm{Db2}}+\Delta\tilde{m}_{\mathrm{Db3}} \tag{20}$$

Where

$$\Delta\tilde{m}_{\mathrm{Dinf}}=\frac{1}{2x_{\mathrm{fD}}}\int_{-x_{\mathrm{fD}}}^{x_{\mathrm{fD}}}K_0\left[\sqrt{u}\sqrt{(x_{\mathrm{D}}-x_{\mathrm{wD}}-\alpha)^2+(y_{\mathrm{D}}-y_{\mathrm{wD}})^2}\right]\mathrm{d}\alpha$$

$$\Delta\tilde{m}_{\mathrm{Db1}}=\frac{1}{x_{\mathrm{eD}}\sqrt{u}}\left\{\begin{array}{l}\mathrm{e}^{[-\sqrt{u}(y_{\mathrm{D}}\pm y_{\mathrm{wD}})]}\\+\mathrm{e}^{[-\sqrt{u}(2y_{\mathrm{eD}}-|y_{\mathrm{D}}\pm y_{\mathrm{wD}}|)]}\end{array}\right\}\cdot\left[1+\sum_{m=1}^{\infty}\mathrm{e}^{(-2my_{\mathrm{eD}}\sqrt{u})}\right]$$

$$\Delta\tilde{m}_{\mathrm{Db2}}=\frac{2}{x_{\mathrm{fD}}}\sum_{n=1}^{\infty}\left\{\begin{array}{l}\dfrac{1}{n\alpha_n}\sin\dfrac{n\pi x_{\mathrm{fD}}}{x_{\mathrm{eD}}}\cos\dfrac{n\pi x_{\mathrm{D}}}{x_{\mathrm{eD}}}\cos\dfrac{n\pi x_{\mathrm{wD}}}{x_{\mathrm{eD}}}\cdot\\\left\{\left[\begin{array}{l}\mathrm{e}^{[-\alpha_n(y_{\mathrm{D}}+y_{\mathrm{wD}})]}\\+\mathrm{e}^{[-\alpha_n(2y_{\mathrm{eD}}-|y_{\mathrm{D}}\pm y_{\mathrm{wD}}|)]}\end{array}\right]\cdot\left[1+\sum_{m=1}^{\infty}\mathrm{e}^{(-2my_{\mathrm{eD}}\alpha_n)}\right]\right.\\\left.+\mathrm{e}^{[-\alpha_n|y_{\mathrm{D}}-y_{\mathrm{wD}}|]}\cdot\sum_{m=1}^{\infty}\mathrm{e}^{(-2my_{\mathrm{eD}}\alpha_n)}\right\}\end{array}\right\}$$

$$\begin{aligned}\Delta\tilde{m}_{\mathrm{Db3}}=&\frac{1}{2x_{\mathrm{fD}}}\int_{-x_{\mathrm{fD}}}^{x_{\mathrm{fD}}}K_0\left[\sqrt{u}\sqrt{(x_{\mathrm{D}}+x_{\mathrm{wD}}-\alpha)^2+(y_{\mathrm{D}}-y_{\mathrm{wD}})^2}\right]\mathrm{d}\alpha\\&+\frac{1}{2x_{\mathrm{fD}}}\sum_{n=1}^{\infty}\int_{-x_{\mathrm{fD}}}^{x_{\mathrm{fD}}}K_0\left[\sqrt{u}\sqrt{(x_{\mathrm{D}}\pm x_{\mathrm{wD}}\pm 2nx_{\mathrm{eD}}-\alpha)^2+(y_{\mathrm{D}}-y_{\mathrm{wD}})^2}\right]\mathrm{d}\alpha\\&-\frac{\pi\mathrm{e}^{[-\sqrt{u}|y_{\mathrm{D}}-y_{\mathrm{wD}}|]}}{x_{\mathrm{eD}}\sqrt{u}}\end{aligned}$$

Where $u=s\tilde{f}(s)$, $\alpha_n=\sqrt{u+n^2\pi^2/x_{\mathrm{eD}}^2}$, and $K_0(x)$ is modified Bessel function.

3.3 Solution for horizontal wells with multiple finite conductivity fractures

The flow pattern in transverse fracture is special because the gas flow is comprised of linear flow and radial flow. Here, we accordingly divided the flow pattern into two parts: convergence flow pattern around wellbore and vertical fractured well flow pattern away from wellbore.

Convergence flow pattern reflects additional pressure drop around wellbore because of occurring of radial flow, which can be described as a kind of skin[23],

$$S_{\mathrm{c}}=\frac{1}{C_{\mathrm{fD}}}\frac{h_{\mathrm{D}}^2}{x_{\mathrm{fD}}}\left(\ln\frac{h_{\mathrm{D}}}{2r_{\mathrm{wD}}}-\frac{\pi}{2}\right) \tag{21}$$

The fracture conductivity is generally finite and the pressure drops along fracture is not uniform, which is closer to reality. The dimensionless fracture conductivity C_{fD} is provided to reflect the characteristic.

$$C_{\mathrm{fD}} = \frac{k_{\mathrm{hf}} w_{\mathrm{f}}}{k_{\mathrm{f}} L_{\mathrm{ref}}} \tag{22}$$

Cinco-Ley[24] presented a semi-analytical solution for a finite-conductivity vertical fractured well. In order to avoid numerical methods, the effect of fracture conductivity on pressure turbulence can be treated as a kind of skin, i. e. "conductivity skin factor", which is the function of dimensionless conductivity and Laplace time variable s. Conductivity skin factor S_{f} can be obtained through modifying the results presented by Wilkinson [25].

$$\tilde{S}_{\mathrm{f}}(C_{\mathrm{fD}}, s) = 2\pi \sum_{n=1}^{\infty} \frac{1}{n^2\pi^2 (C_{\mathrm{fD}}/x_{\mathrm{fD}}) + 2\sqrt{n^2\pi^2 + s}} + \frac{\pi \cdot \Delta S(C_{\mathrm{fD}})}{\pi + 4s \cdot \Delta S(C_{\mathrm{fD}})} \tag{23}$$

$$\Delta S(C_{\mathrm{fD}}) = \frac{C_1}{(C_{\mathrm{fD}}/x_{\mathrm{fD}}) + C_2}, C_1 = \frac{4(\pi - 2)}{\pi} - \frac{\pi}{3}, C_2 = \frac{C_1}{\ln(\pi/2)}$$

Calculative pressure of finite conductivity fracture modified with Equation (23) is seen in Figure 3. The high degree of correlation between the two geometrics shows that finite conductivity skin case mimics the practical flow in hydraulic fracture. Therefore, we can obtain the dimensionless pseudo pressure solution for a single finite conductivity fracture intersecting horizontal well.

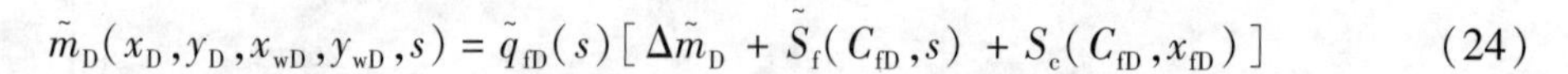

$$\tilde{m}_{\mathrm{D}}(x_{\mathrm{D}}, y_{\mathrm{D}}, x_{\mathrm{wD}}, y_{\mathrm{wD}}, s) = \tilde{q}_{\mathrm{fD}}(s)\left[\Delta\tilde{m}_{\mathrm{D}} + \tilde{S}_{\mathrm{f}}(C_{\mathrm{fD}}, s) + S_{\mathrm{c}}(C_{\mathrm{fD}}, x_{\mathrm{fD}})\right] \tag{24}$$

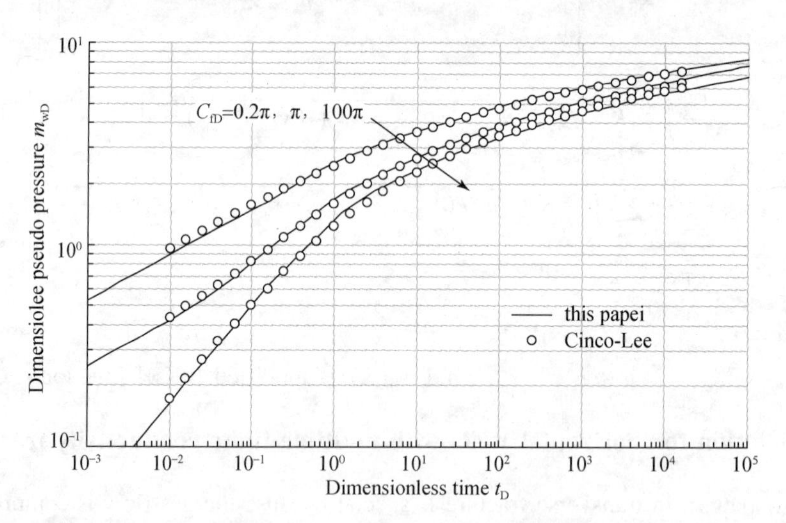

Figure 3　The comparison results of this paper and Cinco-Ley in finite conductivity fracture

According to superposition principle, pressure on the cross point between horizontal wellbore and transverse fracture j is obtained in MFHW.

$$\tilde{m}_{\mathrm{wD}j}(s) = \sum_{i=1}^{n_{\mathrm{f}}} \tilde{q}_{\mathrm{fD}i}(s)\,\Delta\tilde{m}_{\mathrm{D}ji}(x_{\mathrm{wD}j}, y_{\mathrm{wD}j}, x_{\mathrm{wD}i}, y_{\mathrm{wD}i}, s) \tag{25}$$

The assumption of infinite conductivity wellbore means that along the horizontal wellbore pressure at every cross point is equal to each other. The total rate of MFHW is the sum of each fracture rate, and is assumed a constant. These conditions will generate a $n_{\mathrm{f}} \times n_{\mathrm{f}}$ matrix,

$$\begin{bmatrix} s\Delta\tilde{m}_{\mathrm{D11}}(s) & s\Delta\tilde{m}_{\mathrm{D12}}(s) & \cdots & s\Delta\tilde{m}_{\mathrm{D1}n_f}(s) & -1 \\ s\Delta\tilde{m}_{\mathrm{D21}}(s) & s\Delta\tilde{m}_{\mathrm{D22}}(s) & \cdots & s\Delta\tilde{m}_{\mathrm{D2}n_f}(s) & -1 \\ \vdots & \vdots & & \vdots & \vdots \\ s\Delta\tilde{m}_{\mathrm{D}n1}(s) & s\Delta\tilde{m}_{\mathrm{D}n2}(s) & \cdots & s\Delta\tilde{m}_{\mathrm{D}n_fn_f}(s) & -1 \\ s & s & \cdots & s & 0 \end{bmatrix} \cdot \begin{bmatrix} \tilde{q}_{\mathrm{fD1}}(s) \\ \tilde{q}_{\mathrm{fD2}}(s) \\ \vdots \\ \tilde{q}_{\mathrm{fD}n}(s) \\ \tilde{m}_{\mathrm{wD}}(s) \end{bmatrix} = \begin{bmatrix} 0 \\ 0 \\ \vdots \\ 0 \\ 1 \end{bmatrix} \tag{26}$$

In the above system the element $\times 10^{-3}\ \mu\mathrm{m}^2$ represents the effect of fracture i on the location of fracture j. Stehfest algorithm [26] and Newton iteration methods can be used to obtain the solution of the model. The solution provides the dimensionless pseudo pressure response and rate distribution of each fracture.

4 Results and discussion

4.1 Flow regimes

Based on the presented solution, dimensionless pseudo pressure and pressure derivative curves of MFHW are plotted in Figure 4. To fully reflect different flow regimes, parameters used in this section are set without considering the proportion relationship between SRV and MFHW.

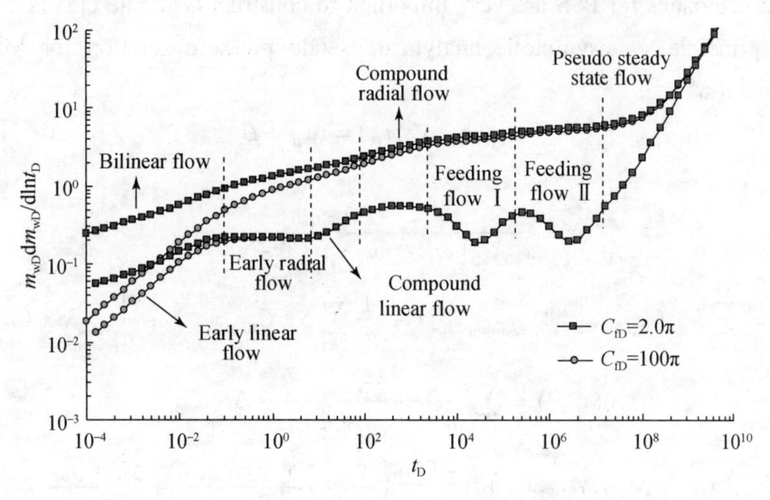

Figure 4 The characteristics of flow regimes throughout whole time domain

It is observed in Figure 4 that the whole flow regimes can be divided into three major parts: transient flow regimes, feeding flow regimes, and pseudo steady state flow regime. The transient regimes cover linear flow with a 1/2 slope trend for high fracture conductivity (1/4 for low conductivity), early radial flow characterized by a horizontal line, compound linear flow identified by 1/2 slope, and compound radial flow characterized by a 0 slope in pressure derivative curves.

In previous flow regimes, neither the free gas stored in matrix blocks nor adsorbed gas stored in sub matrix blocks participate in flowing. When free gases in macro fracture network can not fully supply gas well production, free gases in matrix block will firstly flow into macro fractures, which is called as the first feeding flow regime and characterized by a trough in derivative curves. Simultaneously, pressure in matrix block continues to decrease. When the pressure drop between the matrix and sub matrix reaches the desorbed pressure, adsorbed gases desorb and diffuse into matrix, which is called as the second feeding flow regime and is identified by another trough in derivative curves. The characteristics of these two feeding flow regimes are controlled by the storage capacity of triple continuum and transfer ability of dual mechanism. Finally, gas flow reaches the boundary, pseudo steady state (PSS) flow is attained, which is characterized by a unit slope straight line in both pressure and derivative curves.

It should be noted that not all flow regimes described above exist in shale gas reservoirs, which depends on system properties, such as fracture length, spacing, conductivity, and flowing parameters. Sometimes some regime may be absent.

4.2 Blasingame rate type curves

According to Duhamel principle[26], the dimensionless pressure in constant rate and cumulative rate in constant pressure satisfies the following correlations in Laplace domain.

$$\tilde{m}_{\mathrm{wD}}(s)\tilde{N}_{\mathrm{pD}}(s) = \frac{1}{s^3} \tag{27}$$

Flow characteristics for PSS are very important to construct type rate curves[27]. Employing superposition principle, an asymptotic analytical pseudo pressure solution for MFHW in PSS could be obtained as

$$m_{\mathrm{wD}}(x_{\mathrm{D}}, y_{\mathrm{D}}, t_{\mathrm{D}}) = \alpha t_{\mathrm{D}} + \beta \tag{28}$$

Where

$$\alpha = \frac{2\pi}{x_{\mathrm{eD}} y_{\mathrm{eD}}[\omega_1 + \omega_2 + \sigma(1 - \omega_2)]}$$

$$\beta = \sum_{i=1}^{n_{\mathrm{f}}} b_{\mathrm{pss}i}(x_{\mathrm{D}}, y_{\mathrm{D}}, x_{\mathrm{wD}i}, y_{\mathrm{wD}i}, x_{\mathrm{fD}i}) + \frac{1}{n_{\mathrm{f}}}\sum_{i=1}^{n_{\mathrm{f}}} S_{\mathrm{c}}(C_{\mathrm{fD}i}, x_{\mathrm{fD}i}) + \frac{1}{n_{\mathrm{f}}}\sum_{i=1}^{n_{\mathrm{f}}} S_{\mathrm{f}}(C_{\mathrm{fD}i})$$

$$S_{\mathrm{f}}(C_{\mathrm{fD}i}) = \sum_{n=1}^{\infty} \frac{2\pi}{n^2\pi^2 C_{\mathrm{fD}i} + 2n\pi} + \Delta S(C_{\mathrm{fD}i})$$

$$b_{\mathrm{pss}i}(x_{\mathrm{D}}, y_{\mathrm{D}}, x_{\mathrm{wD}i}, y_{\mathrm{wD}i}, x_{\mathrm{fD}i}) = \frac{2\pi y_{\mathrm{eD}}}{x_{\mathrm{eD}}}\left(\frac{1}{3} - \frac{|y_{\mathrm{D}} \pm y_{\mathrm{wD}i}|}{2y_{\mathrm{eD}}} + \frac{y_{\mathrm{D}}^2 + y_{\mathrm{wD}i}^2}{2y_{\mathrm{eD}}^2}\right)$$

$$+ \frac{2x_{\mathrm{eD}}}{x_{\mathrm{fD}i}\pi}\sum_{n=1}^{\infty} \frac{\cos\frac{n\pi x_{\mathrm{D}}}{x_{\mathrm{eD}}}\cos\frac{n\pi x_{\mathrm{wD}i}}{x_{\mathrm{eD}}}\sin\frac{n\pi x_{\mathrm{fD}i}}{x_{\mathrm{eD}}}}{n^2} \frac{\cosh\frac{n\pi}{x_{\mathrm{eD}}}(y_{\mathrm{eD}} - |y_{\mathrm{D}} \pm y_{\mathrm{wD}i}|)}{\sinh(n\pi y_{\mathrm{eD}}/x_{\mathrm{eD}})}$$

Taking Laplace transform of Equation (28) and using Equation (27), cumulative rate N_{pD} in PSS can be obtain. To eliminate multiple solutions and errors, integral average method of rate was presented instead of raw rate function [28].

$$q_{\mathrm{D}i}(t_{\mathrm{D}}) = \frac{N_{\mathrm{pD}}(t_{\mathrm{D}})}{t_{\mathrm{D}}} = \frac{1}{\alpha t_{\mathrm{D}}}[1 - \mathrm{e}^{(-\frac{\alpha}{\beta}t_{\mathrm{D}})}] \tag{29}$$

According to Equation (29), new variables are redefined to convergence rate decline curves in PSS

$$q_{\mathrm{D}id}(t_{\mathrm{D}}) = \beta \cdot q_{\mathrm{D}i} \tag{30}$$

$$t_{\mathrm{Dd}}(t_{\mathrm{D}}) = \frac{\alpha}{\beta} \cdot t_{\mathrm{D}} \tag{31}$$

This treatment can clearly reflect the rate decline characteristics in unsteady state flow. Meanwhile, rate integral derivative function is given as an auxiliary tool.

$$\frac{\mathrm{d}q_{\mathrm{D}id}(t_{\mathrm{Dd}})}{\mathrm{d}\ln t_{\mathrm{Dd}}} = \frac{\mathrm{d}}{\mathrm{d}\ln t_{\mathrm{Dd}}}\left[\frac{1}{t_{\mathrm{D}}}\int_0^{t_{\mathrm{Dd}}} q_{\mathrm{D}id}(v)\,\mathrm{d}v\right] = q_{\mathrm{D}id}(t_{\mathrm{Dd}}) - q_{\mathrm{Dd}}(t_{\mathrm{Dd}}) \tag{32}$$

4.3 Parameters influence on type curves

There are a number of parameters affecting the shape of type curves, such as MFHW composition, SRV scale, reservoir properties and fluid properties. In this section, we are more concerned with the effect of parameters related to dual mechanism and triple continuum. Here, storage capacity of triple continuum and transfer ability controlled by dual mechanism will be discussed, respectively. Basic data are listed in Table 2.

Table 2 Data used in the discussion

Parameters	Value
Initial reservoir pressure, p_i, Pa	3.457×10^{7}
Reservoir temperature, T, K	333.33
Formation thickness, h, m	20
Reference length, L_{ref}, m	0.1
Reference production, q_{ref}, $\mathrm{m^3/s}$	0.35
Initial gas compressibility, c_{gi}, $\mathrm{Pa^{-1}}$	1.56×10^{-8}
Initial gas viscosity, μ_{gi}, Pa · s	2.96×10^{-5}
Initial gas compressibility factor, Z_i	0.9733
Langmuir concentration constant, V_{L}, $\mathrm{sm^3/m^3}$	18
Langmuir pseudo pressure, m_{L}, Pa	3.62×10^{6}
Matrix porosity, φ_{m}	0.0001
Matrix permeability, k_{m}, $\mathrm{m^2}$	10^{-16}
Macro fracture porosity, φ_{f}	0.01
Macro permeability, k_{f}, $\mathrm{m^2}$	10^{-13}
Number of hydraulic fracture, n_{f}	10
Fracture length, $x_{\mathrm{fD}i}(i=1,2,\cdots,n_{\mathrm{f}})$, m	100
Fracture spacing, D_{f}, m	50
Fracture conductivity, $C_{\mathrm{fD}i}(i=1,2,\cdots,n_{\mathrm{f}})$	100π

4.3.1 Effect of storage capacity of macro fracture

According to definition of storability ratio of macro fracture, ω_1 is positively related to macro fracture porosity, which reflects the capacity of storing free gases in macro fracture network. Figure 5 shows that the effect of ω_1 on rate decline curves of multiple fractured horizontal wells in shale. A bigger ω_1 indicates that massive free gases store in macro fracture network. If the value of ω_1 is large enough compared with storability ratio of matrix ω_2 and adsorption index σ, the production rate can maintain at very high level and the supply from matrix and sub matrix can be neglected (see the case of $\omega_1 = 5$). In this case, the feeding flow regimes are both absent. As the value of ω_1 decreases, the gap between well production and free gases within macro fracture network is supplied by free gases within matrix and desorbed gases within sub matrix. Therefore, there will be one or two troughs in derivative curves, which depends on the scale of production gap. The bigger the gap is, the more the supply from matrix and sub matrix is, the deeper these two troughs are.

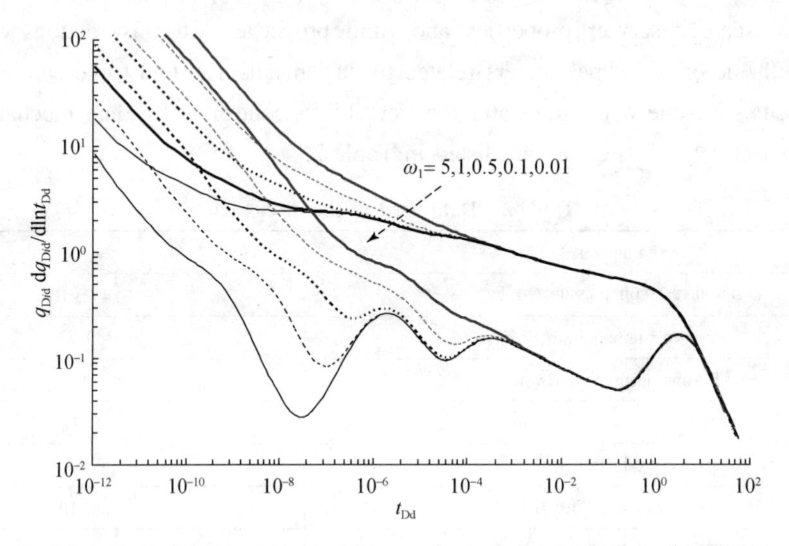

Figure 5　The effect of storage capacity of macro fracture on rate decline of MFHW in shale

4.3.2　Effect of storage capacity of matrix

Storability ratio of matrix block ω_2 reflects storage capacity of free gases within matrix block. Figure 6 shows that the effect of storability ratio of matrix block on rate decline curves of MFHW. After free gases within macro fracture networks are exhausted, free gases within matrix and desorbed gases within sub matrix gradually begin to supply well production. From the viewpoint of supply, storability ratio of matrix block ω_2 can be regarded as a kind of weight coefficient, which reflects the ratio of free gases in matrix to the sum of gases in matrix and sub matrix. A small value of ω_2 indicates a weak storability and slight feeding ability of matrix block. It represents that more desorbed gases from sub matrix will be required to supply well production. The amount of needed desorbed gases is rising continually with ω_2 decreasing, which is characterized by a deeper trough in the second feeding flow regime. When the desorbed gases are absolutely predominant in the process of production, a longer and more intensive feeding flow duration will replace two individual feeding flow regimes. This phenomenon is called as compound feeding flow regime.

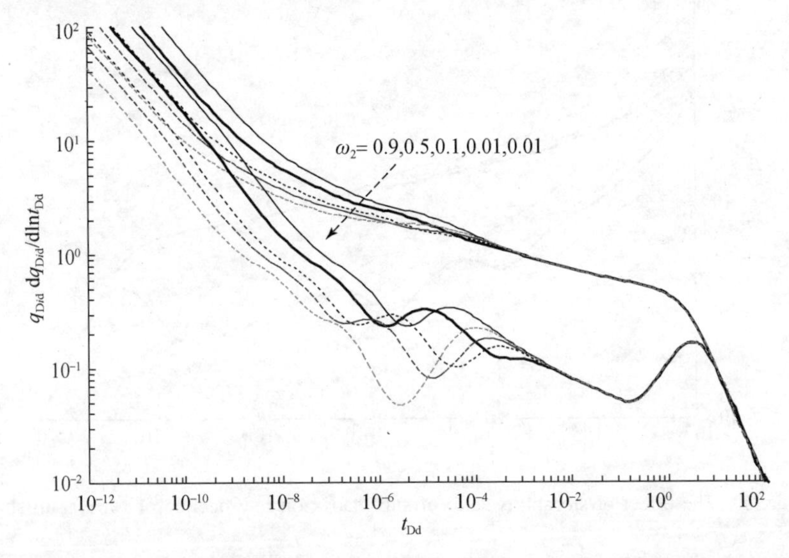

Figure 6 The effect of storability ratio of matrix on rate decline of MFHW in shale

4.3.3 Effect of storage capacity of sub matrix

According to the definition of adsorption index σ in above sections, σ is positive proportional to relative production q_{ref}, Langmuir volume V_L and Langmuir pseudo pressure m_L. Therefore, σ reflects the storage capacity of adsorbed gas in sub matrix. Figure 7 shows that the effect of adsorption index on rate decline curves. When the free gases in macro fracture network and matrix block can not fully feed well production, adsorbed gases desorb from organic material within sub matrix. This phenomenon is identified by a trough in second feeding flow regime. The condition that $\sigma = 0$ indicates that there is no gas source from sub matrix, so the second feeding flow regime is absent in derivative curve. A bigger σ, reflected by a deeper trough, will lead to an earlier arriving of second feeding flow regime and a more fully feeding regime. If σ is large enough (here $\sigma = 100$), the compound feeding flow regime will appear, which is identified by a deeper trough instead of two troughs. It indicates that of the amount of adsorbed gas is more than that of free gas in the production of gas wells.

It is noted that the effect of storage capacity of matrix is similar to that of sub matrix, but these two conditions has some differences in essence. The former is the reflection of matrix block extracting desorbed gases from sub matrix, and the latter is the reflection of sub matrix supplying desorbed gases to well production.

4.3.4 Effect of Fick diffusion transfer ability

Interporosity transmissibility coefficient between matrix and sub matrix λ_2 is positively proportional to desorption time τ, while τ is negatively related to diffusion coefficient D. Therefore, a smaller λ_2 reflects a stronger diffusion ability. Figure 8 shows that the effect of λ_2 on rate decline curves. Feeding procedure is identical to the previous description. Adsorbed gases

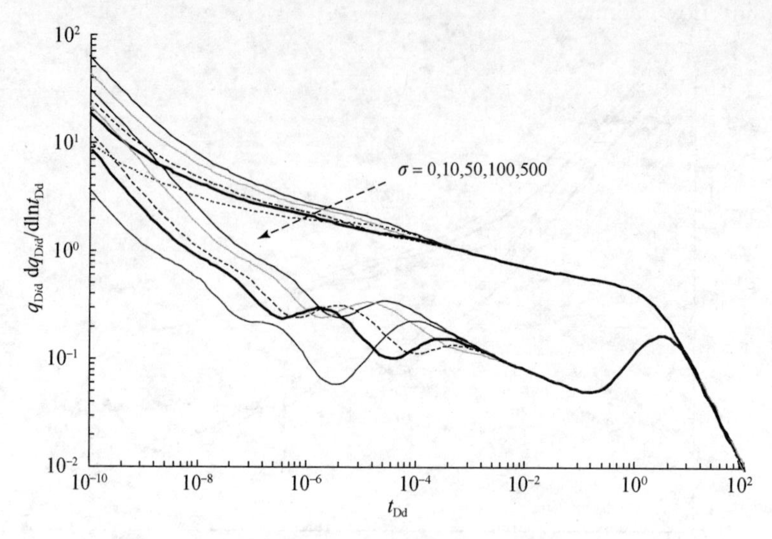

Figure 7 The effect of storability ratio of sub matrix on rate decline of MFHW in shale

desorb from the surface of organic material within sub matrix, and then diffuse into matrix block under concentration difference. A smaller transmissibility coefficient λ_2 indicates desorbed gases could transfer from sub matrix to matrix without too much effort. With enhancing of the transfer ability, the starting time of second feeding flow regime appears in advance. When transfer ability is great enough, the appearance of the second feeding flow regime overlaps with first feeding flow regime. Through the condition that $\lambda_2 = 0$ indicates the diffusion coefficient D is approximately infinite, the second feeding flow regime can't be prior to the first feeding flow regime.

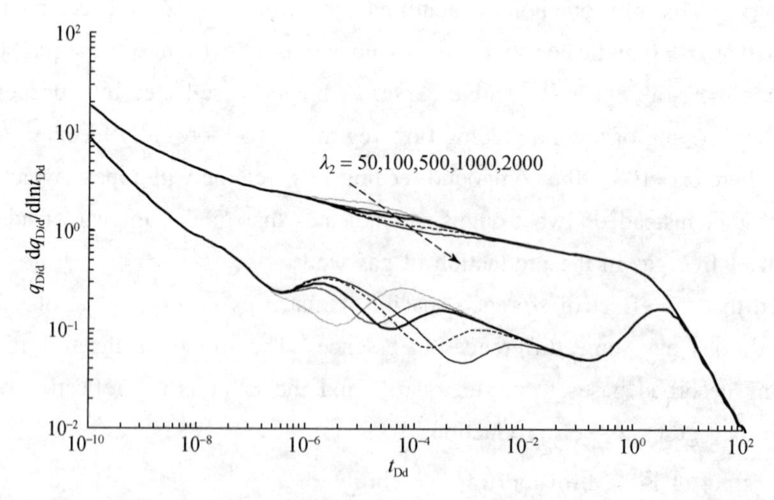

Figure 8 The effect of transfer ability of Fick diffusion on rate decline of MFHW in shale

4.3.5 Effect of Darcy flow transfer ability

Interporosity transmissivity coefficient between macro fracture network and matrix λ_1 is the reflection of gas transfer ability from matrix to macro fracture. The definition of λ_1 is positively proportional to the ratio of the matrix block permeability k_m, and negatively proportional to that of macro fracture k_f. Figure 9 shows that the effect of interporosity transfer coefficient λ_1 on rate decline curves. A bigger value of λ_1 reflects a stronger transfer ability, and results in an earlier appearance of the first feeding flow regime. As λ_1 decreases, the first feeding flow regime is delayed. However, the first feeding flow regime can not always appear prior to the linear flow regime even if the value of λ_1 is large enough.

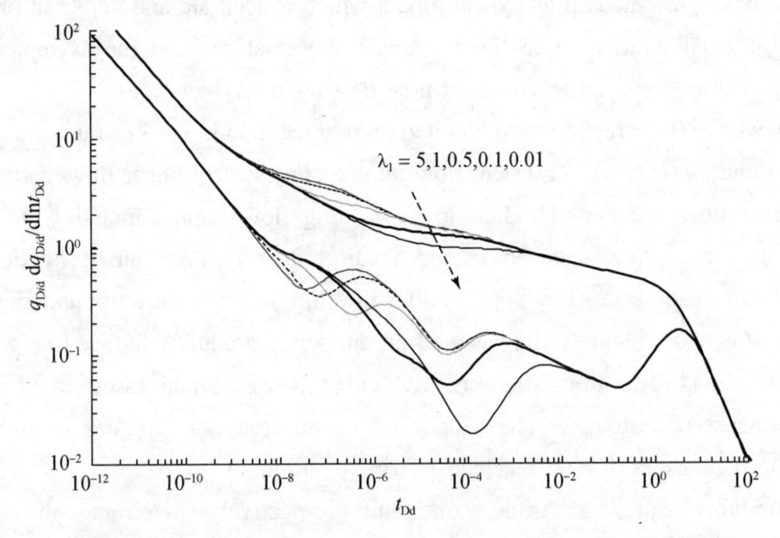

Figure 9 The effect of transfer ability of Darcy flow on rate decline of MFHW in shale

4.3.6 Effect of finite conductivity

Figure 10 shows the effect of fracture conductivity on rate decline curves. The conductivity means the gas flow ability in the hydraulic fracture. Fracture conductivity is considered as a kind

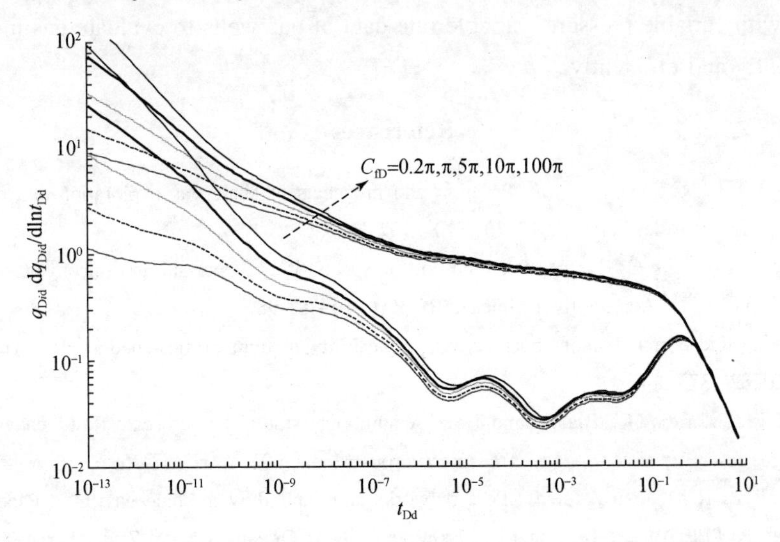

Figure 10 The effect of transverse fracture conductivity on rate decline of MFHW in shale

of transfer ability between macro fracture network and hydraulic fracture. In the entire flow period, a bigger C_{fD} will lead to higher rate, which indicates that horizontal well with high conductivity fractures could exploit more gas resource in shorter producing time. Therefore, it is necessary to keep bigger fracture conductivity to enhance gas well production.

5 Conclusions

(1) This paper presented an analytical model which is accurate and viable in predicting production for horizontal with multiple finite conductivity fractures and the asymptotic results of pseudo steady state is proposed to construct new Blasingame type curves.

(2) The whole flow regimes are divided into three major parts: transient flow, feeding flow, pseudo steady state flow. Transient flow regime covers early linear flow, early radial flow, compound linear flow, compound radial flow. Feeding flow regime includes the first feeding flow caused by Darcy flow and the second feeding flow regime caused by desorption and diffusion. Pseudo steady state flow appears after the flow reaches closed boundaries.

(3) In feeding flow regimes, the gas well production is gradually fed by free gases stored in matrix identified by the first trough in derivative curve, and adsorbed gases stored in sub matrix identified by the second trough. The proportion of different gas sources to well production depends on the parameters of dual mechanism triple continuum.

(4) Storability ratios of triple continuum respectively determine the durations of corresponding feeding flow regimes. Transmissibility coefficients controlled by dual flow mechanism affect the starting time of feeding flow regimes. Except storability ratio of macro fracture network, other sensitive parameters may can determine the appearance of compound feeding flow regime. The effect of fracture conductivity is enhancing gas well productivity.

(5) After incorporating the material balance time function, the presented approach can be used to deal with variable pressure/variable rate data of gas wells to evaluate gas in place within SRV consistently and efficiently.

References

[1] Dong D Z, Zou C N, Yang H, *et al.* Progress and prospects of shale gas exploration and development in China [J]. Acta Petrolei Sinica, 2012, 33(S1): 107-114.

[2] Li J Z, Zheng M, Zhang G S, *et al.* Potential and prospects of conventional and unconventional natural gas resource in China[J]. Acta Petrolei Sinica, 2012, 33(S1): 89-98.

[3] Civan F. Effective correlation of apparent gas permeability in tight porous media [J]. Transport Porous Media, 2010, 82: 375-384.

[4] Freeman C M., Moridis G J, Blasingame T A. A numerical study of microscale flow behavior in tight gas and shale gas reservoir systems [J]. Transport Porous Media, 2011, 90: 253-268.

[5] Wang F P, Reed R M, John A, *et al.* Pore networks and fluid flow in shale gas[C]. Paper SPE 124253 presented at the SPE Annual Technical and Exhibition. New Orleans, USA, 4-7 October 2009.

[6] Loucks R G, Reed R M, Ruppel S C, *et al.* Morphology, genesis, and distribution of nanometer-scale pores

in siliceous mudstones of the Mississippian Barnett shale[J]. Journal of Sediment Resource, 2009, 79: 848-861.

[7] Yao T Y, Zhu W Y, Li J S, *et al.* Fracture mutual interference and fracture propagation roles in production of horizontal gas wells in fractured reservoir[J]. Journal of Central South University, 2013, 44(4): 1487-1493.

[8] Brown M, Ozkan E, Raghavan R, *et al.* Practical solutions for pressure-transient responses of fractured horizontal wells in unconventional shale reservoirs[C]. Paper SPE 125043 presented at the SPE Annual Technical Conference and Exhibition, New Orleans, USA, 4-7 October 2009.

[9] Clarkson C R, Nobakht M, Kaviani D, *et al.* Production analysis of tight-gas and shale-gas reservoirs using the dynamic-slippage concept[C]. Paper SPE 144317 presented at the SPE North American Unconventional Gas Conference and Exhibition. Texas, USA, 12-16 May 2011.

[10] Nobakht M, Clarkson C R, Kaviani D. New and improved methods for performing rate-transient analysis of shale gas reservoirs[J]. SPE Reservoir Evaluation & Engineering. 2012, 12: 335-349.

[11] Zhao Y, Zhang L, Wu F. Pressure transient analysis for multi-fractured horizontal well in shale gas reservoirs[J]. Journal of Petroleum Science and Engineering, 2012, 90: 31-38.

[12] Guo J J, Zhang L H, Wang H T, *et al.* Pressure transient analysis for multi-stage fractured horizontal wells in shale gas reservoirs[J]. Transport Porous Media, 2012, 93: 635-653.

[13] Anderson D M, Nobakht M, Mattar L. Analysis of production data from fractured shale gas wells[C]. Paper SPE 131787 presented at the SPE Unconventional Gas Conference, Pennsylvania, USA, 23-25 February 2010.

[14] Dehghanpour H, Shirdel M. A triple porosity model for shale gas reservoirs[C]. Paper SPE 149501 presented at the SPE Canadian Unconventional Resources Conference, Calgary, Alberta, Canada, 15-17 April 2011.

[15] Yan B C, Wang Y H, Killough J E. Beyond dual-porosity modeling for the simulation of complex flow mechanisms in shale reservoirs[C]. Paper SPE 163651 presented at the SPE Reservoir Simulation Symposium, Texas, USA, 18-20 March 2013.

[16] Wu Y S, Moridis G, Bai B J, *et al.* A multi-continuum model for gas production in tight fractured reservoirs[C]. Paper SPE 118944 presented at the SPE Hydraulic Fracturing Technology Conference, Texas, USA, 19-21 April 2009.

[17] Ozkan E, Raghavan R, Apaydin O G. Modeling of fluid transfer from shale matrix to fracture network [C]. Paper 134830 presented at the SPE Annual Technical Conference and Exhibition, Florence, Italy, 19-22 October 2010.

[18] Javadpour F, Fisher D, Unswort M. Nanoscale gas flow in shale gas sediments[J]. Journal of Canadian Petroleum Technology, 2007, 46(10): 55-61.

[19] Javadpour F. Nanopores and apparent permeability of gas flow in mudrocks (shale and siltstone) [J]. Journal of Canadian Petroleum Technology, 2009, 48(8): 16-21.

[20] Medeiros F, Ozkan E, Kazemi H. Productivity and drainage area of fractured horizontal wells in tight gas reservoirs[J]. SPE Reservoir Evaluation & Engineering, 2008, 10: 902-912.

[21] Song B, Ehlig-Economides C. Rate-normalized pressure analysis for determination of shale gas well performance[C]. Paper SPE 14403 presented at the SPE North American Unconventional Gas conference and Exhibition, Texas, USA, 14-16 May 2011.

[22] Ozkan E, Raghavan R. New solutions for well-test-analysis problems: Part1-Analytical considerations[J]. SPE Formation Evaluation, 1991, 9: 359-371.

[23] Mukherjee H, Economides M J. A parametric comparison of horizontal and vertical well performance[J]. SPE Formation Evaluation, 1991, 6: 209-215.

[24] Cinco-Ley H, Samaniego V F, Dominguez A N. Transient pressure behavior for a well with a finite-conductivity vertical fracture[C]. Paper SPE 6014 presented at the 51th Annual Fall Technical Conference and Exhibition, New Orleans, USA, 3-6 October 1976.

[25] Wilkinson D J. New results for pressure transient behavior of hydraulically fractured wells[C]. Paper SPE 18950 presented at the SPE Joint Rocky Mountain Regional Exhibition, Denver, Colorado, USA, 6-8 May 1989.

[26] Kong X Y. Advanced mechanics of fluids in porous media[M]. Hefei: USTC Press, 1999.

[27] Fetkovich M J. Decline curve analysis using type curves[J]. Journal of Petroleum Technology, 1980, 6: 1065-1077.

[28] Palacio J C, Blasingame T A. Decline-curve analysis with type curves-Analysis of gas well production data [C]. Paper SPE 25909 presented at the SPE Rock Mountain Regional/Low Permeability Reservoirs Symposium, Denver, Colorado, USA, 12-14 August 1993.

非常规油气井产量递减规律分析新模型

齐亚东　王军磊　庞正炼　刘群明

（中国石油勘探开发研究院）

摘要：从非常规油气井生产特征出发，在综合考虑双曲递减和指数递减特点的基础上，构造了具有明确物理意义且量纲齐次的“双曲–指数”混合型递减率表达式，建立了“双曲–指数”混合型产量递减规律分析新模型并绘制了无量纲产量图版。研究结果表明：新模型可以有效地描述油气井的整个生产历史并对未来的生产趋势以及单井控制动态储量进行合理的预测和评价；与目前应用较为普遍的幂律指数递减模型相比，新模型对油气井产量曲线拟合的精度可提高8.1%。

关键词：非常规油气井；产量递减；经验方法；“双曲–指数”混合递减模型

中国非常规油气资源的勘探开发潜力巨大[1]，有评估表明，其可采资源量为890×10^8 ~ 1260×10^8t，大致是常规油气的3倍[2]。随着开发技术的突破与开发模式的创新[2-4]，以致密气、页岩气、致密油为典型代表的非常规油气资源正在逐步得到有效开发，对保障国内油气增储上产发挥着积极的作用。

非常规油气井的生产特征明显有别于常规油气井[2,5-7]：①生产曲线表现出典型的“L”形特征，即产量在投产早期下降十分迅速，第1年递减率多超50%，页岩气井甚至可达80%以上，而中后期递减趋缓；②油气井长期处于低产稳产的状态；③非常规油气井不稳定生产时间很长，很难达到拟稳态或边界控制流阶段。

产量递减规律的研究对单井（或气田）未来产量变化的预测、最终开发指标的制定以及后期开发措施的调整有着重要的现实意义。其分析方法主要有解析法和经验法两类[8]，经验法虽然没有严格的理论基础，但因其简便性和实用有效性而在油气田生产管理中一直发挥着积极的作用。文献[9]首次引入损失率和损失率导数的概念[9]，这是采用经验法建立产量–时间关系曲线进行生产预测的最早尝试；在此基础上，文献[10]提出了经典的产量递减分析经验公式系列，迄今，该公式系列（特别是双曲型递减公式）仍在油气田开发中被广泛应用。但文献[9]中产量递减公式有较为严格的适用条件，它要求定压生产且油气井达到拟稳态流动或边界控制流动状态，依上文所述，对非常规油气而言，此类条件实难满足，若强行使用文献[9]中产量递减分析方法，则常会出现递减指数$b>1$的情况，造成对单井控制储量及未来产量的高估，鉴于此，很多学者提出了相应的改进或替代方法[11-17]。这些方法中，文献[13]基于页岩气生产动态特征所提出的幂律指数递减模型的灵活性和实用性最强，在非常规油气井产量动态预测方面应用效果较好[18]，但该方法也存在明显的瑕疵与不足：①递减率表达式量纲不齐次，物理意义不明确；②对于极典型的“L”形生产曲线（即初期产量在极短的时间内递减极快，尔后以很低的产量保持很长时间的生产）拟合效果不理想。基于上述弊端，本文在综合考虑双曲递减和指数递减特点的基础上构造了全新的递减率表达式，该表达式量纲齐次且物理意义明确，以此为基础建立了“双曲–指数混合型”产量递减规律分析新模型并在页岩气、致密油中进行了成功应用，结果表明，新模型的适用性更强，应用效果更好，产量曲线的拟合精度可提高8.1%。

1　幂律指数递减模型

油气井产量递减的快慢程度多以递减率来表征，其定义式如下：

$$D = -\frac{1}{q}\frac{\mathrm{d}q}{\mathrm{d}t} \tag{1}$$

式中，D 为递减率，d^{-1}；q 为油（气）井产量，t/d（油井）或 $10^4\mathrm{m}^3$/d（气井）；t 为生产时间，d。

文献[13]在对大量页岩气井生产动态数据进行统计分析后发现，在双对数图中“递减率-时间关系曲线”多表现出线性特征，换言之，递减率与时间具有较好的幂律关系，基于此，构造了式（2）所示的幂律型递减率表达式，其特点为生产时间较短时，递减率随时间变化，而当生产时间趋于无穷大时，递减率趋于常数。

$$D = D_\infty + D_1 t^{-(1-n)} \tag{2}$$

式中，D_∞ 为时间趋于无穷大时的递减率，d^{-1}；D_1 为所选拟合时间段内第 1 天的递减率，d^{-1}；n 为时间指数，无量纲。

将式（2）代入式（1）分离变量并积分可得到产量预测模型，如式（3）所示，可以看出，在生产早期，产量特征由 D_1/n 控制，而在生产晚期，即生产达到拟稳态或边界控制阶段，产量特征主要由 D_∞ 控制。

$$q = q_\mathrm{i}\mathrm{e}^{\left(-D_\infty t - \frac{D_1}{n}t^n\right)} \tag{3}$$

式中，q_i 为所选拟合时间段内第 1 天的（初始）产油（气）量，t/d（$10^4\mathrm{m}^3$/d）。

大量的实际应用表明[13,15,16,18]，该模型可描述包括非稳态流、过渡流和拟稳态流（或边界控制流）在内的整个生产过程。然而，该模型并非完美，存在明显的瑕疵。

首先，所构造的递减率式（2）量纲不齐次，等式左边的量纲为 $[\mathrm{T}]^{-1}$，等式右边的量纲为 $[\mathrm{T}]^{-1}+[\mathrm{T}]^{n-2}$，尽管经验法不是建立在严格的理论基础之上的，但在模型中存在如此错误，实为不宜。

其次，递减率表达式的物理意义不够明确，当生产时间趋于无穷大时，递减率趋于定值 D_∞，此处无异议，但当时间为 1 时，即投产之后的第 1 天，递减率理应为 D_1，而按式（2）实际计算的递减率却为 $D_\infty+D_1$，出现了物理意义模糊的问题，也属不妥。

最后，非常规油气井，特别是直井，其生产曲线呈明显的“L”形，即油气井初期产量在极短的时间内快速递减，而后以很缓慢的递减率保持长期的低产量生产。对于这种极典型的“L”形生产曲线，该模型拟合效果似乎并不理想。

以四川盆地一口压裂的页岩气直井 SG1 为例，表 1 和图 1 分别给出了该井的基础数据和生产动态曲线。该气井于 2010 年 10 月投产，截至 2015 年 11 月底已有效生产 1770d，初始日产气量 $0.76\times10^4\mathrm{m}^3$/d，生产 10d 后迅速递减至 $0.15\times10^4\mathrm{m}^3$/d，目前平均日产气量 $0.06\times10^4\mathrm{m}^3$/d，累积产气量 $100.4\times10^4\mathrm{m}^3$。

表 1　SG1 井基础数据

原始地层压力/MPa	产层中深/m	储层厚度/m	渗透率/$\times10^{-3}\mu\mathrm{m}^2$	孔隙度/%	含气饱和度/%
21.8	2106	91.7	1.02×10^{-4}	2.98	70.04

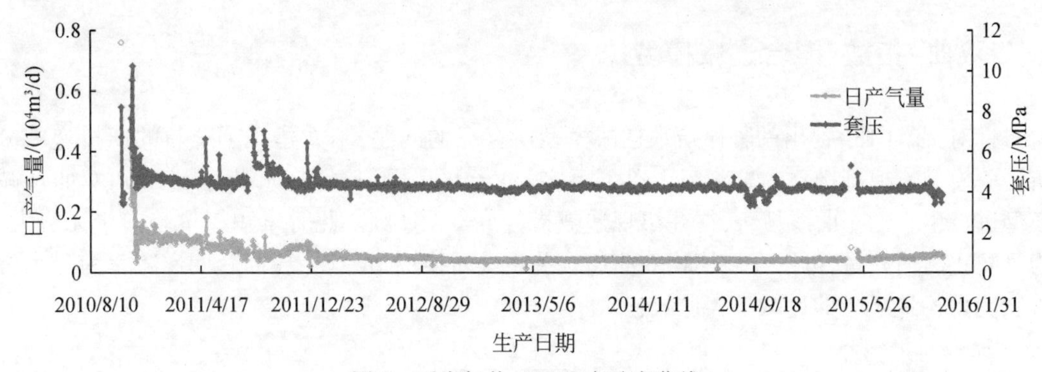

图1 页岩气井 SG1 生产动态曲线

采用文献[13]提出的幂律指数递减模型对生产数据进行拟合，拟合的起始时间取投产之后的第1天，拟合结果见表2、图2。由此得到SG1井的递减率和产量的表达式：

$$D = 1 \times 10^{-5} + 0.152 \times (t-1)^{-0.82}$$

$$q = 0.76e^{-[10^{-5}(t-1)+0.844(t-1)^{0.18}]}$$

式中，产量 q 拟合公式的拟合优度判别系数 $R^2=0.7855$。

表2 SG1井幂律指数递减法拟合参数

初始产量 q_i/(10^4m^3/d)	初始递减率 D_i/d^{-1}	无穷递减率 D_∞/d^{-1}	时间指数 n
0.76	0.152	1×10^{-10}	0.18

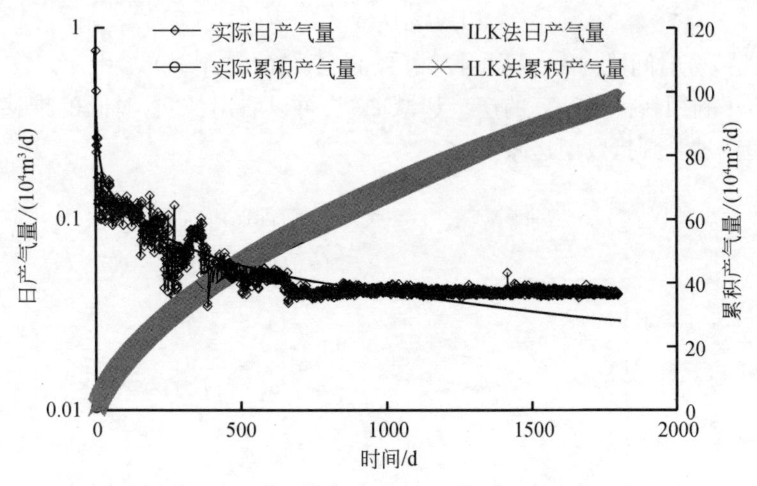

图2 幂律指数递减模型拟合 SG1 井生产数据

通过对上式积分，可以评价气井控制的动态储量，或预测给定时间(或产量)条件下的累积产量。计算可得，该气井控制的动态储量为 $192.8\times10^4m^3$；当产量降至 $0.01\times10^4m^3/d$ 时，累积产气量为 $141.0\times10^4m^3$，所需时间为12.6a，采出量占动态储量的73.1%；而当气井生产20a时，产量递减至 $52m^3/d$，累积产气量为 $160.6\times10^4m^3$，占动态储量的83.3%。

从图2中曲线的拟合程度来看，后期拟合效果并不理想，预测产量低于实际产量，这必将造成对动态储量和废弃产量下可采储量的低估，影响生产决策。

2 “双曲–指数”混合递减模型

针对文献[13]提出的幂律指数递减模型所存在的问题,本文建立了全新的产量递减分析经验模型。模型构建的思路基于以下事实:对于非常规油气井,如果运用 Arps 双曲递减模型进行生产动态曲线拟合,往往出现递减指数 $b>1$ 的情形,此时累积产量趋向于无穷大,为了防止高估储量,可以在后期将双曲递减关系式转换成指数递减(即后期递减率为一个常数),为此构造了新的递减率表达式:

$$D=\frac{D_{\mathrm{i}}-D_{\infty}}{1+mD_{\mathrm{i}}t}+D_{\infty} \tag{4}$$

式中,D 为递减率,d^{-1};D_{i} 为所选拟合时间段内初始时刻的递减率,d^{-1};D_{∞} 为时间趋于无穷大时的递减率,d^{-1},通常取值很小;m 为时间系数,无量纲;t 为生产时间,d。

从形式上看,m 相当于双曲递减模型中的递减指数 b,但此处的 m 并无递减指数的意义,且不局限于 0 ~ 1 的区间,根据实际需要可以是大于零的任意实数。

该递减率表达式兼顾了双曲递减和指数递减的特点,量纲齐次,且物理意义明确:投产时刻($t\to0$ 时),产量递减率等于初始递减率($D=D_{\mathrm{i}}$);开井生产后的早期阶段,递减规律主要受等号右边第一项控制,遵循双曲递减模式;随着生产时间的延长,D_{∞} 的作用越来越明显,当时间足够长后($t\to\infty$ 时),递减规律主要由 D_{∞} 控制。

将式(4)代入式(1),分离变量并进行积分,可以得到单井产量预测表达式

$$q=\frac{q_{\mathrm{i}}}{(1+mD_{\mathrm{i}}t)^{\frac{1}{m}\left(1-\frac{D_{\infty}}{D_{\mathrm{i}}}\right)}\mathrm{e}^{D_{\infty}t}} \tag{5}$$

式中,q_{i} 为所选拟合时间段内第 1 天的产油(气)量,t/d($10^4\mathrm{m}^3$/d)。

图 3 给出了利用新模型计算的产量递减曲线,可以看出,随着时间的推移,产量递减模式由双曲型递减过渡到指数型递减。

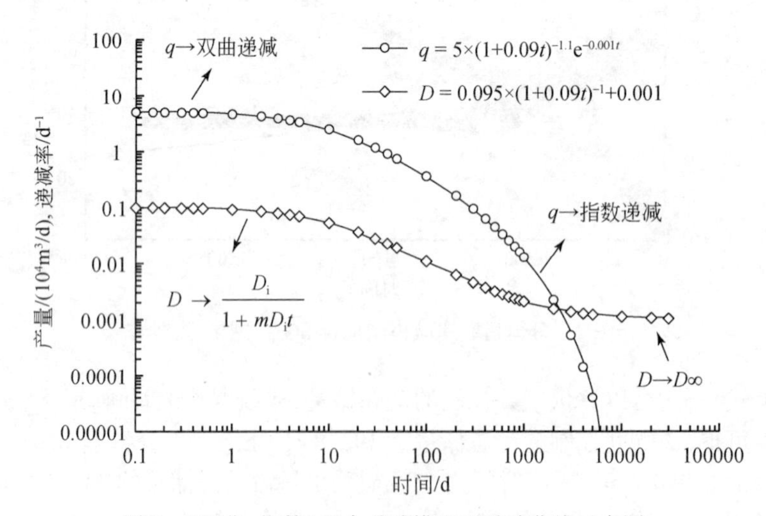

图 3 “双曲–指数”混合递减模型下递减曲线示意图

式(5)中的待定参数包括 q_{i}、D_{i}、D_{∞} 和 m,合理地调整这些参数,使计算出来的产量曲线

与实际生产曲线得到最高程度的拟合，便可最终确定目标井的产量递减规律，预测单井控制可采储量。

实际应用时，需先对待定参数赋初值。q_i初值的给定可参考生产动态数据中的最高日(月)产量；对于D_i，可先根据投产初期实际产量递减的快慢进行初步判断；对于D_∞，致密砂岩油(气)井，多在$10^{-3}\sim10^{-5}$量级，页岩气井多取10^{-5}及以下量级；对于m值，实际生产动态曲线递减越快，m的初始赋值应该越小，非常规油气井，m取值多在0.5~25。

初值给定后，调整参数进行曲线拟合。图4探讨了每个参数对产量曲线形状的影响时机和影响程度，根据实际需要，有些图选用了半对数坐标或双对数坐标，以求清晰展示参数对曲线形态影响的规律：

(1)q_i控制整条曲线的上下移动，如图4(a)所示，增大q_i时，曲线整体呈上移趋势，但早期(t较小)上移幅度大，晚期(t较大)上移幅度小。

(2)D_i主要控制曲线在早期的下降速度和下凹程度，如图4(b)所示，D_i越大，产量曲线在早期下降得越快，下凹程度也越大。

(3)D_∞主要控制曲线在晚期向时间轴靠近的快慢(即晚期产量递减的快慢)，而对早期曲线形状影响甚微，如图4(c)所示，D_∞越大，曲线向时间轴靠近的趋势出现得越早。

(4)m主要控制整条曲线的曲直程度，如图4(d)所示，随着m值的增大，曲线的平均曲率越来越小，产量递减速度越来越慢。

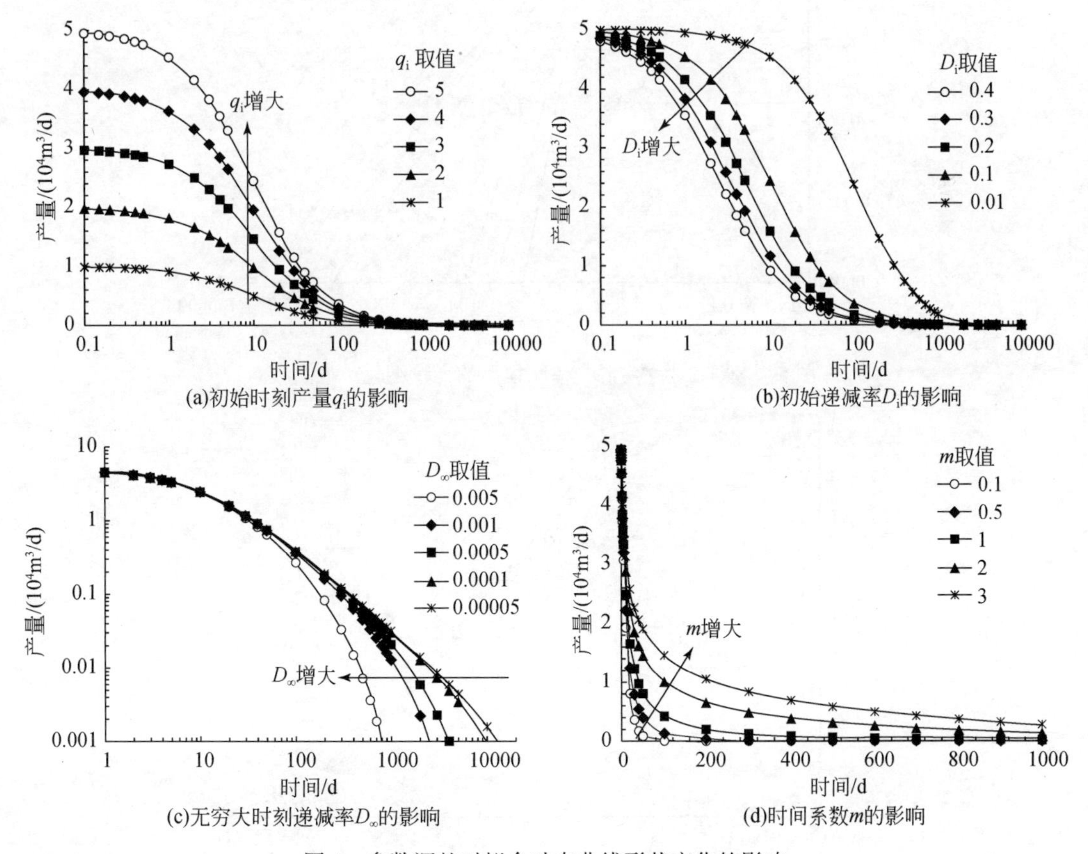

图4 参数调整对拟合动态曲线形状变化的影响

参照上述规律,在应用新模型进行产量曲线拟合时,可快速获得较好的拟合结果。当然,通过编程实现自动寻优拟合更能提高工作效率。

为了更便于应用,可引入无因次产量 q_D、无因次递减率 D_D、无因次时间 t_D 及无因次系数 α,对式(4)和式(5)进行无量纲化,各无量纲量的定义式如下:

$$q_D = \frac{q}{q_i}, \quad D_D = \frac{D}{D_i}, \quad t_D = D_i t, \quad \alpha = \frac{D_\infty}{D_i} \tag{6}$$

新模型无量纲递减率表达式为

$$D_D = \frac{1-\alpha}{1+mt_D} + \alpha (m \text{ 为常数}, m > 0) \tag{7}$$

新模型无量纲产量表达式为

$$q_D = (1+mt_D)^{\frac{1}{m}(\alpha-1)} e^{-\alpha t_D} (m \text{ 为常数}, m > 0) \tag{8}$$

利用上述无因次变量可绘制一系列拟合图版,图 5、图 6 给出的是 α=0.0001 时的情形,其中曲线的形式与初始产量 q_i、初始递减率 D_i 及无穷递减率 D_∞ 均无关,能够适用于所有井的递减分析。

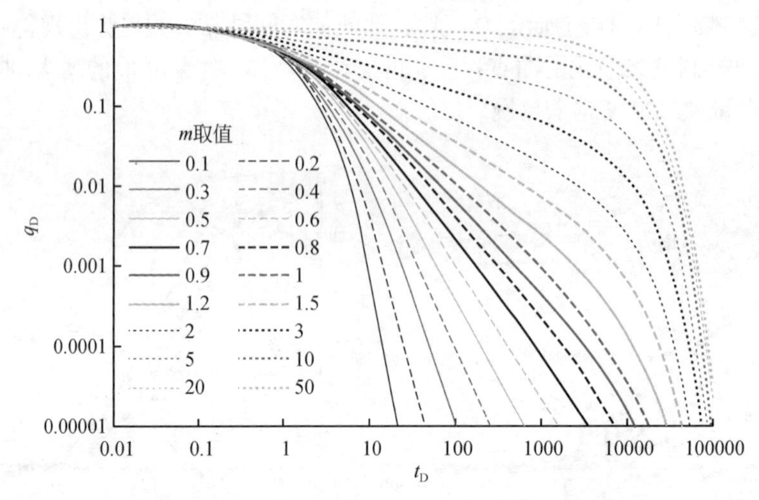

图 5 “双曲-指数”混合递减模型的无量纲化产量图版

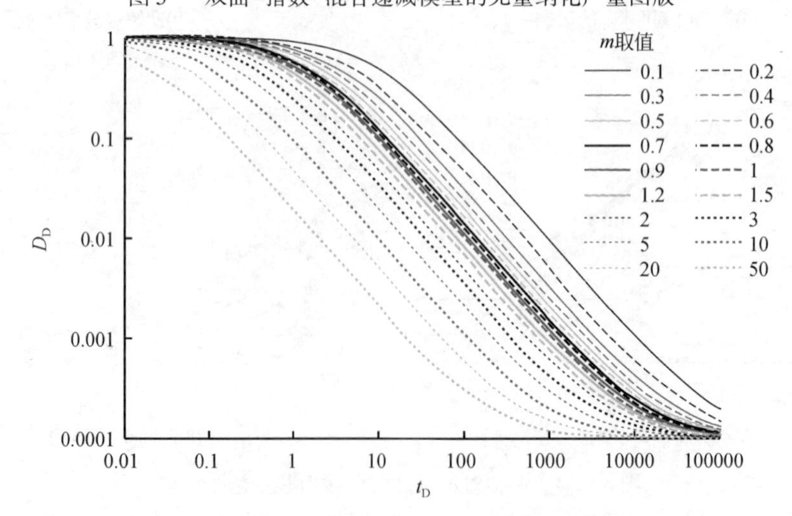

图 6 “双曲-指数”混合递减模型的无量纲化递减率图版

3 应用实例

3.1 页岩气

应用“双曲–指数”混合型产量递减模型对前文所述的 SG1 井生产数据进行拟合，拟合结果见表 3 和图 7，与幂律指数递减模型的拟合效果（图 2）相比，拟合程度得到了很大的改善，拟合精度提高了 8.1%，这充分说明“双曲–指数”混合型产量递减模型可以更为准确地描述页岩气井的产量递减规律。SG1 井递减率和产量的表达式如下：

$$D = \frac{0.85 - 1 \times 10^{-10}}{1 + 2.3205 \times (t - 1)} + 1 \times 10^{-10}$$

$$q = 0.76 \times [1 + 2.3205(t - 1)]^{-0.3663} e^{-10^{-10}(t-1)}$$

式中，产量 q 拟合公式的拟合优度判别系数 $R^2 = 0.8489$。

表 3　SG1 井“双曲–指数”混合递减模型拟合参数

初始产量 $q_i/(10^4 m^3/d)$	初始递减率 D_i/d^{-1}	无穷递减率 D_∞/d^{-1}	时间系数 m
0.76	0.85	1×10^{-10}	2.73

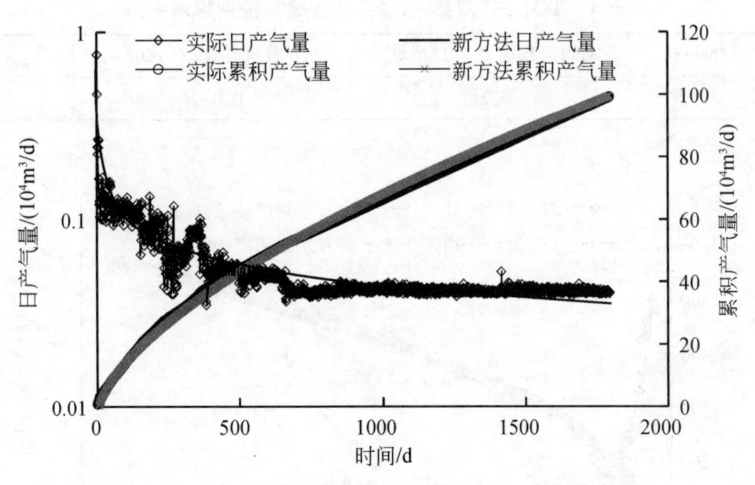

图 7　“双曲–指数”混合递减模型拟合 SG1 井生产数据

由此求得，气井动态储量为 $296.9 \times 10^4 m^3$。当产量降至 $0.01 \times 10^4 m^3/d$ 时，累积产气量为 $216.0 \times 10^4 m^3$，所需时间为 23.1a，采出量占动态储量的 72.7%；而当气井生产 20a 时，产量递减至 117 m^3/d，累积产气量为 $203.6 \times 10^4 m^3$，占动态储量的 68.6%。

3.2 致密油

再以川中一口致密油井 T01 为例，该油井 1996 年 3 月投产，至 2011 年 8 月因井场重建而暂时停产，其间连续生产 188 个月。产层中深 2050m，原始地层压力 24.06MPa，有效储层厚度 36m，储层孔隙度 1.01%，渗透率 $0.07 \times 10^{-3} \mu m^2$，原始含油饱和度 90%，如图 8 所示，投产初期的月产油量最高 593t，投产第 1 年，月产量急剧衰减至 235t，年递减率达 60%；生产

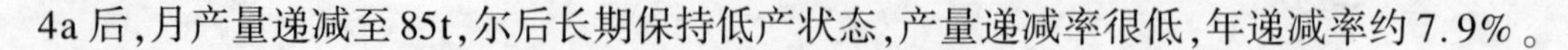

4a 后，月产量递减至 85t，尔后长期保持低产状态，产量递减率很低，年递减率约 7.9%。

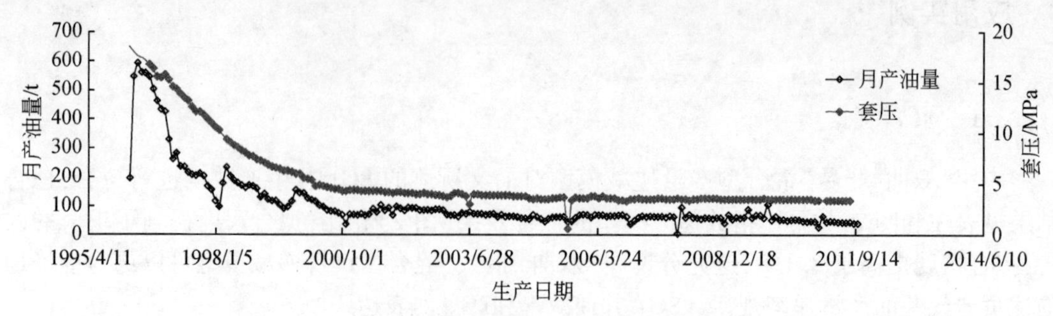

图 8　致密油井 TO1 生产动态曲线

应用“双曲-指数”混合递减模型对 TO1 井的生产动态进行拟合，起始时间取投产后的第 6 个月，拟合所得结果见表 4 和图 9，从曲线的拟合程度看，“双曲-指数”混合递减模型能很好地描述该井的产量递减规律，TO1 井的递减率和产量表达式分别为

$$D = \frac{0.2399}{1 + 0.3912 \times (t - 5)} + 0.0001$$

$$q = 650 \times [1 + 0.3912(t - 5)]^{-0.6132} \times e^{-0.0001(t-5)}$$

表 4　TO1 井“双曲-指数”混合递减模型拟合参数

初始产量 q_i/(t/mon)	初始递减率 D_i/(1/mon)	无穷递减率 D_∞/(1/mon)	时间系数 m
650	0.24	0.0001	1.63

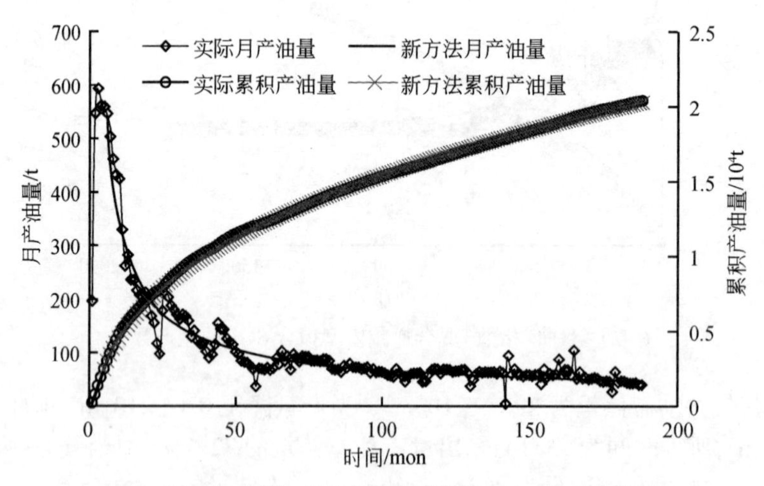

图 9　“双曲-指数”混合递减模型拟合 TO1 井生产数据

由此求得，TO1 井的动态储量为 91500t。当产量降至 3t/mon 时，累积产油量为 72434t，所需时间为 508a，采出量占动态储量的 79.2%；而当油井生产 50a 时，产量递减至 21.6t/mon，累积产油量为 32812×10^4t，占动态储量的 35.8%。

4 结论

(1)文献[13]对页岩气产量递减分析而提出的幂律指数递减模型存在着明显的瑕疵与不足,主要包括:递减率表达式量纲不齐次,物理意义不明确;对于极典型的"L"形生产曲线(即初期产量在极短的时间内递减极快,尔后以很低的产量保持很长时间的生产)拟合效果不理想。

(2)综合考虑双曲递减和指数递减的特征,构造了全新的"双曲–指数"混合递减率表达式,以此为基础建立了"双曲–指数"混合递减分析模型,模型形式如下。

递减率模型:$D = \frac{D_i - D_\infty}{1 + mD_i t} + D_\infty$

产量模型:$q = \frac{q_i}{(1 + mD_i t)^{\frac{1}{m}\left(1-\frac{D_\infty}{D_i}\right)} e^{D_\infty t}}$

(3)将新构建的"双曲–指数"混合递减分析模型成功应用于页岩气和致密油产量递减分析中,与幂律指数递减模型相比,新模型适用性更强,应用效果更好,产量曲线的拟合精度可提高8.1%。

参考文献

[1] 贾承造,郑民,张永峰. 中国非常规油气资源与勘探开发前景[J]. 石油勘探与开发,2012,39(2):129-136.

[2] 邹才能,张国生,杨智,等. 非常规油气概念、特征、潜力及技术:兼论非常规油气地质学[J]. 石油勘探与开发,2013,40(4):385-399.

[3] 马新华,贾爱林,谭健,等. 中国致密砂岩气开发工程与实践[J]. 石油勘探与开发,2012,39(5):572-579.

[4] 李文阳,邹洪岚,吴纯忠,等. 从工程技术角度浅析页岩气的开采[J]. 石油学报,2013,34(6):1211-1224.

[5] 罗瑞兰,雷群,范继武,等. 低渗透致密气藏压裂气井动态储量预测新方法:以苏里格气田为例[J]. 天然气工业,2010,30(7):28-31.

[6] 徐兵祥,李相方,胡小虎,等. 煤层气典型曲线产能预测方法[J]. 中国矿业大学学报,2011,40(5):743-747.

[7] 朱维耀,岳明,高英,等. 致密油层体积压裂非线性渗流模型及产能分析[J]. 中国矿业大学学报,2014,43(2):248-254.

[8] Clarkson C R. Production data analysis of unconventional gas wells: Review of theory and best practices[J]. International Journal of Coal Geology,2013,109: 101-146.

[9] Johnson R H, Bollens A L. The loss ratio method of extrapolating oil well decline curves[J]. Transactions of the American Institute of Mining, Metallurgical and Petroleum Engineers,1927,77: 771.

[10] Arps J J. Analysis of decline curves[J]. Transactions of the American Institute of Mining, Metallurgical and Petroleum Engineers,1945,160: 228-247.

[11] Robertson S. Generalized Hyperbolic Equation[C/OL]. Society of Petroleum Engineers,1988[2015-12-28]. https://www.onepetro.org/general/SPE-18731-MS? sort=&start=0&q=generalized+hyperbolic+equation&from_year=&peer_reviewed=&published_between=&fromSearchResults=true&to_year=&rows=10#.

[12] 俞启泰. Arps 递减指数 $n<0$ 或 $n\geqslant1$ 怎么办[J]. 新疆石油地质,2000,21(5):408-411.

[13] Ilk D,Rushing J,Perego A,*et al*. Exponential vs. hyperbolic decline in tight gas sands: understanding the origin and implications for reserve estimates using Arps' decline curves[C]. Paper SPE 116731 presented at the 2008 SPE Annual Technical Conference and Exhibition, Denver, Colorado, USA, 21-24 September 2008.

[14] Matter L,Moghadam S. Modified power law exponential decline for tight gas[C]. Paper PS 2009-198 presented at the Canadian International Petroleum Conference(CIPC) 2009,Calgary,Alberta,Canada16-18 June 2009.

[15] Ilk D,Curries S,Symmons D,*et al*. Hybrid rate-decline models for the analysis of production performance in unconventional reservoirs[C]. Paper SPE 135616 presented at the SPE Annual Technical Conference and Exhibition,Florence,Italy,19-22 September 2010.

[16] Ilk D. Well performance analysis for low to ultra-low permeability reservoir system[D]. Texas: Texas A&M University,2010.

[17] Valko P P. Assigning value to stimulation in the Barnett Shale: A simultaneous analysis of 7000 plus production histories and well completion records[C]. Paper SPE 119369 presented at the 2009 SPE Hydraulic Fracturing Technology Conference,The Woodlands,Texas,USA,19-21 January 2009.

[18] 段永刚,曹廷宽,王容,等. 页岩气产量幂律指数递减分析[J]. 西南石油大学学报(自然科学版),2013,35(5):172-176.

四、生产应用类

威远页岩气田典型平台生产规律及开发对策

位云生　李易隆　金亦秋　齐亚东　袁　贺

(中国石油勘探开发研究院)

摘要:威远页岩气田是我国页岩气开发最早的区块,总体开发效果越来越好,但同一平台井间存在较大差异。本文选取威远区块 PT2 平台为例,以测试产量为目标,从水平井段钻遇优质页岩比例、水平段轨迹倾向、水平井压裂段长度、改造段数、加砂量以及支撑剂和压裂液类型等多个方面进行上、下半支井的详细对比分析,明确控制页岩气产量的主要因素;以此为基础,通过各井生产动态特征的详细对比,剖析上、下半支井的产量与压力的变化规律,探讨总结引起生产动态变化的根本原因,提出针对性的开发对策,指导威远区块高效开发,并为同类区块的科学开发提供借鉴。

关键词:威远页岩气田;产量主控因素;生产变化规律;临界携液流量;开发对策

页岩气开发已成为目前国内最热门的学术话题之一[1,2]。页岩储层具有横向分布稳定、基质极其致密的特点,平台化钻井、工厂化多段大规模压裂改造已成为目前页岩气效益开发的核心技术[3]。威远区块是国家级页岩气示范区,属四川盆地川中隆起区的川西南低陡褶带,为一大型的穹窿背斜构造[4],构造相对单一,地层分布稳定,储层特征差异较小,但区块内甚至同一平台井间生产特征差异较大,目前尚未明确气井产量的主控因素和合理的开发工艺。本文选取一典型平台,通过水平井地质与工程参数、单井生产规律的分析,明确单井产量的主控因素,并结合实际生产动态,提出具体开发对策。

1　水平井地质与工程特征

1.1　地质特征

威远区块志留系龙马溪组页岩地层自下而上划分为龙一段和龙二段,龙一段为主要目的层,龙一段自下而上又划分为龙一$_1$亚段和龙一$_2$亚段,龙一$_1$亚段自下而上又细分为龙一$_1^1$、龙一$_1^2$、龙一$_1^3$、龙一$_1^4$四个小层(表 1),含气量均较高,但龙一$_1^4$小层脆性矿物含量(硅质、碳酸盐、黄铁矿)较低(表 2),可压性差,因此龙一$_1^1$、龙一$_1^2$、龙一$_1^3$三个小层为优质页岩段;奥陶系五峰组储层在威远地区不发育。

威远区块属海相深水陆棚沉积环境,区域储层地质参数(优质页岩厚度、TOC、孔隙度、脆性指数等)大范围稳定分布,横向变化较小[5](图 1),如威 202 井至威 204 井距离 22km,优质页岩段厚度变化 4.6m,横向变化率 0.2m/km。在一个平台范围内,地质特征横向上变化极小。

表 1 威远区块龙马溪组页岩小层划分表

组	段	亚段	小层	备注	
龙马溪组	龙二段				
	龙一段	龙一$_2$			
		龙一$_1$	龙一$_1^4$	优质页岩段	主要目的层
			龙一$_1^3$		
			龙一$_1^2$		
			龙一$_1^1$		

表 2 威远区块龙一$_1$ 亚段各小层脆性矿物含量

小层	脆性矿物含量/%	
	威 202 井	威 204 井
龙一$_1^4$	49	47
龙一$_1^3$	66	74
龙一$_1^2$	67	58
龙一$_1^1$	82	66

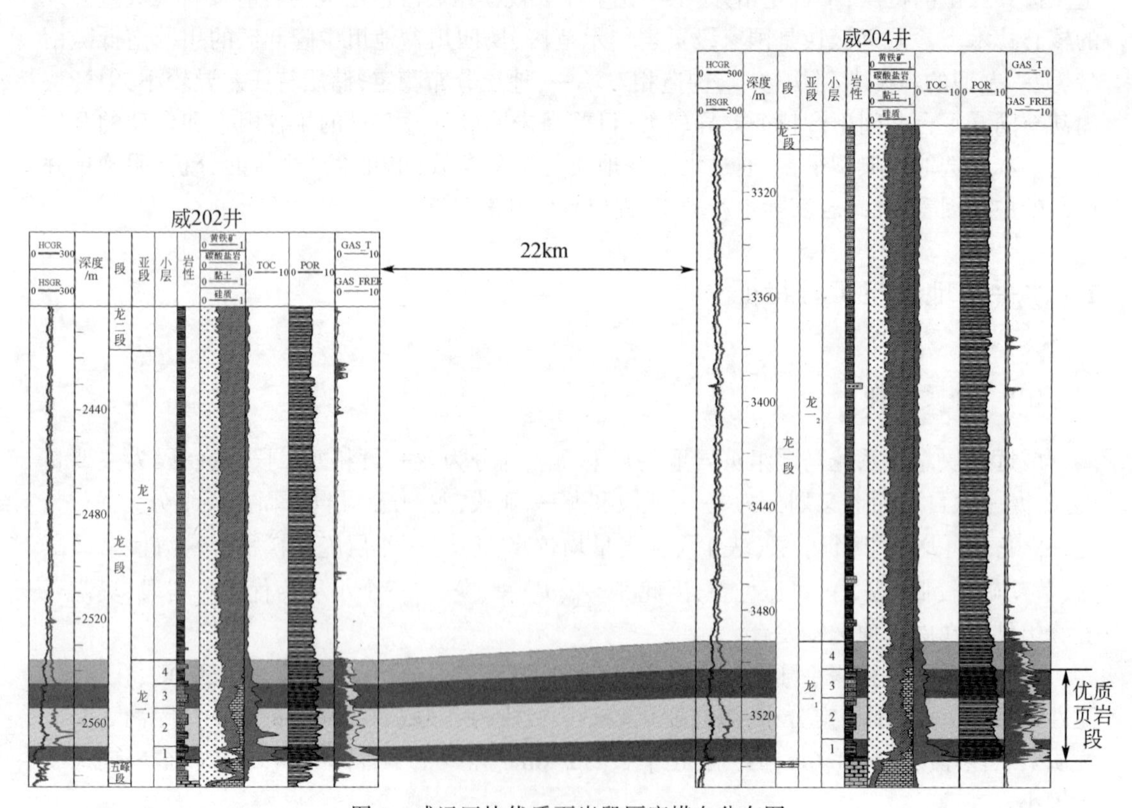

图 1 威远区块优质页岩段厚度横向分布图

1.2 工程参数

威远区块建产区为背斜构造的一翼,水平井钻井方向几乎与埋深等值线垂直(图2),因此普遍存在上半支井上倾、下半支井下倾的情况;上倾井工程施工难度较大,成功率较低,普遍产量低于下半支;据目前的产量和临界携液流量对比判断,2016年之前投产的46口平台井中有31口井存在不同程度的井底积液,且大部分为上半支井。以威远区块PT2平台6口井为例,详细工程参数如图3和表3所示。

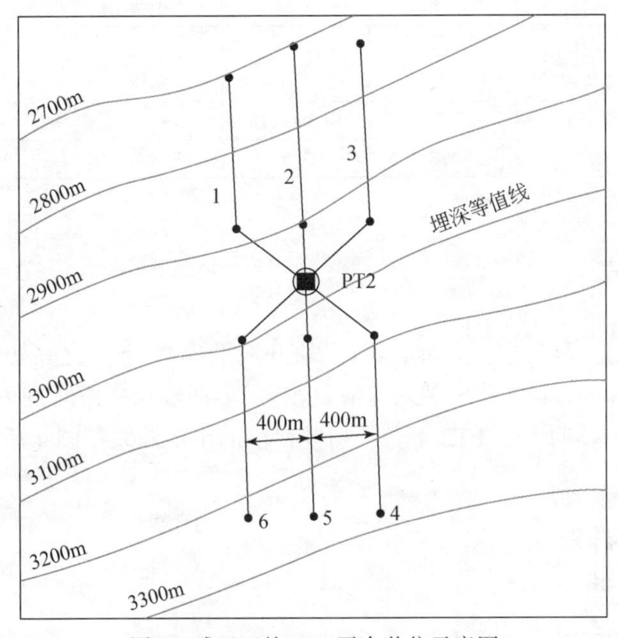

图2 威远区块PT2平台井位示意图

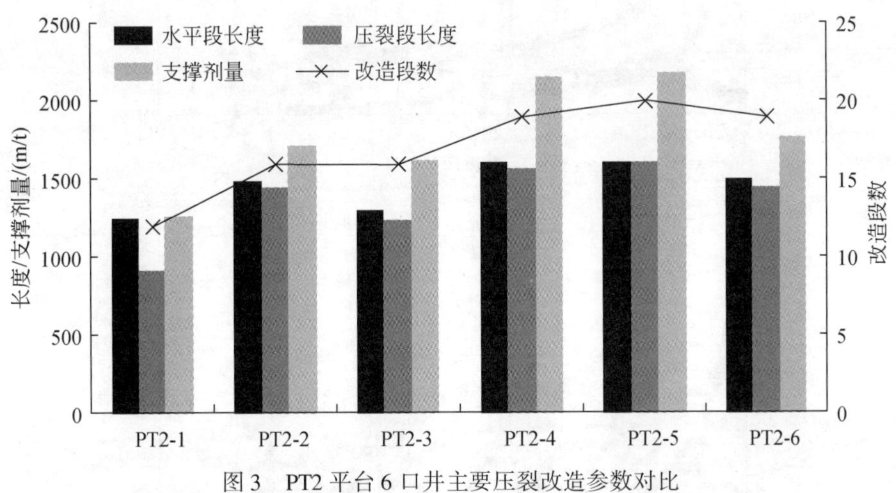

图3 PT2平台6口井主要压裂改造参数对比

表 3　威远区块 PT2 平台 6 口井钻遇地层、工程参数及测试情况

井号		PT2-1		PT2-2		PT2-3		PT2-4		PT2-5		PT2-6	
钻遇地层	厚度/m	钻遇长度/m	钻遇比例/%	钻遇长度/m	钻遇比例/%	钻遇长度/m	钻遇比例/%	钻遇长度/m	钻遇比例/%	钻遇长度/m	钻遇比例/%	钻遇长度/m	钻遇比例/%
龙一$_1^4$	16.73	650	52.42	350	23.65	703	54.08					1423	94.87
龙一$_1^{1+2+3}$	11.84	348	28.07	1130	76.35	545	41.92	1334	83.38	1600	100	77	5.13
五峰	7.90	205	16.53			35	2.68	150	9.38				
宝塔		37	2.98			17	1.31	116	7.25				
水平段倾向		上倾		上倾		上倾		下倾		下倾		下倾	
套变位置距 A 点/m		100		801		—		—		—		—	
支撑剂类型		树脂覆膜砂		树脂覆膜砂		树脂覆膜砂		陶粒		陶粒		陶粒	
压裂液类型		滑溜水		滑溜水		滑溜水+线性胶		滑溜水		滑溜水+线性胶		滑溜水+线性胶	
测试产量/($10^4m^3/d$)		3.4		5.3		3.6		28.8		20.0		6.4	

2　生产动态规律

由于钻遇层位与工程参数的差异,威远区块开发平台上、下半支井的产量与压力变化差异较大[6-10],同时,由于页岩气井压裂入地液量较大,早期较长一段时间内带液生产,返排液对气井产量产生较大影响[11]。PT2 平台 6 口井目前的生产曲线如图 4 所示。

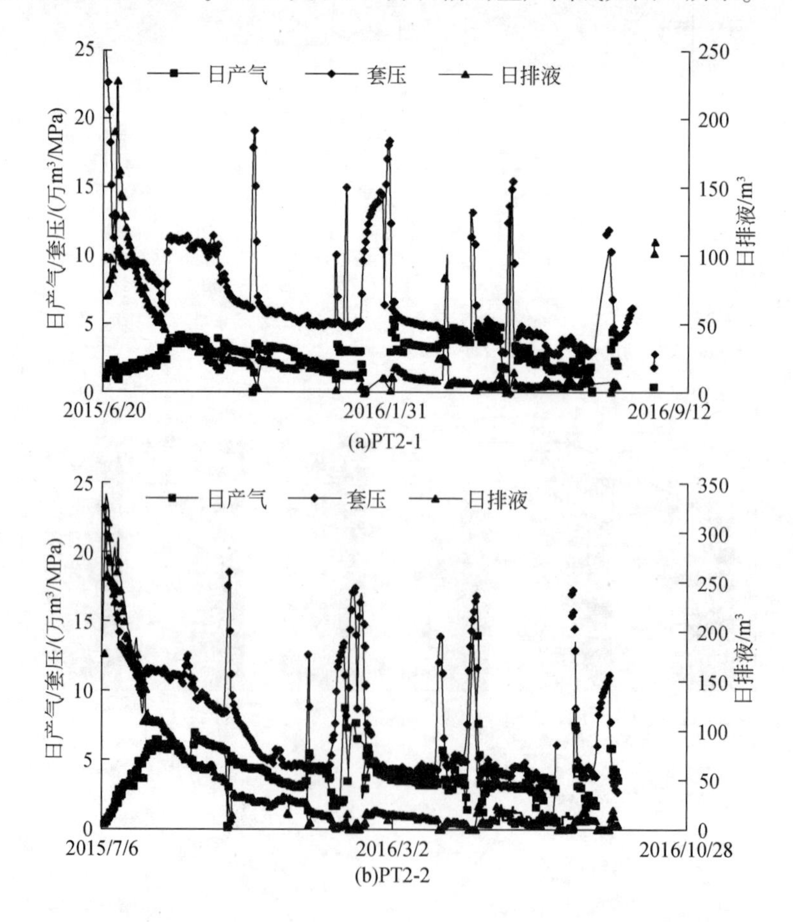

(a)PT2-1

(b)PT2-2

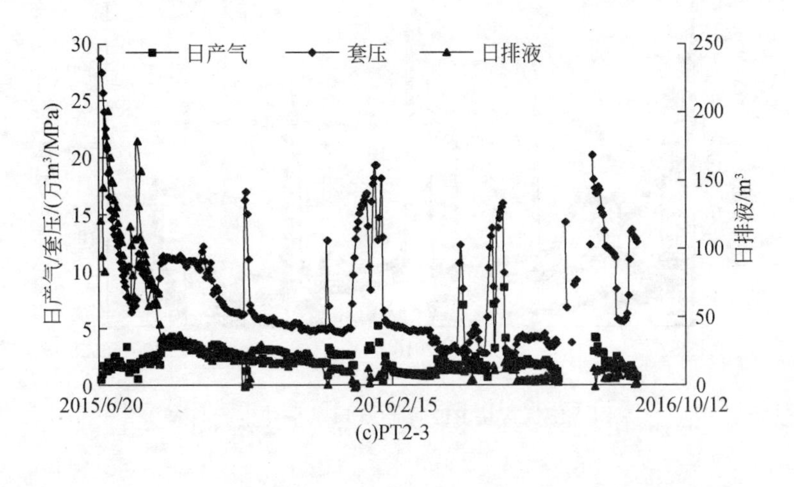

(c)PT2-3

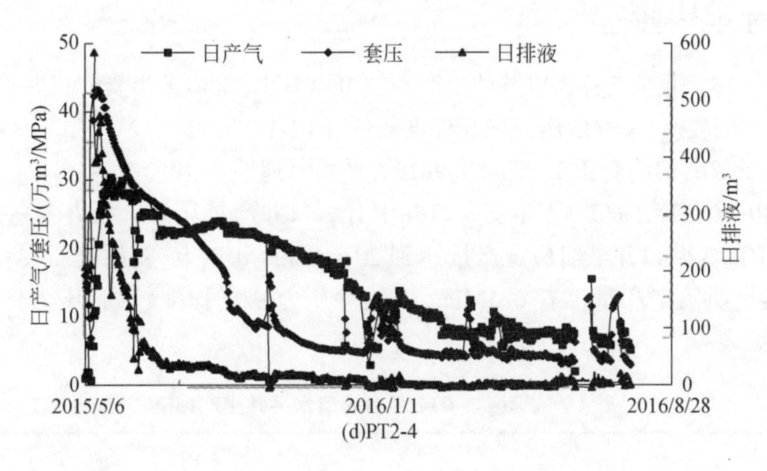

(d)PT2-4

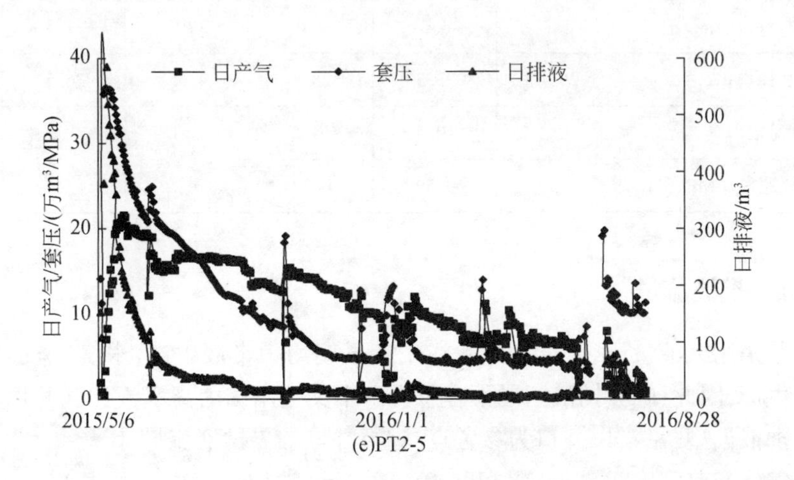

(e)PT2-5

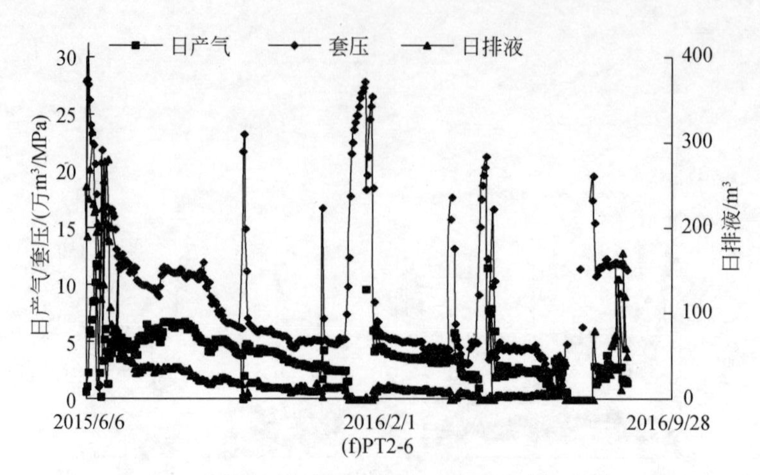

图4　威远区块 PT2 平台 6 口水平井生产曲线

2.1　基本生产特征

对比图 4 中上、下半支各 3 口井的生产曲线可以看出,整体上半支 3 口井产量和井口压力较低,产量递减较快,关井后压力恢复程度较低;PT2-1、PT2-2、PT2-3 三口井早期稳定产量 4×10^4 ~ $6\times10^4 m^3/d$,初始套压 23.3 ~ 28.9MPa,平均月递减率 10.2% ~ 19.4%,关井后压力恢复速率 0.39 ~ 0.52MPa/d。下半支 3 口井中有 2 口井产量和井口压力较高,产量递减较慢,PT2-4 和 PT2-5 两口井早期稳定产量达到 $20\times10^4 m^3/d$ 以上,初始套压在 40MPa 左右;PT2-6 井虽然早期稳定产量仅有 6.5MPa,但关井压力恢复程度较高,压力恢复速率达到 0.96MPa/d,见表 4。

表 4　威远区块 PT2 平台 6 口水平井生产特征

井号	PT2-1	PT2-2	PT2-3	PT2-4	PT2-5	PT2-6
早期稳定产量/($10^4 m^3/d$)	4.2	6.3	4.0	29.7	21.5	6.5
初始套压/MPa	25.1	23.3	28.9	43.5	36.5	28.1
平均月递减率/%	10.2	19.4	14.6	8.8	9.2	14.5
关井压力恢复速率/(MPa/d)	0.39	0.52	0.52	—	—	0.96

2.2　临界携液流量

页岩气井采用大液量压裂改造,因此在生产早期均为带液生产,水气比较大,且普遍采用油层套管生产,因此,临界携液问题不容忽视。本文选用国内高水气比气井比较常用的李闽产水气井携液模型[12-14],以井口为参考点,进行井口临界携液流量计算。

该井区气体相对密度 0.5684,水密度 $1030kg/m^3$,油层套管内径 114.3mm,井口温度平均 310K,气水界面张力取 0.06N/m。根据李闽临界携液流量公式,计算不同井口压力下的临界携液流量,并连同 PT2 平台 6 口井的井口压力和产量数据绘制在图 5 中。

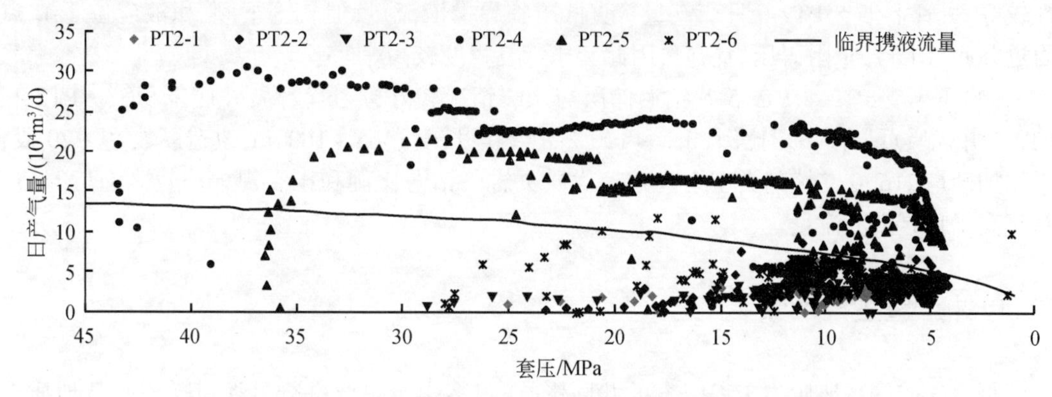

图5　PT2平台6口井不同套压下的产量与临界携液流量对比图

2.3　生产规律分析

通过PT2平台6口水平井优质页岩段的钻遇情况、压裂实施参数、实际生产特征及临界携液流量计算结果综合对比，可以分析出该平台各井产量及压力的变化原因：

(1)由于钻遇地层条件、压裂参数的影响，PT2-1、PT2-2、PT2-3、PT2-6四口井日产气量和初始套压较低。上半支PT2-1、PT2-2、PT2-3井优质页岩段钻遇比例28.1%～76.4%，压裂段长度920～1440m，改造段数12～16段，加砂量1265～1718t，支撑剂类型为树脂覆膜砂，综合判断，压裂规模小造成泄流区的渗透性较差，流动阻力大，是造成日产气量和初始套压低的主要原因，树脂覆膜砂可能对产量有一定影响。而PT2-6井优质页岩段钻遇比例仅为5.1%，压裂段长度1450m，改造段数19段，加砂量1766t，支撑剂类型为陶粒，综合判断，该井压裂规模较大，泄流区的渗透性较好，但靶体钻遇层位大部分位于非优质页岩段内，物质基础较差，因此造成日产气量和初始套压低。

(2)由关井套压恢复速率对比推断，PT2-1、PT2-2、PT2-3三口井泄流区渗透性较差，且日产气量一直低于临界携液流量，因此井底应该存在积液。上半支PT2-1、PT2-2、PT2-3轨迹上倾井积液应位于A点附近井底(图6)，进一步影响套压恢复程度；PT2-6井泄流区渗透

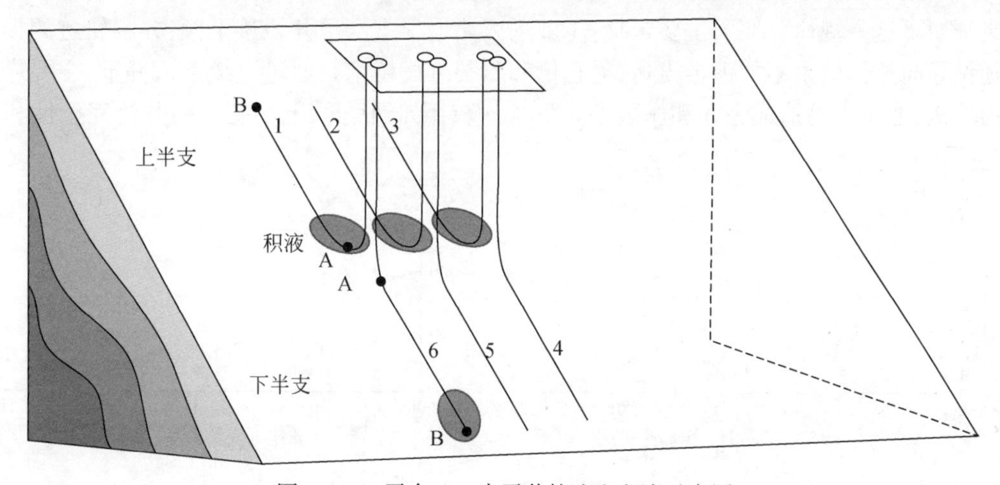

图6　PT2平台6口水平井轨迹和积液示意图

性较好，尽管下半支 PT2-6 日产气量一直低于临界携液流量，但轨迹下倾，积液应位于 B 点附近井底（图 6），未阻碍压力恢复，因此套压恢复程度较高。

（3）下半支 PT2-4、PT2-5 井钻遇储层与压裂情况较好，产量较高，井底未积液。PT2-4、PT2-5 井优质页岩段钻遇比例 83% ~100%，压裂段长度 1565 ~1600m，改造段数 19 ~20 段，平均加砂量 2160t，支撑剂类型为陶粒。优质页岩段钻遇比例和压裂规模均较高，日产气量一直高于临界携液流量。

3　单井产量主控因素

页岩气产量受地质与工程因素的共同影响，很多学者对此进行过探讨[15-19]。从地质上看，储层横向特征变化较小，垂向特征差异是影响页岩气井开发效果的主要地质因素。威远区块水平井目标靶体位置不断优化、下移，目前的靶体位置为龙马溪组底部的龙一$_1^1$小层碳质页岩段，含气量和压力系数最高，通过水平井多段大规模压裂改造，可获得较高的单井产量。但实际钻井过程中，由于目标层位厚度较薄（5m 左右）以及地层倾角的小幅变化和钻井导向的精度限制，实际水平井靶体位置上下有一定的变化，另外，从实际试采动态监测结果看，垂向上水力缝高 40 ~50m，有效支撑缝高 10 ~15m。综合判断，压裂改造后的水平井垂向上可以动用优质页岩段。因此，本文采用优质页岩段储层钻遇比例作为分析控制单井产量的主要地质因素。

在钻遇优质页岩储层的前提下，水平井压裂段长度决定了沿水平井筒方向打开储层的范围。在压裂段内，优化改造段数或簇数，采用“千方砂万方液”的大规模压裂改造，构建复杂的裂缝网络系统，提高泄气面积和改造区储层的渗透性，形成有效的气体流动通道[9]，获得高产气井。因此，水平压裂段长度、改造段数和加砂量是水平井高产的必要手段。

从 PT2 平台 6 口井的数据分析来看，优质页岩段钻遇比例、水平井压裂段长度、改造段数、单井加砂量与测试产量之间有明显的正相关关系（图 7），进一步明确了龙一$_1^{1+2+3}$小层钻遇比例、水平井压裂段长度、改造段数、单井加砂量是控制单井产量的主要因素。同时，上半支井的产量低于下半支井，这是威远区块的一个普遍规律，图 3 产量与其主要控制因素的关系也证明了这一规律。原因主要包括：①威远区块上半支水平井轨迹上倾，井眼轨迹的控制和随钻导向整体效果不如下半支；②射孔枪和压裂分段桥塞是通过电缆下入井下，主要靠水动力输送，上半支易造成完井和压裂工具下入比较困难，因此，上半支井的井筒完整性和改

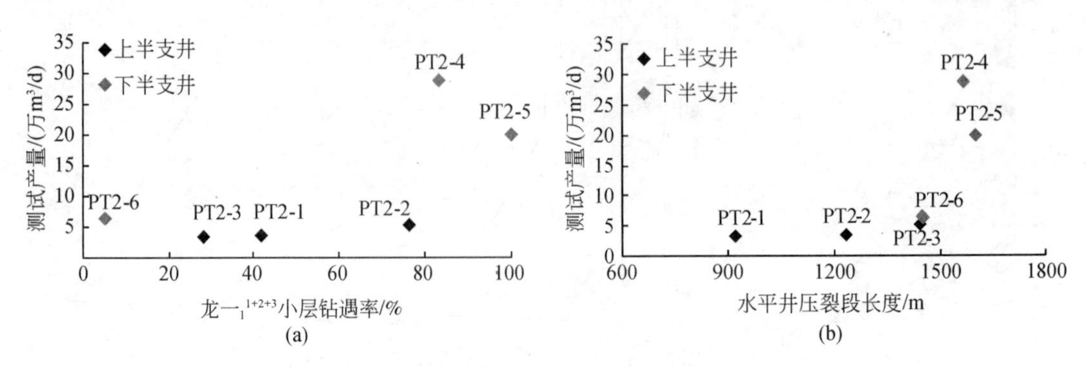

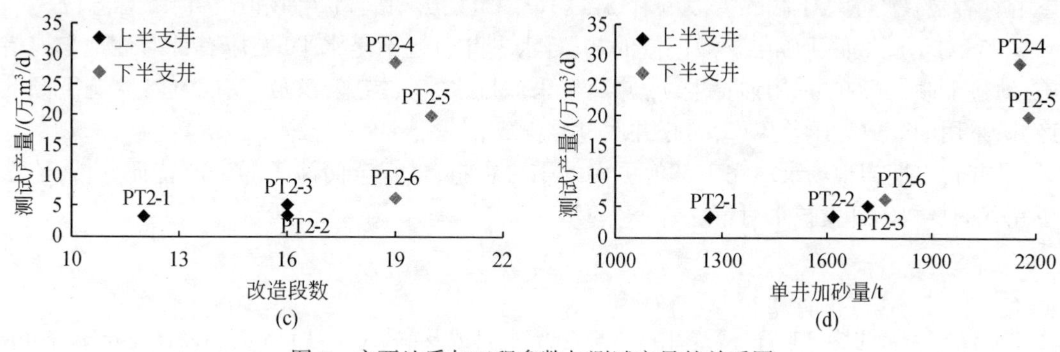

图7 主要地质与工程参数与测试产量的关系图

造段数不及下半支井；③由于滑溜水携砂能力有限，采用滑溜水与陶粒的压裂液与支撑剂组合，上半支井支撑剂泵送较为困难，加砂量一般比下半支要少。另外，支撑剂类型对产量的影响还需要更多井的数据对比论证。

4 开发对策

针对PT2平台6口井实际钻遇、压裂现状及生产规律分析结果，为了更合理、科学地进行平台化开发，提出有针对性的开发对策：

(1)页岩气井钻完井及压裂施工费用较高，因此采用平台化布井模式，现场施工采用“大兵团、工厂化”作业模式，各技术环节和施工效率要求较高，尽可能降低单井投资，因此对平台上的每口井，都应控制水平井靶体位置在优质页岩段内穿行，保证水平井筒完整性和压裂段长度，尽可能增加改造段簇数和加砂规模，提高单井产气量。例如，PT2-4、PT2-5井，钻完井及压裂作业均较好，单井累积产量可达到1亿m^3左右。

(2)针对低产(对比临界携液流量)井，建议采用小油管生产；对于平台上半支轨迹上倾低产井，尽早采取排水采气措施，如PT2-1、PT2-2、PT2-3井，井底积液位于水平段A点附近井底，且压裂液量是有限的，可采用撬装式排水采气工具和措施[20,21]，解放气井产能，发挥气井正常产量；对于平台下半支轨迹下倾低产井，应放开压差生产，尽量保持较高产量生产，防止井底过早积液，如PT2-6井，井底积液位于水平段B点附近井底，排水采气措施较为复杂，因此应放压生产。本文提出的开发对策已在威远页岩气田开发中部分得到推广应用。

5 结论

通过威远区块PT2典型平台6口井的地质、工程及生产动态特征分析，基本明确了控制页岩气井产量的主要因素、不同气井的生产特征和生产规律，并初步提出有针对性的开发对策。

(1)实现页岩气井高产，优质页岩段钻遇比例是物质保障，水平压裂段长度、改造段数和加砂量是必要手段。

(2)气井生产特征除受钻遇储层品质及压裂改造效果影响外，还受地层中压裂液返排的影响。若气井产量低于临界携液流量,存在井底积液,分析井口产量和压力时不容忽视。

(3)对于产量低于临界携液流量的井,若水平井轨迹不是水平的,应有针对性地采取措施:轨迹上倾低产井,采用油管生产,并及早采取排水采气措施,解放气井产能;轨迹下倾低产井,采用油管放大压差生产,防止井底过早积液。

同时,也期望现场录取完善其他更多的平台资料,进行全面深入的生产特征分析,建立更完善的页岩气开发技术对策。

参考文献

[1] 邹才能, 董大忠, 王玉满, 等. 中国页岩气特征、挑战及前景(二)[J]. 石油勘探与开发, 2016, 43(2): 1-13.

[2] 王红岩, 刘玉章, 董大忠, 等. 中国南方海相页岩气高效开发的科学问题[J]. 石油勘探与开发, 2013, 40(5): 574-579.

[3] 李文阳, 邹洪岚, 吴纯忠, 等. 从工程技术角度浅析页岩气的开采[J]. 石油学报, 2013, 34(6): 1218-1224.

[4] 王玉满, 董大忠, 李建忠, 等. 川南下志留统龙马溪组页岩气储层特征[J]. 石油学报, 2012, 33(4): 551-561.

[5] 贾爱林, 位云生, 金亦秋. 中国海相页岩气开发评价关键技术进展[J]. 石油勘探与开发, 2016, 43(6): 1-8.

[6] Wang J L, Jia A L. A general productivity model for optimization of multiple fractures with heterogeneous properties[J]. Journal of Natural Gas Science and Engineering. 2014, 21(2): 608-624.

[7] 王军磊, 贾爱林, 位云生, 等. 有限导流压裂气井拟稳态产能计算及优化[J]. 中国石油大学学报(自然科学版), 2016, 40(1): 100-107.

[8] Wang J L, Yan C Z, Jia A L, *et al.* Rate decline analysis of multiple fractured horizontal well in shale reservoirs with triple continuum[J]. Journal Central South University, 2014, 21:1-10.

[9] 王军磊, 贾爱林, 何东博, 等. 致密气藏分段压裂水平井产量递减规律及影响因素[J]. 天然气地球科学, 2014, 25(2): 278-285.

[10] 王晓泉, 张守良, 吴奇, 等. 水平井分段压裂多段裂缝产能影响因素分析[J]. 石油钻采工艺, 2009, 31(1): 73-76.

[11] 廖开贵, 李颖川, 杨志, 等. 产水气藏气液两相管流动态规律研究[J]. 石油学报, 2009,3(4): 607-612.

[12] 李闽, 郭平, 谭光天. 气井携液新观点[J]. 石油勘探与开发, 2001, 28(5): 105-106.

[13] 李闽, 孙雷, 李士伦. 一个新的气井连续排液模型[J]. 天然气工业, 2001, 21(5): 61-63.

[14] 李闽, 郭平, 张茂林, 等. 气井连续携液模型比较研究[J]. 西南石油学院学报, 2002, 24(4): 30-32.

[15] 张晓明, 石万忠, 徐清海, 等. 四川盆地焦石坝地区页岩气储层特征及控制因素[J]. 石油学报, 2015, 36(8): 926-939, 953.

[16] 郭彤楼, 张汉荣. 四川盆地焦石坝页岩气田形成与富集高产模式[J]. 石油勘探与开发, 2014, 41(1): 28-36.

[17] 李庆辉, 陈勉,Wang F P, 等. 工程因素对页岩气产量的影响——以北美 Haynesville 页岩气藏为例[J]. 天然气工业, 2012, 32(4): 54-59.

[18] 孙海成, 汤达祯, 蒋廷学. 页岩气储层裂缝系统影响产量的数值模拟研究[J]. 石油钻探技术, 2011, 39(11): 63-67.

[19] 王永辉，卢拥军，李永平，等．非常规储层压裂改造技术进展及应用[J]．石油学报，2012，33(1)：149-158.
[20] 张书平，吴革生，白晓弘，等．撬装式小直径管排水采气工艺技术[J]．天然气工业，2008，28(8)：92-94.
[21] 叶长青，熊杰，康琳洁，等．川渝气区排水采气工具研制新进展[J]．天然气工业，2015，35(2)：54-58.

有限导流压裂水平气井拟稳态产能计算及优化

王军磊　贾爱林　位云生

（中国石油勘探开发研究院）

摘要：在拟稳态流动阶段，边界封闭效应会对气井产能计算及优化产生很大影响。以单条人工裂缝为研究单元，在推导有限导流因子基础上，利用积分变换、渐进分析等方法获得单裂缝拟稳态压力基本解，进而使用压降叠加原理、物质平衡方程建立矩形地层有限导流压裂水平井产能计算模型并迭代求解，同时回归产能关于压裂参数的导数极大值获得最优参数的函数关系线。研究表明，气井产能受裂缝条数、长度、间距、导流能力、相对位置及气藏几何形状等因素影响，增大裂缝与地层接触面积、减小缝间干扰、降低边界封闭效应、平衡裂缝与地层流入流出关系能有效提高气井产能；当裂缝系统均分气藏泄流面积时裂缝布局最优，而对应的裂缝最优导流能力关系线则随气藏矩形长宽比、裂缝条数的变化而变化；在最优参数作用下气井能较为显著地达到较高的产能水平，实际使用时应选取最优参数线附近区域作为优化压裂参数的参考范围。

关键词：有限导流；压裂水平井；拟稳态；产能系数；参数优化

对于渗透率小、自然产能低的非常规气藏，利用水平井开发技术辅以水力压裂增产措施，能有效增大泄流面积，减小渗流阻力，增加储量动用程度，提高气井产能。众多理论和实践表明，分段压裂水平井渗流机理复杂、受控因素多，气井产能受水平压裂段长度、裂缝条数、导流能力和裂缝长度等影响显著[1,2]，对其进行参数优化会引起复杂的非线性优化问题[3,4]。寻求简洁合理的产能计算和参数优化方法已成为了提高压裂水平井开采效率的技术关键。

近年来关于压裂水平井产能的计算主要集中在非稳态产能[5-7]和稳态产能[8-10]两大方面，而实际气藏在生产晚期受到井间干扰、断层封闭的影响，通常进入拟稳态流动阶段。在边界封闭效应影响下，气井拟稳态产能公式有别于稳态产能公式[11,12]，影响产能的压裂参数较多且不独立，传统参数优化方法如枚举法、正交实验[13,14]等存在着最优解空间难以全部覆盖、方案数量过大的矛盾，而遗传算法[15,16]等智能技术难以解决由于裂缝条数增加而引起的搜索空间急剧增大的问题。

鉴于此，本文在准确建立分段压裂水平井拟稳态产能计算模型的基础上，深入研究压裂水平井的渗流本质，通过平衡裂缝与地层接触面积、地层边界封闭影响、裂缝间相互干扰、裂缝与地层流入流出动态四种渗流关系来优化气井产能，并借助产能关于压裂参数的导数极大值获得最优参数的函数关系线，同时利用积分平均方法确定压裂参数的优化参考值。

1　气井拟稳态产能计算

对压裂水平井而言，人工裂缝是气体流动的主要通道，首先以单条裂缝作为基本研究单元，进而通过压力叠加原理建立起分段压裂水平井产能计算模型。

1.1 有限导流裂缝拟稳态压力模型

引入气体拟压力 m、拟时间 t_a 函数能够将气体渗流问题等效为液体渗流问题。

$$m = \frac{\mu_{gi} Z_i}{p_i} \int_0^p \frac{\xi}{\mu_g(\xi) Z(\xi)} d\xi \tag{1}$$

$$t_a = \int_0^t \frac{\mu_{gi} c_{gi}}{\mu_g(\xi) c_g(\xi)} d\xi \tag{2}$$

假设垂直裂缝完全穿透地层,平面上平行于短轴边界 x_e(相当于1/2 排距),长轴边界为 y_e(相当于水平井段长),裂缝流量为 q_{fsc},位于(x_w, y_w)的长度为 d_{xw} 微元对应流量为 $q_{fsc} \times d_{xw}/(2x_f)$,则相应微元在地层中引起的无量纲拟压力控制方程满足式(3),

$$\nabla^2 m_D + \frac{\pi q_{fD}(t_{aD}) \Delta x_{wD}}{x_{fD}} \delta(x_D - x_{wD}) \delta(y_D - y_{wD}) = \frac{\partial m_D}{\partial t_{aD}} \tag{3}$$

其中,气藏带有 $x_{eD} \times y_{eD}$ 的矩形封闭外边界,初始时刻压力分布均匀:

$$m_D(x_D, y_D, 0) = 0 \tag{4}$$

$$\frac{\partial m_D(0, y_D, t_D)}{\partial x_D} = \frac{\partial m_D(x_{eD}, y_D, t_D)}{\partial x_D} = 0 \tag{5a}$$

$$\frac{\partial m_D(x_D, 0, t_D)}{\partial y_D} = \frac{\partial m_D(x_D, y_{eD}, t_D)}{\partial y_D} = 0 \tag{5b}$$

SI 单位制下的无量纲量定义为

$$t_{aD} = \frac{3.6 \times 10^{-3} k_m}{\phi \mu_{gi} c_{gi} L_{ref}^2} t_a \tag{6}$$

$$m_D = \frac{k_m h(m_i - m)}{1.842 q_{ref} \mu_{gi} B_{gi}} \tag{7}$$

$$q_{fD} = \frac{q_{fsc}}{q_{ref}} \tag{8}$$

$$\xi_D = \frac{\xi}{L_{ref}} (\xi = x, y, x_e, y_e, x_f, h, r_w) \tag{9}$$

$$C_{fD} = \frac{k_f w_f}{k_m x_f} \tag{10}$$

式中,μ_g 为气体黏度,mPa · s; c_g 为气体压缩系数,MPa^{-1}; Z 为气体偏差因子,无量纲; k_m 为基质渗透率,$\times 10^{-3}\ \mu m^2$; k_f 为裂缝渗透率,$\times 10^{-3}\ \mu m^2$; w_f 为裂缝宽度,m; x_f 为裂缝半长,m; h 为气藏厚度,m; y_e 为气藏长轴,m; x_e 为短轴,m; L_{ref} 为参考长度,m; t 为时间,h; q_{fsc} 为标准状况下裂缝产量,m^3/d; q_{ref} 为参考变量,m^3/d; C_{fD} 为无量纲裂缝导流能力。

利用 Laplace 变换、Fourier 有限余弦积分变换及反变换处理式(3)~式(5),可以得到 Laplace 空间下的微元压力基本解[17,18]:

$$\tilde{m}_{D0} = \frac{\tilde{q}_{fD}(s) \Delta x_{wD}}{2x_{fD}} \frac{\pi}{x_{eD}} \left\{ \frac{\cosh\sqrt{s}(y_{eD} - |y_D \pm y_{wD}|)}{\sqrt{s} \sinh\sqrt{s}\, y_{eD}} + 2\sum_{n=1}^{\infty} \frac{\cos\frac{n\pi x_D}{x_{eD}} \cos\frac{n\pi x_{wD}}{x_{eD}} \cosh\alpha_n(y_{eD} - |y_D \pm y_{wD}|)}{\alpha_n \sinh\alpha_n y_{eD}} \right\} \tag{11}$$

式中，

$$\cosh\alpha_n(y_{eD} - |y_D \pm y_{wD}|) = \cosh\alpha_n(y_{eD} - |y_D - y_{wD}|) + \cosh\alpha_n(y_{eD} - |y_D + y_{wD}|) \tag{12}$$

$$\alpha_n = \sqrt{s + n^2\pi^2/x_{eD}^2} \tag{13}$$

利用线性叠加原理，沿裂缝面积分获得均匀流量裂缝压力解：

$$\tilde{m}_D = \int_{x_{wD}-x_{fD}}^{x_{wD}+x_{fD}} \tilde{m}_{D0} dx_{wD} = \tilde{q}_{fD}(s)(HT + HV) \tag{14}$$

式中，

$$HT = \frac{\pi}{x_{eD}} \frac{\cosh\sqrt{s}(y_{eD} - |y_D \pm y_{wD}|)}{\sqrt{s}\sinh\sqrt{s}\, y_{eD}} \tag{15}$$

$$HV = \frac{2}{x_{fD}} \sum_{n=1}^{\infty} \left[\begin{array}{l} \dfrac{1}{n\alpha_n} \sin\dfrac{n\pi x_{fD}}{x_{eD}} \cos\dfrac{n\pi x_{wD}}{x_{eD}} \\ \cos\dfrac{n\pi x_D}{x_{eD}} \dfrac{\cosh\alpha_n(y_{eD} - |y_D \pm y_{wD}|)}{\sinh\alpha_n y_{eD}} \end{array} \right] \tag{16}$$

Cinco-Ley 等[19]、Al-Kobaisi 等[20]利用不同的数值方法获得了有限导流裂缝的压力动态，但解法复杂、计算量大、不易推广使用。借助相关文献[21,22]研究思路，将裂缝导流能力看作一种表皮，起到增加额外压力降落的作用，有限导流裂缝分解可无限导流裂缝解 $\tilde{m}_{Dinf}$ 与有限导流函数 $\tilde{S}_{fD}$ 的复合解：

$$\tilde{m}_D = \tilde{q}_{fD}(\tilde{m}_{Dinf} + \tilde{S}_{fD}) \tag{17}$$

Blasingame 和 Poe[21]给出了三线性流模型（相当于有限导流函数）与无限导流模型的复合解，三线性流模型能模拟早期的裂缝线性流和双线性流阶段，无限导流解能够较好模拟地层线性流和拟径向流阶段，但复合解难以模拟有限导流裂缝从双线性流过渡到拟径向流时缺失的地层线性流阶段。

针对这个问题，Wilkinson[23]首先给出了低导流裂缝压力解，随后将裂缝置于裂缝端点处存在不渗透边界的地层中，利用连续 Fourier 变换解析出低导流裂缝的早期流动特征函数 $\tilde{m}_{inner}$（裂缝线性流、双线性流），处理变换后形式为

$$\tilde{m}_{inner} = \frac{\pi}{2\sqrt{s}} + 2\pi\left[\sum_{n=1}^{\infty} \frac{1}{n^2\pi^2 C_{fD} + 2\sqrt{n^2\pi^2 + s}}\right] \tag{18}$$

基于渐进拟合分析法[24]给出拟合公式 $\tilde{m}_{match}$，用以修正 $\tilde{m}_{inner}$ 晚期双线性流与 $\tilde{m}_{inner}$ 地层线性流耦合过程中的过渡流阶段；基于数值模拟结果，同时引入校正函数 $\tilde{m}_{corr}$ 用以改进复合解的精度。

$$\tilde{m}_{match} = \frac{\pi}{2\sqrt{s}} \tag{19}$$

$$\tilde{m}_{corr} = \frac{\pi\Delta s(C_{fD})}{\pi + 4s \cdot \Delta s(C_{fD})} \tag{20}$$

因此，最终“改进”的有限导流函数可表述为

$$\tilde{S}_f(C_{fD,s}) = \tilde{m}_{inner} - \tilde{m}_{match} + \tilde{m}_{corr} \tag{21}$$

其中

$$\Delta S(C_{\mathrm{fD}}) = \frac{C_1}{C_{\mathrm{fD}} + C_2} \tag{22}$$

$$C_1 = \frac{4(\pi - 2)}{\pi} - \frac{\pi}{3} \tag{23}$$

$$C_2 = \frac{C_1}{\ln(\pi/2)} \tag{24}$$

利用无限导流与均匀流量裂缝间的转换关系[25]，令式(14)中的 $x_{\mathrm{D}}=x_{\mathrm{wD}}+0.732x_{\mathrm{fD}}$，结合式(21)即可得到有限导流裂缝的不稳态空间压力分布：

$$\tilde{m}_{\mathrm{D}} = \tilde{q}_{\mathrm{fD}}(s)[\mathrm{HT} + \mathrm{HV} + \tilde{S}_{\mathrm{f}}(C_{\mathrm{fD}})] \tag{25}$$

研究表明，定产条件下的压裂水平井在拟稳态阶段单裂缝流量趋于稳定[26]，不随时间变化。在拟稳态阶段，式(14)中的 Laplace 产量退化为 q_{fD}/s，利用 s 趋近 ∞ 渐进分析式(15)、式(16)，结果如下：

$$\frac{\mathrm{HT}_{\mathrm{pss}}}{s} = \left[\frac{2\pi}{x_{\mathrm{eD}}y_{\mathrm{eD}}s^2} + \frac{2\pi y_{\mathrm{eD}}}{x_{\mathrm{eD}}s}\left(\frac{1}{3} - \frac{|y_{\mathrm{D}} \pm y_{\mathrm{wD}}|}{2y_{\mathrm{eD}}} + \frac{y_{\mathrm{D}}^2 + y_{\mathrm{wD}}^2}{2y_{\mathrm{eD}}^2}\right)\right] \tag{26}$$

$$\frac{\pi x_{\mathrm{fD}}}{2x_{\mathrm{eD}}}\mathrm{HV}_{\mathrm{pss}}(x_{\mathrm{D}}, y_{\mathrm{D}}; x_{\mathrm{wD}}, y_{\mathrm{wD}}) = \sum_{n=1}^{\infty} \frac{\sin\frac{n\pi x_{\mathrm{fD}}}{x_{\mathrm{eD}}}\cos\frac{n\pi x_{\mathrm{wD}}}{x_{\mathrm{eD}}}\cos\frac{n\pi x_{\mathrm{D}}}{x_{\mathrm{eD}}}}{n^2} \frac{\mathrm{e}^{\frac{-n\pi|y_{\mathrm{D}} \pm y_{\mathrm{wD}}|}{x_{\mathrm{eD}}}} + \mathrm{e}^{\frac{-n\pi(2y_{\mathrm{eD}} - |y_{\mathrm{D}} \pm y_{\mathrm{wD}}|)}{x_{\mathrm{eD}}}}}{1 - \mathrm{e}^{\left(-\frac{2n\pi y_{\mathrm{eD}}}{x_{\mathrm{eD}}}\right)}} \tag{27}$$

对式(26)、式(27)做 Laplace 反变换[27]，得到拟稳态压力解：

$$\tilde{m}_{\mathrm{D}} = q_{\mathrm{fD}}\left\{L^{-1}\left[\frac{\mathrm{HT}_{\mathrm{pss}}}{s}\right] + L^{-1}\left[\frac{\mathrm{HV}_{\mathrm{pss}}}{s}\right] + S_{\mathrm{f}}(C_{\mathrm{fD}})\right\} \tag{28}$$

同时在 Laplace 空间内对式(21)做拟稳态流动渐进分析(s 趋近∞)，得到有限导流因子拟稳态表达式：

$$S_{\mathrm{f}}(C_{\mathrm{fD}}) = \sum_{n=1}^{\infty} \frac{2}{n[n\pi(C_{\mathrm{fD}}) + 2]} + \Delta S(C_{\mathrm{fD}}) \tag{29}$$

Wang 等[28]利用边界元数值方法(BEM)计算了有限导流裂缝的半解析压力解，通过数据回归给出了拟稳态阶段的有限导流因子：

$$S_{\mathrm{f}}(C_{\mathrm{fd}}) = \frac{0.95 - 0.56\omega + 0.16\omega^2 - 0.028\omega^3 + 0.0028\omega^4 - 0.00011\omega^5}{1 + 0.094\omega + 0.093\omega^2 + 0.0084\omega^3 + 0.001\omega^4 + 0.00036\omega_5} \tag{30}$$

式中 $\omega=\ln C_{\mathrm{fd}}$。对比分析式(29)，从低导流能力到高导流能力两种方法拟合效果良好(图1)，证明了本文推导有限导流因子的可靠性。

1.2 有限导流压裂水平井拟稳态产能模型

在拟稳态阶段，定容气藏物质平衡方程满足：

$$\frac{p_{\mathrm{avg}}}{Z_{\mathrm{avg}}} = \frac{p_{\mathrm{i}}}{Z_{\mathrm{i}}}\left(1 - \frac{G_{\mathrm{p}}}{G}\right) \tag{31}$$

式中，G_{p}为气藏累积产量，G 为气藏地质储量。方程两侧关于 t 求导，结果如下(图1)：

$$\int_{p_{\mathrm{i}}}^{p_{\mathrm{avg}}} \frac{Z_{\mathrm{g}}(p_{\mathrm{i}})\mu_{\mathrm{g}}(p_{\mathrm{i}})}{p_{\mathrm{i}}} \frac{p_{\mathrm{avg}}}{Z_{\mathrm{g}}(p_{\mathrm{avg}})\mu_{\mathrm{g}}(p_{\mathrm{avg}})}\partial p_{\mathrm{avg}} = -\frac{N_{\mathrm{f}}}{G}\frac{1}{c_{\mathrm{g}}(p_{\mathrm{i}})}\int_0^t \frac{c_{\mathrm{g}}(p_{\mathrm{i}})\mu_{\mathrm{g}}(p_{\mathrm{i}})}{c_{\mathrm{g}}(p_{\mathrm{avg}})\mu_{\mathrm{g}}(p_{\mathrm{avg}})}\partial t \tag{32}$$

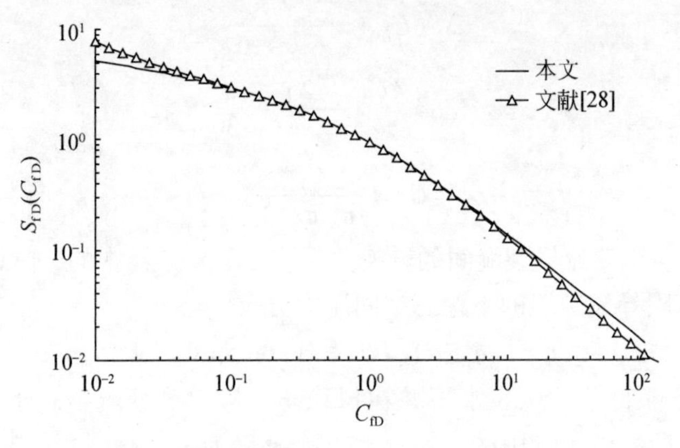

图 1　有限导流因子验证分析

引入拟压力定义式(1)、拟时间定义式(2),并做无量纲化处理,最终得到无量纲形式的物质平衡方程:

$$\frac{m_{\mathrm{D}}(p_{\mathrm{avg}})}{q_{\mathrm{scD}}} = 2\pi \frac{t_{\mathrm{D}}}{x_{\mathrm{eD}} y_{\mathrm{eD}}} \tag{33}$$

式中,q_{scD}为压裂水平井的无量纲产量。将式(33)代入式(28)获得单条有限导流垂直裂缝的拟稳态压力公式:

$$m_{\mathrm{D}} = q_{\mathrm{fD}} \begin{Bmatrix} \dfrac{m_{\mathrm{avgD}}}{q_{\mathrm{scD}}} + L^{-1}\left[\dfrac{\mathrm{HV}_{\mathrm{pss}}}{s}\right] + S_{\mathrm{f}}(C_{\mathrm{fD}}) + \\ \dfrac{2\pi y_{\mathrm{eD}}}{x_{\mathrm{eD}}}\left(\dfrac{1}{3} - \dfrac{|y_{\mathrm{D}} \pm y_{\mathrm{wD}}|}{2y_{\mathrm{eD}}} + \dfrac{y_{\mathrm{D}}^{2} + y_{\mathrm{wD}}^{2}}{2y_{\mathrm{eD}}^{2}}\right) \end{Bmatrix} \tag{34}$$

与垂直裂缝中单一的线性流相比,横向压裂缝在井筒附近会产生一个附加压力降,通常用聚流表皮因子[7]修正:

$$S_{\mathrm{c}} = \frac{1}{C_{\mathrm{fD}}} \frac{h_{\mathrm{D}}}{x_{\mathrm{fD}}}\left[\ln \frac{h_{\mathrm{D}}}{2r_{\mathrm{wD}}} - \frac{\pi}{2}\right] \tag{35}$$

根据压力叠加原理,地层任意点处的压降等于各裂缝单独工作时在该点引起的压降总和(图 2)。这样可获得井筒与裂缝 i 交叉点处压力值 $m_{\mathrm{wD}i}$:

$$m_{\mathrm{wD}i} = m_{\mathrm{avgD}} + \sum_{j=1}^{n_{\mathrm{f}}} q_{\mathrm{scD}j} \Delta m_{\mathrm{D}ij} \tag{36}$$

假设井筒具有无限导流能力,即 $m_{\mathrm{wD1}} = m_{\mathrm{wD2}} = \cdots = m_{\mathrm{wDnf}}$,结合流量约束条件 $\sum q_{\mathrm{fD}i} = q_{\mathrm{scD}} = 1$ 可以获得 $n_{\mathrm{f}} \times n_{\mathrm{f}}$阶的 n_{f}段压裂水平井产能计算模型:

$$\begin{cases} 1 = \mathrm{PI}_1 \Delta m_{\mathrm{D11}} + \mathrm{PI}_2 \Delta m_{\mathrm{D12}} + \mathrm{PI}_3 \Delta m_{\mathrm{D13}} + \cdots + \mathrm{PI}_{n_{\mathrm{f}}} \Delta m_{\mathrm{D1}n_{\mathrm{f}}} \\ 1 = \mathrm{PI}_1 \Delta m_{\mathrm{D21}} + \mathrm{PI}_2 \Delta m_{\mathrm{D22}} + \mathrm{PI}_3 \Delta m_{\mathrm{D23}} + \cdots + \mathrm{PI}_{n_{\mathrm{f}}} \Delta m_{\mathrm{D2}n_{\mathrm{f}}} \\ \quad \vdots \\ 1 = \mathrm{PI}_1 \Delta m_{\mathrm{D}n_{\mathrm{f}}1} + \mathrm{PI}_2 \Delta m_{\mathrm{D}n_{\mathrm{f}}2} + \mathrm{PI}_3 \Delta m_{\mathrm{D}n_{\mathrm{f}}3} + \cdots + \mathrm{PI}_{n_{\mathrm{f}}} \Delta m_{\mathrm{D}n_{\mathrm{f}}n_{\mathrm{f}}} \end{cases} \tag{37}$$

式中,PI_i为第 i 条裂缝的无量纲产能系数;$\Delta m_{\mathrm{D}ij}$为第 j 条裂缝在第 i 条裂缝上引起的单位产量下压降。

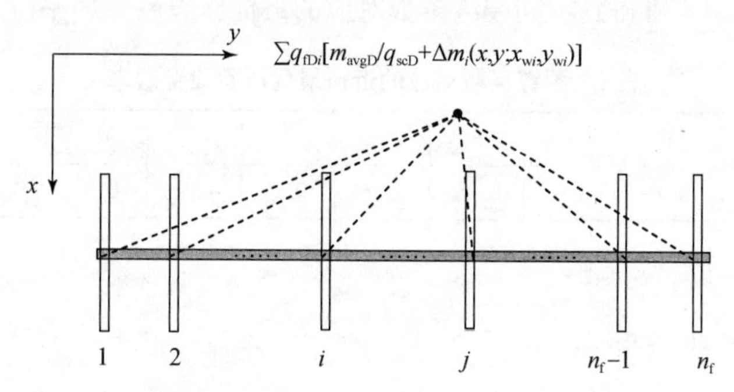

图 2　压裂水平井压降叠加示意图(俯视面)

$$\mathrm{PI}_i = \frac{q_{fDj}}{m_{wD} - m_{avgD}} \tag{38}$$

$$\Delta m_{Dij} = \frac{2\pi y_{eD}}{x_{eD}}\left(\frac{1}{3} - \frac{|y_{wDi} \pm y_{wDj}|}{2y_{eD}} + \frac{y_D^2 + y_{wDj}^2}{2y_{eD}^2}\right) + L^{-1}\left[\frac{\mathrm{HV}_{pss}(x_{wi},y_{wi};x_{wj},y_{wj})}{s}\right] + \gamma_{pss} \tag{39}$$

$$\gamma_{pss} = \begin{cases} S_f(C_{fD}) + S_c(C_{fD},x_{fD}), i = j \\ 0, i \neq j \end{cases} \tag{40}$$

式(40)表明裂缝 i 的有限导流因子和聚流表皮影响只体现在流体从裂缝 i 流入井筒过程中形成的附加压降,对其他裂缝的流动过程不产生影响[18,26]。式(37)可以借助 Newton 迭代算法数值求解。从式(37)可知,裂缝的无量纲导流能力 C_{fD}、条数 n_f、半长 x_f、相对位置(x_{wD},y_{wD})都会影响气井产能;压裂参数与给定地层间存在着最佳匹配关系,优化这种关系能够降低缝间干扰,减小地层封闭影响,增大裂缝与地层接触面积,平衡裂缝流入流出动态,提高气井产能。为方便讨论,文中以后涉及的压裂水平井各裂缝参数均一致,即裂缝长度、导流能力、裂缝间距等相等。

2　气井产能优化

2.1　优化裂缝布局

裂缝布局主要包括裂缝条数(n_f)、裂缝穿透率($I_x=2x_f/x_e$)、压裂段穿透率($I_y=L_f/y_e$,L_f 为最外侧两条裂缝间距),这些压裂参数决定着裂缝系统与地层的接触面积、与封闭边界的相互作用以及裂缝间的相互干扰。图 3、图 4 反映了气井产能($\mathrm{PI}=\sum \mathrm{PI}_i$)随裂缝条数、压裂段穿透率的变化规律,可从两种角度进行分析。

(1)固定压裂段穿透率 I_y:增加裂缝条数增大了压裂水平井与地层接触面积,同时也加剧了裂缝间相互干扰,但整体上减小了渗流阻力,提高了气井产能。

(2)固定裂缝条数 n_f:当 $n_f \geqslant 14$ 时裂缝条数较多,有效增加了压裂水平井与地层的接触面积,“掩盖”了裂缝间相互干扰的影响。在此基础上增大压裂段穿透率,可以有效缓解裂缝间相互干扰,进一步减小渗流阻力,所以压裂段穿透率越大对应的气井产能越高;当 $n_f \leqslant 13$ 时裂缝条数较少,接触面积增大产生的“正”影响不足以完全弥补裂缝间相互干扰产生的“负”影响,此时压裂参数间存在最佳匹配问题。

定量分析图3、图4可得到不同裂缝条数对应的最优压裂段穿透率(I_{yopt}),结果见表1。

表1 裂缝条数对应的近似最优压裂段穿透率

n_f	2	3	4	5	6	7	8	9	10	11	12	13
I_{yopt}	0.5	0.65	0.75	0.8	0.85	0.85	0.9	0.9	0.9	0.9	0.9	0.9

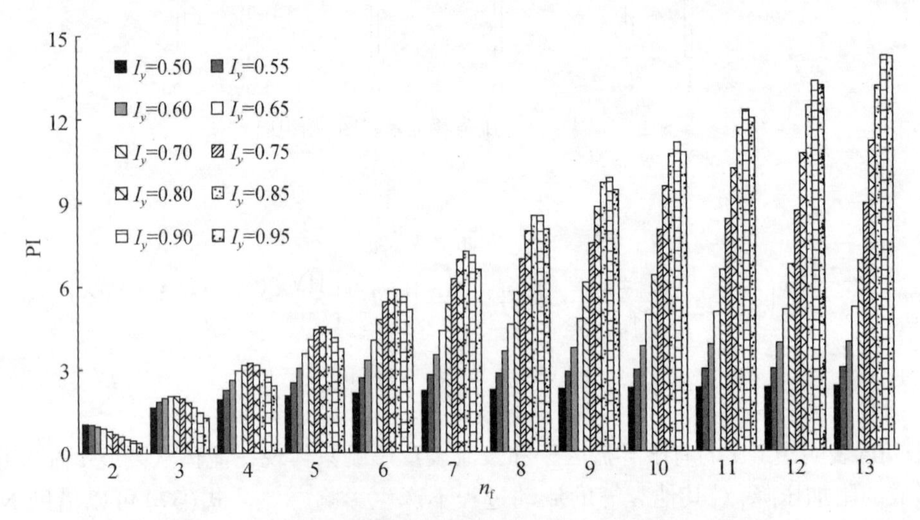

图3 气井产能随压裂段穿透率及裂缝数变化规律($n_f \leqslant 13$)

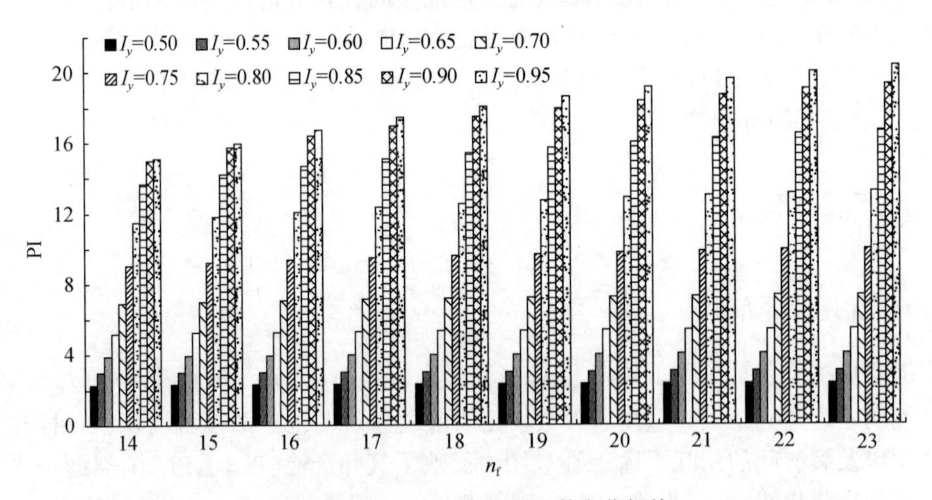

图4 气井产能随压裂段穿透率及裂缝数变化规律($n_f \geqslant 14$)

为进一步分析,计算不同裂缝条数对应产能随压裂段穿透率的变化规律(图5)。图5反映不同裂缝条数对应不同最优压裂段穿透率,拟合裂缝条数、最优压裂段穿透率离散点可获得二者间的近似关系式:

$$I_{yopt} = -6 \times 10^{-5} n_f^4 + 0.0025 n_f^3 - 0.0365 n_f^2 + 0.2547 n_f + 0.1708 \tag{41}$$

经过渐进分析可知,式(41)满足近似关系式:

$$I_{yopt} = 1 - 1/n_f \tag{42}$$

式(42)有明确物理意义:在缝间干扰与边界封闭作用的共同影响下,相邻裂缝间形成分流线,将裂缝系统间隔成一系列具有不同泄流面积的单缝[29]。当每条裂缝对应的泄流面积相等时,裂缝间干扰最小、边界封闭影响最低、单裂缝产能相同、气井产能最大。$I_y<I_{yopt}$时外侧裂缝对应泄流面积较大、单缝产能较高,$I_y>I_{yopt}$情况则与之相反(图5)。

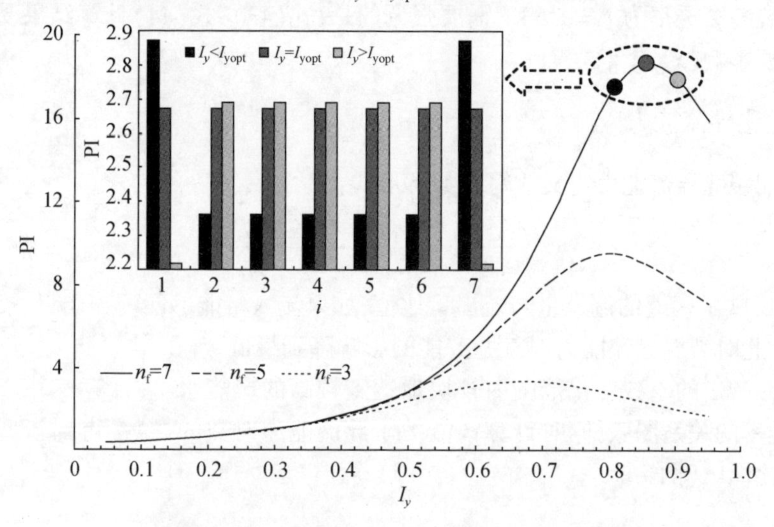

图5　气井产能随压裂段穿透率的变化规律

以苏里格压裂水平井为例说明确定最优裂缝条数的流程。目前典型的苏里格气藏南北向水平井段长1000m,东西向排距600m(即$x_{eD}/y_{eD}=0.3$),按均匀布缝原理[式(42)]沿水平井筒进行压裂。计算不同裂缝穿透率的气井产能关于裂缝条数的导数值(图6)。其中导数最大值对应的裂缝条数为最优裂缝条数n_{fopt},气井能够在此范围内较为显著地达到较高的产能。同样通过数据拟合可获得裂缝最优条数n_{fopt}与I_x的函数关系式:

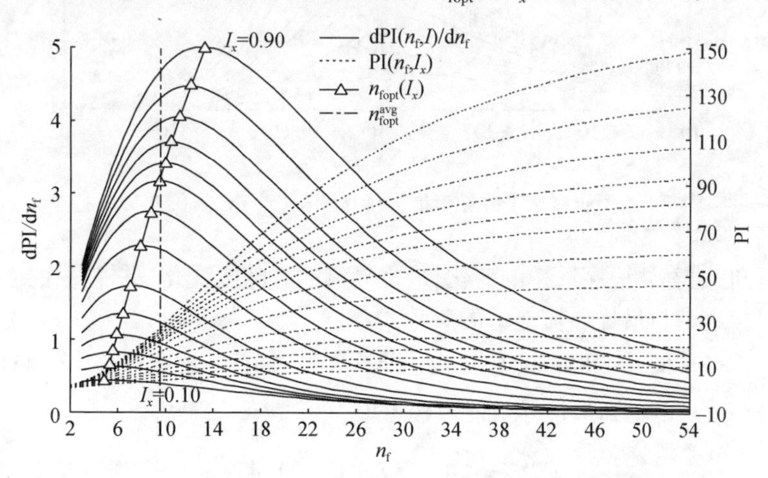

图6　裂缝均匀分布下的气井产能优化图版($x_{eD}/y_{eD}=0.3$)

当$I_x<0.64$时,

$$n_{fopt}=-511.99I_x^5+970.24I_x^4-648.77I_x^3+183.05I_x^2-16.438I_x+5.3403 \quad (43a)$$

当$I_x\geqslant 0.64$时,

$$n_{\mathrm{fopt}} = 76512I_x^5 - 300071I_x^4 + 469183I_x^3 - 365481I_x^2 + 141818I_x - 21921 \tag{43b}$$

对式[43(a)、(b)]进行积分平均可得到最优裂缝条数参考值：

$$n_{\mathrm{fopt}}^{\mathrm{avg}} = \int_0^1 n_{\mathrm{fopt}}(I_x)\,\mathrm{d}I_x \approx 8 \tag{44}$$

对应的裂缝穿透率为 $I_x=0.53$。需要强调的是，式(43)、式(44)计算结果受气藏泄流面积的几何规模影响($x_{\mathrm{eD}}/y_{\mathrm{eD}}$)。

2.2　优化导流能力

将无量纲裂缝导流能力定义改写为如下形式：

$$C_{\mathrm{fD}} = \frac{k_{\mathrm{f}}w_{\mathrm{f}}}{k_{\mathrm{m}}x_{\mathrm{f}}} = \frac{w_{\mathrm{f}}h(k_{\mathrm{f}}/\mu_{\mathrm{g}})}{x_{\mathrm{f}}h(k_{\mathrm{m}}/\mu_{\mathrm{g}})} = \frac{q_{\mathrm{inf}}}{q_{\mathrm{outf}}} \tag{45}$$

式(45)反映了裂缝的流入量与流出量比值，如果流入量能够匹配流出量，裂缝将达到最佳导流状态，此时裂缝导流能力对气井产能的影响降到最低。

为了能在一定的参数变化范围内较快地达到较高的产能水平，计算气井产能与裂缝导流能力、穿透率的关系图版，同时计算产能关于导流能力对数的导数，得到新型分段压裂水平井产能优化图版(图7)。

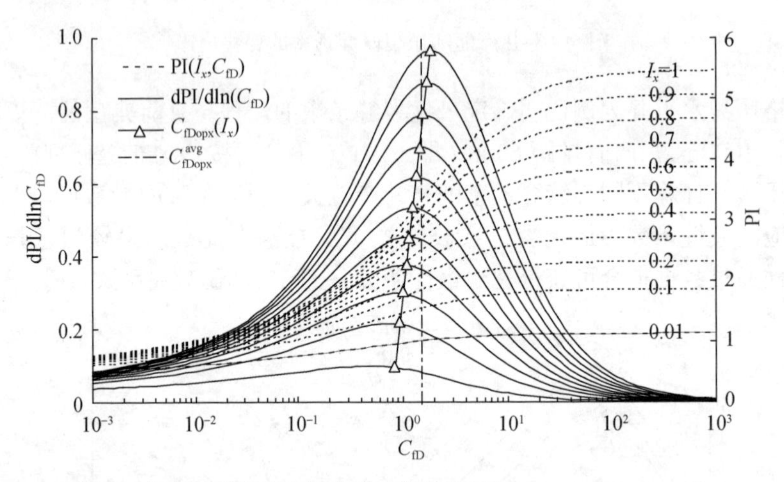

图7　新型压裂水平井导流能力优化图版($x_{\mathrm{eD}}/y_{\mathrm{eD}}=0.3$，$n_{\mathrm{f}}=8$)

由图7可知，随着裂缝穿透率的变化，最优导流能力随之变化，并非一个定值，通过回归可以得到裂缝穿透率与最优导流能力的关系式：

$$C_{\mathrm{fDopt}} = 36.92I_x^5 - 88.303I_x^4 + 72.824I_x^3 - 23.86I_x^2 + 3.9491I_x + 0.4519 \tag{46}$$

对 C_{fDopt} 取积分平均，可得到最优导流能力参考值：

$$C_{\mathrm{fDopt}}^{\mathrm{avg}} = \int_0^1 C_{\mathrm{ftopt}}(I_x)\,\mathrm{d}I_x = 1.472 \tag{47}$$

式(47)计算值与 Economides 等[30]给出的最佳无量纲导流能力值1.6相近，验证了本文算法的合理性。需要强调的是，在裂缝布局最优的情况下，压裂水平井可等效为一系列的单裂缝问题，对应单缝泄流面积的几何形状为 $x_{\mathrm{e}}\times(y_{\mathrm{e}}/n_{\mathrm{f}})$，此时裂缝导流能力的优化函数关系[式(46)]随地层长度与宽度比例 $y_{\mathrm{e}}/x_{\mathrm{e}}$、裂缝条数 n_{f} 的变化而发生改变。

综上所述,合理使用式(42)、式(43)、式(44)、式(46)、式(47)的步骤如下:

(1)根据气井的有效控制面积确定气藏矩形长宽比。

(2)利用式(42)中压裂段穿透率与裂缝条数的最优关系,获得裂缝穿透率与裂缝条数的优化关系线[式(43),图5],进而利用式(44)得到最优裂缝条数参考值和对应的裂缝穿透率。

(3)根据特定的矩形长宽比和裂缝条数参考值确定裂缝穿透率与导流能力的优化关系线[式(46),图6],同时利用式(47)得到最优导流能力参考值。

(4)实际应用时,可以选取最优参数关系线附近的一个小条带区域作为优化压裂参数的参考区域。在最优参数的作用下气井能够较为显著地达到较高的产能水平。

3 结论

(1)利用积分变换、渐进分析、压降叠加、附加导流因子等方法能有效地建立起考虑边界封闭影响的有限导流压裂水平井拟稳态产能计算模型,其中裂缝条数、长度、导流能力、压裂段长度以及地层几何规模、裂缝相对分布均会对气井产能产生影响。

(2)提高气井产能主要通过减小缝间干扰、降低边界封闭作用、增大裂缝与地层接触面积、平衡裂缝与地层的流入流出关系实现,当裂缝系统中各裂缝对应的泄流面积相同时,缝间干扰最小、边界封闭影响最低。

(3)对于实际生产而言,气井能在一定压裂参数变化范围内较快地达到较高的产能水平更具现实意义。通过求解产能关于压裂参数的导数极大值得到最优压裂参数分布线,利用积分平均的方法获得压裂参数的优化参考值。

参考文献

[1] 郝明强,王晓冬,胡永乐.压敏性特低渗透油藏压裂水平井产能计算[J].中国石油大学学报(自然科学版),2011,35(6):99-104.

[2] 王本成,贾永禄,李友全,等.多段压裂水平井试井模型求解新方法[J].石油学报,2013,34(6):1150-1156.

[3] 周明,孙树栋.遗传算法原理及应用[M].北京:国防工业出版社,1999.

[4] 樊冬艳,姚军,姚婷,等.基于自适应遗传算法的压裂水平井参数优化[J].油气地质与采收率,2011,18(5):85-89.

[5] Ranghvan R S, Chen C C, Agaewal B. An analysis of horizontal wells intercepted by multiple fractures[J]. SPE Journal, 1997, 2: 235-245.

[6] Chen C C, Raghavan R S. A multiply fractured horizontal well in a rectangular drainage region[J]. SPE Journal, 1997, 2: 455-465.

[7] 郝明强,胡永乐,李凡华.特低渗透油藏压裂水平井产量递减规律[J].石油学报,2012,33(2):269-273.

[8] 姚军,刘丕养,吴明禄.裂缝性尤其藏压裂水平井试井分析[J].中国石油大学学报(自然科学版),2013,37(5):107-114.

[9] 宁正福,韩树刚,程林松,等.低渗透油气藏压裂水平井产能计算方法[J].石油学报,2002,23(2):68-72.

[10] Wang X D, Li G H, Wang F. Productivity analysis of horizontal wells intercepted by multiple finite-con-

ductivity fractures[J]. Petroleum Science, 2010, 7: 367-371.

[11] 王晓冬, 张义堂, 刘慈群. 垂直裂缝井产能及导流能力优化研究[J]. 石油勘探与开发, 2004, 31(5): 78-81.

[12] 罗万静, 王晓冬, 李凡华. 分段射孔水平井产能计算[J]. 石油勘探与开发, 2009, 36(1): 97-102.

[13] 曾凡辉, 郭建春, 何颂根, 等. 致密砂岩气藏压裂水平井裂缝参数的优化[J]. 天然气工业, 2012, 32(11): 54-60.

[14] 曲占庆, 曲冠政, 何利敏, 等. 压裂水平井裂缝分布对产能影响的电模拟实验[J]. 天然气工业, 2013, 33(10): 1-7.

[15] 柳毓松, 廉培庆, 同登科, 等. 利用遗传算法进行水平井水平段长度优化设计[J]. 石油学报, 2008, 29(2): 296-299.

[16] 刘珊, 同登科. 水平井分段采油优化模型[J]. 计算力学学报, 2010, 27(2): 342-246.

[17] Ozkan E, Raghavan R. New solutions for well-test analysis problems: part1: Analytical consideration [R]. SPE 18303, 1991.

[18] 王晓冬, 罗万静, 侯晓春, 等. 矩形油藏多段压裂水平井不稳态压力分析[J]. 石油勘探与开发, 2014, 41(1): 74-78.

[19] Cinco-Ley H, Samaniego V F, Dominguez A N. Transient pressure behavior for a well with a finite conductivity vertical fractures[C]. Paper SPE 6014 presented at the 51th Annual Fall Technical Conference and Exhibition, New Orleans, USA, 3-6 October 1976.

[20] Al-Kobaisi M, Ozkan E, Kazemi H. A hybrid numerical/analytical model of finite-conductivity vertical fracture intercepted by a horizontal well[C]. SPE 92040, 2006.

[21] Blasingame T A, Poe B D. Semianalytic solutions for a well with a single finite-conductivity vertical fracture[C]. Paper SPE 26424 presented at the 68th Annual Technical Conference and Exhibition, Houston, Texas, USA, 3-6 October 1993.

[22] Brown M, Ozkan E, Raghavan R, *et al.* Practical solutions for pressure-transient responses of fractured horizontal wells in unconventional shale reservoirs[C]. Paper SPE 125043 presented at the SPE Annual Technical Conference and Exhibition, New Orleans, USA, 4-7 October 2009.

[23] Wilkinson D J. New results for pressure transient behavior of hydraulically fractured wells[C]. Paper SPE 18950 presented at the SPE Joint Rocky Mountain Regional Exhibition, Denver, Colorado, USA, 6-8 May 1989.

[24] Van D M. Perturbation methods in fluid mechanics[M]. California: Parabolic Press, 1975.

[25] Gringarden A C, Ramey H J, Raghavan R. Unsteady-state pressure distributions created by a well with a single infinite-conductivity vertical fracture[C]. Paper SPE 4051 presented at the SPE-AIME 47th Annual Fall Meeting, San Antonio, Texas, USA, 8-11 October 1972.

[26] Zerzar A, Bttam Y. Interpretation of multiple hydraulically fractured horizontal wells in closed systems [C]. Paper SPE 84888 presented at the SPE International Improved Oil Recovery Conference in Asia Pacific, Kuala Lumpur, Malaysia, 20-21October 2003.

[27] 孔祥言. 高等渗流力学 [M]. 安徽: 中国科学技术大学出版社, 2010.

[28] Wang L, Wang X D, Ding X M, *et al.* Rate decline curves analysis of a vertical fractured well with fracture face damage[J]. Journal of Energy Resource Technology, 2012, 134: 1-9.

[29] 王军磊, 贾爱林, 何东博, 等. 致密气藏分段压裂水平井产量递减规律及影响因素[J]. 天然气地球科学, 2014, 25(2): 278-285.

[30] Economides M, Oligney R, Valko P. Unified fracture design: Bridge the gap between theory and design [M]. Texas: Orsa Press, 2002.

页岩气井压后返排规律研究

刘乃震[1]　柳　明[1,2]　张士诚[2]

(1. 中国石油集团长城钻探工程有限公司;2. 中国石油大学(北京)石油工程教育部重点实验室)

摘要:页岩气藏通常需要进行大规模的水力压裂才具有工业开采价值,但是页岩气井压后返排率普遍较低。针对这一问题,采用数值模拟和实验相结合的方法,研究了天然裂缝间距、裂缝导流能力、压裂规模、压力系数和关井时间等因素对返排的影响,并从机理上分析了页岩气井压后返排困难的原因。结果表明,返排率随天然裂缝间距、裂缝导流能力和压力系数的增加而增加,随压裂规模和关井时间的增加而减少。从微观机理上分析,水通过毛细管自吸作用进入微裂纹,页岩基质中矿物颗粒间原有的氢键被羟基取代进而发生水化作用,造成新的微裂纹的产生和主裂缝的扩展,形成复杂的裂缝网络,使大部分水难以返排,返排率低;对于页岩气井压裂,一般裂缝间距和裂缝导流能力较小、压裂规模大,很大一部分注入水存在于比表面积极大、形态极为复杂的裂缝网络中,以致无法返排。研究表明,页岩气井压后返排率的高低受多种因素的影响,不应该刻意追求返排率;地返排率的页岩气井的产量一般较高。

关键词:页岩气井;返排;数值模拟;裂缝间距;压裂规模;压力系数;自吸水化

页岩储层具有脆性高、渗透率低、天然微裂缝发育等特点[1,2],通过"高排量、高液量、高砂量、低黏度、低砂比"的大规模水力压裂可以形成复杂的裂缝网络,增大裂缝面与页岩基质的接触面积,从而实现页岩气的工业开采。美国 Barnett 页岩的开发实践证明,清水压裂不但能大量减少施工费用,而且有利于提高页岩气的最终采收率[3]。然而,与常规低渗气藏不同的是,页岩气藏压后返排率比较低(一般为 10% ~40%)[4],大量残留于地层中的水对基质和裂缝系统中气体的运移必然产生不容忽视的影响。因此,认识水在页岩储层中的分布对掌握返排规律具有重要意义。

1　模型的建立

由于页岩气的渗流规律极为复杂,目前对于页岩气的数值模拟主要基于煤层气渗流理论,即采用基质-裂缝双重介质模型,同时考虑气体的解吸附、扩散和应力敏感等因素,模型基本参数见表1。

表1　模型的基本参数

参数	数值	参数	数值
气藏体积/(m×m×m)	1400×1000×50	压裂规模/m^3	10000 ~ 25000
网格数量	383×205×10	朗缪尔压力/MPa	4.5
基质渗透率/(10^{-3} μm^2)	0.0001	朗缪尔体积/(m^3/t)	2.3
基质孔隙度/%	5	水平段长/m	1000
主裂缝导流能力/(μm^2 · cm)	2 ~ 5	裂缝半长/m	200
次裂缝导流能力/(μm^2 · cm)	0.5	簇间距/m	40

续表

参数	数值	参数	数值
原始地层压力/MPa	30	次裂缝间距/m	20,40,80
压裂级数/簇数	12 级/25 簇	地层初始含水饱和度/%	0.25

(1)裂缝网络的处理。页岩气藏压后形成的裂缝网络极为复杂,由于计算机内存和求解的收敛性限制,模拟中必须进行简化。模型中分为主裂缝和次裂缝(天然裂缝),根据等效导流能力理论进行主、次裂缝的尺寸和属性设置,以保证裂缝对气液两相的运移能力不变,主裂缝和次裂缝分别垂直和平行于水平井筒方向(图 1)。

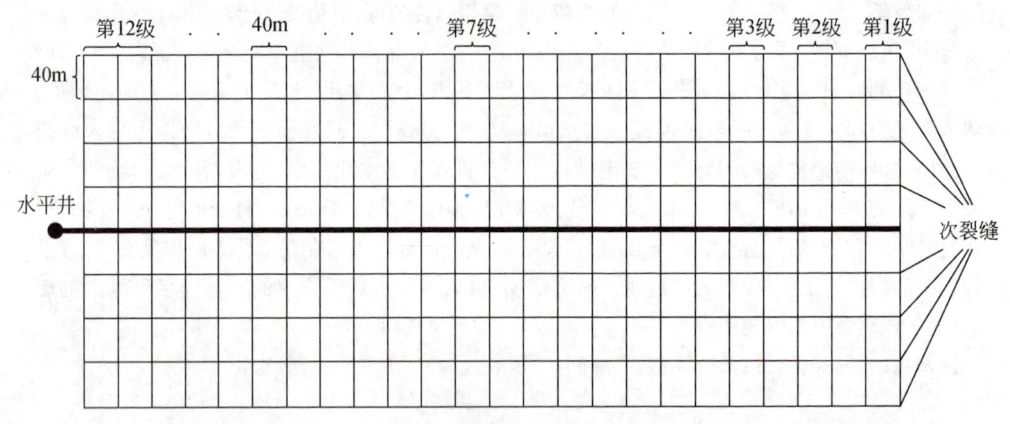

图 1　裂缝网络示意图

(2)渗流方式的处理。由于页岩基质的渗透率极低,认为气体在基质中不发生流动,只随压力和气体分子浓度的变化发生吸附、解吸和扩散,解吸出的气体流入裂缝遵循达西定律,并最终流入井底。

(3)模型的初始化。采用注水井注水的方法,根据压裂规模对模型的初始含水饱和度和地层压力进行预处理,从而改变地层的含水饱和度和压力(图 2)。

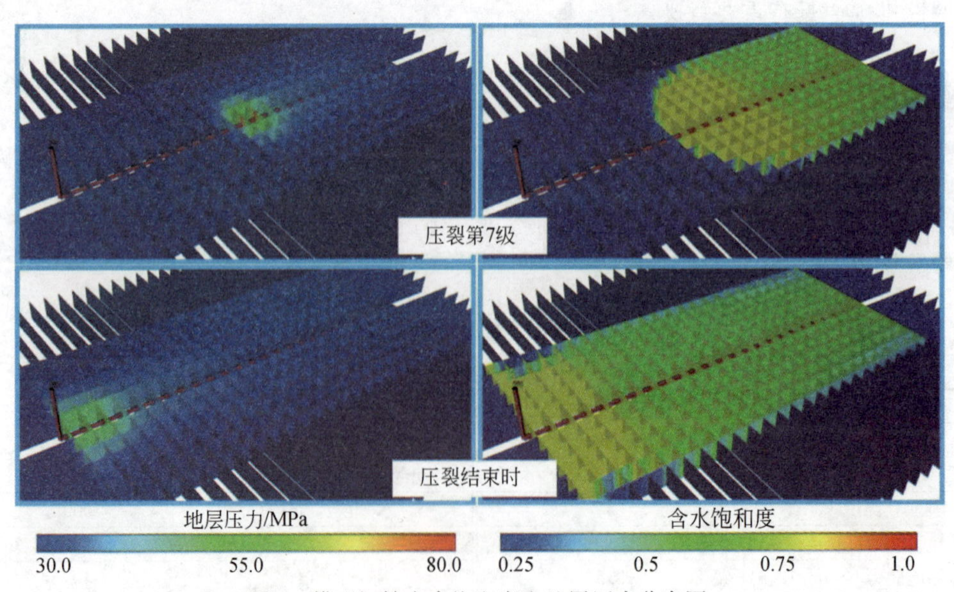

图 2　模型初始含水饱和度和地层压力分布图

2 返排规律影响因素分析

页岩气压后返排与产能通常呈现出反比的关系。压裂过程中，一部分压裂液（滑溜水）存在于压开的主裂缝和次裂缝中，另一部分滤失进入页岩基质中；压裂结束后，在压差的驱动下，滑溜水从裂缝流入井筒，在这个过程中，裂缝间距及导流能力、压裂规模、压力系数和关井时间会对返排产生较大的影响。

2.1 导流能力

导流能力比定义为主裂缝与次裂缝的导流能力之比，模拟中次裂缝的导流能力为 $0.5\mu m^2 \cdot cm$ 保持不变。图 3 所示为不同次裂缝间距时五年返排率随导流能力比的变化。可以看出，随导流能力比和间距的增加，返排率增加。由于次裂缝的导流能力为恒定，导流能力比增加意味着主裂缝的导流能力增加，存在于其中的滑溜水更易返排；次裂缝间距越大意味着天然裂缝密度越小，存在于其中，以及通过次裂缝滤失进地层的滑溜水越少，大量滑溜水存在于主裂缝中，因此更易于返排。可以推断，页岩储层脆性越高、天然裂缝越发育，即压后形成的裂缝网络越发达，返排率一般越低，但产量一般较高。

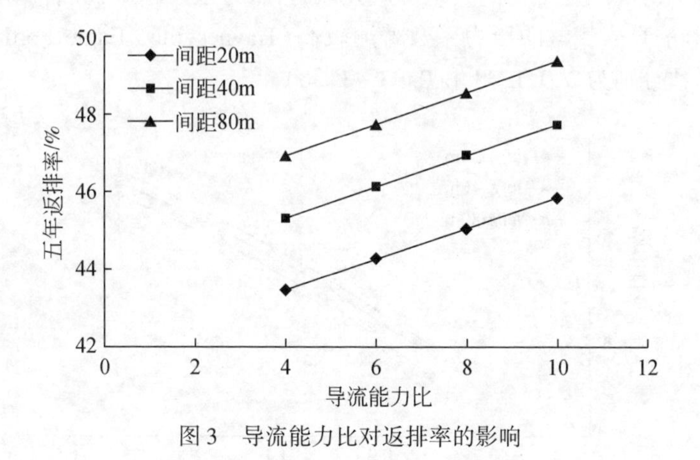

图 3 导流能力比对返排率的影响

2.2 压裂规模

统计国内外页岩气井压裂规模现状[4-7]，发现单簇规模多在 400 ~ 1000m^3，因此对于 12 级 25 簇的水平井来说，压裂规模在 10000 ~ 25000m^3。从图 4 可以看出，压裂规模越大，即单簇吸水量越大，返排率越低。这是因为，单簇吸水量越大，存在于次裂缝和滤失入基质的滑溜水越多，而次裂缝和基质的流动能力较弱，因此较难返排。若保持单簇吸水量不变，模拟不同的压裂规模对返排率的影响，发现其规律与图 4 相似。这主要是因为压裂规模越大，压裂级数和簇数越多，裂缝网络越发达，返排率也越低，产量一般越高。

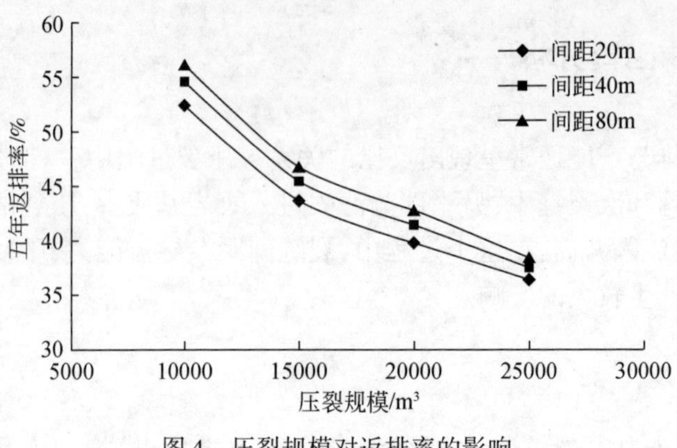

图 4　压裂规模对返排率的影响

2.3　压力系数

压力系数是一个对返排率和产能均非常敏感的因素。相同条件下,压力系数越高,返排时基质与裂缝之间的压差越大,地层提供的返排能量越大,返排率越高(图 5)。生产时,压力系数越高,一方面吸附气含量越多,另一方面生产压差越大,产能越好。可以看出,压力系数表现出的返排率与产能是正比的关系。研究认为,压力系数越高,页岩储层的开发效果越好,美国产量较高的页岩气藏的压力系数普遍较高(Haynesville, Eagle Ford 和 Marcellus 页岩气藏的压力系数分别为 2.0,1.33 和 0.93 ~ 1.56)[8,9]。

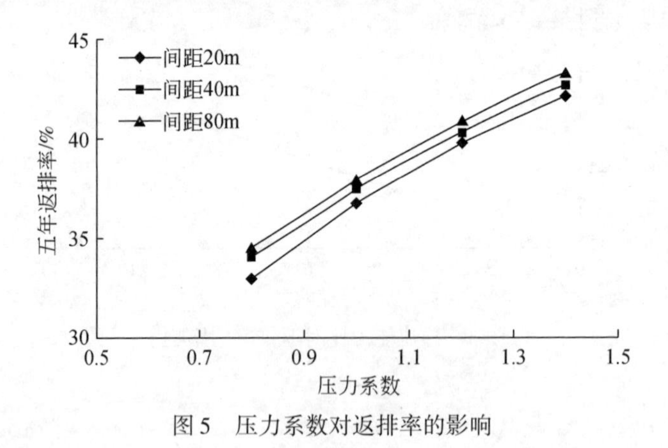

图 5　压力系数对返排率的影响

2.4　关井时间

页岩气井压后一般经历返排、关井和投产三个阶段。返排后的关井期(10 ~ 30 天)主要用于输气管道铺设,这段时间的长短对后期投产也有一定的影响。从图 6 可见,关井时间越长,相同次裂缝间距条件下的返排率越低,页岩气井的产量也越低。关井期间,主裂缝中的滑溜水进一步向次裂缝和基质中滤失,关井时间越长,滤失量越大,造成返排率越低。因此,对于可工业开采的页岩气藏,宜尽量缩短返排后的关井时间,条件允许的情况下,应提前完成管道铺设工作,不关井直接投产。

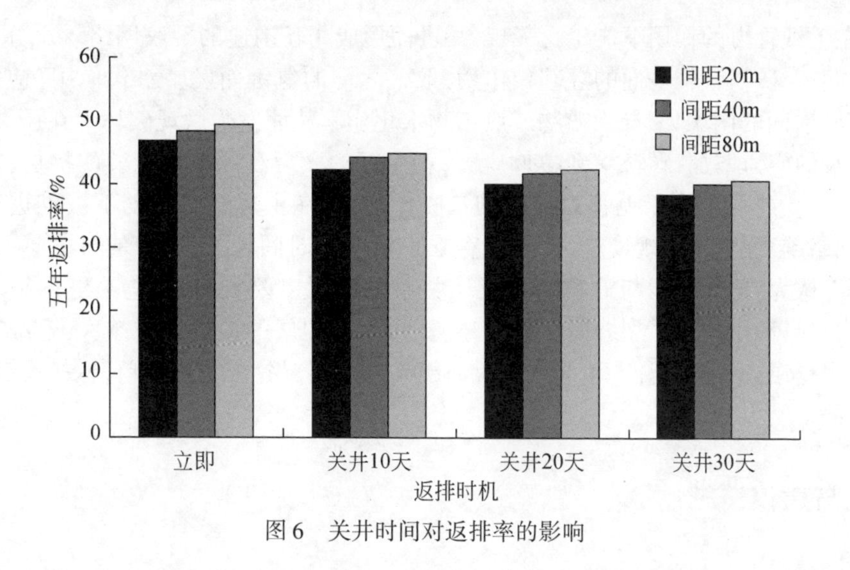

图6　关井时间对返排率的影响

3　页岩自吸水化机理

页岩的自吸水特征可以作为解释页岩压后返排率低的机理之一。页岩气藏压裂所用压裂液中含水一般都大于90%[10],因此水在页岩中的存在方式与返排率的高低密切相关。将3个页岩露头样本分别置于盛有5mm水的容器中,24h后观察到如图7所示的毛细管自吸现象。

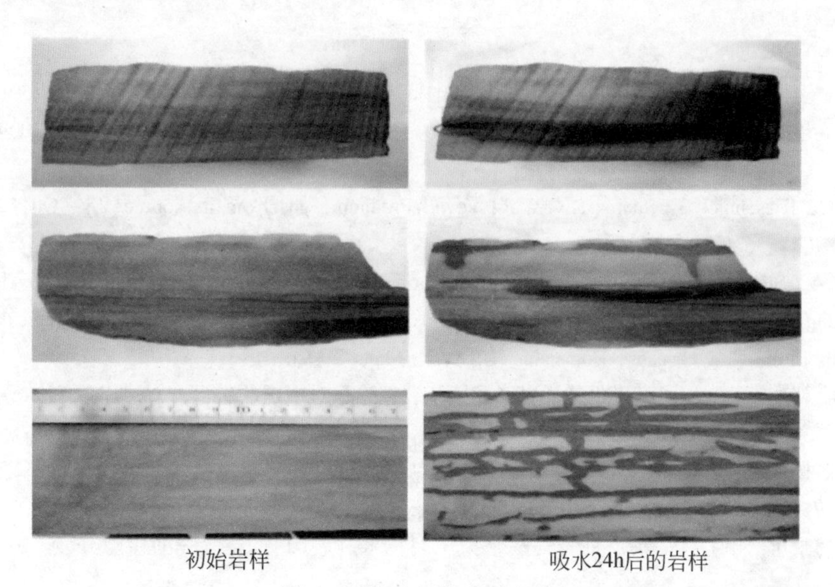

图7　页岩毛细管自吸现象

脆硬性页岩是矿物颗粒相互胶结在一起的集合体,矿物颗粒间或颗粒与胶结物间存在弱结合点或面,水化作用导致这些地方成为初始微观损伤点或面。页岩在成岩作用下因脱水形成了大量的微裂纹,进入地层的水首先进入较大的微裂纹,之后进入与之相连通的细小微裂纹,在毛细管力的作用下产生自吸现象,进而形成相互沟通的裂纹网络。水分子吸附于矿物颗粒间,与矿物颗粒表面的羟基相作用,取代之前的氢键而发生水化作用(应力腐蚀作用),削弱颗粒间的内聚力,导致矿物或胶结物水化崩解从而发生微观破坏。在压裂过程中,微裂纹尖端的应力增加,颗粒受到拉伸应力,产生新的微裂纹和较大裂缝的扩展,从而产生宏观破坏[11,12]。因此,由于毛管力控制的自吸作用,水在主裂缝的带动下被推进至远井区域,裂缝网络越复杂、压裂规模越大、水残存在地层中的时间越长,返排的难度越大,返排率很低。可以认为,单井产能与返排率呈反比关系的原因主要有两点:①裂缝网络的复杂程度;②页岩自吸水化作用。裂缝网络越复杂,注入水与地层的接触面积越大,毛细管力产生的自吸作用越强,返排率越低,单井产能反而越高;反之,裂缝网络越简单,返排率越高,单井产能却越低。

4　结论与建议

(1)返排率与裂缝导流能力、裂缝间距和压力系数呈正比关系,与压裂规模和关井时间呈反比关系。

(2)水在微裂纹中的毛细管自吸现象导致基质颗粒间内聚力减小,进而在压裂过程中产生汇合贯通的复杂裂缝网络,使大部分水难以返排,返排率低。

(3)页岩气井压后返排率的高低受多种因素的影响,但不应刻意强调返排率。压裂液在页岩中发生的自吸水化作用能很大程度地增加裂缝网络的复杂程度、提高储层与井筒的接触面积,使大量的吸附气解吸,因此低返排率的页岩气井的产量一般较高。

参考文献

[1] 蒋裕强, 董大忠, 漆麟, 等. 页岩气储层的基本特征及其评价[J]. 天然气工业, 2010, 30(10): 7-12.

[2] 张金川, 姜生玲, 唐玄, 等. 我国页岩气富集类型及资源特点[J]. 天然气工业, 2009, 29(12): 109-114.

[3] Boyer C, Kieschnick J, Suarez-Rivera R, *et al.* Producing gas from its source[J]. Oil field Review, 2006, 18(3):36-49.

[4] Ahmad A, Steven M, Robert A W. Estimation of effective fracture volume using water flowback and production data for shale gas wells[C]. Paper SPE166279 presented at SPE Annual Technical Conference and Exhibition, New Orleans, Louisiana, USA, 30 Sept. -2 Oct. 2013.

[5] 曾凡辉, 郭建春, 刘恒, 等. 北美页岩气高效压裂经验及对中国的启示[J]. 西南石油大学学报(自然科学版), 2013, 35(6): 90-98.

[6] 曾雨辰, 杨保军, 王凌冰. 涪页 HF-1 井泵送易钻桥塞分段大型压裂技术[J]. 石油钻采工艺, 2012, 34(5): 75-79.

[7] 叶静, 胡永全, 叶生林, 等. 页岩气藏水力压裂技术进展[J]. 天然气勘探与开发, 2012, 35(4): 64-68.

[8] Robert G L, Robert M R, Stephen C R, *et al.* Morphology, genesis, and distribution of nanometer-scale pores in siliceous mudstones of the Mississippian Barnett shale [J]. Journal of Sedimentary Research, 2009, 79: 848-861.

[9] 黄金亮, 邹才能, 李建忠, 等. 川南志留系龙马溪组页岩气形成条件与有利区分析[J]. 煤炭学报, 2012, 37(5): 782-787.

[10] 高树生，胡志明，郭为，等．页岩储层吸水特征与返排能力[J]．天然气工业，2013，33(12)：71-76.
[11] 李果，张诗通.用石英玻璃模拟低渗砂岩毛细管自吸效应的研究.重庆科技学院学报(自然科学版)，2010，12(3)：1-3.
[12] 石秉忠，夏柏如，林永学，等.硬脆性泥页岩水化裂缝发展的CT成像与机理[J].石油学报，2012，33(1)：137-142.

基于施工曲线的页岩气井压后评估新方法

卞晓冰　蒋廷学　贾长贵　王海涛　李双明　苏　瑗　卫　然
（中国石油化工股份有限公司石油工程技术研究院）

摘要：目前页岩气井压后评估手段较少，如何基于有限资料进行压后评估分析对于压裂方案的持续改进具有重要意义。为了再认识储层及人工裂缝参数，把压裂施工曲线分为前置液注入和主压裂施工两个阶段：①通过统计前置液注入阶段的地层破裂次数、平均压力降幅和平均压力降速，定性判断地层脆塑性；根据压裂施工中的能量区域可定量化计算综合脆性指数。②真三轴大型物理模拟实验结果显示，主压裂施工阶段曲线压力波动频率和幅度反映了裂缝的复杂程度，结合地层脆塑性可综合诊断远井裂缝形态。以渝东南某页岩气 P 井为例进行压后评估分析，结果表明，水平井筒延伸方向的页岩地层非均质性较强，以自然伽马值 260API 作为该区地层脆塑性的界限，偏脆性地层更易形成复杂裂缝系统。P 井有一半裂缝为单一缝，为了进一步改善开发效果，应进一步采取精细分段、转向压裂等措施。

关键词：页岩气；施工曲线；破裂压力；压后评估；页岩脆性；露头；裂缝复杂性

页岩气广泛分布于页岩烃源岩地层中，是一种重要的非常规清洁能源，页岩基质孔渗极低，且富含吸附气，具有开采寿命长和生产周期长的特点[1-4]。渝东南某页岩气区块单井钻完井及压裂开发成本高达 7000 万～9000 万元，稳定产量为 1×10^4～$2\times10^4 m^3/d$。为了提高经济效益，许多辅助性措施如常规测井、裂缝形态监测及产气剖面测试等做得较少[5-7]。基于成熟软件模型的压后评估方法具有多解性，且需提供模型可解释的数据，如基于停泵压降曲线的 G 函数分析方法需要长时间的停泵后压力测试数据，若现场测压时间较短则无法进行解释，因此压后评估结果的可信度与评估者的经验直接相关[8-10]。如何基于有限的资料对页岩地层进行更为准确的认识，是一直困扰广大科技工作者的难题。本文提出了基于压裂施工曲线反演储层参数与定性评估压后裂缝形态的新方法，通过对已施工井资料进行充分挖掘，进行储层及裂缝参数再认识，并以渝东南某页岩气井 P 井为例进行压后评价，为进一步改进该区压裂井设计、提高压裂有效改造体积提供理论依据。

1　已施工井概况

P 井位于渝东南某页岩气区块，完钻斜深 4190m，水平段长 1260m，目的层为龙马溪组。该井总共压裂 22 段，2～3 簇/段，最高施工排量保持在 13～14.5m^3/min，平均施工压力 60～70MPa，施工总液量 46542m^3，总砂量 2108m^3。

按照地层瞬时停泵压力梯度（ISIP）及地层是否渗漏（前 6 段钻遇漏失层），将施工压力曲线分为 4 种类型：①类型 1。第 1～2 段，渗漏地层，ISIP＝0.021～0.023MPa/m。②类型 2。第 3～7 段，渗漏地层，ISIP＝0.016～0.017MPa/m。③类型 3。第 8～13 段，非渗漏地层，ISIP＝0.018～0.022MPa/m。④类型 4。第 14～22 段，非渗漏地层，ISIP＝0.023～0.026MPa/m。如图 1 所示，P 井在趾端前 100m 附近地应力较高，之后到跟端地应力逐渐增加。

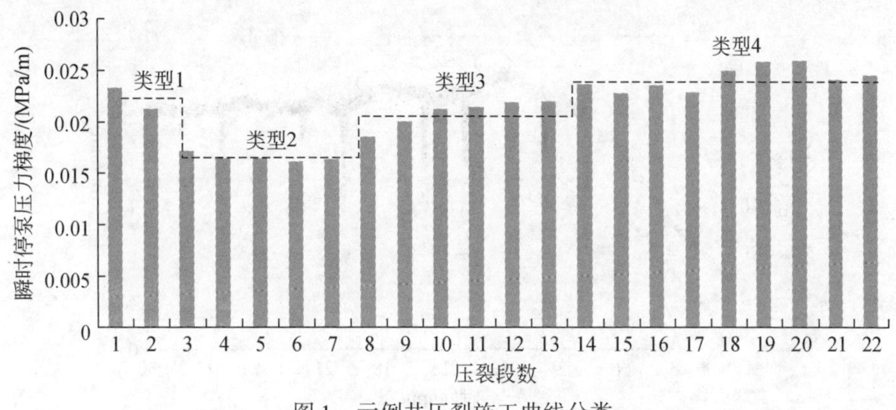

图1 示例井压裂施工曲线分类

2 地层脆塑性识别

在页岩气井压裂施工过程中,随着前置液逐渐泵入地层,裂缝不断起裂并随之延伸扩展。基于P井破裂压力点压力降幅和降速,进行页岩地层脆塑性识别。

2.1 地层破裂特征表征

选取类型2第5段压裂施工曲线(图2),对矩形红色虚线圈闭的前置液阶段进行地层破裂压力特征分析。该段为低地应力渗漏层,在升排量过程中及较大排量下地层均发生破裂,较为明显的3处破裂压力降幅2.1~5.2MPa,压力降速为1.68~6.67MPa/min。地层发生破裂后压力降幅较大、降速较快,说明地层脆性好、滤失大,天然裂缝较发育。

同理,统计了P井22段压裂施工在升排量阶段的地层破裂次数、平均压力降幅及降速,见表1。其中前6段天然裂缝发育,压力降幅和降速较大,整个排量过程发生多次明显破裂,地层偏脆性;第7~11段压力降幅和降速小,相对低排量发生2~3次微小破裂,地层偏塑性;第12~22段受较高的地应力影响,压力降速有所降低,发生明显破裂的次数减少,地层脆塑性居中。

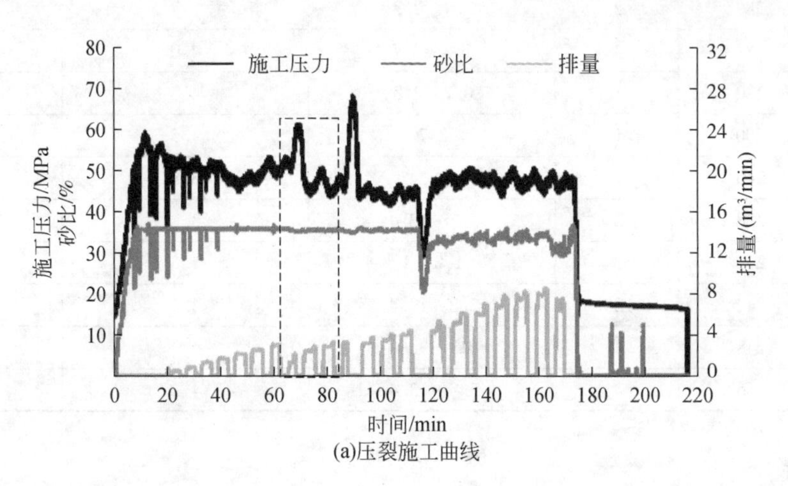

(a)压裂施工曲线

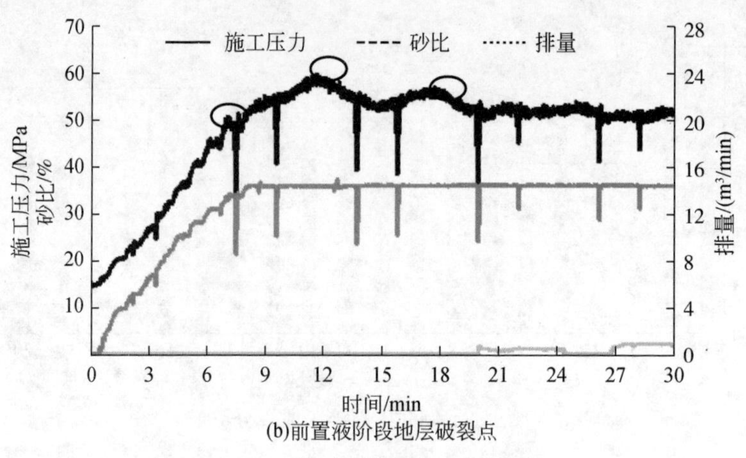

(b)前置液阶段地层破裂点

图2　第5段压裂施工数据

表1　P井22段地层破裂特征数据统计表

射孔段	伽马值/API	破裂次数	施工排量/(m^3/min)	平均压力降幅/MPa	降速/(MPa/min)
1	243	7	5.5～11	3.7	17.4
2	244	6	4～12.2	4.7	34.5
3	259	12	0.9～13.2	4.0	29.5
4	245	7	5.7～13	1.9	8.4
5	210	3	10～14.4	4.2	4.7
6	248	5	4～14.1	3.6	17.4
7	370	3	3.5～6	0.9	15.0
8	431	2	5～9.3	1.8	8.2
9	388	2	7.4～9.5	2.4	6.9
10	413	3	2.7～6.6	2.4	9.1
11	235	4	2.7～13.3	4.3	25.8
12	231	3	2.5～4.8	2.2	8.2
13	389	2	1.4～2.3	2.7	9.5
14	187	4	1.8～2.7	4.0	18.4
15	180	3	0.7～2.8	4.2	6.8
16	200	1	2.7	1.5	10.0
17	240	1	2.3	8.8	7.3
18	228	2	5.7～13.4	3.8	2.4
19	242	2	1.8～3	2.6	8.0
20	242	2	2.8～13.4	4.3	1.2
21	248	4	1～13.6	4.2	15.7
22	156	3	5.8～13.4	4.2	7.4

2.2 地层脆塑性定量评价

塑性页岩地层发生破裂后施工压力几乎不变，但持续的变形导致较大的能量消耗；脆性页层地层破裂后压力快速下降，能量消耗相对较小。根据相关文献[11]中提出的方法，利用压裂施工中地层发生破裂时的能量区域来表征施工过程中的综合脆性指数：

$$BI = \frac{(T_c - T_0)(p_{max} + p_h - p_f) - \int_{T_0}^{T_c} [p(t) + p_h - p_f] dt}{(T_c - T_0)(p_{max} + p_h - p_f)} \quad (1)$$

式中，BI 为页岩的综合脆性指数；$p(t)$为井口压力，MPa；p_{max}为页岩井口破裂压力，MPa；p_h为静液柱压力，MPa；p_f为井筒沿程摩阻压力，MPa；T_c为地层破裂变形后压力下降到最低值时的时间，min；T_0为地层变形后压力上升到最高值时的时间，min。

计算 P 井施工井段的综合脆性指数，结果见表 2，脆性指数范围为 30.9% ~54.3%，水平井段穿行页岩地层非均质性较强。

表 2　P 井各段脆性指数

段数	1	2	3	4	5	6	7	8	9	10	11
脆性指数/%	36.2	38.5	39.8	33.1	48.9	38.9	30.9	34.4	34.6	43.9	35
段数	12	13	14	15	16	17	18	19	20	21	22
脆性指数/%	33.7	35.4	32.4	33.6	47	48.5	35.3	42.4	43.7	54.3	39.9

2.3 地层脆塑性综合评价

综合看来，P 井水平段射孔位置自然伽马值以及各压裂段施工压力曲线地层破裂特征具有如下对应关系：①自然伽马值 150 ~260API，平均破裂次数 4.1 次，平均压力降幅 3.9MPa，平均降速 13.1MPa/min，平均脆性指数 40.1%；②自然伽马值>260API，平均破裂次数 2.4 次，平均压力降幅 2.0MPa，平均降速 9.7MPa/min，平均脆性指数 35.8%。针对该区块页岩气井，可根据自然伽马值对水平井段穿越地层的脆塑性进行判断，进而为每段压裂设计思路提供依据。

3　裂缝复杂形态综合评价方法

3.1 大型物理模拟实验

针对该区块进行了大量真三轴压缩条件下的 300mm×300mm×300mm 页岩露头水力压裂物理模拟实验，压后监测显示复杂缝和单一缝均有形成。以复杂缝典型岩样为例（图 3），预制模拟井眼的露头试样完整性良好，仅有少量沿天然层理面方向的未贯穿天然裂缝。三向应力分别设定为 $\sigma_v = 20.4$MPa，$\sigma_H = 18.4$MPa，$\sigma_h = 14.7$MPa，泵压排量为 0.5mL/s，在压裂液中加入红色示踪剂以记录裂缝延伸状态。压裂后试样剖切显示，压裂后形成垂直层理面的裂缝，与开裂的天然层理面相互交汇形成缝网，图 4 的声发射实时监测结果也验证了该区块页岩具备形成复杂裂缝的基础。

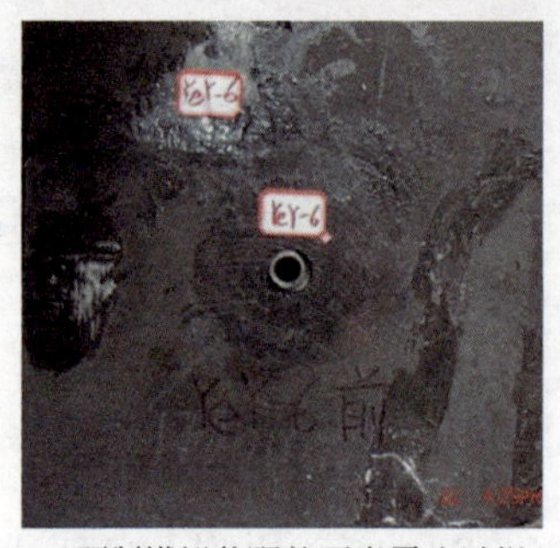

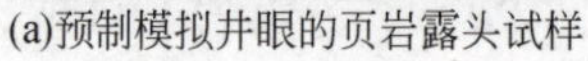

(a)预制模拟井眼的页岩露头试样

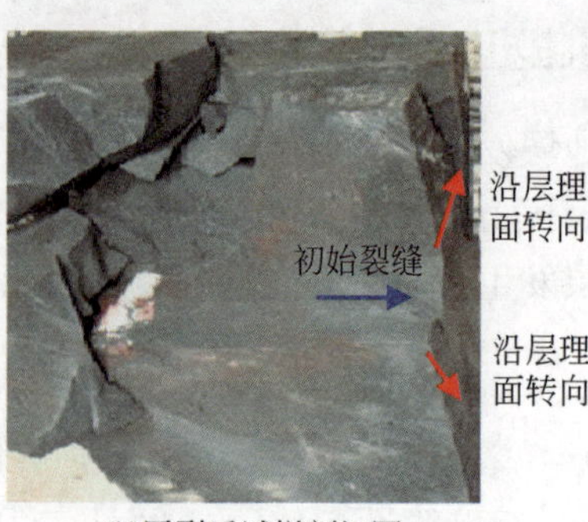

(b)压裂后试样剖切图

图3　页岩露头物理模拟实验

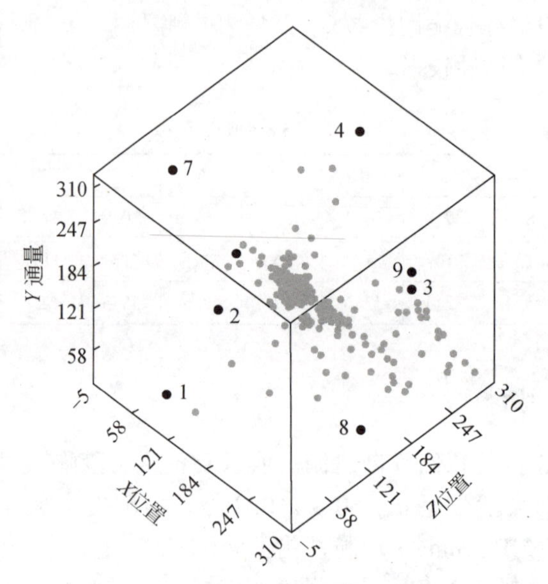

图4　页岩露头声发射实时监测结果

3.2　压力曲线形态

P井仅有4段曲线进行了停泵后压力降测试,因此无法通过G函数分析诊断各段裂缝形态。大型物理模拟实验表明,施工过程中压力曲线波动频率越大、压力降幅越高,则声发射监测到的信号分布范围越广,裂缝形态越复杂。基于此,统计了P井滑溜水压裂施工阶段消除携砂液密度差影响的井底压力波动频率和平均压力波幅,井底压力计算公式见式(2)[12]。

$$p_b = p_w + p_h - p_f \tag{2}$$

式中,p_b为井底压力,MPa;p_w为井口压力,MPa。

以4种压力类型施工曲线为例,分别选取第2、第5、第8、第18段进行压力波动频率和幅度统计,结果见表3。P井22段施工曲线的压力波动频率和幅度统计结果如图5所示。综合而言,第5、第6、第18、第20段远井裂缝发育程度较好、分布范围均较大,压裂后易形成复杂裂缝。

表3 P井代表段压力波动频率和幅度

类型	代表段	压力波动频率	压力波动幅度/MPa	裂缝发育程度
1	2	10	5个波幅6~8MPa,其余3~5MPa	远井裂缝相对较发育,天然裂缝分布范围大
2	5	18	9个在4MPa以上(有两个达9MPa和22MPa),其余为1.5MPa左右的小波幅	远井裂缝相对较发育,天然裂缝分布范围大
3	8	7	仅有5个4MPa以上波幅,其余波幅在2MPa以下	裂缝相对不发育,天然裂缝分布有限
4	18	12	6个波幅在4~5MPa,其余波幅在3MPa左右	地层均质,裂缝发育,分布范围有限

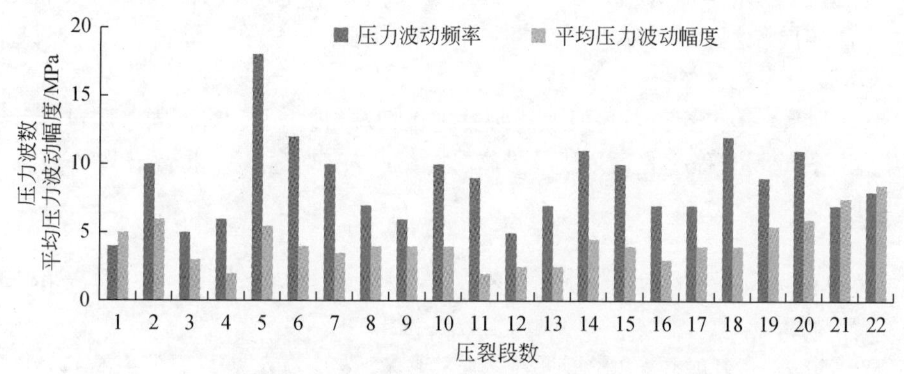

图5 P井22段施工曲线的压力波动频率和幅度统计图

3.3 裂缝形态综合诊断

结合页岩脆塑性、典型页岩压裂实验,以及压力曲线形态分析结果,进行P井裂缝形态综合诊断,见表4,P井单一裂缝比例高达50%。总体而言,对于施工分段的4种类型,类型1和类型2裂缝发育程度较好、分布范围均较大,压裂后易形成天然层理缝与水力裂缝相交的复杂裂缝。类型3塑性强、天然裂缝不发育,易形成单一缝。类型4整体压力波动情况稍差于类型1和类型2,有形成复杂缝的可能。

表4 P井裂缝形态诊断结果

裂缝形态	单一裂缝	复杂裂缝	网络裂缝
压裂段	7~11、13~16、21、22段	1、3、4、12、17~20段	2、5、6段
比例/%	50	36.4	13.6

4 提高裂缝复杂性的工艺措施优化

页岩气井压后产量高低取决于两个要素[13-15]:①压裂段簇是否处于优质甜点区;②压裂施工是否形成复杂裂缝。针对页岩地层非均质性强、压裂改造复杂缝形成比例低的问题,为了改善区块开发效果,应进一步采取如下工艺措施:①精细分段。根据新井的伽马测井数据初步判断沿水平井筒方向的地层脆塑性,进而优选含气性高及天然裂缝发育的脆性地层为地质甜点区,针对性设置段簇分布。②提高裂缝复杂程度。适当提高施工排量及滑溜水黏度,配合加砂浓度、时机、段塞量及压裂液交替注入等工艺措施,增加施工过程中的净压力;还可借鉴美国的转向压裂技术,进行缝内及缝口暂堵以增加页岩改造体积。

5　结论

(1)针对页岩储层非均质性较强的特征,利用施工曲线对地层进行脆塑性识别,初步建立自然伽马值与页岩脆塑性的关联性,伽马值在260API以上地层脆性相对变差。

(2)基于大型物理模拟实验结果,通过统计施工过程中井底压力曲线波动频率及压力降幅,同时结合地层脆塑性分析结果综合判断裂缝形态。P井单一缝占50%,复杂裂缝占36.4%,网络裂缝占13.6%。

(3)考虑到水平段的穿行轨迹及已压裂井施工情况,提出了针对不同小层进行针对性精细分段及提高裂缝复杂程度的工艺技术措施,以增加有效改造体积、提高开发效果。

参考文献

[1] 唐颖, 张金川, 张琴, 等. 页岩气井水力压裂技术及其应用分析[J]. 天然气工业, 2010, 30(10): 33-38.

[2] 陈尚斌, 朱炎铭, 王红岩, 等. 中国页岩气研究现状与发展趋势[J]. 石油学报, 2010, 31(4): 689-694.

[3] 刘玉章, 修乃岭, 丁云宏, 等. 页岩储层水力裂缝网络多因素耦合分析[J]. 天然气工业, 2015, 35(1): 61-66.

[4] Gulen G, Ikonnikova S, Browning J, *et al.* Fayetteville shale-production outlook [J]. SPE Economics & Management, 2014, 7(2): 47-59.

[5] 郭小哲, 周长沙. 基于扩散的页岩气藏压裂水平井渗流模型研究[J]. 西南石油大学学报(自然科学版), 2015, 37(3): 38-44.

[6] 叶登胜, 李建忠, 朱炬辉, 等. 四川盆地页岩气水平井压裂实践与展望[J]. 钻采工艺, 2014, 37(3): 42-44.

[7] 钱斌, 张俊成, 朱炬辉, 等. 四川盆地长宁地区页岩气水平井组"拉链式"压裂实践[J]. 天然气工业, 2015, 35(1): 81-84.

[8] Soliman M Y, East L E, Augustine J R. Fracturing design aimed at enhancing fracture complexity [C]. Paper SPE 130043 presented at SPE EUROPEC/EAGE Annual Conference and Exhibition, Barcelona, Spain, 14-17 June 2010.

[9] 吴奇, 胥云, 王腾飞, 等. 增产改造理念的重大变革——体积改造技术概论[J]. 天然气工业, 2011, 31(4): 7-12.

[10] 高世葵, 朱文丽, 殷诚. 页岩气资源的经济性分析——以Marcellus页岩气区带为例[J]. 天然气工业, 2014, 34(6): 141-148.

[11] 蒋廷学, 卞晓冰, 苏瑗, 等. 页岩可压性指数评价新方法及应用[J]. 石油钻探技术, 2014, 42(5): 16-20.

[12] 蒋廷学. 压裂施工中井底压力的计算方法及其应用[J]. 天然气工业, 1997, 17(5): 82-84.

[13] 侯磊, Elsworth D,孙宝江,等. 页岩支撑裂缝中渗透率变化规律实验研究[J]. 西南石油大学学报(自然科学版), 2015, 37(3): 31-37.

[14] Kennedy R L, Gupta R, Kotov S V, *et al.* Optimized shale resource development: Proper placement of wells and hydraulic fracture stages [C]. Paper SPE 162534 presented at Abu Dhabi International Petroleum Conference and Exhibition, Abu Dhabi, UAE, 11-14 November 2012.

[15] Willian J, Clarkson C R. Stochastic modeling of Two-Phase flowback of multi-fractured horizontal wells to estimate hydraulic fracture properties and forecast production [C]. Paper SPE 164550 presented at SPE Unconventional Resources Conference, The Woodlands, Texas, USA, 10-12 April 2013.

页岩气控压生产的理论认识与现场实践

齐亚东　贾爱林　位云生　王军磊

（中国石油勘探开发研究院）

摘要：随着我国南方海相页岩气产能建设工作的稳步推进，业界对页岩气生产规律的认识逐步加深，气井生产制度优化问题备受关注。针对龙马溪组页岩储层的“大甜点”分布特征及超高压属性，本文从机理分析、理论模型计算和现场实例解剖三个角度明确了页岩气井控压生产的必要性：首先从岩心和气藏两个尺度剖析了页岩储层强应力敏感性产生的机理；在此基础上，引入渗透率模量模型表征应力敏感性并建立了地层-裂缝耦合的多段压裂水平井生产理论模型，通过模型实例计算论证了控压方式对气井生产的积极影响；最后通过现场实例解剖进一步明确了页岩气控压生产的重要性。综合研究认为，控压对气井生产潜力的充分发挥十分有利，可有效提高单井最终累积产气量30%左右。

关键词：页岩气；应力敏感；生产制度；控压；最终累积产量

通过“十二五”期间的理论研究和技术攻关，我国已经在重庆涪陵、四川长宁—威远和滇黔北昭通建成了3个海相页岩气示范区，截至2015年年底产能规模已达$77\times10^8m^3$[1]，页岩气产业开始步入规模开发阶段。此阶段，制定科学合理的开发技术政策是页岩气高效开发的关键，其中一个重要的核心议题便是生产制度优化。国内外业界有关页岩气是否应采取控压方式投产的讨论由来已久，就美国经验来看，Barnett、Marcellus等常压页岩气（压力系数分别为1.04～1.13和1.01～1.34）多采用放压方式生产，而Haynesville页岩气因其地层压力高（压力系数1.6～2.0），储层应力敏感性不容忽视，气井多采用控压限产的方式生产；我国目前初步开发的页岩气区块地层压力普遍较高，借鉴Haynesville的开发经验[2-5]，国内对控压生产的认可程度逐渐增强[6,7]。本文从龙马溪组页岩储层的“大甜点”分布特征及超高压属性出发，首先从页岩储层强应力敏感性产生机理的角度阐述了控压生产的理论依据，在此基础上，引入渗透率模量模型表征应力敏感性，建立起地层-裂缝耦合的多段压裂水平井控压生产理论模型并分析探讨了控压对气井生产的积极影响，最后通过现场实例分析进一步明确了控压生产的必要性。

1　储层应力敏感机理分析

气井生产制度优化时之所以有放压、控压的讨论，其根本原因是油气储层普遍具有应力敏感性。

储层应力敏感性强弱的评判主要基于真实岩心压敏测试结果。大量的室内实验测试表明，致密气、页岩气普遍具有岩心尺度上的强应力敏感性，且储层越致密，其储渗孔隙越小，有效应力作用下越容易失去渗透能力，表现出的应力敏感性越强[8-11]。如果室内测试结果应用在实际生产中可以发现，强应力敏感性会对气田开发造成不利影响。

以苏里格致密砂岩气为例，图1给出了典型的真实岩心应力敏感性测试曲线，从测试结果看，以原始地层条件下的渗透率为基础，气藏开采后期，储层有效渗透率降低 15.30% ~ 48.57%，平均 32.57%；基于测试结果，对不同应力敏感程度下的气田采收率进行了理论计算（表 1），当应力敏感指数为 0.14 时，储层应力敏感造成采收率降低 4.76%；当应力敏感指数为 0.26 时，储层应力敏感造成采收率降低 10.54%。其中，应力敏感指数的表达式为

$$K=K_0\mathrm{e}^{-S(\sigma-\alpha P)}$$

式中，S 为应力敏感指数，$\mathrm{MPa^{-1}}$；K_0 为有效应力为 0 时的岩石渗透率，$\times10^{-3}\,\mu\mathrm{m}^2$；$\sigma$ 为上覆地层压力，MPa；α 为有效应力系数，无因次。

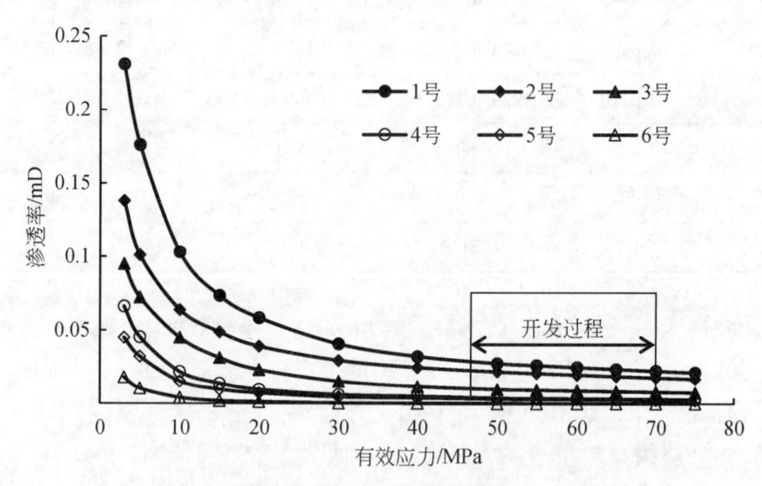

图 1　苏里格气田应力敏感性典型测试曲线

表 1　苏里格气田应力敏感指数对采收率影响程度分析结果

理论计算内容	应力敏感指数 $S/\mathrm{MPa^{-1}}$		
	0	0.14	0.26
稳产结束压力/MPa	4.84	6.27	7.99
井控区内理论采收率/%	84.11	79.35	73.57
应力敏感对采收率影响程度/%	0	4.76	10.54

注：极限产量取值 $1400\mathrm{m^3/d}$

取四川盆地龙马溪组页岩进行应力敏感测试，为了更客观地反映压裂施工后的储层状况，测试之前用压裂液浸泡，应力敏感测试结果见表 2 和图 2，采用相关文献[12]关于页岩储层应力敏感性的表征模型及理论对实验结果进行分析，其中，应力敏感系数表达式为

$$\gamma=-\frac{1}{K}\frac{\mathrm{d}K}{\mathrm{d}\sigma}$$

式中，γ 为应力敏感系数，$\mathrm{MPa^{-1}}$；K 为渗透率，$\times10^{-3}\,\mu\mathrm{m}^2$；$\sigma$ 为有效应力，MPa。

表 2　四川龙马溪组页岩应力敏感性测试结果

有效压力/MPa	K/K_0			平均值	
	S1 号样	S2 号样	S3 号样	K/K_0	Φ/Φ_0
3	1	1	1	1	1

续表

有效压力/MPa	K/K_0			平均值	
	S1 号样	S2 号样	S3 号样	K/K_0	Φ/Φ_0
5	0.5902	0.8310	0.7850	0.7354	0.9854
7	0.3130	0.5577	0.3710	0.4139	0.9581
10	0.1746	0.2506	0.2000	0.2084	0.9255
15	0.0702	0.0733	0.0786	0.0740	0.8763
20	0.0295	0.0293	0.0378	0.0322	0.8368
30	0.0110	0.0118	0.0120	0.0116	0.7883
40	0.0041	0.0044	0.0046	0.0044	0.7418
50	0.0023	0.0028	0.0028	0.0026	0.7178
应力敏感系数 γ	0.128	0.132	0.128	0.130	

从评价结果看，四川龙马溪组页岩储层表现出强应力敏感性，其应力敏感系数平均值为 0.130MPa^{-1}，与致密砂岩储层相比，高出一个数量级；将应力状况回归到原始地层条件，模拟储层生产情况，有效应力从20MPa增至50MPa时，储层渗透率损失率达91.8%；从图2回归的孔渗关系看，龙马溪组页岩的孔渗幂指数平均值为17.961，依文献[12]中的理论分析可以推断，储层中微裂缝的尺度远大于基质孔隙，微裂缝是储层的主要渗流通道，而微裂缝对有效应力的变化更为敏感。

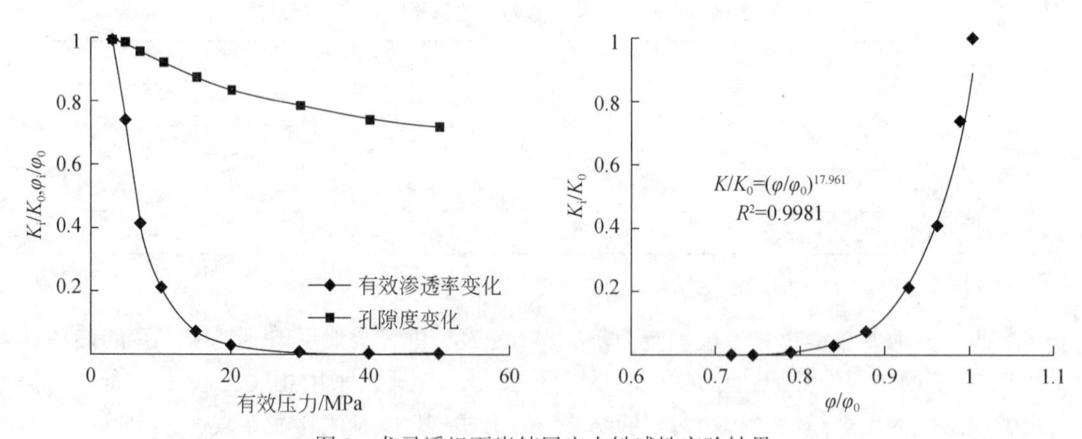

图2　龙马溪组页岩储层应力敏感性实验结果

从上述测试结果可以看出，岩心尺度上，致密气和页岩气储层均具有较强的应力敏感性，基于这一事实，人们也普遍认为开发技术政策制定时储层应力敏感因素不可忽略；但岩心尺度上的实验现象在实际生产过程中能否得到充分体现还受限于储层的内部结构，储层内部结构不同，生产过程中产气层位的受力状况也会有很大的差异。

苏里格气田的储集体具有强非均质性[13]，在大面积连片分布的宏观背景上，气藏内部具有明显“二元”结构特征，即“主力含气砂体”+“基质储集层”；其中，主力含气砂体为辫状河心滩相沉积，基质储集层为辫状河河道充填相沉积，受沉积相控制，主力含气砂体孤立状分布在连续展布的基质储集层中,如图3所示,这种结构的地层,透镜状砂体与基质储集层之间的穹拱状接触面上存在压力拱效应[14-16]。尽管气井投产后透镜状砂体内的孔隙压力降低,但压力拱效应使穹拱状接触面承受住了大部分的上覆岩层压力,阻碍了上覆压力对透镜状砂体的压缩,再考虑到苏里格气田低压的特征(压力系数0.87),可以推断:与实验过程中岩心的受压缩程度相比,投产后实际地层中透镜状砂体内部的岩石受压缩程度并不显著,苏里格型致密气田整体应力敏感性不强。此外,截至目前,苏里格实际生产资料也未明显反映出放压生产和控压生产对采收率的影响存在显著差别。

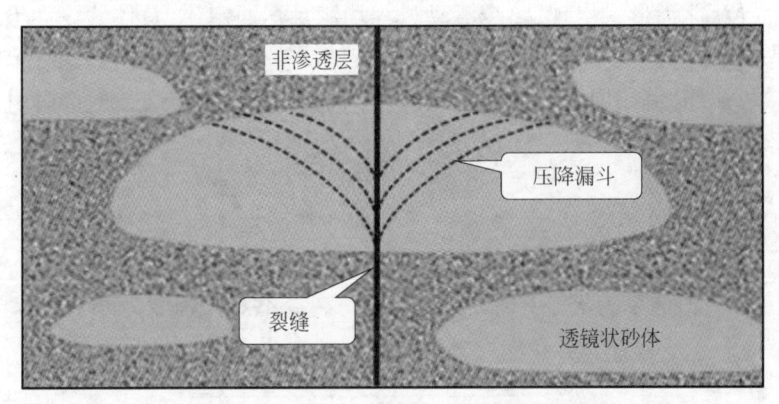

图3 致密砂岩气藏气井生产对储层的影响

页岩储层则不同,中国南方海相优质页岩储集层主要位于深水陆棚相,地势平、展布广,较大区域内保持一致的古环境、古气候和古地理特征,从储层内部结构上看,储集层厚度在平面上分布极为稳定,变化率很小,表现出明显的"大甜点"特征,单个甜点范围多在几十到上百平方千米[1],再考虑到南方海相页岩储层的超高压特征(压力系数1.3~2.1),可以推断,这种结构的地层不存在压力拱效应,投产后孔隙流体压力的降低会直接导致作用在生产层上的有效应力显著增大,产层表现出强应力敏感性。这种情况下,页岩储层一旦放压生产(图4),水力裂缝内会迅速泄压,造成裂缝区和近缝区的储层渗透率急剧下降,快速形成储层伤害区,过早阻挡外围气体进入主裂缝系统,进而导致单井累积产量减少。

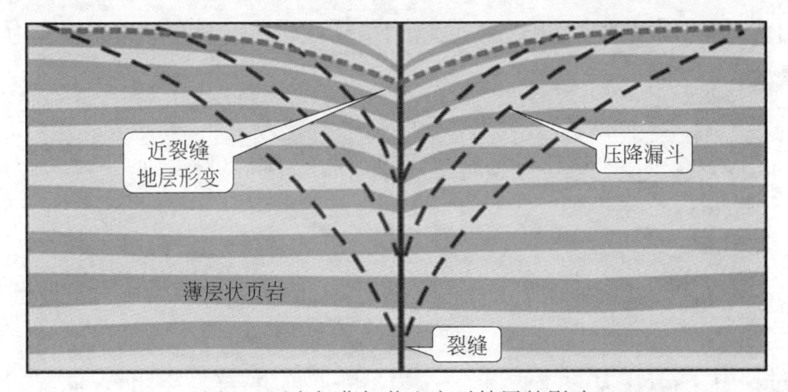

图4 页岩气藏气井生产对储层的影响

2　控压生产理论模型分析

气井完钻后进行水力压裂改造,改造后的地层气体从地层基质到井筒的不稳定渗流过程分为地层到裂缝及裂缝内部流动两部分。

通过裂缝与地层在接触面上的流量及压力耦合条件可以将两个流动过程耦合为一个统一的整体。本文物理量采用 SI 单位制。假定地层内渗透率随压力变化满足指数形式,压裂缝中以无量纲导流能力形式出现,整个流动过程均满足达西定律,具体假设如下。

(1)地层均质、等厚、上下封闭且水平无限大,储层厚度为 h,孔隙度为 φ_m,渗透率为 K_m,储层和流体微可压缩,储层内综合压缩系数为 C_{tm},流体黏度为 μ;气井生产时引入拟压力:

$$p_p = \frac{\mu_{gi} Z_{gi}}{p_i} \int_{p_t}^{p} \frac{\zeta}{\mu_g(\zeta) Z_g(\zeta)} d\zeta$$

(2)压裂裂缝纵向上穿透地层,裂缝垂直于水平井筒且关于井筒呈不对称延伸,多裂缝沿水平井筒方向排列,裂缝中点坐标为(x_{w0},y_w),裂缝与井筒交点坐标为(x_{fw},y_w),裂缝半长为 x_f,宽度为 w_f,渗透率为 K_f,压裂段间距为 D_f,裂缝不对称程度为 θ(图 5)。

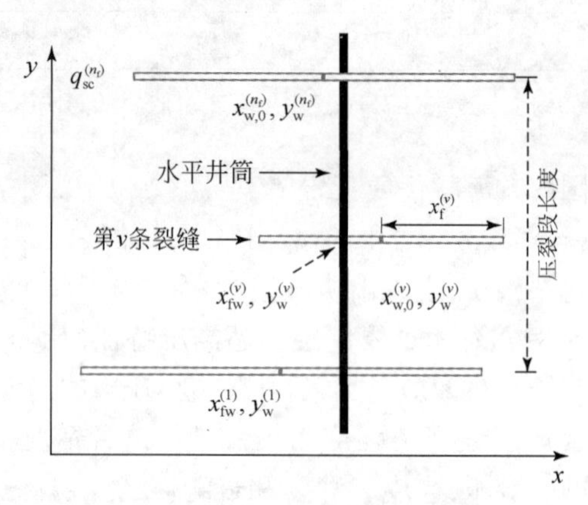

图 5　分段压裂水平井模型

(3)相对渗透率变化,通常忽略孔隙度应力敏感性[17],这里设定渗透率应力敏感与孔隙压力间呈指数递减关系:

$$K_\zeta = K_\zeta e^{[-\gamma_\zeta(p_\zeta - p_\zeta)]} \tag{1}$$

而在裂缝系统内,渗透率应力敏感性通常用等效导流能力 C'_{fD} 来表征:

$$C'_{fD}(p_f) = \frac{K_{fi} w_f}{K_{mi} x_f} e^{[-\gamma_f(p_i - p_f)]} \tag{2}$$

式中,p_i 为原始压力,Pa;K_{fi} 为原始压力下裂缝渗透率,m^2;K_{mi} 为原始压力下地层渗透率,m^2;w_f 为裂缝宽度,m;x_f 为裂缝半长,m;γ_f 为裂缝渗透率模量,Pa^{-1};γ_m 为地层渗透率模量,Pa^{-1}。

在生产过程中,随着压力的降低,有效应力不断增加,裂缝发生闭合、储层流动通道发生变形(图 6)。从数学模型的角度看,引入应力敏感效应后渗流控制方程表现为强非线性,本文引入 Pedrosa 变换及相应的扰动方法处理非线性方程,其中 Pedrosa 变换中无量纲扰动量与无量纲压力转换关系为

$$p_{\zeta D}=-\frac{1}{\gamma_{\zeta D}}\ln(1-\gamma_{\zeta D}\eta_{\zeta D}),\zeta=f \text{ and } m \tag{3}$$

关于式(3)需要强调的是,为了保证转换关系的有效性需满足前提条件 $0<\gamma_{\xi D}\eta_{\xi D}<1$。

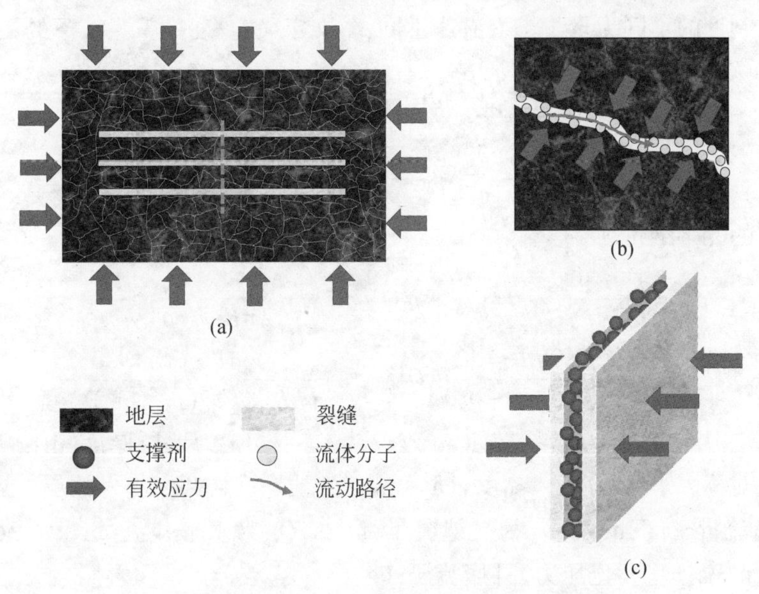

图 6　储层流动通道闭合和裂缝导流闭合模型

2.1　裂缝内部流动模型

地层内流体沿裂缝壁面不断流入裂缝,并沿裂缝流向井筒。基于以上假设条件,裂缝与地层流动过程为独立且连续的两部分,因此裂缝与地层内的空间坐标相互独立(图 7)。

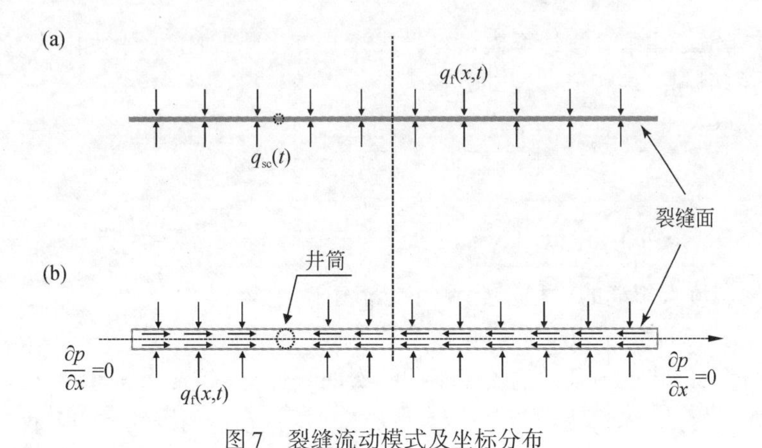

图 7　裂缝流动模式及坐标分布

这里将裂缝内的流动可以简化为一维流动,结合式(3)获得第 v 条裂缝内部不稳定流动方程为

$$
-\gamma_{\mathrm{fD}}^{(v)} \mathrm{e}^{(-\gamma_{\mathrm{fD}}^{(v)} p_{\mathrm{fD}}^{(v)})} \left(\frac{\partial p_{\mathrm{fD}}^{(v)}}{\partial x_{\mathrm{D}}}\right)^2 + \mathrm{e}^{(-\gamma_{\mathrm{fD}}^{(v)} p_{\mathrm{fD}}^{(v)})} \frac{\partial^2 p_{\mathrm{fD}}^{(v)}}{\partial x_{\mathrm{D}}^2} - \frac{\pi}{C_{\mathrm{fD}}^{(v)}} q_{\mathrm{fD}}^{(v)}(x_{\mathrm{D}}) + \frac{2\pi}{C_{\mathrm{fD}}^{(v)}} q_{\mathrm{scD}}^{(v)} \delta(x_{\mathrm{D}} - \theta^{(v)}) = 0 \tag{4}
$$

根据生产制度的不同无量纲定义有所不同,本文定义定/变产量生产下的无量纲量为

$$
p_{\mathrm{fD}} = \frac{2\pi K_{\mathrm{mi}} h(p_{\mathrm{i}} - p_{\mathrm{f}})}{q_{\mathrm{ref}}\mu}, \gamma_{\mathrm{fD}} = \frac{q_{\mathrm{ref}}\mu\gamma_{\mathrm{f}}}{2\pi K_{\mathrm{mi}} h}, t_{\mathrm{D}} = \frac{K_{\mathrm{mi}} t}{\varphi_{\mathrm{m}}\mu c_{\mathrm{tm}} L_{\mathrm{ref}}^2}, L_{\mathrm{D}} = \frac{L}{L_{\mathrm{ref}}}, \theta = \frac{|x_{\mathrm{w},0} - x_{\mathrm{fw}}|}{x_{\mathrm{f}}},
$$

$$
C_{\mathrm{fD}} = \frac{K_{\mathrm{fi}} w_{\mathrm{f}}}{K_{\mathrm{mi}} L_{\mathrm{ref}}}, q_{\mathrm{scD}} = \frac{q_{\mathrm{sc}}}{q_{\mathrm{ref}}}, q_{\mathrm{fD}} = \frac{2 q_{\mathrm{f}} L_{\mathrm{ref}}}{q_{\mathrm{ref}}}, Q_{\mathrm{cumD}} = \frac{k Q_{\mathrm{cum}}}{\varphi_{\mathrm{m}}\mu c_{\mathrm{tm}} L_{\mathrm{ref}}^2 q_{\mathrm{ref}}}
$$

定压力下的无量纲量为

$$
p_{\mathrm{fD}} = \frac{p_{\mathrm{i}} - p_{\mathrm{f}}}{p_{\mathrm{i}} - p_{\mathrm{w}}}, \gamma_{\mathrm{fD}} = (p_{\mathrm{i}} - p_{\mathrm{w}})\gamma_{\mathrm{f}}, q_{\mathrm{scD}} = \frac{\mu q_{\mathrm{sc}}}{2\pi K_{\mathrm{mi}} h(p_{\mathrm{i}} - p_{\mathrm{w}})},
$$

$$
q_{\mathrm{fD}} = \frac{\mu(2 q_{\mathrm{f}} L_{\mathrm{ref}})}{2\pi K_{\mathrm{mi}} h(p_{\mathrm{i}} - p_{\mathrm{w}})}, Q_{\mathrm{cumD}} = \frac{Q_{\mathrm{cum}}}{2\pi h L_{\mathrm{ref}}^2 \varphi_{\mathrm{m}} c_{\mathrm{tm}}(p_{\mathrm{i}} - p_{\mathrm{w}})}
$$

式中,p_{f}为裂缝压力,Pa;q_{ref}为参考流量,$\mathrm{m^3/s}$;t 为时间,s;h 为地层厚度,m;x 为裂缝坐标,m;θ 为裂缝非对称因子;μ 为流体黏度,Pa · s;φ_{m}为地层孔隙度;c_{tm}为地层综合压缩系数,$\mathrm{Pa^{-1}}$;q_{f}为裂缝流量密度,$\mathrm{m^2/s}$;q_{sc}为裂缝流量,$\mathrm{m^3/s}$;Q_{sc}为水平井流量,$\mathrm{m^3/s}$;Q_{cum}为(定压力)累积产量,$\mathrm{m^3/s}$;p_{w}为(定压力)井底流压,Pa。

引入式(3)处理式(4),可以得到扰动变量形式下的渗流控制方程:

$$
\frac{\partial^2 \eta_{\mathrm{fD}}^{(v)}}{\partial x_{\mathrm{D}}^2} - \frac{\pi}{C_{\mathrm{fD}}^{(v)}} q_{\mathrm{fD}}^{(v)}(x_{\mathrm{D}}, t_{\mathrm{D}}) + \frac{2\pi}{C_{\mathrm{fD}}^{(v)}} q_{\mathrm{scD}}^{(v)}(t_{\mathrm{D}}) \delta(x_{\mathrm{D}} - \theta^{(v)}) = 0, x_{\mathrm{wD},0}^{(v)} - x_{\mathrm{fD}}^{(v)} \leqslant x_{\mathrm{D}} \leqslant x_{\mathrm{wD},0}^{(v)} + x_{\mathrm{fD}}^{(v)} \tag{5}
$$

2.2　地层流动模型

根据物质守恒定律结合渗透率模量定义式(2),在上下封闭、水平无限大、各向同性的储层中由 n_{f}条裂缝引起的单相流体流动方程为

$$
\gamma_{\mathrm{mD}} \mathrm{e}^{(-\gamma_{\mathrm{mD}} p_{\mathrm{mD}})} \left[\left(\frac{\partial p_{\mathrm{mD}}}{\partial x_{\mathrm{D}}}\right)^2 + \left(\frac{\partial p_{\mathrm{mD}}}{\partial y_{\mathrm{D}}}\right)^2\right] + \mathrm{e}^{(-\gamma_{\mathrm{mD}} p_{\mathrm{mD}})} \nabla^2 p_{\mathrm{mD}} + \pi \Delta x'_{\mathrm{D}} \sum_{v=1}^{n_{\mathrm{f}}} q_{\mathrm{fD}}^{(v)}(x'_{\mathrm{D}}, t_{\mathrm{D}}) \delta(x_{\mathrm{D}} - x'_{\mathrm{D}}, y_{\mathrm{D}} - y_{\mathrm{wD}}^{(v)}) = \frac{\partial p_{\mathrm{mD}}}{\partial t_{\mathrm{D}}} \tag{6}
$$

式中,流体从地层流入裂缝的过程用点源函数 δ 表征。

将转换关系式(3)代入式(6),同时将扰动量表示为幂级数的形式($\eta = \eta_0 + \alpha\eta_1 + \alpha^2\eta_2 + \cdots$)。根据扰动法原理可以获得一系列的关于不同阶次的线性方程组,当 $\eta < 0.2$ 时 0 阶解近似可以满足计算精度,因此式(6)可以近似转换为关于地层扰动量的线性控制方程:

$$
\frac{\partial^2 \eta_{\mathrm{mD}}}{\partial x_{\mathrm{D}}^2} + \frac{\partial^2 \eta_{\mathrm{mD}}}{\partial y_{\mathrm{D}}^2} + \pi \Delta x'_{\mathrm{D}} \sum_{v=1}^{n_{\mathrm{f}}} q_{\mathrm{fD}}^{(v)}(x'_{\mathrm{D}}, t_{\mathrm{D}}) \delta(x_{\mathrm{D}} - x'_{\mathrm{D}}, y_{\mathrm{D}} - y_{\mathrm{wD}}^{(v)}) = \frac{\partial \eta_{\mathrm{mD}}}{\partial t_{\mathrm{D}}} \tag{7}
$$

2.3　模型耦合

在裂缝与地层的接触面上存在着流量及压力耦合条件,而不是扰动变量的耦合,这与 Chen 等[18]的耦合条件不同。基于式(3),可以获得地层与裂缝间的扰动量耦合条件:

$$\eta_{\mathrm{fD},j}^{(v,m)} = \{1 - [1 - \gamma_{\mathrm{mD}}\eta_{\mathrm{mD},j}^{(v,m)}]^{\gamma_{\mathrm{fD}}^{(v)}/\gamma_{\mathrm{mD}}}\}/\gamma_{\mathrm{fD}}^{(v)} \tag{8}$$

同时裂缝与井筒结合处的压力耦合,满足耦合条件:

$$p_{\mathrm{wD}}^{(m)} = -\frac{1}{\gamma_{\mathrm{fD}}^{(1)}}\ln(1 - \gamma_{\mathrm{fD}}^{(1)}\eta_{\mathrm{wD}}^{(1,m)}) = \cdots = -\frac{1}{\gamma_{\mathrm{fD}}^{(n_f)}}\ln(1 - \gamma_{\mathrm{fD}}^{(n_f)}\eta_{\mathrm{wD}}^{(n_f,m)}) \tag{9}$$

2.4 模拟计算

基于上述理论模型,以某南方海相页岩气建产区已投产井的平均静态参数为基础,对放压和控压两种生产制度下的气井产量进行模拟研究。

优质页岩储层厚度21m,原始地层压力36MPa;裂缝半长200m,裂缝宽度1mm;原始地层条件下裂缝渗透率$100\times10^{-3}\,\mu m^2$,裂缝孔隙度0.5%,基质渗透率$0.000102\times10^{-3}\,\mu m^2$,基质孔隙度5%;水平井长度1450m,水力压裂18段;每段射孔3簇。

放压生产情形,井口压力取外输管线压力1.5MPa;控压生产情形,投产初期定产$4\times10^4 m^3/d$,根据气井的稳产情况,产量依次调整为$3.0\times10^4 m^3/d$、$2.0\times10^4 m^3/d$、$1.5\times10^4 m^3/d$和$1.0\times10^4 m^3/d$,而后产量自然递减;两种情形的废弃产量均取$1000m^3/d$。

模拟计算的结果(表3和图8)表明,生产制度的选择对气井生产潜力的发挥程度影响显著。

表3　考虑应力敏感时放压与控压条件下的产量预测

生产时间/a	1	2	3.6	5	10	15	20	25	EUR
放压累积产量/$10^4 m^3$	2214	3145	4161	4718	5944	6576	6964	7226	7346
控压累积产量/$10^4 m^3$	1460	2582	4161	5132	7405	8290	8752	9033	9151

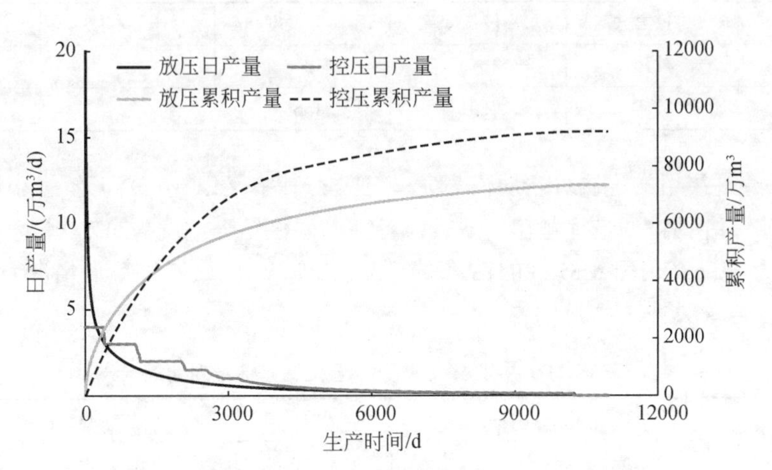

图8　考虑应力敏感时放压与控压条件下的产量预测

从气井产量剖面看,放压生产初期产量高,但产量递减急剧,投产首月平均日产量$19.7\times10^4 m^3/d$,第二月平均日产量仅为$4.08\times10^4 m^3/d$,月递减率达60.5%;生产至243天时(即投产8个月后),放压生产的日产量开始低于控压生产的日产量($4.0\times10^4 m^3/d$),换言之,与控压生产相比,放压生产的日产优势持续时间只有8个月。

从累积产量曲线看,放压生产首年累积产量$2214\times10^4 m^3$,控压生产首年累积产量1460×

10^4m^3,放压生产的首年产量是控压生产的 1.5 倍;生产 3.6 年时,两种生产制度下的累积产量相当,为 $4161×10^4m^3$,而后控压生产的累积产量逐渐超过放压生产,换言之,与控压生产相比,放压生产的累积产量优势持续时间为 3.6 年。

最终,放压生产 EUR 为 $7346×10^4m^3$,控压生产 EUR 为 $9151×10^4m^3$,控压生产比放压生产可有效提高单井最终累积产量 25%。

由此可见,控压生产对提高页岩气井最终累积产量是有利的;如果考虑到成本回收周期、气价变化等经济因素,则需要进一步论证何种配产方案最为合理,以获得特定时限内的最大经济效益。

3 控压生产现场实例分析

随着我国页岩气建产规模的逐步扩大,有关南方海相页岩气的动静态资料日趋丰富,相应地,业内对页岩气生产规律的认识也在不断加深,再加之对美国页岩气成功开发经验的跟踪和借鉴,目前,各建产区针对控压生产和放压生产的探讨和试验正在有序开展。

现选取某建产区工程参数和地质参数相近,但生产制度不同的两口井(HD 井和 MM 井)进行剖析,以深入认识控压和放压两种制度对气井生产动态的影响。

工程上,大量的实践数据表明,投产井的生产效果与压裂段长度、压裂段数和加砂量之间具有良好的正相关性。从表 4 所示的工程参数角度看,HD 井的压裂段长度和压裂段数略次于 MM 井,但其压裂加砂量是 MM 井的 1.5 倍,明显占优;优劣相抵,可以认为两口井的储层改造程度相近。

表 4 HD 井和 MM 井工程参数对比表

井号	生产方式	水平段长度/m	压裂段长度/m	压裂段数	加砂量/t
HD 井	放压	1510	1318	18	1673
MM 井	控压	1605	1568	22	1101

地质上,龙马溪组的优质页岩层位为龙一$_1$亚段,其中又以龙一$_1^1$和龙一$_1^3$小层为最优,这两个层位是水平井钻进的最优靶位。从表 5 各井的层位钻遇情况看,HD 井龙一$_1^1$和龙一$_1^3$小层的钻遇率为 47.53%,而 MM 井的钻遇率为 33.12%,从优质层位钻遇情况看,HD 井优于 MM 井。

表 5 HD 井和 MM 井页岩层位钻遇情况对比表

层位	HD 井		MM 井	
	钻遇长度/m	钻遇比例/%	钻遇长度/m	钻遇比例/%
龙一$_1^4$	403	26.69		
龙一$_1^3$	519.5	34.40	531.6	33.12
龙一$_1^2$	389.3	25.78	1073.4	66.88
龙一$_1^1$	198.2	13.13		

综合上述信息不难判断,若无其他因素影响,HD 井和 MM 井的生产效果应较为相似,甚至 HD 井的开发效果应略优于 MM 井;但从生产动态信息(图 9、图 10 和表 6)看,情况恰恰相反,分析认为,造成此种差异的一个重要原因在于生产制度选择的不同:HD 井放压投产(high drawdown),而 MM 井控压投产(managed drawdown)。

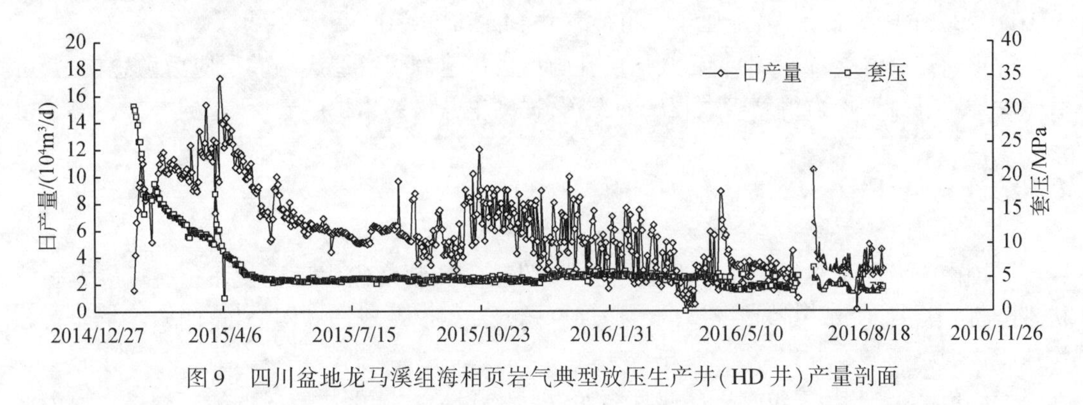

图 9 四川盆地龙马溪组海相页岩气典型放压生产井(HD 井)产量剖面

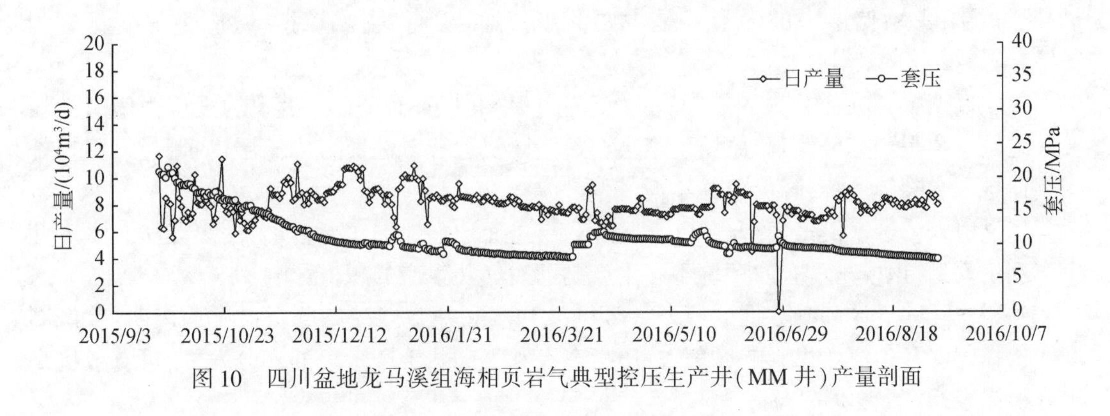

图 10 四川盆地龙马溪组海相页岩气典型控压生产井(MM 井)产量剖面

表 6 HD 井和 MM 井生产动态参数对比表

井号	生产方式	前三个月压降速率/(MPa/d)	前三个月单位压降产量/(10^4m^3/MPa)	首年单位压降产量/(10^4m^3/MPa)	预测累积产量/(10^8m^3)
HD 井	放压	0.278	37.8	102	0.61
MM 井	控压	0.130	64.7	206	0.83

从图 9 可知,HD 井放压投产后,早期产量随压裂液的返排而逐渐上升,最高值 $17.1\times10^4m^3/d$,而后产量快速下降,4 个月后降至 $5.0\times10^4m^3/d$;套压自投产日起便急剧衰减,仅 4 个月时间便从初始的 30.5MPa 降至管线输压 5MPa。从图 10 可知,MM 井自投产之初便以较为恒定的产量 $8.0\times10^4m^3/d$ 生产,生产一年多以来,产量一直保持平稳;套压衰减相对缓慢,1 年时间,由初期的 21.0MPa 降至目前的 7.8MPa。

表 6 给出了更为量化的生产动态特征参数:①HD 井前三个月的压降速率为 0.278MPa/d,而 MM 井压降速率为 0.130MPa/d,仅为 HD 井的 46.8%,控压生产井的压力降落明显趋缓,这将有助于延缓储层应力敏感伤害;②HD 井前三个月的单位压降产量为 $37.8\times10^4m^3/$

MPa，而 MM 井前三个月的单位压降产量是 HD 井的 1.7 倍，达 64.7×10^4m^3/MPa；随着生产时间的延长，控压生产井的这种优势越来越明显，生产时间达 1 年时，控压生产的 MM 井单位压降产量已经达到放压生产的 HD 井的 2.0 倍；③采用多种生产动态分析技术对两口气井的最终累积产量进行预测，HD 井 EUR 为 6100×10^4m^3，而 MM 井的 EUR 为 8300×10^4m^3，是 HD 井的 1.4 倍，换言之，控压生产可使气井 EUR 提高 30% 以上。

综上所述，控压生产有助于延缓储层孔隙压力快速衰减所造成的储层应力敏感伤害，可有效保障气井生产潜力得到较大程度的发挥，提高单井最终累积产气量。

4　结论与认识

中国南方海相页岩气产能建设已经初具规模，生产制度优化日受关注。通过应力敏感机理分析、考虑应力敏感性情况下的生产理论模型计算以及现场实例分析，明确了页岩气井控压生产的必要性。

岩心尺度上，龙马溪组页岩储层具有强应力敏感性，实际生产情况下储层渗透率损失率最高可达 91.8%；储层中的微裂缝是主要的渗流通道，而微裂缝对有效应力的变化更为敏感。

气藏尺度上，储层的“大甜点”特征及超高压属性决定了岩心尺度上的强应力敏感性在实际地层中会得到充分体现，放压生产会在裂缝区及近缝带快速形成储层伤害，过早阻挡外围气体进入主裂缝系统，降低单井累积产量。

通过引入渗透率模量模型表征应力敏感性，进而建立地层-裂缝耦合的多段压裂水平井生产理论模型，该模型可有效评价不同生产制度对气井生产的积极影响。

基于实际建产区基础数据的模型计算结果和实际井生产动态分析对比均表明：生产制度的选择对气井生产效果影响显著，控压生产有利于气井生产潜力的充分发挥，可有效提高单井最终累积产量 30% 左右。

参考文献

[1] 贾爱林，位云生，金亦秋．中国海相页岩气开发评价关键技术进展[J]．石油勘探与开发，2016，43(6)：1-7.

[2] Okouma V, Guillot F, Sarfare M, *et al.* Estimated ultimate recovery (EUR) as a function of production practices in the Haynesville shale [C]. Paper SPE 147623 presented at the SPE Annual Technical Conference and Exhibition, Denver, Colarado, USA, 30 October-2 November 2011.

[3] Akande J, Spivey J P. Consideration for pore volumes stress-effects in over-pressured shale gas under controlled drawdown well management strategy[C]. Paper SPE 162666 presented at the SPE Canadian Unconventional Resources Conference, Calgary, Alberta, Canada, 30 October-1 November 2012.

[4] Wilson K. Analysis of drawdown sensitivity in shale reservoirs using coupled-geomechanics models[C]. Paper SPE 175029 presented at the SPE Annual Technical Conference and Exhibition, Houston, Texas, USA, 28-30 Septermber 2015.

[5] Thompson J W, Fan L, Martin R B, *et al.* An overview of horizontal well completions in the Haynesville Shale[C]. Paper CSUG/SPE 136875 presented at the Canadian Unconventional Resources & International Petroleum Conference, Calgary, Alberta, Canada, 19-20 October 2010.

[6] 刘洪林，王红岩，方朝合，等. 中国南方古老海相页岩气超压富集特征及勘探开发意义[J]. 非常规油气，2014，1(1)：11-16.
[7] 杨波，罗迪，张鑫，等. 异常高压页岩气藏应力敏感及其合理配产研究[J]. 西南石油大学学报(自然科学版)，2016，38(2)：115-121.
[8] 杨朝蓬，高树生，郭立辉，等. 致密砂岩气藏应力敏感性及其对产能的影响[J]. 钻采工艺，2013，36(2)：58-61.
[9] 钟高润，张小莉，杜江民，等. 致密砂岩储层应力敏感性实验研究[J]. 地球物理学进展，2016，31(3)：1300-1306.
[10] 朱维耀，马东旭，朱华银，等. 页岩储层应力敏感性及其对产能影响[J]. 天然气地球科学，2016，27(5)：892-897.
[11] 张睿，宁正福，杨峰，等. 页岩应力敏感实验研究及影响因素分析[J]. 岩石力学与工程学报，2015，34(S1)：2617-2622.
[12] 张睿，宁正福，杨峰，等. 页岩应力敏感实验与机理[J]. 石油学报，2015，36(2)：224-231.
[13] 何东博，贾爱林，冀光，等. 苏里格大型致密砂岩气田开发井型井网技术[J]. 石油勘探与开发，2013，40(1)：79-89.
[14] 李乐忠，李相方. 储层应力敏感实验评价方法的误差分析[J]. 天然气工业，2013，33(2)：48-51.
[15] 王钒潦，李相方，韩彬，等. 考虑压力拱效应的应力敏感实验[J]. 科技导报，2013，31(19)：26-32.
[16] 王钒潦，李相方，汪洋，等. 考虑压力拱效应的苏里格气田上覆压力计算[J]. 大庆石油地质与开发，2013，32(5)：61-66.
[17] Berumen S. Evaluation of fractured wells in pressure-sensitive formations [D]. University of Oklahoma，1995.
[18] Chen Z M，Liao X W，Zhao X L，*et al.* Performance of horizontal wells with fracture networks in shale gas formation. Journal of Petroleum Science and Engineering，2015，133：646-664.

页岩气水平井井间干扰分析及井距优化

位云生　王军磊　齐亚东　贾成业

（中国石油勘探开发研究院）

摘要:页岩气井距设计与优化是评价页岩气开发效果的重要指标之一。根据类比法、数值模拟、经济评价方法论证,本文形成了从井间干扰模拟、动态数据诊断到多井生产模拟、井距优化的完整工作流程:①通过建立压力探测边界传播模型,模拟不同沟通条件下井间干扰响应程度;②基于井间干扰响应规律,根据气井生产动态数据演绎识别、诊断井间干扰;③以地质解释和动态分析结果为基础参数,建立气藏体积压裂多井数值模型,模拟气田生产动态,结合净现值模型优化井距。模拟表明,减小井距使井间干扰提前发生,同时也提高区块整体采收率;基于目前的参数体系,400~500m井距有进一步缩小至300~400m的空间,单位面积内井数增加30%,区块储量采收率提高10%左右;区块整体净现值随着生产年限不断增加,但对应的最优井距结果不随生产周期而改变。同时本文讨论了井间裂缝沟通、交错布缝模式、生产制度优化等对合理井距论证产生的影响,为下一步的科研工作提供了攻关方向。

关键词:分段压裂水平井;裂缝沟通;井间干扰;井距优化;净现值模型

近些年,随着长宁—威远、昭通、焦石坝等页岩气的大规模开采,以工厂化钻井、体积压裂为核心的工程技术得到了广泛应用[1,2]。为了获得最优的开发效果,需要厘清气井/区块的生产动态特征。影响气井生产动态的因素主要分为不可控因素和可控因素:不可控因素包括孔隙度、含水饱和度、初始压力、渗透率和天然裂缝分布等,可控因素包括开发井距、完井方式、工作制度等,可通过优化可控因素提高气井产能。其中,可控因素中,页岩气水平井开发井距最为关键,直接影响页岩气采收率、最终累积产量(EUR)和经济效益,一旦不合理,造成的经济后果是无法挽救的,因此合理井距/井网密度一直以来是气藏高效开发的关键指标。非常规气藏的开发经验表明,在开发过程中,当存在井间干扰时,单井最终累积产量会随着井距减小而减小,同时井间储量将得到有效动用,气藏整体采收率增加[3],但投入随之增加,因此,页岩气开发同样存在最优井距/井数[4,5]。

开发井距决定着井间干扰方式及干扰程度,合理井距能够平衡区块采收率和单井累积产量的关系,而合理井距主要通过气藏模拟、现场试验和经济评价来确定[6]。例如,Marcellus页岩气田,数值模拟结果表明,开发井距从640m调整到320m时,可提高10%的气藏可采储量,但单井EUR将降低57%,结合经济评价模型确定640m是最优井距[7]。而目前中国石油长宁—威远、昭通等区块400~500m[1]的开发井距主要利用现场试验如干扰试井、微地震监测等手段,获得有效缝长与微地震监测缝长之间的关系,进行同区类比,再综合考虑水平两向应力、天然裂缝发育程度等因素确定,论证方法不严谨,且生产动态表明开发井距偏大。为了充分论证合理井距,提高静态、动态资料利用率,本文形成了综合井间干扰模拟、井间干扰响应识别、多井生产动态模拟、经济评价等方法的井距优化工作流程(图1),并做了详细演绎模拟,力图提供一种考虑全面、方便可行的井距优化方法。

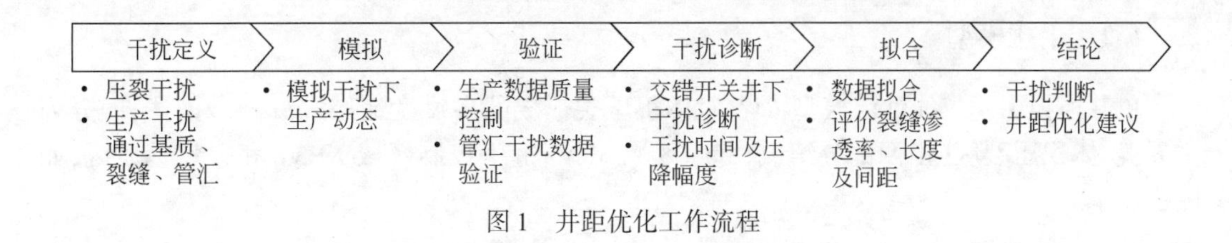

图1　井距优化工作流程

1　井间干扰模拟

1.1　干扰模式

不同学科对井间干扰有不同的定义,如压裂领域中的井间裂缝击穿(fracture hit),即所谓的压穿[8,9],产生压裂干扰。本文中的干扰主要指不同井生产引起的压力扰动在介质中传播时的相互干扰,即生产干扰。根据压力扰动在不同传播介质类型中的差别,井间(生产)干扰方式可进一步分为两种:通过裂缝和基质产生干扰。

图2反映了不同井距下的两种沟通情况,其中灰色区域指体积改造的有效区域(SRV),即裂缝覆盖区域,白色区域为未改造区域,即天然基质区域。这里,裂缝沟通主要指相邻两口井的主裂缝相互沟通,形成高速导流通道,压力波在裂缝介质中向外传播,如图2(a)、(c)所示,通过这种沟通方式井间极易发生生产干扰;基质沟通主要指压力扰动通过基质传播,根据复合区域模型的概念[10],压力波依次通过主裂缝、体积改造区域、未改造区域向外传播,如图2(b)、(d)所示,这种情况下在有限的生产周期内井间发生生产干扰概率较低(或井间干扰强度较弱)。需要强调的是,不同沟通方式下,井距对单井生产动态的影响程度不同,直接影响井距优化结果。同时对于图2(c)、(d)的情况,由于井距足够小,井间裂缝可以认为全部沟通(灰色区域可视为一系列裂缝排列,缝长等于区域宽度),此时裂缝沟通与基质沟通几乎等效。本文重点研究基质沟通情况下的井间干扰及井距优化,裂缝沟通情况作为对比进行研究。

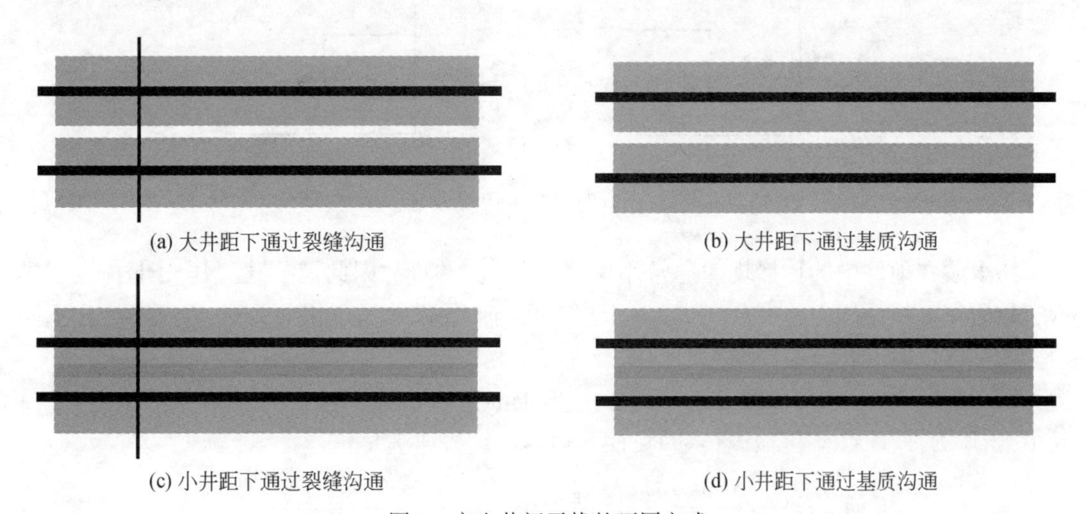

(a) 大井距下通过裂缝沟通　(b) 大井距下通过基质沟通

(c) 小井距下通过裂缝沟通　(d) 小井距下通过基质沟通

图2　产生井间干扰的不同方式

1.2　干扰模型

单井模拟引用三线性流模型[10,11]对体积压裂形成的多尺度地层进行分布描述，即分为主裂缝、体积压裂区（由两部分组成，缝间区域 SRV+缝尖控制区域 XRV）和天然基质区，如图 3 所示。

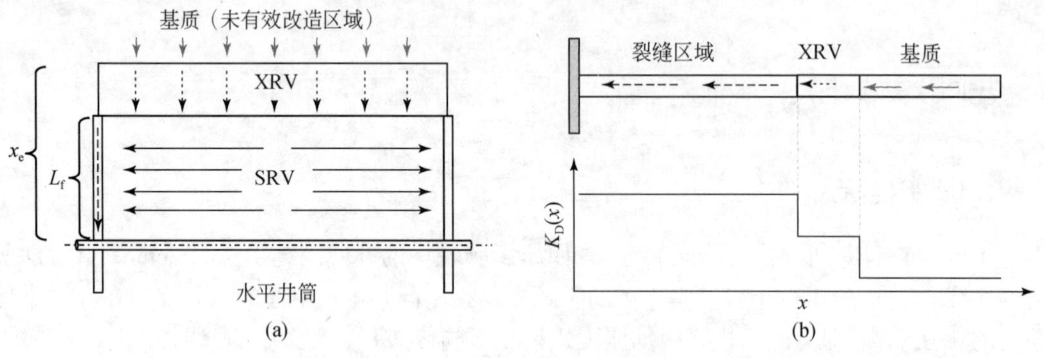

图 3　一维线性流分区流动模型

在同一渗流系统中多口井同时开井/关井时，所引起的压力扰动在介质中传播，当压力扰动发生相互干扰时，表现为某井的工作制度改变影响邻井的井底压力或产量，导致地层中的能量在井间进行重新分配，如图 4 所示。

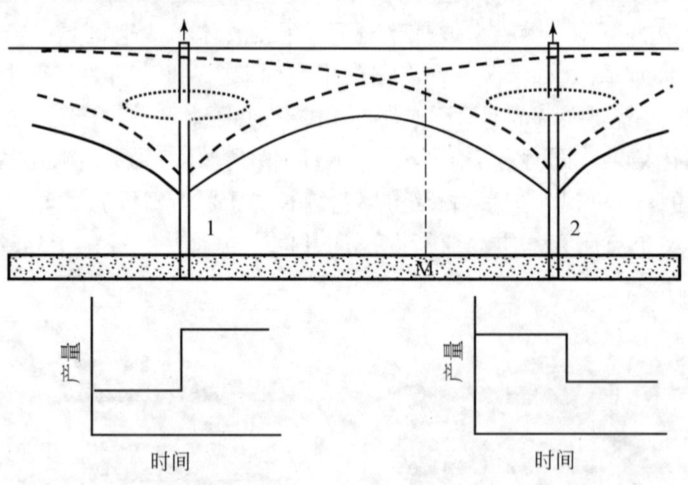

图 4　井间存在干扰时邻井生产动态响应对应关系

压力波在地层中的传播规律通常用探测半径进行模拟，因此需要建立压力干扰模型进行定量研究。根据压力传播公式[12,13]，干扰时间与传导系数呈正相关关系，当压力波在不同的介质中传播时干扰响应的发生时间为

$$\mathrm{Time_{total}} = \mathrm{Time_{fracture}} + \mathrm{Time_{XRV}} + \mathrm{Time_{matrix}} \tag{1}$$

其中，当压力波在 XRV 区域内传播时，探测边界移动规律满足：

$$t_{\mathrm{XRV}} = \frac{\phi_{\mathrm{f}}\,\mu_{\mathrm{gi}}\,c_{\mathrm{gi}}\,L_{\mathrm{ref}}^{2}}{3.6\times 6\times 10^{-3}K_{\mathrm{f}}}\left(\frac{K_{\mathrm{XRV}}}{K_{\mathrm{f}}}\right)^{2}\left(\frac{x_{\mathrm{f}}^{3}}{L_{\mathrm{f}}+(K_{\mathrm{XRV}}/K_{\mathrm{f}})(x_{\mathrm{f}}-L_{\mathrm{f}})}\right) \tag{2}$$

当压力波在基质内传播时,探测边界移动规律满足:

$$t_m = \frac{\phi_f \mu_{gi} c_{gi} L_{ref}^2}{3.6 \times 6 \times 10^{-3} K_f} \left(\frac{K_m}{K_f}\right)^2 \left(\frac{x_f^3}{L_f + (K_m/K_f)(x_f - x_e)}\right) \quad (3)$$

国际上通常是把单位压力脉冲引起的最大响应位置定义为探测边界,可用探测边界内地层对气井流量供给比例进行表征。对比文献[13]和文献[14]可见,经典公式在定产条件下的计算结果为68.3%,定压条件下为63.9%,本文解分别对应为91.68%,95.02%。这说明近似解增加了压力扰动的有效波及范围,提高了在探测边界内使用物质平衡方程的精度,更为重要的是本文可解决气井变产量生产和复合介质条件下的探测边界移动规律。

利用探测边界移动规律公式和压力分布公式可以获得不同介质参数下的干扰发生时间及压力波的传播规律。这里压力干扰观测点取两口井中间位置,其中通过裂缝沟通时观测点位于裂缝内部,通过基质沟通时观测点位于基质中,如图5所示。

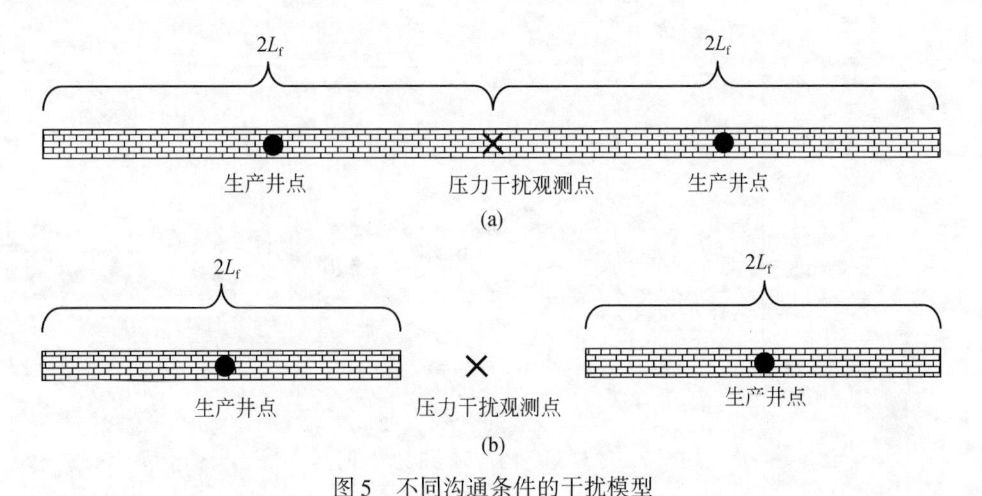

图5 不同沟通条件的干扰模型

通过裂缝沟通模型使用参数见表1,假设压裂水平井共15段,每段3簇,共45条有效主裂缝,日产量设定为4.5万m^3/d,裂缝半长150m,井距300m,即裂缝恰好沟通。根据探测半径公式可知,裂缝渗透率很高且传播时间很短,因此本模型未考虑基质内气体对气井产量的贡献,认为在基质供给前裂缝内压力波即发生干扰。

表1 裂缝沟通情况使用参数

P_i/MPa	S_g/%	S_w/%	C_t/MPa^{-1}	L_f/m	w_f/m	h/m	D/m	Q/($10^4 m^3$/d)
50	80	20	10^{-4}	150	0.01	30	300	0.1

模拟结果表明,根据探测边界传播公式计算结果与压力观测点记录结果拟合程度较高,也验证了本文模型的可靠性。利用模型模拟10%、20%、30%三种裂缝孔隙下不同裂缝渗透率与干扰发生时间的对应关系,如图6所示。图7反映了通过基质沟通时不同基质渗透率下干扰时间的变化规律,模型中裂缝半长为50m。

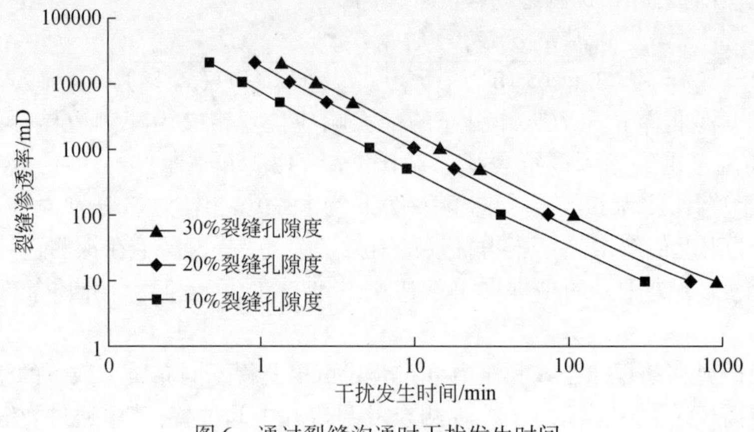

图6　通过裂缝沟通时干扰发生时间

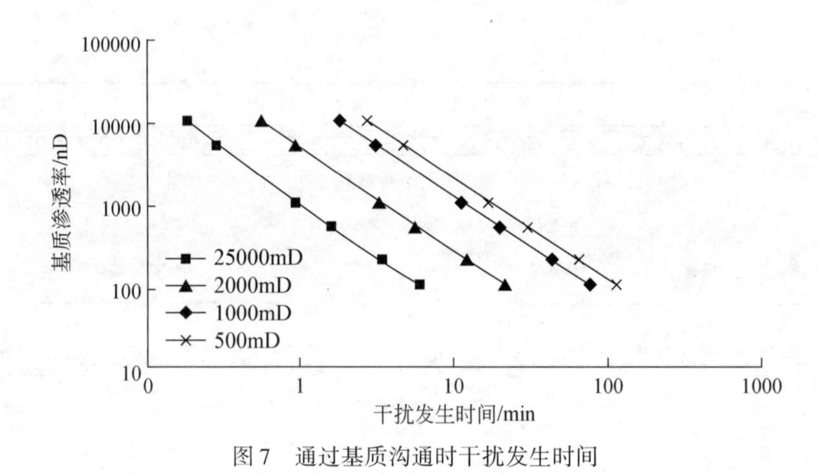

图7　通过基质沟通时干扰发生时间

2　井间干扰诊断和分析

基于井间干扰模拟的结果，充分利用实际动态数据，诊断井间干扰及干扰程度。当邻井进行开关井时，通过分析自身动态(产量、压力)响应可以较为直观地判断是否发生井间干扰。同时为了定量化表征井间干扰对生产动态的影响，使用产量递减、动态分析和产能评价等方法。

2.1　生产动态诊断

当邻井工作制度发生改变时，观察分析气井产量、压力变化规律是判断井间干扰最简单的方法，以 X 平台为例进行定性的诊断分析。

X 平台上半支共有三口井，进行了干扰试井测试。其中 X-1 井与 X-2 井距离 300m，X-2 井与 X-3 井距离 400m。相邻井间干扰测试表明井距小于 400m 时，井间存在部分干扰。图 8 反映了 X-1 井与 X-2 井的生产动态响应：在干扰测试期间，X-2 井提前开井，压降幅度为 4. 2MPa，同时引起邻井 X-1 井 2. 16MPa 的压降，以及相邻 X-3 井 0. 42MPa 的压降，说明 300m 井距条件下有较为明显的井间干扰。

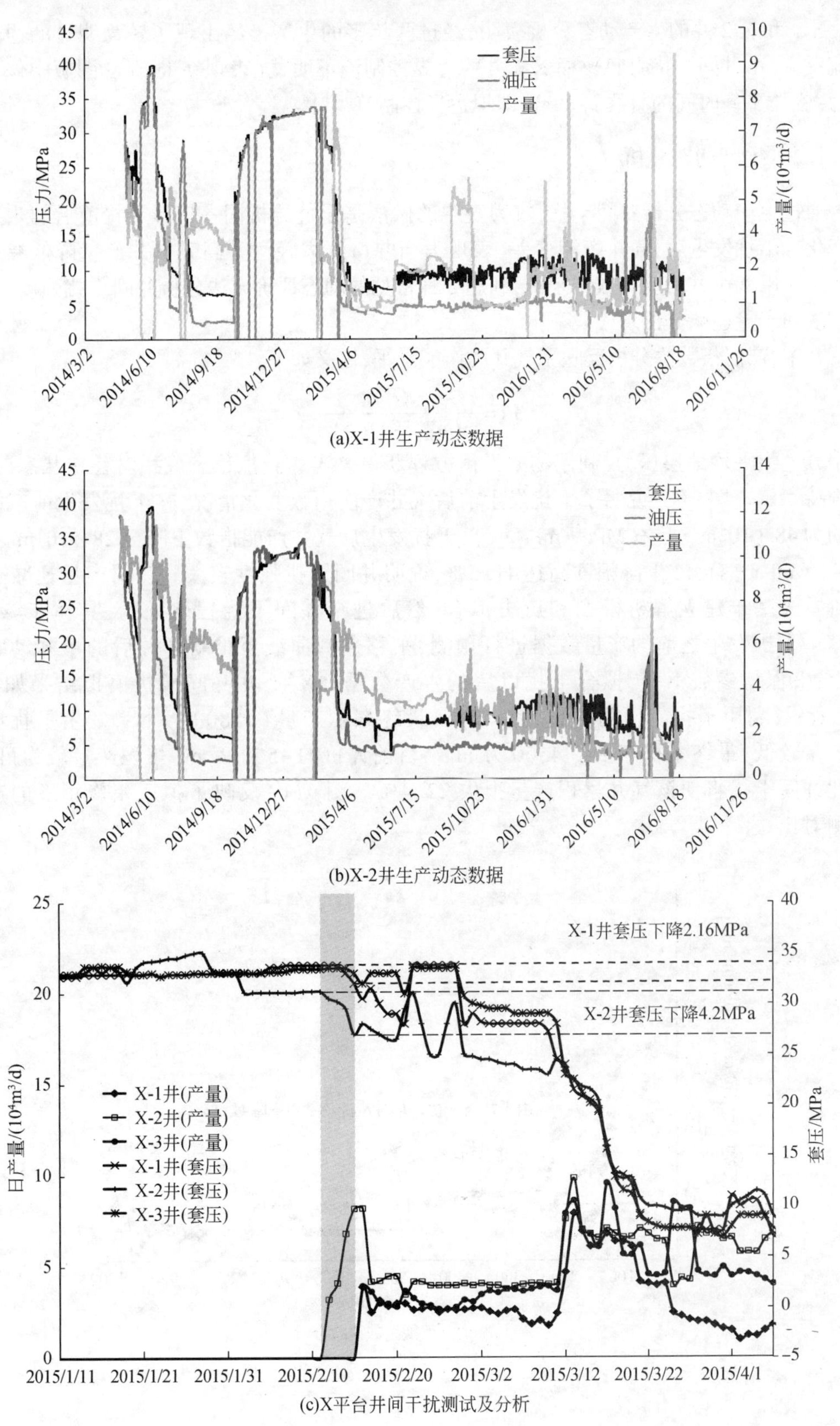

图 8 X 平台生产动态数据及分析

X-1 和 X-2 井的生产动态分析表明,经过两年多的生产已经出现了较为明显的边界控制流反应(由井间干扰引起),而 X-3 井的边界控制流不明显(表现在 log-log 图版中修正拟压力–物质平衡时间的直线斜率<1,即拟边界控制流[15])。

2.2　生产动态分析

通过生产动态分析井间干扰,较为直观的做法是通过分析同一口井干扰前后的生产动态变化。干扰发生后,气井产能变差,表现为当井口压力轻微升高时气井产量降低幅度较大。本文以 X-1 井为基础数据来源,利用 2 种常见的动态评价方法分析井间干扰强度:产能系数法和产量递减分析。

其中,产能指数法分析反映了气井的产能水平,定义如下:

$$PI=\frac{Q_{g}}{m(p_{avg})-m(p_{wf})} \tag{4}$$

式中,p_{avg}为平均地层压力,利用物质平衡方程获得。当达到拟稳态或相对稳定状态时,PI 值维持稳定。图 9(a)说明了干扰发生前后气井产能指数变化情况,在干扰发生前产能指数约为 48 $(10^4 m^3/d)/(MPa^2/mPa \cdot s)$,干扰发生后气井产能指数下降至 28 $(10^4 m^3/d)/(MPa^2/mPa \cdot s)$,PI 下降幅度高达 41.6%,说明井间干扰强度较大。使用产量递减分析评价首先基于建立解析模型和历史拟合,然后进行井间干扰程度评价,具体步骤是:①基于干扰发生之前的产量数据进行预测,平移递减曲线至重新开井后的单井动态数据上;②保持参数不变,拟合干扰以后的生产数据。X–1 井井间干扰模拟结果如图 9(b)所示,利用干扰前预测产量相对较高,最终累积产量约 5800 万 m^3,发生干扰后预测产量较低,最终累积产量约 4500 万 m^3。目前评价的 4500 万 m^3 累积产量较为可靠,因此井间干扰将使该气井累积产量损失 22.4%。图 9(c)反映了拟压差修正下的产量预测对比。

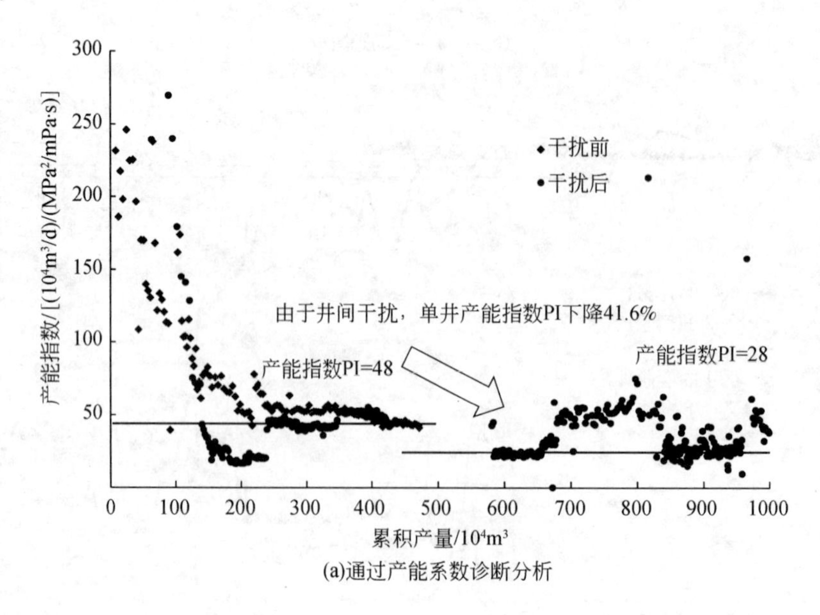

(a)通过产能系数诊断分析

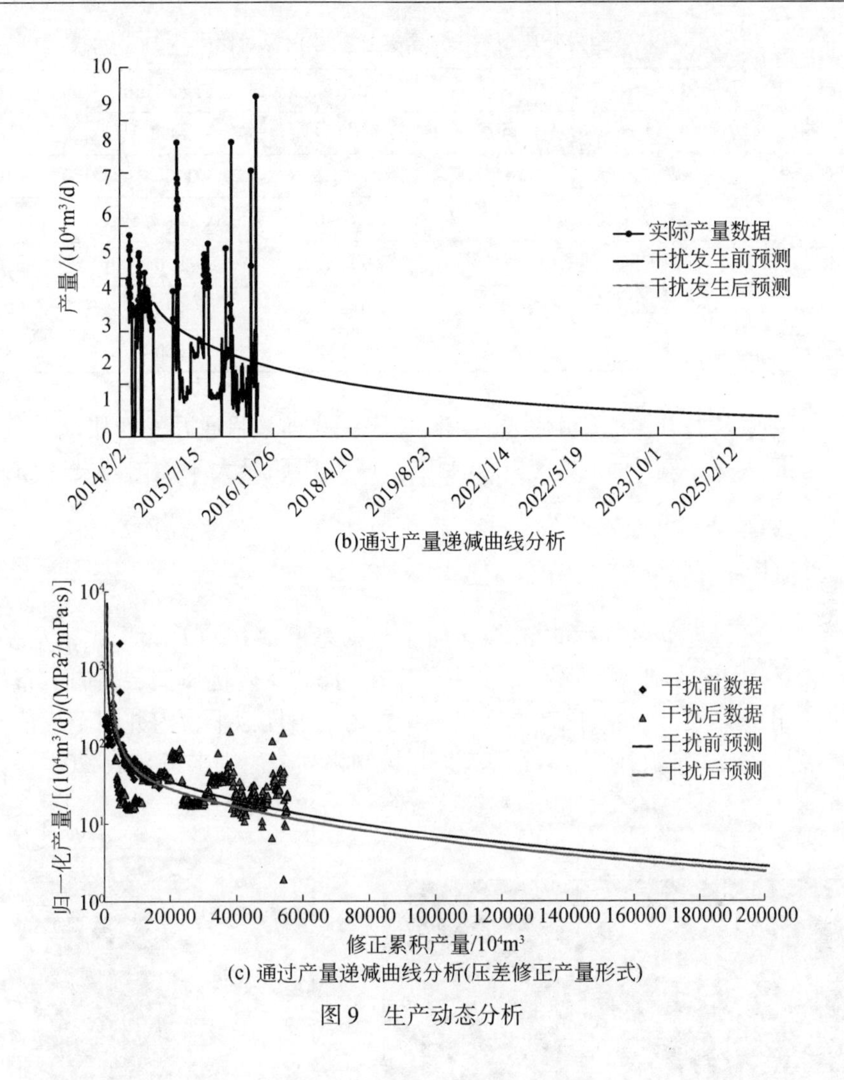

(b)通过产量递减曲线分析

(c) 通过产量递减曲线分析(压差修正产量形式)

图9 生产动态分析

3 合理井距论证

影响合理井距的因素较多,总体上分为可控因素和不可控因素,不可控因素包括地质参数、流体参数等,可控参数包括水平井、裂缝参数及生产制度等。井距论证方法主要分为类比法、数值模拟和经济评价三种方法。

3.1 类比法

类比法主要建立在相似的地质、工程背景下,借鉴已开发的成熟气田井距参数作为开发早期井距设计的依据,见表2。类比法的局限性在于当地质或工程参数近似度不高时,对比结果具有很高的不确定性,因此,类比法只能作为一种初步井距设计的参考。但通过与多个同类气田的详细类比,能够有效降低不确定性,能大致确定合理井距的范围。

表 2　国内外典型气田主要开发参数指标对比[15-19]

参数	Barnett	Haynesville	Marcellus	Eagle Ford	长宁—威远—昭通
水平段长度/m	1219	1402	1128	1494	1465
单段支撑剂量/t	129. 7	162. 3	181. 2	112. 6	101. 2
井控面积/km^2	0. 24 ~ 0. 65	0. 16 ~ 2. 27	0. 16 ~ 0. 65	0. 32 ~ 2. 59	0. 39 ~ 1. 0
平均井控面积/km^2	0. 45	0. 5	0. 42	0. 6	0. 6
裂缝半长/m	91 ~ 122	91	91 ~ 122	107	90 ~ 120
平均井距/m	280	260	260	300	400 ~ 500

从表 2 中影响气井井距的影响参数上看,长宁—威远—昭通开发区块气井单段加砂量明显低于美国四大页岩气开发区块,相反,目前开发井距明显大于美国开发区块,因此开发井距存在优化空间。

3. 2　数值模拟法

数值模拟法通过建立全过程的单井/区块生产动态模型,以 EUR、采收率及经济指标为目标函数,综合考虑多参数对井距评价结果的影响。海相页岩优质储层横向展布相对稳定,通常采用平台化批量布井开发,因此本文以三口井为一组模拟平台内井间干扰情况,分布模式及相关参数如图 10 所示,使用参数见表 1,相关网格划分原则参考文献[20]。

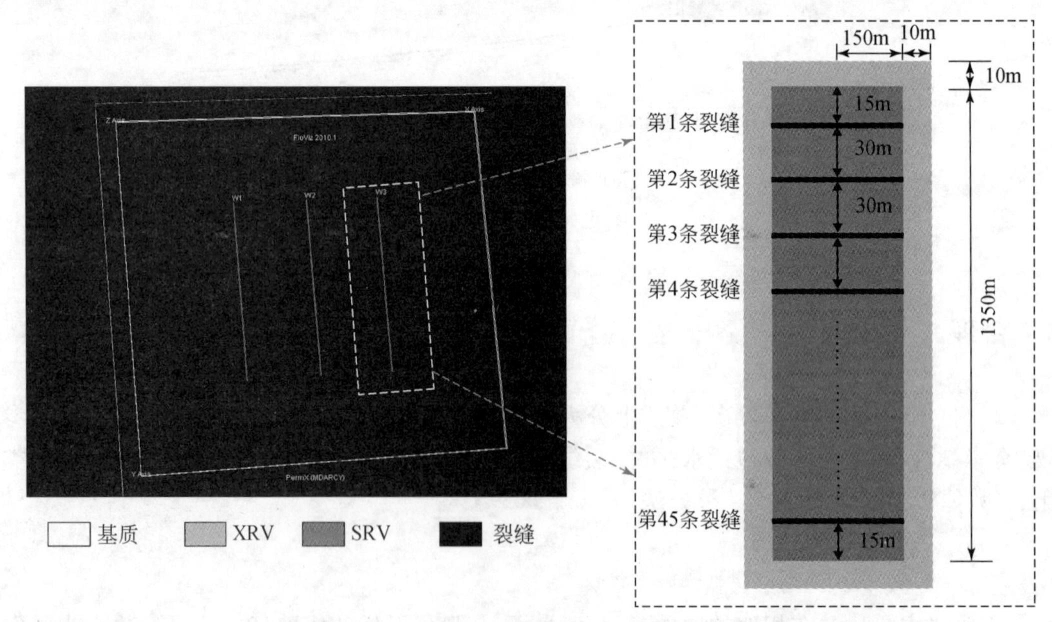

图 10　平台水平井分布模式(W2 代表中心井)

为了对比井间干扰对单井生产动态的影响,提出两种模拟方案:①模拟单井生产动态(无井间干扰);②模拟平台生产动态(有井间干扰),通过比较单井模式和多井模式下的开发指标(主要是单井 EUR)来量化井距对生产动态的影响程度。

图 11 和图 12 展示了两种方案下压力场分布及井间干扰引起的中间井产量降低模

拟，这里定义 ΔG 表征累积产量降低量。随着生产时间的增加，ΔG 值逐渐增大，说明井间干扰强度越来越大，因此缩小井距提高了区块储量平面动用程度，但单井 EUR 在不断降低。

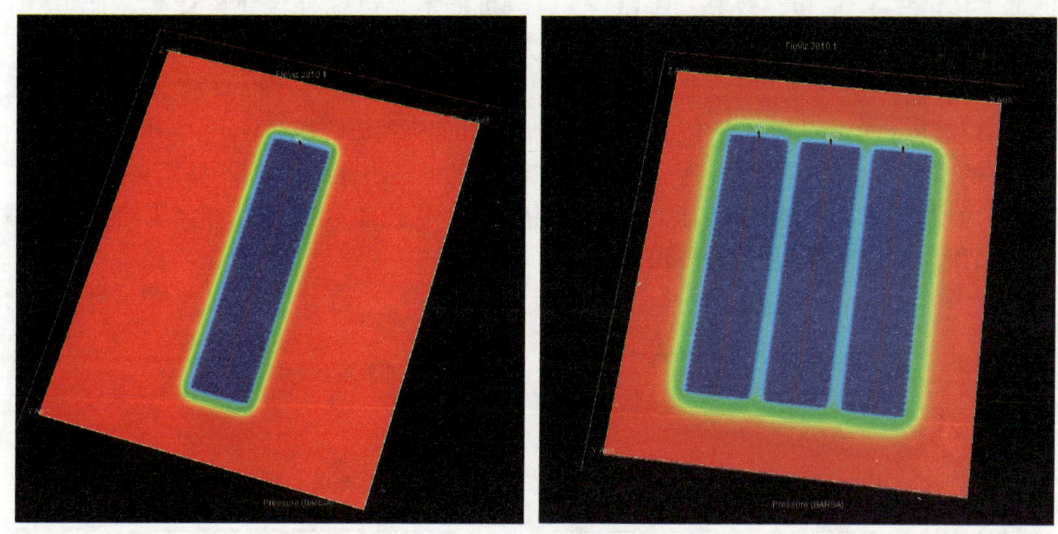

图 11 平台水平井分布模式(W2 代表中心井)

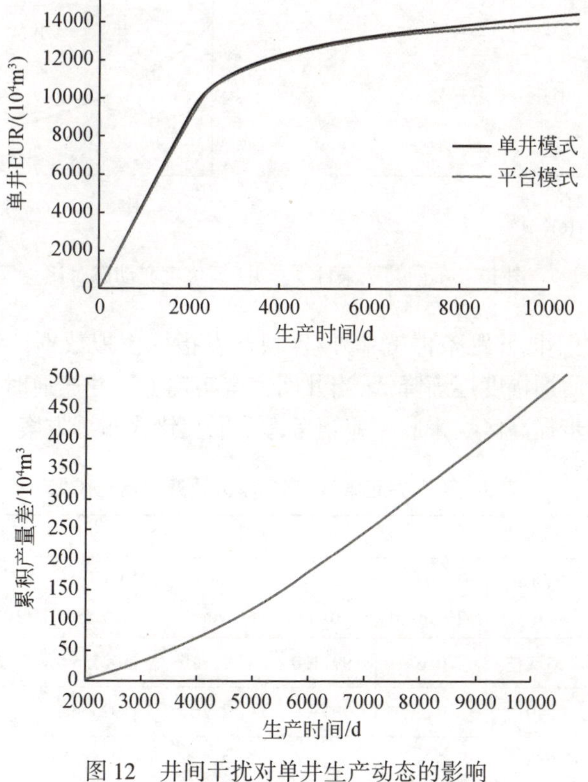

图 12 井间干扰对单井生产动态的影响

从产能系数角度分析，在同一区块内减小井距可以增加井数，即增加了裂缝与地层的接触面积，可以不断提高区块整体的产能系数，因此区块累积产量增加幅度较快，在有限的生产周期内可以采出更多的气量，如图13(a)所示；增大井距，井数相应减少，井间干扰强度降低，单井(累积)产量相对较高[图13(c)]，但可能不足以弥补区块整体生产能力，如图13(b)所示。在井间干扰发生前，不同井距方案下单井生产动态相同[图13(d)]，但小井距方案中井数较多，因此区块的整体产量较高；当井间干扰发生后[图13(c)、(d)中曲线偏离点对应时刻]，在大井距方案中，区块总体产量受井间干扰影响相对较小，单井可以保持相对较高的产量，但由于井数的差距，区块累积产量仍然低于小井距方案；小井距条件下，随着单井压力控制范围的不断扩大，压力干扰出现较早，干扰强度不断加强，相当于储量基础不断减小，因此单井产量递减速度较快，如图13(d)所示。

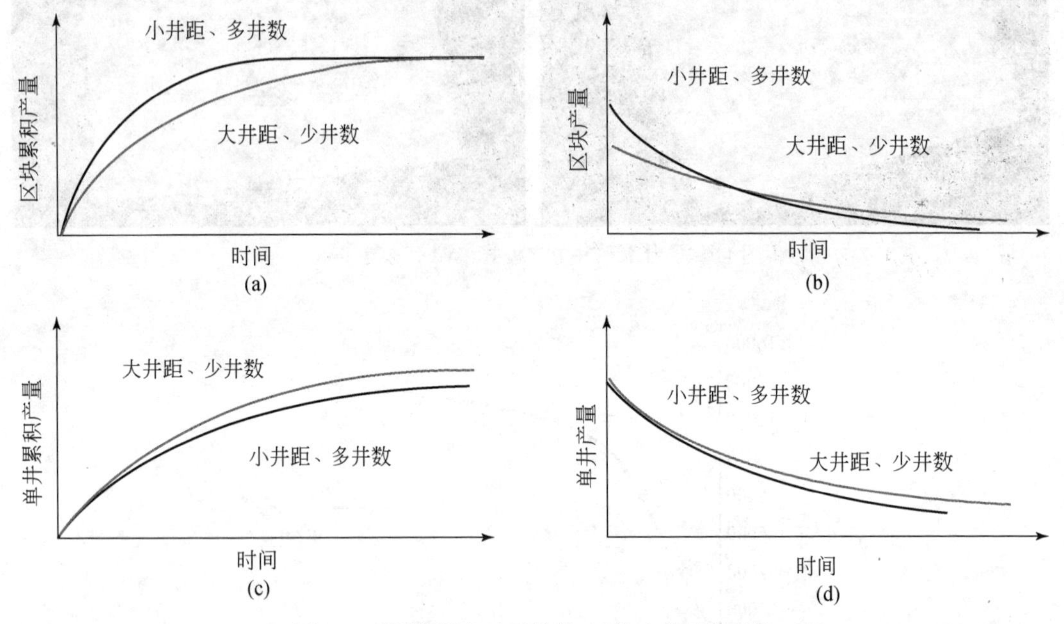

图13　不同井距条件下单井、区块生产动态对比

表3是同一区块不同井距条件下单井/区块开发指标模拟数据。结果表明随着井距的减小，区块累积产量增加幅度逐渐降低，当井间完全实现了裂缝连通时区块累积产量达到最大值，此时若要进一步提高区块采收率应当考虑采取交错式布缝方案。

表3　不同井距条件下单井/区块开发指标模拟

单井					平台				
井数	井控面积/km^2	井间距/m	初始产量/(10^4m^3/d)	EUR/10^4m^3	总面积/km^2	OGIP/10^4m^3	初始产量/(10^4m^3/d)	EUR/10^4m^3	采收率/%
4	0.68	511	0.6	9911.0	2.72	192556	2.5	39643.8	21
6	0.45	340	0.6	9911.0	2.72	192556	3.5	59465.7	31
8	0.34	255	0.6	9911.0	2.72	192556	4.6	79287.6	41

续表

单井					平台				
井数	井控面积/km^2	井间距/m	初始产量/(10^4m^3/d)	EUR/10^4m^3	总面积/km^2	OGIP/10^4m^3	初始产量/(10^4m^3/d)	EUR/10^4m^3	采收率/%
10	0.27	204	0.6	9911.0	2.72	192556	6.0	99109.5	51
12	0.23	170	0.6	9344.6	2.72	192556	7.1	110436.3	57
14	0.19	146	0.6	8495.1	2.72	192556	8.1	116099.7	60
16	0.17	128	0.5	7362.4	2.72	192556	8.5	118931.4	62

表4为川南—黔北地区海相页岩气区块典型井井距优化前后指标对比，优化前井距为400m，根据有效缝长和控制面积内的穿透比进行井距优化，可以看出优化后单井EUR略有降低，但能够较大幅度地提高井点储量采出程度，达到充分利用资源的目的。

表4 川南—黔北地区海相页岩气区块典型井井距优化前后单井开发指标对比

井号	优化前			优化后		
	有效裂缝半长/m	EUR/10^8m^3	井点储量采出程度/%	井距/m	EUR/10^8m^3	井点储量采出程度/%
X1	130.2	1.03	25.6	289	0.98	33.7
X2	176.3	0.68	19.3	392	0.67	19.5
Y1	167.5	0.99	28.9	373	0.96	30.3
Y2	122.1	0.72	28.6	271	0.71	40.7
Z1	147.2	0.93	26.5	327	0.90	29.7
Z3	129.5	0.72	21.4	288	0.68	28.9

3.3 经济评价法

净现值模型是一种广泛用于油气行业评价投资项目经济效益的模型，净现值是按行业基准折现率或其他设定的折现率计算的各年净现金流量现值的代数和。其中现值P与未来值F关系满足：

$$P=F/(1+i)^n \tag{5}$$

对于气藏开发，实际井距/井数优化结果主要受经济效益影响。井距减小、井数增加整体提高了气田储量采出程度，但同时也增加了钻完井成本。因此本文引入净现值(NPV)模型优化井距/井数，形式如下：

$$\mathrm{NPV}=\sum_{j=1}^{n}\frac{V_{\mathrm{F},j}}{(1+i)^j}-\left[\mathrm{FC}+\sum_{k=1}^{N}(C_{\mathrm{well}}+C_{\mathrm{fracture}})\right] \tag{6}$$

式中，V_F为产气所获收益，主要受产气量和气价影响；FC为固定总投入；C_{well}为单口水平井钻井成本；$C_{fracture}$为单口水平井压裂成本；N为总投产井数；i为年利率；n为生产年限。

基于区块生产动态模拟结果，结合式(6)评价5年生产周期和30年生产周期时的经济评价结果。基本经济指标参数参考文献[20]，见表5。

表5 主要经济评价指标参数

水平井长度/m	成本/(万元/口)	裂缝半长/m	成本/(万元/口)	参数	值
304. 8	1366	76. 2	68. 3	气价	0. 72 元/m^3
609. 6	1434	152. 4	85. 4	利率	10%
914. 4	1503	228. 6	102. 5	开采费	12. 5%
1219. 2	1571	304. 8	119. 5		

设定所有井均部署在1. 6km×3. 2km的区块内,所有井均同时开井生产,采用早期配产,后期转定压的生产制度,结合目前工程参数,按当前单井综合投资5500万元计算,设计6种井距方案,对比结果见表6。随着生产年限的增加,区块净现值不断增加,在相同生产时间内存在着最优井距,即井距260m、部署6口井最优,最优值不随生产周期的改变而改变。

表6 不同井距经济评价结果

方案	井数	30年NPV/亿元	20年NPV/亿元	10年NPV/亿元	5年NPV/亿元
1	3	2. 4588	1. 96704	1. 59822	1. 2294
2	4	2. 9369	2. 41782	2. 02851	1. 6392
3	5	3. 2784	2. 67736	2. 22658	1. 7758
4	6	3. 515	2. 8686	2. 4588	2. 049
5	7	3. 4833	2. 85494	2. 38367	1. 9124
6	8	3. 6199	2. 90958	2. 37684	1. 8441

4 讨论

本文研究是建立在概念模型的基础上,即以分段多簇压裂水平井为基本井型,主裂缝等长、均匀分布,与实际开发中存在一定的差异。本文重点研究井间距对生产动态的影响,而未考虑其他影响因素(包括裂缝长度、基质渗透率、有效裂缝间距、裂缝导流能力及天然裂缝等),这里主要讨论影响合理井距的几个可控因素。

4.1 裂缝沟通

井间裂缝沟通后,增加了裂缝与地层的接触面积,提高了单井产能,因此对提高单井EUR有一定的积极作用。其中,影响裂缝沟通的主要因素是裂缝导流能力,在生产早期,主要流动区域集中在裂缝控制区域(SRV),因此井间裂缝导流能力对生产动态影响较小;到生产中后期,井间裂缝导流能力影响变得明显。当井间裂缝导流能力较强时,对单井生产动态影响较大,见表7。

表7 (无限导流)裂缝沟通对单井指标影响

年限/年	无裂缝沟通/$10^4 m^3$	存在裂缝沟/$10^4 m^3$	提高程度/%
1	3768.993	4275.867	12
5	8302.544	9879.801	16
10	10012.89	12380.19	19
60	13753.57	19700.14	31

随着井距的减小,连通区域中井间裂缝长度和其能沟通的基质面积不断降低,因此,井间裂缝的作用将逐渐降低。裂缝沟通下的平台压力场如图14所示。

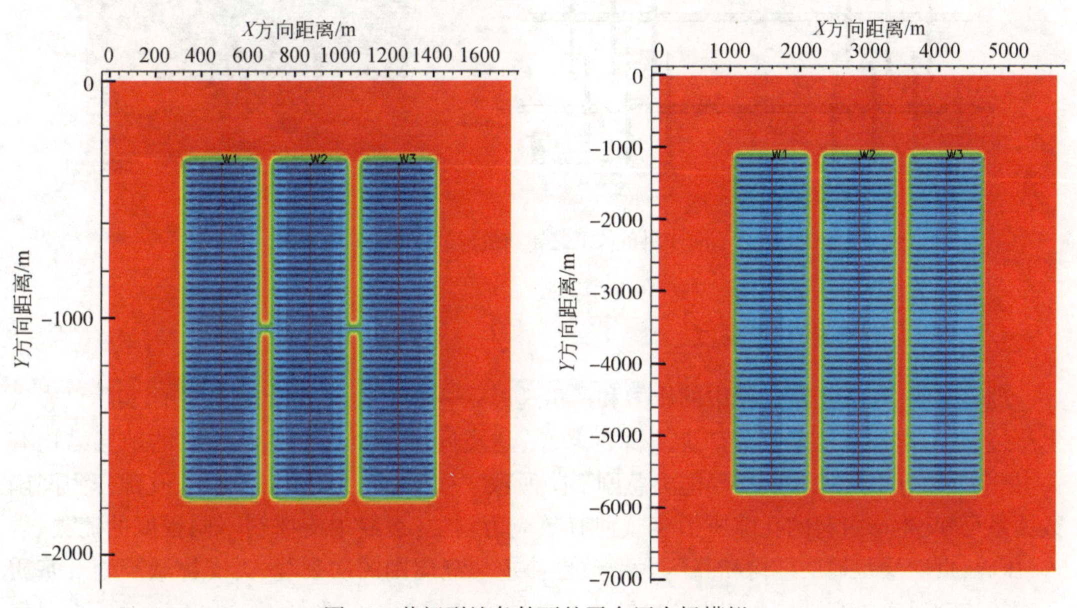

图14 井间裂缝条件下的平台压力场模拟

4.2 交错布缝模式

采用交错布缝可以有效动用井间储量,同时在一定程度上可以减缓井间干扰。研究表明:对于裂缝间距较小的情况,交错布缝优势不明显;但当裂缝间距较大时,交错布缝可以显著提高单井产能[21]。主要原因在于裂缝间距较小时,缝间干扰强度远大于井间干扰,因此,井间干扰的作用相对较弱;裂缝间距较大时情况则恰好相反,同时,交错布缝可以形成更大的有效动用面积,提供了更多的有效储量,如图15(a)、(b)所示。此外,交错布缝为进一步缩小井距提供了空间,如图15(c)、(d)所示。常规布缝时,当井间距等于裂缝长度时,区块产能达到最高值,无缩小井距的空间(进一步减小井距相当于减小裂缝的有效长度)。而交错布缝时,当井距小于裂缝长度时,裂缝“嵌入”邻井裂缝间,在保证总的裂缝-地层接触面积不变条件下,相当于减小裂缝间距,缩短了气体从基质运移至裂缝的距离,提高了气藏整体流动能力。同时在保证单井产能的基础上,减少了井间和缝间储量动用“盲区”,提高了区块整体的资源利用率。

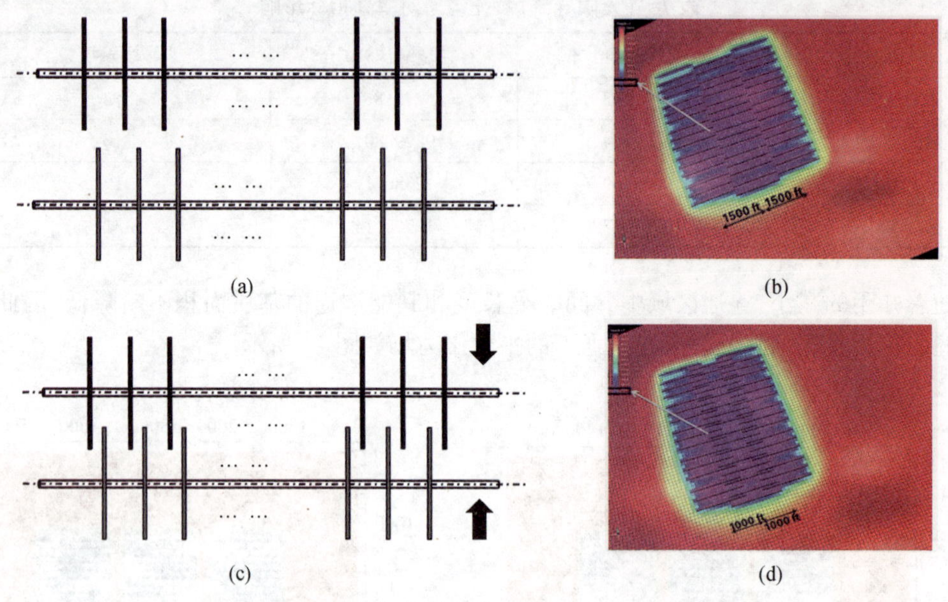

图 15　不同井距条件下交错布缝模式及生产动态模拟

4.3　应力敏感和生产制度

理论上讲，气藏开发过程中的压力和产量递减是一个连续、缓慢的过程。而实际页岩气生产过程中，出现了很多气井产能突降的现象，这类问题与页岩薄层状分布的地层结构有关，是应力敏感引起的介质传导能力急剧下降所致。根据有效应力与总应力、孔隙压力间函数关系可知，生产引起的孔隙压力衰减使有效应力增加，裂缝中支撑剂不同程度的流失、嵌入、压碎，页岩层状地层结构和高压力系数也使主裂缝周围的次裂缝、无支撑微裂缝及基质孔喉等流动通道更容易发生变形[22]，导致流动通道渗透能力的降低，如图 16 所示。页岩气开发实践也表明，与“放压生产”相比，“控压限产”除了在增加吸附气解吸量、降低出砂风险等方面存在优势外，更重要的作用在于能保持人工裂缝(网络)的长期开启，特别是延长主裂缝周围无支撑微裂缝及基质孔喉等流动通道的“寿命”，缓解裂缝及基质渗流场的应力敏感效应。

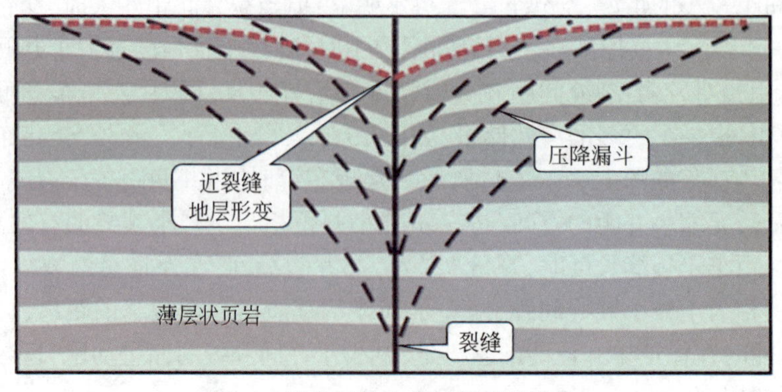

图 16　页岩气井生产对储层的影响

以本文模型为基础,考虑应力敏感效应,对比放压和控压两种生产制度下气井产量。表8中的模拟计算结果表明,生产制度的选择对气井生产潜力的发挥程度影响显著。而不同井距条件下,压裂过程形成的应力场也不同,对应生产过程中的应力敏感程度也不同,同时,单井的有效泄流面积也将发生变化,这将直接影响井距优化结果,此时存在最优井距:若小于最优井距,单井产量将显著降低,若高于最优井距,单井产量增加幅度极小,具体结论需要下一步工作进行论证。

表8 考虑应力敏感时放压与控压条件下的EUR预测

生产时间/a	1	2	3.6	5	10	15	20	25	EUR
放压累积产量/10^4m^3	2214	3145	4161	4718	5944	6576	6964	7226	7346
控压累积产量/10^4m^3	1460	2582	4161	5132	7405	8290	8752	9033	9151

5 结论

本文通过引入井间干扰模拟、井间干扰响应识别、多井生产动态模拟、经济评价等方法,建立起完整的井距优化工作流程,并给出川南—黔北地区海相页岩气区块井距优化的结果。

(1)井间干扰存在基质和裂缝沟通两种形式,通过裂缝沟通时井间干扰快速发生,通过基质干扰时至少需要数小时才会发生较弱的压力响应。利用井间干扰测试资料,通过定量分析生产制度(开关井)改变对邻井生产动态的影响,可以识别井间是否存在有效连通。

(2)井间干扰特征直接影响井距优化结果,主要受裂缝长度、基质渗透率控制。基质渗透率较高时,可以适当增加井距;裂缝长度较短时,可以适当缩小井距。

(3)当井间裂缝沟通时,可使井间干扰提前发生,裂缝导流能力大小决定区块累积产量提高程度,同时会影响经济评价结果,对应的最优井距结果相对较大。

(4)缩小井距、增加井数可以提高区块整体产能水平,增加储量平面动用程度,但相应单井累积产量将降低,同时导致整体投资增加。当井距设计合理时,在充分利用资源的前提下,可获得较高的净现值。

(5)根据川南—黔北地区海相页岩气区块的地质和工程参数条件,目前采用的400~500m开发井距有进一步缩小至300~400m的空间,优化后区块储量采出程度可提高5%~10%。

参考文献

[1] 贾爱林, 位云生, 金亦秋. 中国海相页岩气开发评价关键技术进展[J]. 石油勘探与开发, 2016, 43(6): 949-955.

[2] 王志刚. 涪陵页岩气勘探开发重大突破与启示[J]. 石油与天然气地质, 2015, 36(1): 1-6.

[3] Suarez M, Pichon S. Completion and well spacing optimization for horizontal wells in pad development in the Vaca Muerta shale[A]. SPE 180956-MS, 2016.

[4] Sahai V, Jackson G, Rai R. Optimal well spacing configurations for unconventional gas reservoirs[A]. SPE 155751-MS, 2012.

[5] Baker M, Mazumder S, Sharma H, *et al.* Well design and well spacing optimization in unconventional

plays[A]. SPE 159325-MS, 2012.

[6] 何东博, 王丽娟, 冀光, 等. 苏里格致密砂岩气田开发井距优化[J]. 石油勘探与开发, 2012, 39(4): 458-464.

[7] Malayalam A, Bhokare A, Plemons P, *et al.* Multi-disciplinary integration for lateral length, staging and well spacing optimization in unconventional reservoirs[A]. URTeC: 1922270, 2014.

[8] Marongiu-Porcu M, Lee D, Shan D. Advanced modeling of interwell-fracturing interference: an Eagle Ford shale-oil study[J]. SPE Journal, 2016, 1-15.

[9] Awada A, Santo M, Lougheed D, *et al.* Is that interference? A work flow for identifying and analyzing communication through hydraulic fractures in a multiwall pad[A]. SPE 178509-MS, 2015.

[10] Brown M, Ozkan E, Raghavan R, *et al.* Practical solutions for pressure-transient responses of fractured horizontal wells in unconventional shale reservoirs[C]. Paper SPE 125043 presented at the SPE Annual Technical Conference and Exhibition, New Orleans, USA, 4-7 October 2009.

[11] Song B, Ehlig-Economides C. Rate-normalized pressure analysis for determination of shale gas well performance[C]. Paper SPE 14403 presented at the SPE North American Unconventional Gas conference and Exhibition, Texas, USA, 14-16 May 2011.

[12] 王军磊, 李波, 赵凡, 等. 致密气藏压力扰动传播规律[J]. 大庆石油地质与开发, 2015, 34(1): 63-67.

[13] Nobakht M, Clarkson C. A new analytical method for analyzing linear flow in tight/shale gas reservoirs: constant-flow-pressure boundary condition [C]. Paper SPE 143989 presented at the Americans Unconventional Gas Conference, Texas, USA, 12-16 June 2011.

[14] Wattenbarger R A, EI-Banbi A H, Villegas M E, *et al.* Production analysis of linear flow into fractured tight gas wells[C]. Paper SPE 39931 presented the 1998 Rocky Mountain Regional Symposium and Exhibition, Denver, Colorado, USA, 5-8 April 1998a.

[15] Pratikno H, Reese D E, Maguire M M, *et al.* Spacing pilots performance study using public data in the Barnett shale[A]. SPE 168937-MS, 2013.

[16] Boulis A, Jayakumar R, Rai R. Application of well spacing optimization workflow in various shale gas resource: lessons learned[A]. IPTC 17150, 2013.

[17] Krisanne L, Sean W, Josh J, *et al.* Marcellus shale hydraulic fracturing and optimal well spacing to maximize recovery and control costs[A]. SPE 140463-MS, 2011.

[18] Douglas H, Hector B, Mark M, *et al.* Searching for the optimal well spacing in the Eagle Ford shale: a practical tool-kit[A]. SPE 168810-MS, 2013.

[19] Hemali P, John J, Michael F. Borehole microseismic, completion and production data analysis to determine future wellbore placement, spacing and vertical connectivity, Eagle Ford shale, South Texas [A]. SEG 2014-0572, 2014.

[20] Svjetlana L. Fracutre face interference of finite conductivity fractured wells using numerical simulation[M]. University of Oklahoma, 2008.

[21] Yu W, Sepehrnoori K. Optimization of well spacing for Bakken tight oil reservoirs[A]. URTeC: 1922108, 2014.

[22] Britt L K, Smith M B, Klein H H, *et al.* Production benefits form complexity-effects of rock fabric, managed drawdown, and propped fracture conductivity[A]. SPE 179159-MS, 2016.

工程因素对页岩气产量的影响
——以北美 Haynesville 页岩气藏为例

李庆辉[1,2]　陈　勉[1]　Fred P. Wang[2]　金　衍[1]　李志猛[1]

(1. 中国石油大学(北京)石油工程学院;2. University of Texas at Austin)

摘要:我国页岩气开发尚处于前期探索阶段,分析国外成功的开发案例具有重要的借鉴作用和参考价值。在对北美 Haynesville 页岩气藏生产数据进行统计和分析的基础上,总结了工程因素对该页岩气藏开发的影响规律。研究表明,页岩气井水平段长度、压裂级数、加砂量、油嘴尺寸、井底压力和是否采取重复压裂施工对累积产量有显著影响;水平段长度在 1500～1650m、压裂级数 12～17 级、加砂量 1500～3000t、油嘴尺寸 8～9mm 为该地区常用且效果较好的设计方案;井底压力高的区域页岩气产量相应较高,这与井底超压改善页岩脆性、优化压裂效果有关;重复压裂能够显著提高页岩气总产量,作业时机的选择与增产效果关系密切。

关键词:页岩气;工程因素;生产能力;水平井;压裂(岩石)规模;重复压裂;增产效果

北美地区已圈定页岩气和页岩油盆地超过 70 个。2010 年美国页岩气产量约为 $1270\times10^8m^3$,预计到 2020 年将增至 $5000\times10^8m^3$。在当前能源供需矛盾突出的情况下,页岩气的高效开发可有效缓解国家能源的对外依存情况,降低 CO_2 排放。因此,页岩气的开采对我国乃至世界各国均具有重要意义[1-3]。

含气页岩的品位、供气能力和产能持久性将决定页岩气开发的成功与否。页岩的品质和供气能力主要由总有机碳含量、热成熟度、含气量、厚度、脆性、压力梯度和矿物组成决定。而产能的持久性除与储层特性有关,还同时受到水平钻井和分段压裂等关键工程因素的显著影响。由于页岩具有低孔、低渗的特征,在生储共生的情况下,理论上的渗流理论和规律尚未建立起来。研究钻井和压裂工程因素对产能的影响,对了解产气规律、优化钻完井设计具有重要意义。

本文在前人研究的基础上,以北美 Haynesville 页岩为例,结合实际资料和测试成果,综合运用产能分析和模拟方法,分析水平钻井和压裂施工等关键因素对页岩气产能的影响,研究了该地区最新的钻井和压裂趋势,为我国页岩气开发过程中把握关键设计因素、借鉴相关经验提供了依据。

1　Haynesville 页岩气藏勘探开发现状

Haynesville 页岩气盆地位于美国得克萨斯州东部与西路易斯安那州交界,是北美地区少数产能巨大的页岩气盆地之一。Haynesville 页岩广泛分布于路易斯安那州的 Bienville、Bossier、Caddo、De Soto、Red River、Sabine 等郡和得克萨斯州的 Harrison、Nacogdoches、Panola、San Augustine、Shelby 等郡[4]。

晚侏罗世 Haynesville 页岩是一套在相对半封闭沉积环境下发育的高碳泥页岩,南西方向由碳酸盐岩台地包围，北东方向为大陆架构造。Haynesville 页岩与下伏的 Smackover、Buckner 石灰岩、碳酸盐岩呈不整合接触，与下伏的硅（砂质）Bossier 地层呈局部整合接触。构造形态上受石炭纪 Sabine 台地隆升和北路易斯安那州—得克萨斯州东部盐丘盆地形成的影响，该套页岩形成 NE-SW 向倾斜产状[5,6]。

古地理构造及环境影响着现今岩层的矿物组成。Haynesville 页岩在盆地的北及北西方向为高黏土–硅质页岩，越往南西方向过渡碳酸盐矿物和总有机碳含量逐渐升高。气藏产层的孔隙度在 8% ~14%，束缚水饱和度和总有机碳含量相对较低，绝大多数游离气储存在非有机质骨架。页岩总体矿物组成中方解石和石英等脆性矿物含量为 14% ~35%，天然裂缝及地层超压增强了页岩的脆性[1,6,7]。

自 2007 年发现到 2011 年 10 月为止，已钻页岩气井约 1500 口，日产气约 $1.56\times10^8m^3$，累积产气量已超过 $710\times10^8m^3$。与开发历史 30 年（1981 ~2011 年），累积产能 $2550\times10^8m^3$ 的 Barnett 页岩（14900 口井）相比，Haynesville 页岩气藏无疑是开发潜力极大的非常规能源。

由于缺乏共享数据，不同作业公司在该地区开发页岩气时采用的工艺技术有很大差距，已完钻的 1500 多口井中水平段长度范围在 427.6 ~2220.8m，压裂级数 3 ~20 段，支撑剂用量 460.7 ~3413.0t，油嘴尺寸 3 ~25mm。

考虑到目前 Haynesville 页岩尚处于开发早期，绝大多数生产井未停止生产，因此最终产量也无法确定。在分析上述工程因素对页岩气产能的影响时，采用初始产量、1 年和 2 年累积产量进行研究。

从图 1（a）、（b）可以看到，初始产量和单井累积产量之间呈现大致的相关关系，初始产量越高，最终产量也相应越高。从图 1（b）、（c）看，1 年和 2 年累积产量与单井累积产量的相

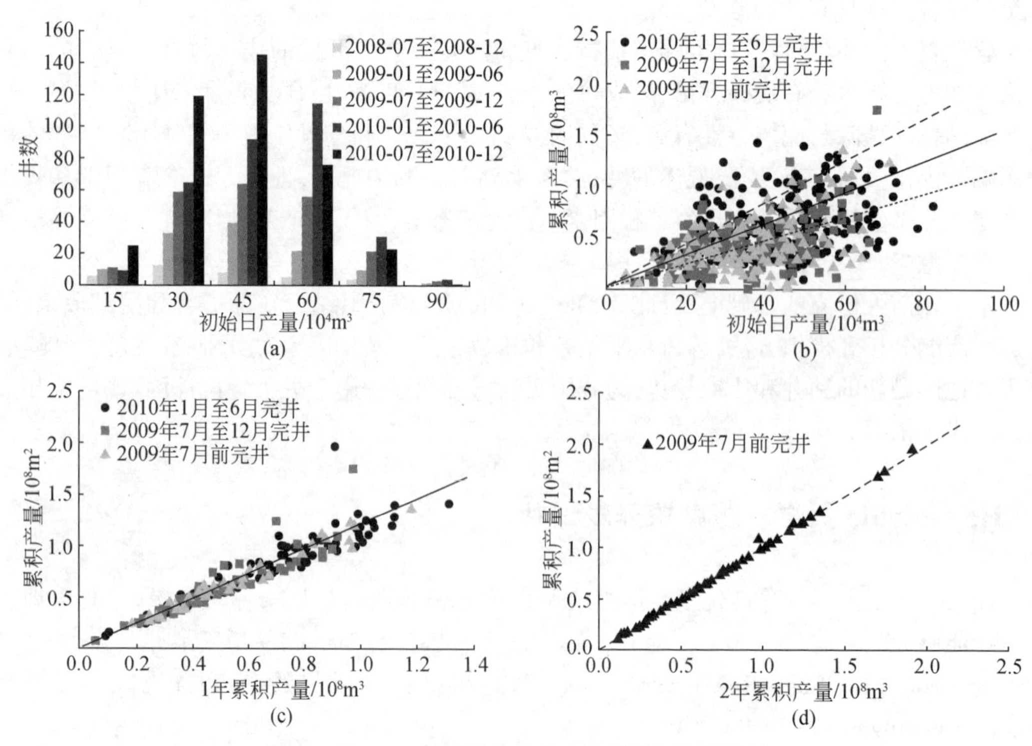

图 1　Haynesville 页岩气井生产数据的规律分析图

关关系非常明显,1 年和 2 年累积产量越高,单井累积产量也越高。因此,对于尚处生产期的页岩气井来说,采用初始产量、1 年和 2 年累积产量作为指标,分析工程因素对产能的影响具有可行性和借鉴性。

2 工程因素对产能的影响

2.1 水平段长度影响

Haynesville 页岩开发过程中极大地借鉴了临近 Barnett 页岩气开采过程积累下的经验。水平钻井和多段压裂是其普遍采用的开发技术。

一般页岩气井的水平段越长采气面积越大,储量的控制和动用程度越高。但是水平井的设计长度并不是越长越好,水平段越长施工难度越大,脆性页岩垮塌和破裂等复杂问题越突出。同时,由于井筒压差的存在,水平段越长抽吸压力越大,总体页岩气产量反而降低。此外,从经济技术的角度考虑,水平段越长,钻井及开发耗费资金越多,成本越高(图 2)。

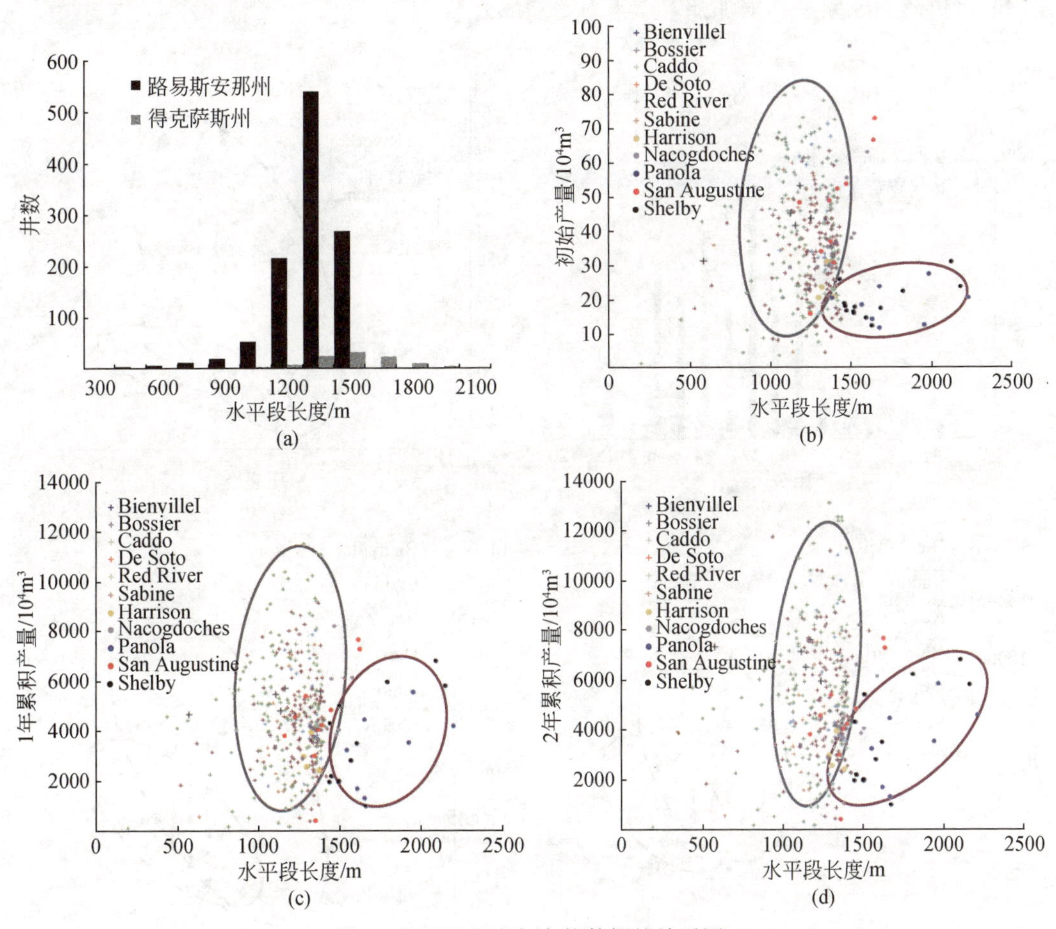

图 2 水平段长度与产能数据的关系图

统计路易斯安那州和得克萨斯州页岩气井水平段的长度发现,两个州分别倾向于选择

1500m 和 1650m 水平段长度完井，路易斯安那州集中在 1200 ~ 1500m，且以 1500m 为最多，两侧近似对称分布；得克萨斯州集中在 1350 ~ 1800m，以 1650m 为最多，两侧近似对称分布[图 2(a)]。

分析显示，两州页岩气井产能情况有差异，1500m 以上产能与水平段长度线性关系变差。虽然两州 1000 ~ 1500m 水平段长度井数均最多，初始产量、1 年和 2 年累积产量最高的生产井也在此长度段范围内，该长度段内产量差异性非常明显，反映了页岩气钻井过程中水平段长度不是决定产能的唯一因素。得克萨斯州平均水平段长度(1650m)长于路易斯安那州(1500m)，但是产量却低于后者。

2.2 压裂级数影响

两个州的页岩气井压裂级数集中在 2 ~ 20 级。路易斯安那州倾向于 12 ~ 14 级压裂，得克萨斯州倾向于 15 ~ 17 级压裂[图 3(a)]。从生产情况看，压裂级数越多，产能倾向于越高[图 3(b)、(c)、(d)]。路易斯安那州的页岩气井产能情况优于得克萨斯州，初始产量一般高于得克萨斯州，但得克萨斯的 San Augustine 郡页岩气产能趋势良好，此为例外，其 2 年累积产量仍落后于路易斯安那州。

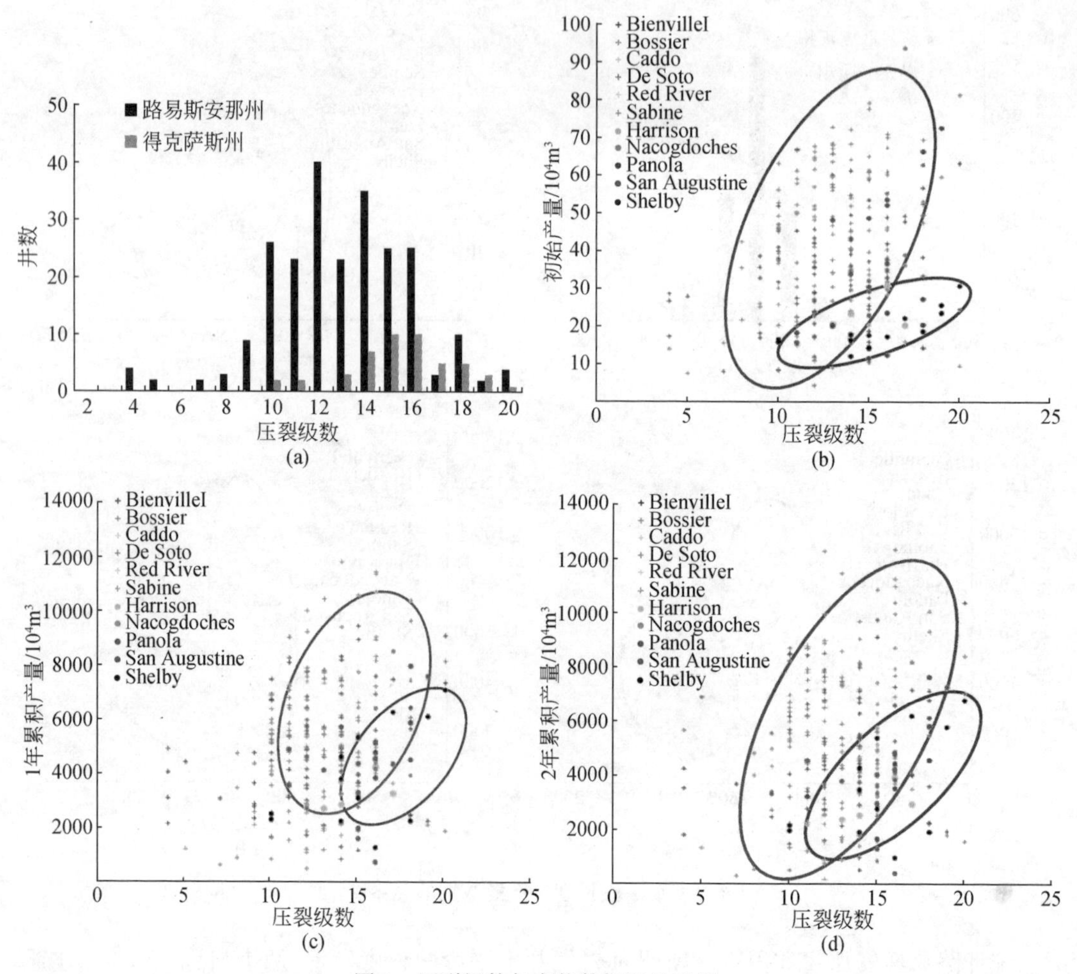

图 3 压裂级数与产能数据的关系图

2.3 加砂量影响

上述两州的加砂量情况上也不相同,路易斯安那州页岩气井加砂量多在 1500 ~ 2500t,得克萨斯州则为 1500 ~ 3000t。总加砂量的差异主要和页岩气井的水平段长度、压裂级数和压裂规模有关。总体上看,加砂量越多产量不一定越好,加砂量与产量的线性关系最差(图 4)。

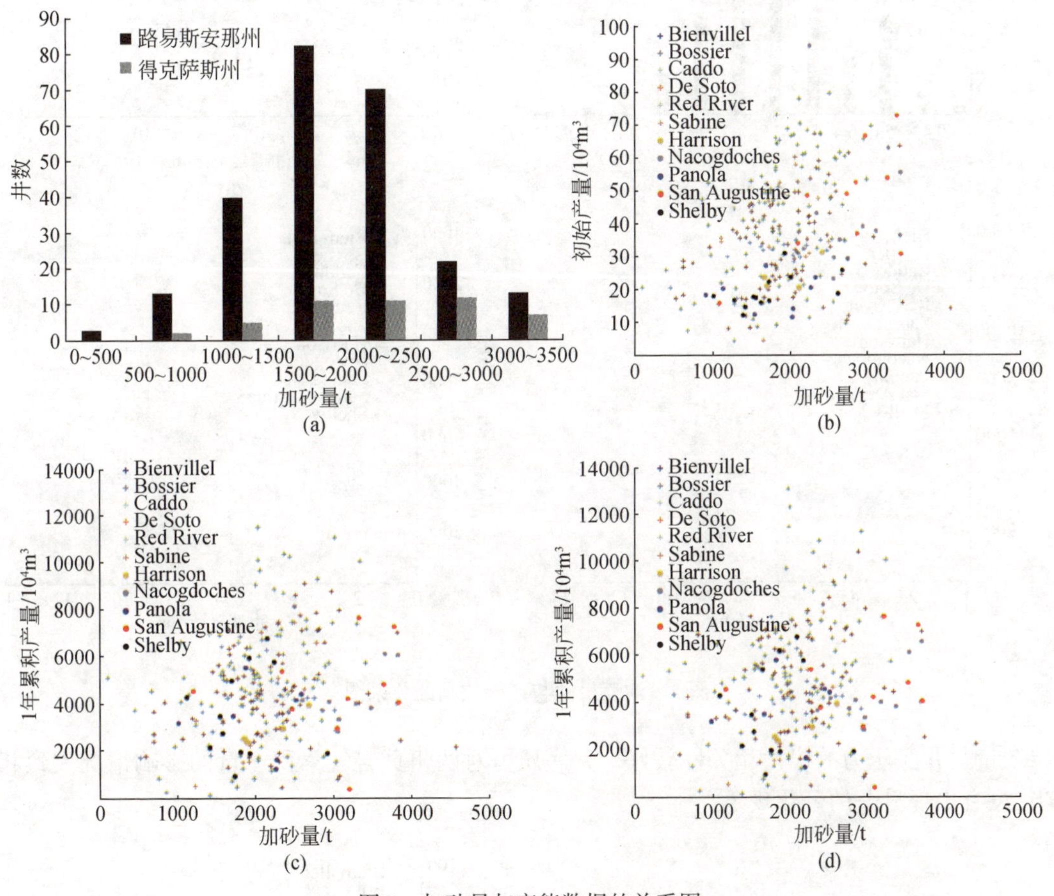

图 4 加砂量与产能数据的关系图

2.4 油嘴尺寸影响

油嘴尺寸方面,路易斯安那州和得克萨斯州分别倾向于 9mm 和 8mm 油嘴求产[图 5(a)]。产能方面,油嘴尺寸选取越大,初始产量越高,相关关系明显[图 5(b)]。1 年和 2 年累积产量与油嘴尺寸关系仍大致可见[图 5(c)、(d)]。新的趋势是:路易斯安那州采用 6mm 油嘴的案例逐渐增多,可能与经济环境、成本需求有关,具体原因有待分析。

2.5 井底压力影响

井底压力方面,路易斯安那州井底压力大致呈正态分布,得克萨斯州井底压力则无明显趋势,这反映了两个地区页岩气品质上的差异(图 6)。井底压力与产能的关系较为明显,初

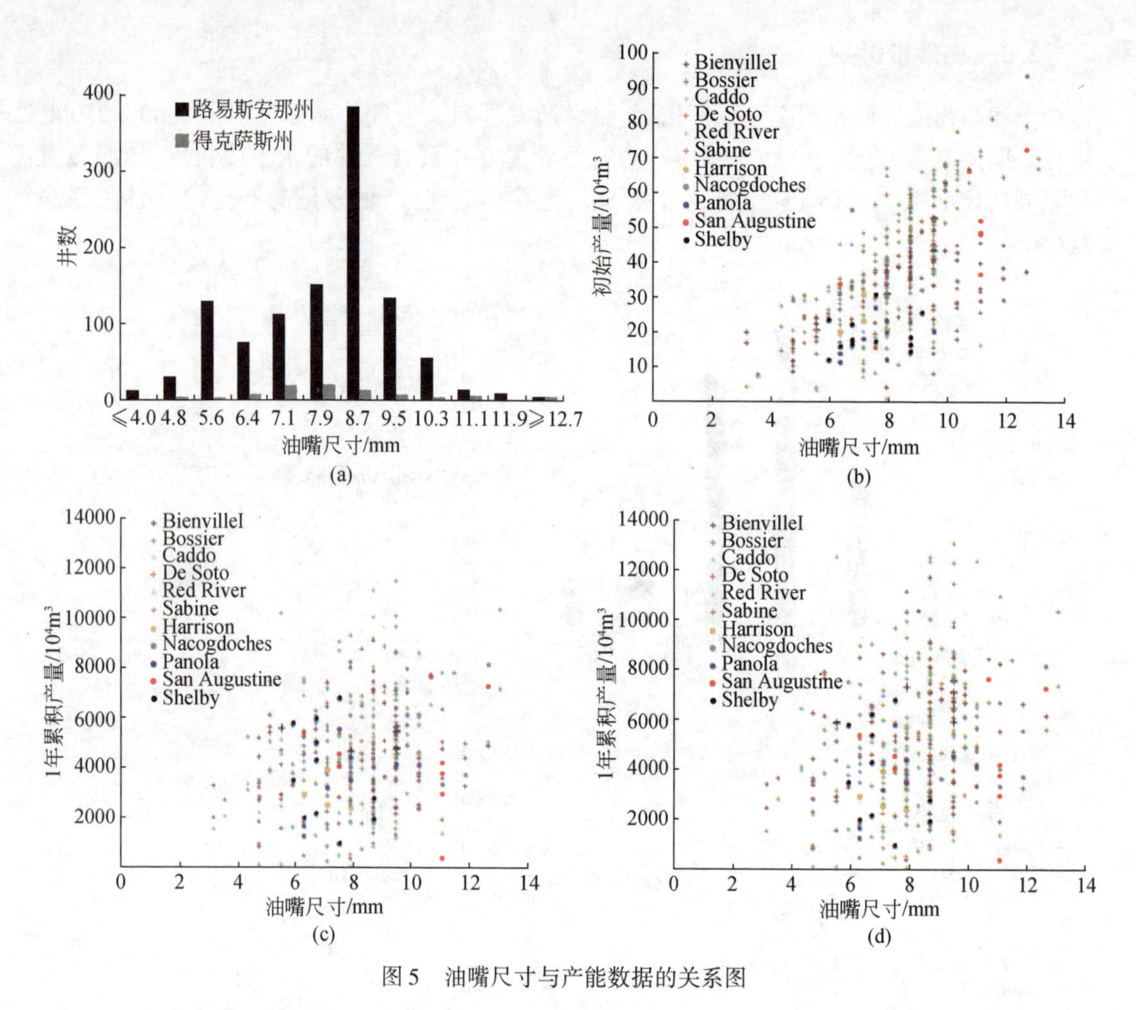

图5　油嘴尺寸与产能数据的关系图

始产量与井底压力的线性相关性较好。井底压力对两州页岩气累积产量的影响情况大致相当，没有出现显著的分化现象。

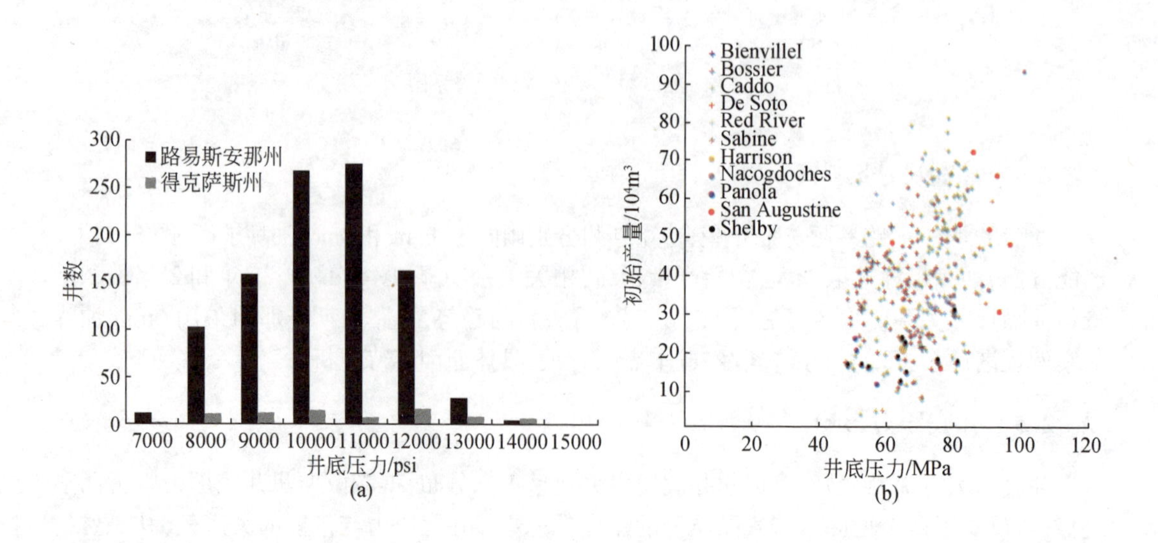

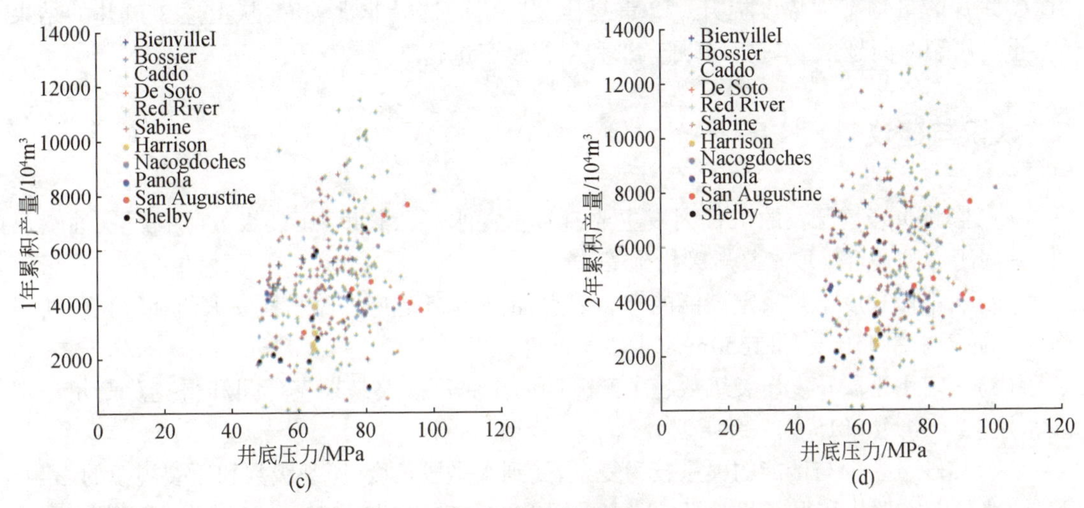

图6 井底压力与产能数据的关系图(1psi = 6.89476kPa)

进一步研究发现,较高的压力梯度增加了孔隙度、含气量(来自于孔隙度和页岩气密度增加)和含气页岩的表观脆性。大量数据显示,Haynesville 页岩的内在脆性实际上比 Barnett 页岩更差,异常的压力作用在增加孔隙度和含气量的同时,降低了骨架有效应力,从而使页岩储层容易压裂,且效果显著。通过对以 Haynesville 页岩层为目的层的 1100 多口井的井底压力进行估算,看到井底压力在 70MPa 的井位以 SW-NE 为主,包括 Red River、Sabine、Nacogdoches 和 San Augustine 等郡县。结合产能统计可见,这几个郡县的产量均较高。其中,以 Red River 为代表,井底压力均值较高,为 76 ~ 83MPa,页岩气井初始产量普遍较高,均值为 $42\times10^4m^3$。这与井底超压改善页岩的脆性,从而激化压裂增产效果有关。

2.6 重复压裂时机的影响

重复压裂是有效提高页岩气井产能的增产方法,压裂时机的选择对最终产量的影响显著。选取 A 井、B 井、C 井这 3 口页岩气井进行分析,3 口井完井时间相近:A 井、C 井 2009 年 12 月完井,B 井 2009 年 11 月完井。求产时油嘴尺寸也相近:A 井 8mm,B 井和 C 井 9mm。首月产量接近:A 井 $1774\times10^4m^3$,B 井 $1803\times10^4m^3$,C 井略高于前两者为 $2034\times10^4m^3$。A 井和 B 井分别在投产后第 7 个月开展重复压裂,C 井则在第 12 个月进行重复压裂(图 7)。分析压裂结果可得以下结论:①重复压裂能够显著增加页岩气产量,重复压裂后,A 井、B

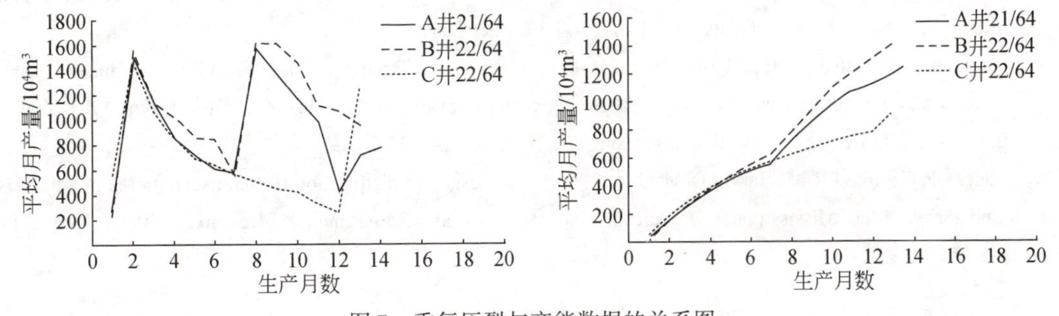

图7 重复压裂与产能数据的关系图

井、C 井月产量均有大幅度提升。②重复压裂时机的选择非常关键，从上述 3 口井的结果看，重复压裂时间越早，作业后单月产量和累积总产量越高。

3　认识与建议

(1)影响页岩气开发的工程因素主要有水平井水平段长度、压裂级数、加砂量、油嘴尺寸、井底压力和是否采取重复压裂措施等。

(2)水平段长度不是决定产能的唯一因素，也非越长越好。路易斯安那州和得克萨斯州分别趋向于采用 1500m 和 1650m 水平段生产井开发页岩气。

(3)一定水平段长度压裂级数越多产能倾向于越高。路易斯安那州和得克萨斯州分别多采用 12～14 级和 15～17 级压裂。

(4)加砂量多少与产能的关系较为复杂，受到水平段长度、压裂级数和压裂规模的影响。目前，路易斯安那州和得克萨斯州分别多采用 1500～2500t 和 1500～3000t 两种加砂量。

(5)油嘴尺寸与初始产量之间关系显著，与 1 年和 2 年累积产量也有一定相关关系。路易斯安那州和得克萨斯州分别倾向于采用 9mm 和 8mm 两种油嘴尺寸。此外，路易斯安那州有倾向于采用 6mm 油嘴的趋势。

(6)井底压力对产能的影响显著，超压范围与高产区域吻合良好。这与超压改善页岩脆性，优化压裂改造效果有关。

(7)重复压裂能够显著提高压裂效果，最佳压裂时机的选择可以通过产能分析与预测进行判断。实例显示压裂越早效果越好，但具体何时最好，仍需进一步分析。

参 考 文 献

[1] Wang F P, Reed R M. Pore networks and fluid flow in gas shales[C]. Paper SPE 124253-MS presented at the SPE Annual Technical Conference and Exhibition, New Orleans, Louisiana, USA, 4-7 October 2009.

[2] 翟光明，何文渊，王世洪. 中国页岩气实现产业化发展需重视的几个问题[J]. 天然气工业，2012，32(2)：1-4.

[3] 邱中建，邓松涛. 中国非常规天然气的战略地位[J]. 天然气工业，2012，32(1)：1-5.

[4] Stoneburner R. The Haynesville Shale：A look back at the first year[R]. Dallas, Texas：8th Gas Shales Summit, 2009.

[5] Pope C, Peters B, Benton T, *et al.* Haynesville Shale-one operator's approach to well completions in this evolving play [C]. Paper SPE 125079-MS presented at the SPE Annual Technical Conference and Exhibition, New Orleans, Louisiana, USA, 4-7 October 2009.

[6] Hammes U, Eastwood R, Hamlin H S. Influence of faces variations on exploration, production, and resource assessment in gas-shale plays：a geologic and petrophysical evaluation of the Haynesville Shale, East Texas, USA [C]. New Orleans, Louisiana, USA, AAPG, 2010.

[7] Loucks R G, Reed R M, Jarvie D M. Morphology, genesis and distribution of nano scale pores in siliceous mudstones of the Mississippian Barnett Shale[J]. Journal of Sedimentary Research, 2009, 79(12)：848-861.